COBALAMIN

George R. Minot

William B. Castle

H. A. Barker

Dorothy C. Hodgkin

COBALAMIN

BIOCHEMISTRY AND PATHOPHYSIOLOGY

EDITED BY

BERNARD M. BABIOR, M.D.
NEW ENGLAND MEDICAL CENTER HOSPITAL

A Wiley-Interscience Publication

JOHN WILEY & SONS, New York • London • Sydney • Toronto

Library of Congress Cataloging in Publication Data:

Babior, Bernard M

Cobalamin : biochemistry and pathophysiology

"A Wiley-Interscience publication."

Includes bibliographical references and index.

1. Cyanocobalamine deficiency. 2. Cyanocobalamine metabolism. I. Title. [DNLM: 1. Vitamin B 12. 2. Vitamin B 12–Deficiency. QU195 B114c]

RC620.5.B3 616.3′99 74-32499

ISBN 0–471–03970–5

Printed in the United States of America.

10 9 8 7 6 5 4 3 2 1

AUTHORS

Babior, Bernard M., M.D., Ph.D.
New England Medical Center Hospital
Boston, Massachusetts

Beck, William S., M.D.
Harvard University School of Medicine
Cambridge, Massachusetts
Hematology Research Laboratory of the Massachusetts General Hospital
Boston, Massachusetts

Castle, William B., M.D.
Harvard University School of Medicine
Cambridge, Massachusetts

Donaldson, Robert M., Jr., M.D.
Yale University School of Medicine
New Haven, Connecticut

Ellenbogen, Leon, Ph.D.
Lederle Laboratories
American Cyanamid Company
Pearl River, New York

Friedmann, Herbert C., Ph.D.
Department of Biochemistry
University of Chicago
Chicago, Illinois

Hogenkamp, H. P. C., Ph.D.
Department of Biochemistry
University of Iowa
Iowa City, Iowa

Linnell, John C., Ph.D.
Vincent Square Laboratories
Westminster Hospital
London, England

Mahoney, Maurice J., M.D.
Department of Human Genetics
Yale University School of Medicine
New Haven, Connecticut

Poston, Michael J., Ph.D.
National Institutes of Health
Bethesda, Maryland

Rosenberg, Leon E., M.D.
Department of Human Genetics
Yale University School of Medicine
New Haven Connecticut

Stadtman, Thressa C., Ph.D.
National Institutes of Health
Bethesda, Maryland

PREFACE

In the 50 years since its discovery, cobalamin has been extensively investigated by two groups of scientists. Clinical researchers have studied its absorption and transport in man and animals, and have puzzled over why the vitamin is required in man, as well as more specifically over the nature of its effects on the hematopoietic and nervous systems. At the same time, basic scientists have studied it to understand how its subtle and complex chemistry is related to its mode of action in a series of remarkable biochemical processes. Though there is a great deal that these two groups of scientists can learn from each other, there seems to have been relatively little exchange of ideas between them. The purpose of this book, edited by someone with a foot in each camp, is twofold: to survey the field of corrinoids at a biochemical and cellular level, and to provide clinical and basic researchers with a feeling for what each group is doing.

To this end, the book is divided (arbitrarily) into two sections, preceded by a historical introduction. In the first section, the chemistry and biochemistry of corrinoids are reviewed. A chapter outlining the major aspects of corrinoid chemistry is followed by chapters on the biosynthesis of corrinoids and on the biochemistry of the reactions catalyzed cobamide-requiring enzymes. The second section, dealing with what may loosely be termed the pathophysiology of the vitamin, begins with two chapters on the absorption and fate of cobamides in man and animals, continues with a discussion of the defects in absorption and transport of the vitamin followed by a survey of the recently described familial defects in cobalamin metabolism, and concludes with a chapter revealing how little is known about the biochemical basis for the clinical manifestations of cobalamin deficiency.

Outside the small circle of cobalamin aficionados, the corrinoid field seems to be regarded as a riddle wrapped in a mystery inside an enigma, an area of investigation only one step removed from alchemy. This widely held view may be attributed in part to the nomenclature which has afflicted the field for many years, a nomenclature so specialized that the field possesses aspects of a secret society, the members communicating with each other in what to an outsider must seem like unintelligible gibberish, lacking only a secret handshake to complete the analogy. To remedy this situation, IUPAC-IUB have recently issued a revised set of recommendations for corrinoid nomenclature, including suggestions for constructing rational symbolic representations of corrinoids which will do away with such terms as B_{12a}, factor A, factor III_m, DMBC, and others similarly obscure. By and large, the terminology and abbreviations used in this book are in accord with the recommendations of IUPAC-IUB, and the recommendations themselves are included in Appendix I. In addition, the book contains illustrations of certain corrinoid structures printed on the end-papers so they can be referred to at any time without turning pages.

I wish to acknowledge my indebtedness to many people involved, both directly and indirectly, in the preparation of this book. I am grateful to Robert Badger and Mary Conway, both of John Wiley & Sons, for patient and cheerful guidance. Michal Walke and Madeleine Mack eased my editorial burdens immeasurably. Professors Konrad E. Bloch and James H. Jandl, both of Harvard University, and Ephraim Y. Levin, of the National Institutes of Health, provided me by instruction and example with whatever skills I may possess as a scientist and an educator. Finally, for their encouragement and enthusiasm during the preparation of this book, and for their tolerance during many hours of writing and revising, I thank my family—Shirley, Greg, and Jill.

BERNARD M. BABIOR

Boston, Massachusetts
May 1974

#ACKNOWLEDGMENT

When a figure in this book is a reproduction of a figure published elsewhere, the original source of the figure is cited in the legend. We thank the publishers of these sources for permission to reproduce the figures.

B. M. B.

CONTENTS

COBALAMIN

INTRODUCTION

The History of Corrinoids

WILLIAM B. CASTLE, M.D.

Francis W. Peabody Faculty Professor of Medicine,
Emeritus, Harvard University
Former Distinguished Physician, Veterans Administration

CONTENTS

EARLY CLINICAL OBSERVATIONS

History has the special fascination that with its help, from any moment in time-past the sequence of subsequent events may be foreseen as with the gift of prophecy. This is especially true of the often complex unfolding of scientific discovery. Thus we now can see that the story of corrinoids began in 1824 with Combe's (1) account of a fatal case of anemia in Edinburgh, provided as he suggested, that its cause was "some disorder of the digestive and assimilative organs." More certainly related to subsequent medical events is the description in 1855 of an "idiopathic anaemia" by Thomas Addison (2) of Guys Hospital in London. However, for the recognition of the particular kind of anemia of which he wrote somewhat incidentally in this famous monograph, *On the Constitutional and Local Effects of Disease of the Suprarenal Capsules,* we depend largely on the sagacity of medical contemporaries. Indeed, only 5 years later in the United States the elder Austin Flint, finding that autopsies of patients that he recognized as similar cases had disclosed no gross abnormalities, made a shrewd surmise. Aware that the microscopic examination of 100 stomachs by Handfield Jones had found atrophy of the secretory glands in 14, he predicted (3) that similar microscopic inspection would disclose similar findings in the idiopathic anemia described by Addison. A decade later Samuel Fenwick (4) of the London Hospital saw with the aid of the microscope the glandular atrophy of the stomach of a patient considered to have Addison's "idiopathic anaemia" and demonstrated the inability of acidified scrapings of the withered mucosa to digest hard-boiled white of egg. In 1886 Cahn and von Mehring (5) reported for the first time during life the typical lack of hydrochloric acid that in 1900 was found by Faber and Bloch (6) in 33 cases. The introduction of the fractional test meal and the flexible gastric tube of small diameter allowed in 1921 confirmation of the constancy of a lack of hydrochloric acid in over 100 cases (7); and following the introduction of the gastric biopsy tube by Wood et al. (8) in 1949 the progressive atrophic process was clearly shown during life to spare only the pyloric gland region (9).

It is understandable that the early suggestions (3, 4) that the anemia was due to defective nutrition were neglected in the era dominated by the concept of the infectious or toxic etiology of disease, so successfully illustrated by the discoveries of Pasteur, Koch, and their successors. Thus in 1900 William Hunter (10) emphasized the hemolytic features of the anemia and expressed his conviction concerning the septic origin of the disease. Shortly thereafter he claimed to have found streptococci as the cause of the inflamed tongue and gastrointestinal symptoms. Indeed, as late as 1924 Arthur Hurst (11) presumed that the invariable achlorhydria permitted streptococcal colonization of the intestine with resulting elaboration of a hemolytic toxin. For this interpretation an analogy seemed to be found in several reports (12) after 1885 of a relation of the broad tapeworm to a form of pernicious anemia occurring in Finland and northern Russia that was cured when the parasite was expelled from its intestinal location.

CONCEPT OF NUTRITIONAL DEFICIENCY

It was only with the beginning of the twentieth century and the pioneer work of Eijkman in Java on beri-beri, followed by experiments with purified diets in animals in England by F. Gowland Hopkins and in this country by Osborne and Mendel, MacCollum, and others that a different concept of disease was established. Thus the idea gained credence that deficiencies of traces of accessory dietary factors, termed "vitamins" by Casimir Funk in 1913, might cause human disease under restricted dietary conditions. The idea that good food makes good blood is a lay belief of long standing. It was now to receive controlled experimental examination when George Whipple, working first at the Hooper Institute in San Francisco, began shortly before 1920 to study in chronically bled dogs the ability of various additions to a basal dietary regimen to enhance hemoglobin formation. In 1922, after his appointment as Professor of Pathology and Dean of the new medical school at Rochester, New York, Whipple (13) reported favorable effects with spinach, pork muscle, beef heart, and beef liver. By 1925 he and his principal research associate Dr. F. S. Robscheit-Robbins (14) with more prolonged experiments concluded that beef liver should head the list. Moreover it was fortunate that not until 1936 did it become apparent that the principal effective component of these items was their available iron content (15). Consequently, when these early results came to be known to the medical profession several clinicians tried dietary supplementation with meat and liver in various kinds of anemia in man including pernicious anemia (16, 17). Although it was found that a positive nitrogen balance could be achieved in that condition and some improvement in blood levels was observed to follow the feeding of such improved diets similar inexplicable "spontaneous remissions" were well known to experienced clinicians, despite the eventual fatal outcome of the disease. Consequently, when clinical improvement appeared it was not regarded as necessarily related to the modest amounts of meat or liver given; and the sage advice offered was not to abandon the well-established use of iron and arsenic (18).

On the other hand, George Minot was prepared to believe in the possible significance of Whipple's experiments for pernicious anemia. This Boston physician was Assistant Professor of Medicine at Harvard and had a special interest in diseases of the blood. While still a medical "house pupil" at the Massachusetts General Hospital he (19) had gained the impression from the cross-questioning of patients with pernicious anemia that they had often subsisted for some time on abnormal diets. Therefore, Whipple's work seemed to confirm his clinical suspicion; soon after 1922 Minot began to advise pernicious anemia patients to include liver in their diets. He became even more enthusiastic a little later when marked improvement occurred in the blood levels of one patient who really enjoyed eating liver in plentiful amounts. At that point he invited an assistant in private practice, Dr. William P. Murphy, to join him in an all-out trial of a special diet contaning "an abundance of food rich in complete proteins and iron—particularly liver—and relatively low in fat." As a result, in 1926 Minot and Murphy (20) were able to report consistent clinical improvement and gain in red blood cell levels in 45 patients who had successfully partaken of the special diet containing at least 120 to 240 g of liver daily. Shortly thereafter Minot consulted Edwin J. Cohn, Professor of Physical Chemistry at Harvard, as to the possibility of preparing a clinically effective extract of liver and of initiating a search for the identity of the active principle or principles involved. Protein of high biological quality was thought by Minot to be the basis of the therapu-tic success of liver feeding. Therefore he presumably approached Cohn because of the latter's well-known research interest in proteins. A collaborative program was agreed upon: Cohn and his associates were to proceed with the preparation of chemical fractions of liver, each of which was to be tested for potency in an untreated patient with pernicious anemia. For this clinical evaluation daily percentage counts of newly formed red cells (reticulocytes) were made during successive contiguous 10-day periods, respectively, of uniform daily administration of the liver fractions to be tested. If, in a first period, no reticulocyte increase appeared, the fraction was provisionally judged to be inactive, until in a second period of administration of a different liver fraction a reticulocyte rise indicated its activity and so confirmed the lack of potency of the first liver fraction (21). In this way each patient was his own control and the necessity of waiting for substantial increases of circulating red cells was avoided.

SEARCH FOR THE ACTIVE PRINCIPLE IN LIVER

The pioneering fractionation procedures of Cohn and his associates (22) began with the extraction of minced liver with water at pH 5 and subsequent heating of the extract to 70°C. This coagulated most of the bulky and disagreeable-tasting proteins, which happily were found almost at once to be inert on clinical trial. Next, after extraction with ether, alcohol was added to 70% concentration and the resulting clinically inert precipitate discarded. When the concentration of alcohol in the filtrate was increased to 95% another precipitate formed, the so-called fraction G, and was found to be clinically active. At this point the help of Eli Lilly and Company was invited through collaboration with its medical director, Dr. G. H. A. Clowes, for the commercial production of fraction G. By 1928 this liver fraction in the form of a yellow powder known as Liver Extract 343 (Lilly) became available for testing in several clinics in this country and abroad under the auspices of a Harvard Committee (23). This preparation conserved most of the hematopoietic activity of 300 g of beef liver in a daily oral dose of 12.75 g.

Attempts in Cohn's laboratory to produce a more purified fraction for parenteral use encountered great losses of potency in getting rid of an active vasodepressor principle. In 1930 Gänsslen (24) in Germany described the surprisingly great clinical activity of a nearly protein-free extract of liver when injected in daily doses of material derived from only 5 g of liver and probably representing about 0.35 g of dry material. It was then found (25) that a sterile neutralized aqueous solution of Liver Extract 343 possessed on intravenous injection as much as 30 times the activity of the oral preparation in pernicious anemia. The vasodepressor effects (probably due to histamine) were evanescent and could be minimized by slow injection. This "crude liver extract" was also highly effective in the treatment of tropical sprue in Puerto Rico (26).

Progress in further fractionation of liver, by then also begun in other laboratories, was slow. The clinical testing required untreated patients with pernicious anemia who were not always available, especially as by now medical practitioners had available to them commercial liver extracts. Chemical techniques originally applied to the concentration of water-soluble vitamins were further utilized, among them metal salts to precipitate impurities (22). More useful was precipitation of activity with phosphotungstic acid and later with the help of Reinecke's salt, a method introduced by Dakin and West (27) in New York. Their work was also effectively aided by salting out techniques employing ammonium or magnesium sulfate. By 1935 this led to the development of a so-called purified liver extract commercially available for parenteral use of which the daily dose of the dried material present had become as small as 0.015 g. A new impetus to progress came in 1936 when by means of initial extraction of minced liver with 50% acetone, followed by partition between water and phenol and adsorption on charcoal, LaLand and Klem (28) in Norway produced material active clinically in a single parenteral dose of 0.7 mg. In 1939 drops of an even more refined preparation, described as semicrystalline and orange-yellow, were placed on microscopic slides under sealed coverglasses. The German invasion interrupted their work, but 19 years later, when examined by others (29), the nearly forgotten slides displayed the deep red crystals by then known to be typical of vitamin B_{12}.

In 1945 Spies and his collaborators (30) briefly astonished the hematological world by demonstrating clear-cut activity in pernicious anemia with synthetic pteroylmonoglutamic (folic) acid in daily doses of 29 mg. This was an unexpected result because concentrates of purified liver extract were effective in much lower dosage and in fact by microbial analysis contained little or no folic acid. The explanation is now known to depend on the biochemical relationships of both folic acid and vitamin B_{12} to hematopoiesis to be discussed in another chapter.

The goal was now close at hand. In 1946 Emery and Parker (31) reported that a complete hematological remission could be obtained from a single injection of 1 mg of their material. In 1948 Randolph West (32), who from the early years had worked toward its isolation, was the first to demonstrate the clinical activity of crystalline vitamin B_{12} when injected in the form of cyanocobalamin in pernicious anemia patients. The precious material had been supplied to him by Karl Folkers and his associates (33) at the Merck Laboratories in the United States and was also isolated within a few weeks thereafter by Smith and Parker (34) at Glaxo Laboratories in England. In retrospect, according to E. Lester Smith (35) the reasons for the protracted effort were several. In the first place the losses in fractionation were sc great that large quantities of starting material were required as well as the processing facilities of industrial laboratories. Preoccupation with the notion that pernicious anemia was a unique human disease limited testing of fractions to suitable patients and inhibited the search for methods of assay involving growth rates in laboratory animals or microbial species until almost the end. Finally, the classical methods of fractionating water-soluble substances were inadequate to the task. Only with the development of adsorption and partition chromatography could the necessary separations be effected. An illustration of this last point is to be found in the detailed accounts of the isolation procedures used by the Glaxo Laboratories team (36).

MICROBIOLOGICAL GUIDANCE

The main trunk of the tree of knowledge that flowered with the discovery of vitamin B_{12} was the search for the active principle of liver effective in pernicious anemia. However, other at first seemingly unrelated scientific explorations helped either to reach that goal or afterward to aid in understanding the biological origin and functions of vitamin B_{12} and other corrinoids. Thus belatedly study of the growth requirements of *Lactobacillus lactis* Dorner by Mary Shorb (37) in 1947 led to recognition that crude and slightly purified liver extracts contained an essential growth factor. This suggestive finding provided a biological assay of great assistance to the Merck scientists in the final stages of the isolation of vitamin B_{12}, which was indeed the factor required by this particular lactobacillus. Since then other and more specific assays have been devised. *E. coli* mutants were introduced for this purpose in 1950 (38). Lactobacilli, including *Lactobacillus leichmannii* used officially by the United States Pharmacopoeia in 1955, have the disadvantage for this purpose that they respond to thymidine as well as to other deoxyribonucleosides. In 1949 Hutner et al. (39) had found photosynthesis by the flagellate *Euglena gracilis* to require vitamin B_{12}; after heating to liberate the vitamin from binding proteins the assay of human serum by this organism (40) has only recently begun to be replaced by isotope dilution assays (41) using serum protein or other binders with charcoal or Sephadex to provide proportionate exclusion.

MECHANISM OF VITAMIN B_{12} DEFICIENCY

The success of liver feeding in 1926 made it clear that pernicious anemia was either the result of a nutritional defect or metabolic aberration. The clinical knowledge that achlorhydria was known sometimes to precede by many years the development of the anemia and that it persisted after liver therapy had corrected all other aspects of the disease suggested to Castle that the causative mechanism might be, as he defined it in 1929, "an inability to carry out some essential step in the process of gastric digestion" (42). This supposition seemed to be confirmed when it was shown by the use of the double reticulocyte response method described above that daily administration of 200 g of ground beef muscle was ineffectual unless given simultaneously with 150 cc of normal human gastric juice. The unknown component of the beef muscle was termed "extrinsic factor" and that of the gastric juice "intrinsic factor" (43). Because it contained both factors, dessicated hog stomach was found to be therapeutically active by Sturgis and Isaacs (44). Over the next 20 years the nature of each factor was repeatedly sought by various investigators and the conditions of their supposed interaction were defined to some extent purely by observations on patients. The natural supposition that the gastric instrinsic factor was an enzyme could not be substantiated. The clue as to a possible substrate provided by Reimann's (45) demonstration in 1931 that liver or oral liver extract could be potentiated when given with gastric juice was unfortunately not followed up until 1948. Then Berk et al. (46), after finding that small amounts of highly purified liver extract were enhanced in activity upon daily oral administration with gastric juice, showed the same thing to be true for as little as 5 μg of the newly isolated active principle of liver extract, vitamin B_{12}. In 1952 Glass and his associates (47) reported that the acid glandular mucoprotein of human gastric juice possessed intrinsic factor activity with vitamin B_{12}. The same year, after Rosenblum and Woodbury (48) had made available to investigators isotopically labeled ^{60}Co-cyanocobalamin, Heinle, Welch, and their co-workers (49) demonstrated that in pernicious anemia simultaneous administration of gastric intrinsic factor decreased fecal excretion of the label, shown the following year by Schilling (50) to be absorbed and in part excreted in the urine as the vitamin. It thus appeared that the function of intrinsic factor was simply to promote the assimilation of the vitamin in food without essential modification. However, when large amounts of vitamin B_{12} are fed, as with 240 g of liver, passive diffusion alone without intrinsic factor suffices for assimilation.

In 1955 Watson and Florey (51) demonstrated that removal of the secretory portion of the rat stomach abolished the animal's ability to assimilate ^{60}Co-B_{12}. This function could be restored by a source of rat, but not of human or hog, intrinsic factor. Subsequent observations by many authors using isotopically labeled vitamin B_{12} in patients, in gastrectomized animals, and in perfused or everted segments of their distal small bowel or with mucosal scrapings have defined the normal process of vitamin B_{12} assimilation as follows. In the stomach the native vitamin of animal food, mostly in the coenzyme form linked to its specific protein, is released by peptic digestion and is at once strongly bound to intrinsic factor (52). This binding phenomenon was first postulated by Ternberg and Eakin (53) in 1949 from the observed *in vitro* inhibitory effect of gastric juice upon the growth of bacteria requiring vitamin B_{12}. The intrinsic factor, a glycoprotein secreted as such exclusively by the gastric parietal cells in man (54), is partially protected in this bound form from intestinal enzymes and adventitious parasites (55). The complex, in which the nucleotide portion of the vitamin B_{12} molecule relates structurally to the intrinsic factor (56), traverses the small bowel to the distal ileum where, by a specific but purely physical process (57), it becomes attached to the microvilli of the epithelial cells in the presence of calcium ion when, as normally, the local pH is 6.4 or above (58). Next, by an energy-requiring process, the vitamin is released into the intestinal cell from the intrinsic factor, which seems to remain active on the intestinal surface for a time (59). After a delay of some hours in the intestinal cell (60) the vitamin B_{12} enters the bloodstream where it is bound briefly to a plasma globulin, transcobalamin II. Later it is found attached to what may be a storage plasma globulin probably derived from leukocytes, transcobalamin I (61). A hereditary lack of transcobalamin II has recently been shown (62) to be associated with defective intestinal assimilation and to prevent transfer of the plasma vitamin to the erythropoietic cells. Significant knowledge of these plasma transport proteins began with the work of Hall and Finkler (63) in 1963.

The finding by Schwartz (64) in 1958 and the next year by Taylor (65) of anti-intrinsic factor antibodies in the serum of some patients with pernicious anemia at first seemed to provide an additional mechanism for vitamin B_{12} malabsorption. However, it is now recognized that this is significant only when such serum antibodies leak or are secreted into the alimentary tract. Instead, a suggestion of an immunological basis for the chronic atrophic gastritis was provided, especially when antiparietal cell antibodies were found in association with gastritis, sometimes preceding the anemia (66). More recent evidence that cellular rather than humoral immunity is more likely to damage tissue cells (67) and the increased incidence of pernicious anemia reported in younger subjects with hypogammaglobulinemia (68) first reported in 1961, seem to favor a chronic cellular immune reaction expressed by an infiltration of the gastric mucosa with lymphocytes and macrophages and reflected by the presence of circulating antibodies specifically directed against intrinsic factor and parietal cell cytoplasmic antigens.

ANIMAL PROTEIN FACTOR

Isolation from liver would never have provided a practical source of vitamin B_{12} for therapeutic purposes or of other corrinoids without industrial fermentations with streptomyces as initially employed in the production of antibiotic (69). This technique came to be used because of difficulties during World War II in obtaining satisfactory growth rates in hogs and in chickens fed vegetable rations lacking in the usual meat scrap supplementation. At maturity the hens laid eggs many of which failed to hatch (70). It was soon discovered that these abnormalities could be corrected by the addition of animal or fish wastes to the vegetable rations, which were thus deemed to be lacking in an unknown "animal protein factor" (APF). Experiments with rats on similar restricted diets disclosed that "factor X" and "zoopherin" were identical with APF. When laying hens were raised under natural conditions, rather than the seemingly more hygienic way of being kept on wire mesh, egg hatchability improved. Rubin et al. (71) found that poultry droppings caused fermentation of spilled food and, like dried cow manure, supplied APF as liver extracts were already known to do. Fermentation of an artificial medium with an organism isolated from hen feces was found to yield APF; and after the isolation of vitamin B_{12} from liver this APF was shown to be effective in pernicious anemia (72).

COBALT DEFICIENCY IN RUMINANTS

A further unexpected relation to synthesis of corrinoids by fermentation emerged from the successful analysis of the cause of a wasting disease of sheep and cattle when pastured in certain localities in several parts of the world. Recognized by veterinarians under different names such as "bush sickness" and "pining," loss of appetite is followed by loss of weight and eventually anemia. In South Australia the so-called coast disease was at first found to be curable by commercial iron salts, but later not by pure iron salts. Eventually the curative impurity turned out to be cobalt, which could be supplied by oral administration or by addition in trace amounts to the soil of the grazing areas (73). The relative inefficacy of cobalt when given to sheep by injection led to consideration of its possible role as a growth requirement for the microorganisms of the rumen that play an essential role in ruminant nutrition by producing fatty acids from cellulose. No search for a cobalt-containing metabolite in rumen contents was made until after the isolation of vitamin B_{12} when rumen contents were found to be a good source of that vitamin and especially of related corrinoids (74). The discovery of cobalt in the vitamin B_{12} molecule then suggested an essential synthetic role for that element in rumen microbiology (75). Although this was rendered doubtful when the usual human doses of vitamin B_{12} were found to be ineffective in sheep upon parenteral injection, subsequently larger dosages were found to be fully effective in restoring appetite, growth, and blood formation (76). The explanation of this greater requirement is still uncertain.

METABOLIC FUNCTIONS OF VITAMIN B_{12}

Such were the diverse researches, termed "convergent" by E. Lester Smith (35), prior to the isolation of vitamin B_{12} in 1948. Since then much has been learned of its biogenesis, structure, and biological functions in the body in coenzyme forms. Much of this information has been derived from studies of its metabolic actions in bacteria requiring it for growth. The discovery of an orange-yellow, photosensitive coenzyme form of an analog of vitamin B_{12} was announced by Barker et al. (77) in 1958, as a result of studies of an unusual metabolic pathway for glutamate present in the organism *Clostridium tetanomorphum*. This led shortly to the discovery of the adenosyl coenzyme form of vitamin B_{12} itself. The complete spatial configuration of cyanocobalamin having been specified as a result of several years of chemical investigations culminating in the X-ray crystallographic work and exhaustive calculations of Dorothy Hodgkin's group in 1957 (78), she and Lenhert (79) were able to report the structure of the adenosyl coenzyme in 1961 only 3 years after its isolation. The vitamin B_{12} molecule was shown to consist of the nearly planar corrin ring joined through its central cobalt atom almost at a right angle to the so-called nucleotide portion of the molecule. In the coenzyme, the cyanide group, attached above the macrocycle to cobalt in cyanocobalamin, is replaced by the adenosine moiety attached by the C-5′ of the pentose.

The original supposition that vitamin B_{12} was involved in the synthesis of nuclear DNA was soon confirmed by Roberts et al. (80) in 1949. Within a decade attention had become focused upon the synthesis of the deoxyriboside moiety. Beck et al. (81) in 1962 proposed that the adenosyl coenzyme might be involved in the reduction of all four ribose nucleotides in man, as they showed to be the case in *L. leichmannii*, preparatory to DNA synthesis. On the other hand, there is suggestive evidence that in man, as in the mutant strains of *E. coli* requiring vitamin B_{12}, a different pathway is concerned in which B_{12} is only indirectly involved, and then as the methyl, not the adenosyl, derivative.

Early experiments on the growth of chicks and rats showed that vitamin B_{12} exerted a sparing effect upon their methionine requirement and could reduce somewhat their need for dietary choline (82). It was also discovered that a few mutant strains of *E. coli* need vitamin B_{12} for methionine synthesis unless supplied with 10,000 times as much of the amino acid. Around 1960 Woods' and Buchanan's groups independently concluded that the transfer of the methyl group from *N*-5-methyltetrahydrofolic acid, the principal form in serum, requires a different coenzyme of vitamin B_{12} than the adenosyl. When synthetic methylcobalamin became available it was shown that its methyl group was donated to convert homocysteine to methyl homocysteine (methionine), though in a somewhat complex system demonstrated by Foster et al. (83) in 1964. The cooperative role of methylcobalamin in transferring the one-carbon fragment from N-5-methyltetrahydrofolic acid to homocysteine in man was originally suggested in 1962 by Herbert and Zalusky (84) as an explanation of the "pile up" of the serum folate in clinical vitamin B_{12} deficiency. Experiments by Killman (85) in 1964 with human bone marrow deficient in vitamin B_{12} and by Metz et al. (86) in 1968 tend to confirm the suggestion that lack of vitamin B_{12} prevents the formation of the specific N-5,10-methylene form of tetrahydrofolate required for the methylation of deoxyuridylic to thymidylic acid in the synthesis of DNA. However, whether this or another system is involved in man remains to be determined.

A firmly established function of vitamin B_{12} in man is the involvement of the adenosyl coenzyme in the major pathway of propionate metabolism demonstrated by Beck and Ochoa in 1958. The final step involves the isomerization of methylmalonyl CoA to succinyl CoA and is catalyzed by the vitamin B_{12} coenzyme as was shown in 1961 by Marston et al. (87) in vitamin B_{12}-deficient sheep. In the human deficiency Cox and White (88) demonstrated the next year a large excretion of methylmalonic acid in the urine that was corrected by cyanocobalamin but not by folic acid. Recently this abnormality has been found to be associated with the synthesis of abnormal fatty acids with odd-numbered carbon chain lengths both *in vivo* (89) and *in vitro* (90) and so is possibly related to the specific demyelination of the spinal cord sometimes occurring in vitamin B_{12} deficiency.

BIOGENESIS AND SYNTHESIS OF VITAMIN B_{12}

The history of the discovery of the nature of the biogenesis of vitamin B_{12} and of corrinoid analogs is too complex a subject for detail here. British workers (91) engaged in research related to the dairy industry discovered in 1955 the existence of vitamin B_{12} analogs after microbial assays for vitamin B_{12} in rumen contents indicated less than the expected growth rates in vitamin-deficient chicks given such material. Further understanding has come largely from the use of labeled precursors in vitamin B_{12}- or analog-producing microorganisms. In 1950 Sahashi et al. (92) showed that preformed 5,6-dimethylbenzimidazole can be incorporated into the nucleotide portion of the vitamin B_{12} molecule. Fantes and O'Callaghan (93) found in 1955 than in *Streptomyces griseus* phenylenediamine was a precursor of an analog, whereas dimethylphenylenediamine increased the yield of vitamin B_{12} proper. Shemin and his associates (94) in 1956, utilizing ^{14}C-labeled glycine, δ-aminolevulinic acid, or porphobilinogen showed that the corrin ring is constructed in a fashion similar to that of the porphyrin ring of heme. Synthesis of the corrin ring was achieved by Eschenmoser in Zurich in 1964, and recently Eschenmoser and Woodward, working in collaboration, have effected total synthesis of the vitamin (95). So much for the historical highlights of biosynthesis. Man's ability to achieve with immense labor what the most primitive bacteria accomplish with ease is, judged by human standards, a superb achievement.

CONCLUSION

The history of the discovery of the corrinoids is a classical illustration of the seminal importance of clinical investigation to medical and biological science. The empirical discovery of the efficacy of liver feeding in pernicious anemia led eventually to the isolation, structural analysis, and synthesis of vitamin B_{12}. Without the first clinical steps, however, no one can say how long it would have been before the time came for the present elegant revelations of the nature and functions of the corrinoids. Indeed, at the service of the clinical investigator, as for no other biological scientist, is the boundless ingenuity of nature in providing the experiments created by disease. Once understanding of their proximate meaning reaches a certain point the torch can be passed to basic science for further and more rapid progress.

REFERENCES

1. Combe, J. S. (1824). *Trans. Med-Chirurg. Soc. Edinburg* **1**, 194–203.
2. Addison, T. (1855). *On the Constitutional and Local Effects of Disease of the Suprarenal Capsules,* Samuel Highley, London, pp. 2–4.
3. Flint, A. (1860). *Am. Med. Times* **1**, 181–186.
4. Fenwick, S. (1870). *Lancet* **2**, 78–80.
5. Cahn, A., and von Mehring, J. (1886). *Dtsch. Arch. Klin. Med.* **39**, 233–253.
6. Faber, K., and Bloch, C. E. (1900). *Z. Klin. Med.,* 113–114.
7. Levine, S. A., and Ladd, W. S. (1921). *Bull. Johns Hopkins Hosp.* **32**, 254–266.
8. Wood, I. J., Doig, R. K., Motteram, R., and Hughes, A. (1949). *Lancet* **1**, 18–21.
9. Siurala, M., Eräman, E., and Nyberg, W. (1960). *Acta Med. Scand.* **166**, 213–223.

10. Hunter, W. (1900). In *Pernicious Anemia: Its Pathology, Septic Origin, Symptoms, Diagnosis and Treatment,* Griffin, London, pp. 240–246.
11. Hurst, A. F. (1924). *Brit. Med. J.* **1**, 93–100.
12. Cited by Birkeland, I. W. (1932). *Medicine* **11**, 79–83.
13. Whipple, G. H. (1922). *Arch. Intern. Med.* **29**, 711–731.
14. Whipple, G. W., and Robscheit-Robbins, F. S. (1925). *Am. J. Physiol.* **72**, 408–418.
15. Whipple, G. H., and Robscheit-Robbins, F. S. (1936). *Am. J. Med. Sci.* **191**, 11–24.
16. Barker, L. F., and Sprunt, T. P. (1917). *JAMA* **69**, 1919–1927.
17. Mosenthal, H. O. (1918). *Bull. Johns Hopkins Hosp.* **29**, 129–134.
18. Gibson, R. B., and Howard, C. P. (1923). *Arch. Intern. Med.* **32**, 1–6.
19. Minot, G. R. (1935). *Lancet* **1**, 361–364.
20. Minot, G. R., and Murphy, W. P. (1926). *JAMA* **87**, 470–476.
21. Minot, G. R., and Castle, W. B. (1935). *Lancet* **2**, 319–330.
22. Cohn, E. J. Minot, G. R., Alles, G. A., and Salter, W. T. (1928). *J. Biol. Chem.* **77**, 325–358.
23. Castle, W. B. (1966). *Clin. Pharmacol. Ther.* **7**, 147–161.
24. Gänsslen, M. (1930). *Klin. Wochenschr.* **9**, 2099–2102.
25. Strauss, M. B., Taylor, F. H. L., and Castle, W. B. (1931). *JAMA* **97**, 313–314.
26. Castle, W. B., and Rhoads, C. P. (1931). *Trans. Assoc. Am. Physicians* **47**, 245–246.
27. Dakin, H. D., and West, R. (1935). *J. Biol. Chem.* **109**, 489–522.
28. LaLand, P., and Klem, A. (1936). *Acta Med. Scand.* **88**, 620–623.
29. Jorpes, J. E., and Standell, B., (1960). *Acta Med. Scand.* **168**, 325–327.
30. Spies, T. D., Vilter, C. F., Koch, M. B., and Caldwell, M. H. (1945). *South. Med. J.* **38**, 707–709.
31. Emery, W. B., and Parker, L. F. J. (1946). *Biochem. J.* **40**, iv.
32. West, R. (1948). *Science* **107**, 398.
33. Rickes, E. L., Brink, N. G., Koniuszy, F. R., Wood, T. R., and Folkers, K. (1948). *Science* **107**, 396–397.
34. Smith, E. L., and Parker, L. F. J. (1948). *Biochem.* **43**, viii.
35. Smith, E. L. (1965). *Vitamin B_{12},* 3rd Ed., Methuen, London.
36. Fantes, K. H., Page, J. E., Parker, L. F. J., and Smith, E. L. (1949). *Proc. Roy. Soc. (B)* **136**, 592–613.
37. Shorb, M. S. (1947). *J. Biol. Chem.* **169**, 455–456.
38. Davis, B. D., and Mingiolo, E. S. (1950). *J. Bacteriol* **60**, 17–28.
39. Hutner, S. H., Provasoli, L., Stokstad, E. L. R., Hoffman, C. E., Belt, M., Franklin, A. L., and Jukes, T. H. (1949). *Proc. Soc. Exp. Biol. Med.* **70**, 118-120.
40. Ross, G. I. M. (1952). *J. Clin. Pathol.* **5**, 250–256.
41. Lau, K., Gottlieb, C., Wasserman, L. R., and Herbert, V. (1965). *Blood* **26**, 202–214.
42. Castle, W. B. (1929). *Am. J. Med. Sci.* **178**, 764–777.

43. Castle, W. B., Townsend, W. C., and Heath, C. W. (1930). *Am. J. Med. Sci.* **180**, 305–335.
44. Sturgis, C. C., and Isaacs, R. (1929). *JAMA* **93**, 747–749.
45. Reimann, F. (1931). *Med. Klin.* **27**, 880–881.
46. Berk, L., Castle, W. B., Welch, A. D., Heinle, R. W., Anker, R., and Epstein, M. (1948). *N. Engl. J. Med.* **239**, 911–913.
47. Glass, G. B. J., Boyd, L. J., Rubenstein, M. A., and Svigals, C. S. (1952). *Science* **115**, 101.
48. Rosenblum, C., and Woodbury, D. T. (1950). *Science* **111**, 601–602.
49. Heinle, R. W., Welch, A. D., Scharf, V., Meacham, G. C., and Prusoff, W. H. (1952). *Trans. Assoc. Am. Physicians* **65**, 214–222.
50. Schilling, R. F. (1953). *J. Lab. Clin. Med.* **42**, 860–866.
51. Watson, G. M., and Florey, H. W. (1955). *Brit. J. Exp. Pathol.* **36**, 479–486.
52. Cooper, B. A., and Castle, W. B. (1960). *J. Clin. Invest.* **39**, 199–214.
53. Ternberg, J. L., and Eakin, R. E. (1949). *J. Am. Chem. Soc.* **71**, 3858.
54. Hoedemaeker, P. J., Abels, J., Wachters, L. J., Arends, A., and Nieweg, H. O. (1964). *Lab. Invest.* **13**, 1394–1399.
55. Nyberg, W. (1960). *Acta Med. Scand.* **167**, 185–192.
56. Bunge, M. B., Schloesser, L. I., and Schilling, R. F. (1956). *J. Lab. Clin. Med.* **48**, 735-744.
57. Strauss, E. W., and Wilson, T. H. (1960). *Am. J. Physiol.* **198**, 103–107.
58. Herbert, V., and Castle, W. B. (1961). *J. Clin. Invest.* **40**, 1978–1983.
59. Hines, J. D., Rosenberg, A., and Harris, J .W. (1968). *Proc. Soc. Exp. Biol. Med.* **129**, 653–658.
60. Doscherholmen, A., and Hagen, P. S. (1957). *J. Clin. Invest.* **36**, 1551–1557.
61. Hom, B. L. (1967). *Scand. J. Haematol.* **4**, 321–332.
62. Hakami, N., Neiman, P. E., Canellos, G. R., and Lazerson, J. (1971). *N. Engl. J. Med.* **285**, 1163–1170.
63. Hall, C. A., and Finkler, A. E. (1963). *Biochim. Biophys. Acta* **78**, 234–236.
64. Schwartz, M. (1958). *Lancet* **2**, 61–62.
65. Taylor, K. B. (1959). *Lancet* **2**, 106–108.
66. Fisher, J. M., and Taylor, K. B. (1965). *N. Engl. J. Med.* **272**, 499–503.
67. Perlmann, P., Perlmann, H., and Holm, G. (1968). *Science* **160**, 306.
68. Twomey, J. J., Jordan, P. H., Jarrold, T., Trubowitz, S., Ritz, N. D., and Conn, H. O. (1968). *Am. J. Med.* **47**, 340–350.
69. Rickes, E. L., Brink, N. G., Koniuszy, F. R., Wood, T. R., and Folkers, K. (1948). *Science* **108**, 634–635.
70. Wilson, D., Titus, H. W., and Bird, H. R. (1946). *Poultry Sci.* **25**, 143–147.
71. Rubin, M., Groschke, A. C., and Bird, H. R. (1947). *Proc. Soc. Exp. Biol. Med.* **66**, 36–38.
72. Stokstad, E. L. R., Page, A., Jr., Pierce, J., Franklin, A. L., Jukes, T. H., Heinle, R. W., Epstein, J., and Welch, A. D. (1948). *J. Lab. Clin. Med.* **33**, 860–864.
73. Underwood, E. J., and Filmer, J. F. (1935). *Aust. Vet. J.* **11**, 84–92.

74. Marston, H. R. (1952). *Physiol. Rev.* **32**, 66–121.
75. Ford, J. E., Kon, S. K., and Porter, J. W. G. (1951). *Biochem. J.* **50**, ix.
76. Marston, H. R., and Lee, H. J. (1952). *Nature* **170**, 791.
77. Barker, H. A., Weissbach, H., and Smyth, R. D. (1958). *Proc. Natl. Acad. Sci. U.S.A.* **44**, 1093–1097.
78. Hodgkin, D. C., Pickworth, J., Robertson, J. H., Trueblood, K. N., Prosen, R. J., White, J. G., Bonnett, R., Cannon, J. R., Johnson, A. W., Sutherland, I., Todd, Sir Alexander, and Smith, E. L. (1955). *Nature* **176**, 325–328.
79. Lenhert, P. G., and Hodgkin, D. C. (1961). *Nature* **192**, 937–938.
80. Roberts, I. Z., Roberts, R. B., and Abelson, P. H. (1949). *J. Bacteriol.* **58**, 709–710.
81. Beck, W. S., Hook, S., Barnett, B. H., and Goulian, M. (1962). *Biochem. Biophys. Acta* **55**, 455–469.
82. Jukes, T. H., and Stokstad, E. L. R. (1951). *J. Nutr.* **43**, 459–467.
83. Foster, M. A., Dilworth, M. J., and Woods, D. D. (1964). *Nature* **201**, 39–42.
84. Herbert, V., and Zalusky, R. (1962). *J. Clin. Invest.* **41**, 1263–1276.
85. Killman, S. A. (1964). *Acta Med. Scand.* **175**, 483–488.
86. Metz, J., Kelly, A., Swett, V. C., Waxman, S., and Herbert, V. (1968). *Brit. J. Haematol.* **14**, 575–592.
87. Marston, H. R., Allen, S. H., and Smith, R. M. (1961). *Nature* **190**, 1085–1087.
88. Cox, E. V., and White, A. M. (1962). *Lancet.* **2**, 853–856.
89. Frenkel, E. P. (1971). *J. Clin. Invest.* **50**, 33a.
90. Barley, F. W., Sato, G. H., and Abeles, R. H. (1972). *J. Biol. Chem.* **247**, 4270–4276.
91. Ford, J. E., Holdsworth, E. S., and Kon, S. K. (1955). *Biochem. J.* **59**, 86–93.
92. Sahashi, Y., Mikata, M., and Sakai, H. (1950). *Bull. Chem. Soc. Japan* **23**, 247–249.
93. Fantes, K. H., and O'Callaghan, C. H. (1955). *Biochem. J.* **59**, 79–82.
94. Shemin, D., Corcoran, J. W., Rosenblum, C., and Miller, J. M. (1956). *Science* **124**, 272.
95. Woodward, R. B. (1973). *Pure Appl. Chem.* **33**, 145–177.

BIOCHEMISTRY

CHAPTER ONE

THE CHEMISTRY OF COBALAMINS AND RELATED COMPOUNDS

H. P. C. HOGENKAMP

Department of Biochemistry
University of Iowa
Iowa City, Iowa

CONTENTS

INTRODUCTION

In 1948 cyanocobalamin was isolated as a red crystalline compound almost simultaneously in two laboratories (1, 2), while 10 years later adenosylcobalamin, one of the coenzymatically active forms of the vitamin, was isolated by Barker and co-workers (3, 4). Both compounds are now recognized as members of a larger family of corrinoids.

All corrinoids contain the corrin nucleus (*1*) (Figure 1-1), a metal atom, usually cobalt, coordinated in the center of the corrin ring by the nitrogens of the four pyrroline rings, and ligands in the fifth and the sixth coordination positions of the metal. The basic skeleton of the corrinoids is cobyrinic acid (*2*) (Figure 1-2), which has three acetic acid, four propionic acid, and eight methyl groups on the periphery of the corrin ring. In cobyric acid (*3*) all carboxyl groups except the one at position *f* are amidated with ammonia, while in cobinic acid (*4*) the carboxyl group at position *f* is amidated with (R)-1-amino-2-propanol and the other carboxyl groups are free. In cobinamide (*5*) the *f*-propionic acid group is amidated with (R)-1-amino-2-propanol and the other carboxyl groups are amidated with ammonia. In cobamide (*6*) the hydroxyl-function of aminopropanol is esterified with α-D-ribofuranose-3-phosphoric acid. The corresponding hexacarboxylic acid of cobamide is called cobamic acid (*7*). Many cobamides contain a substituted imidazole linked by way of an *N*-α-glycosidic bond to the ribose moiety; the second nitrogen of the imidazole can then be coordinated to cobalt in the α coordination position. Among the imidazole bases found in cobamides are benzimidazole, dimethylbenzimidazole, hydroxybenzimidazole, and methoxybenzimidazole. Cobamides in which 5,6-dimethylbenzimidazole is bound to ribose in an α-glycosidic linkage are called cobalamins. For instance, cyanocobalamin is a cobalamin (i.e., a cobamide with 5,6-dimethylbenzimidazole linked to the ribose) with a cyanide ligand in the β coordination position. Named as a cobamide, it would be called *Co*α- (5,6-dimethylbenzimidazolyl)-*Co*β-cyanocobamide. The cobalamin with a 5′-deoxyadenosyl moiety linked to cobalt (Figure 1-3) is properly named adenosylcobalamin (AdoCbl) or *Co*α-(5,6-dimethylbenzimidazolyl)-*Co*β-adenosylcobamide. On the other hand, corrinoids possessing bases other than 5,6-dimethylbenzimidazole are called cobamides. For example, the cyanocobamide containing adenine bound to the ribosyl moiety by way of a 7-α-glycosidic linkage is named *Co*α-adenyl-*Co*β-cyanocobamide [(Ade)CN-Cba] (5, 6). Corrinoid nomenclature is discussed fully in the Appendices.

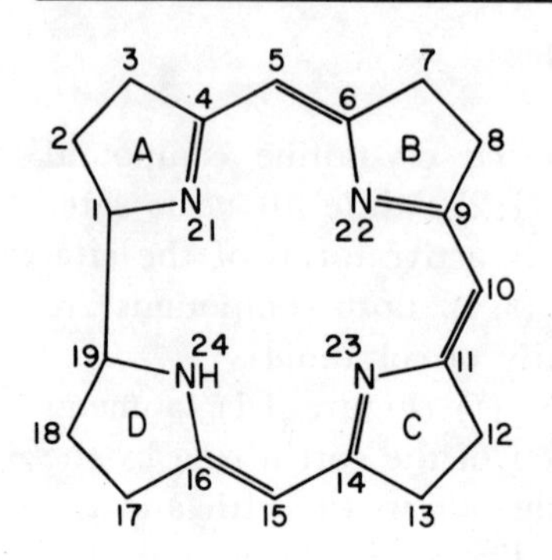

Fig. 1-1. Corrin (*1*). The basic ring of all corrinoids. The capital letters denote the four pyrroline rings. The number 20 is omitted so that the numbering system will correspond to that of the porphyrin nucleus.

STRUCTURE AND REACTIONS

The structure determination of cyanocobalamin and adenosylcobalamin followed very similar patterns: for both compounds detailed chemical studies were done in several laboratories (7–9) but the final elucidation of the structure was accomplished by X-ray analysis in the laboratory of Hodgkin (10). The structural relation of the coenzymes to the vitamins was first recognized by the formation of dicyanocobalamin on exposure of the coenzyme to alkaline cyanide (3, 4, 11). Treatment of adenosylcobalamin with alkaline cyanide resulted in a change from a red to a purple color. The purple compound had an absorption spectrum typical of the spectrum of dicyanocobalamin.

The most precise evidence for the structure of the corrin nucleus comes from the X-ray analysis of cobyric acid (12). This corrinoid, isolated by Bernhauer et al. (13) from sewage sludge, crystallized from water in the presence of acid as the aquocyano form, (CN)AqCby. The four rings of the corrin moiety (see Figure 1-2) form a nearly planar nucleus. Three of the pyrroline rings are bridged through a carbon atom, while rings *A* and *D* are connected directly through their α-carbons. The corrin nucleus contains six conjugated double bonds, a level of unsaturation fixed by methyl groups on the periphery of the ring. Three acetic acid side chains and four propionic acid side chains are attached to the β-carbons of the pyrroline rings and project to opposite sides of the corrin ring. The X-ray data also show that the corrin ring is buckled with the two methyl groups C-35 and C-53 on opposite sides of the central cobalt-nitrogen plane (Figure 1-4).

The four nitrogens alternate 0.05 Å above and below the central plane. Furthermore, the cyanide ligand bound to cobalt in the α coordination position (the side of the propionic acid side chain) is pushed

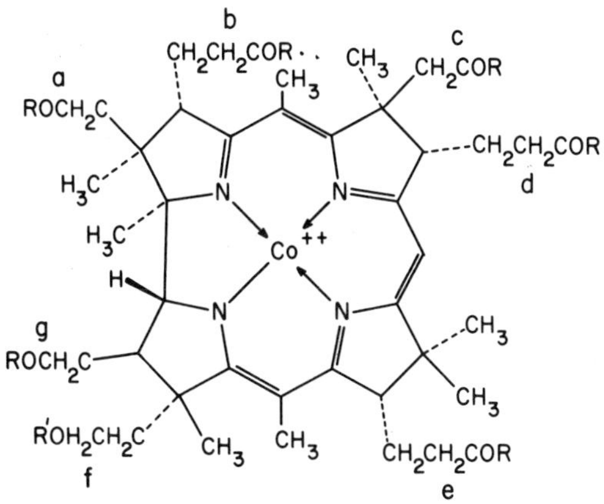

Fig. 1-2. Nomenclature of the corrinoids: *2* Cobyrinic acid, R = R′ = OH. *3* Cobyric acid, R = NH_2, R′ = OH. *4* Cobinic acid, R = OH, R′ = $NHCH_2CHOHCH_3$. *5* Cobinamide, R = NH_2, R′ = $NHCH_2CHOHCH_3$. 6 Cobamide, R = NH_2,

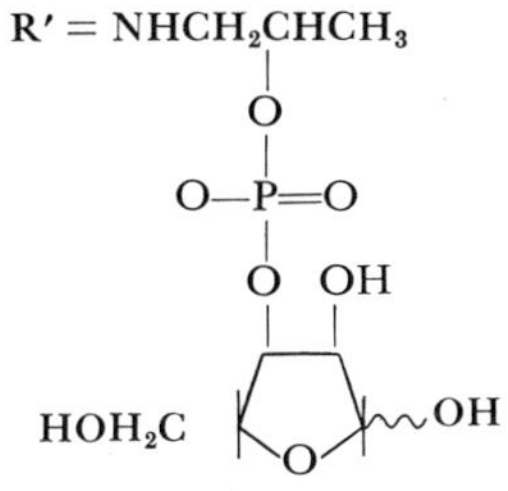

R′ = $NHCH_2CHCH_3$

O

O—P=O

O OH

HOH_2C O OH

7 Cobamic acid, R = OH, R′ as in *6*.

away from normal by the C-20 methyl group. The water molecule attached to cobalt in the β coordination position is hydrogen bonded to the oxygen of the acetamide side chain of ring *B* and is also connected to the carboxylate ion of the propionic acid side chain on ring *D* by way of an intervening water molecule.

In the cobalamins the distortion of the corrin nucleus is quite different (Figure 1-4). The nucleus is now buckled so that the two methyl groups C-35 and C-53 are on the same side of the cobalt-nitrogen plane. Like the cyanide ligand in cyanoaquocobyric acid, the benzimidazole moiety in the cobalamins is tilted away from rings *A* and *D* by the C-20 methyl group. In adenosylcobalamin the 5′-carbon of the nucleoside in the β coordination position is attached to the cobalt atom by a sigma bond normal to the cobalt-nitrogen plane. This organometallic bond is

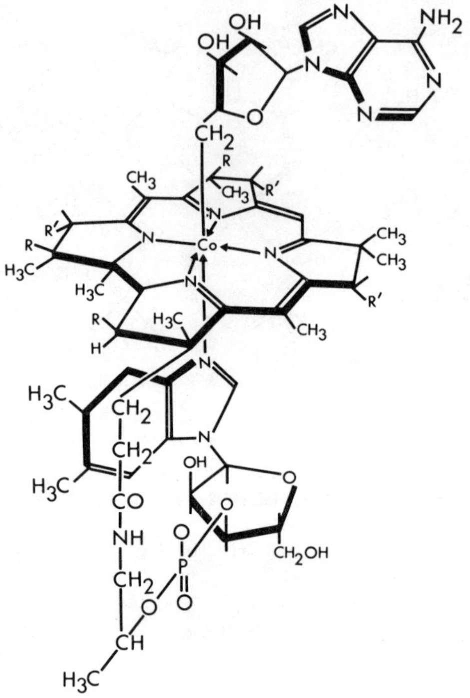

Fig. 1-3. Adenosylcobalamin. R = CH_2COHN_2, R′ = $CH_2CH_2CONH_2$.

surrounded by two methyl groups (C-46 and C-54) and by two methylene groups (C-26 and C-37) at distances of 4 to 5 Å from the 5′-methylene carbon of the nucleoside. Thus these groups and the deoxyadenosine moiety itself serve to protect the unique organometallic bond from the approach of reagents. The bond length of the carbon-cobalt bond is 2.05 ± 0.05 Å, while the Co–C-5′–C-4′ angle (125°) is larger than that observed for a tetrahedral or even trigonal carbon atom (14) (Figure 1-5). Free rotation of the nucleoside ligand about the carbon-cobalt bond is hindered by the interaction between the nucleoside and the corrin ring. For instance, the 6-amino group of the adenine moiety is hydrogen bonded to the carbonyl oxygen of the acetamide side chain of ring *B*. Thus the adenine moiety is positioned partly over ring *C* in a plane at an angle of 20.5° with the plane of the four corrin ring nitrogen atoms. The ribofuranosyl group lies in a plane at an angle of 72.5° with the plane of the corrin ring.

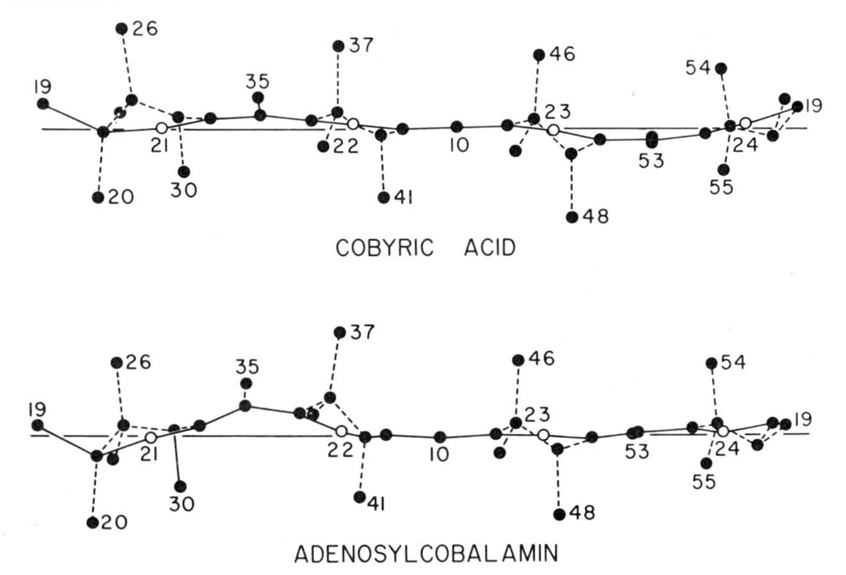

Fig. 1-4. Cylindrical projection of cobyric acid and adenosylcobalamin. The molecules are shown as they would be seen when viewed from the cobalt atom outward ([16)]. The pyrrolidine rings are seen edge on. All four rings are shown, the *A* ring lying on the left. The line running through the structure from left to right represents the least-squares plane through the four pyrrolidine nitrogen atoms (open circles).

The Corrin Nucleus

The nucleus of the corrinoids is resistant to vigorous hydrolysis in acid or alkali. Drastic conditions such as oxidation with chromic acid (15) or with permanganate (16), or distillation with sodium hydroxide (17), are required to disrupt the corrin nucleus and release the cobalt atom. These degradative procedures have been very useful in the elucidation of the structure of cyanocobalamin (7, 9). Attempts to exchange the cobalt atom with radioactive cobaltous salts under a variety of conditions and over prolonged periods of time have been unsuccessful (18, 19).

Acid Hydrolysis of the Peripheral Amide Group

Acid hydrolysis of corrinoids yields mixtures of mono- to heptacarboxylic acids (20-22), all of which have been isolated as homogeneous preparations (23). Mild acid hydrolysis of cyanocobalamin yields a mixture of three monobasic and three dibasic acids as well as one tribasic acid, all derived from the propionamide side chains. One of the monocarboxylic

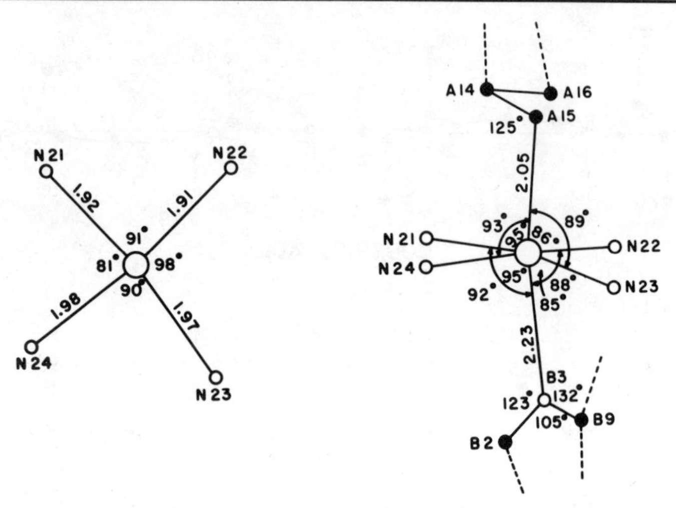

Fig. 1-5. Bond distances and angles of the cobalt ligands in adenosylcobalamin [16]. A14, A15, and A16 correspond to C-4′, C-5′, and the furanose ring oxygen of the 5′-deoxyadenosyl moiety; B2, B3, and B9 correspond to C2, N-3, and C-9 of the 5,6-dimethylbenzimidazole nucleotide.

acids constitutes more than half of the monocarboxylic acids fraction and has been identified as cyano[α-(5,6-dimethylbenzimidazolyl)cobamic acid-*a,b,c,d,g*-pentaamide] (cyanocobalamin-*e*-monocarboxylic acid) by X-ray analysis (24, 25). The acetamide groups are more resistant to hydrolysis. The ease of hydrolysis of the amide groups of cyanocobalamin decreases in the following order: propionamide groups $e >> b, d >$ acetamide groups $c > a, g$. The susceptibility of the *e*-propionamide moiety to acid hydrolysis may be due to intramolecular catalysis by phosphate of the nucleotide ligand. Indeed, an aqueous solution of cyanocobinamide phosphate slowly hydrolyzes to the corresponding *e*-carboxylic acid on standing at room temperature. Conversely, the *e*-carboxyl group catalyzes the hydrolysis of the phosphate ester. Of course, acid hydrolysis under conditions severe enough to cleave the acetamide groups also causes the release of the nucleotide moiety and the isopropanolamine group at the propionic acid side chain *f*. The course of these hydrolysis reactions is very much dependent on the ligands attached to cobalt in the α and β coordination positions. For example, the order of acetamide hydrolysis in aquocobalamin is $g > a, c$, different from that seen with cyanocobalamin (see above). In cyanocobalamin the nucleotide is more susceptible to acid hydrolysis than in aquocobalamin. Acid hydrolysis of aquocobalamin, but not of cyanocobalamin, leads in part to lactone formation;

vigorous acid hydrolysis of cyanocobalamin (2 *N* hydrochloric acid, 100°, 16 hr) yields predominantly the hexa- and heptacarboxylic acids.

Oxidative Cyclization at the B *Ring*

Alkaline hydrolysis of cyanocobalamin with 30% aqueous sodium hydroxide at 150° gives a mixture of corrinoids containing predominantly the nucleotide-free penta- and hexacarboxylic acids, with only a trace of the tetracarboxylic acids. However, these carboxylic acids differ from those obtained by acid hydrolysis. No heptacarboxylic acid can be obtained by further treatment of the hexacarboxylic acid with alkali. The purified hexacarboxylic acid crystallizes as red prisms which are amenable to X-ray analysis. The X-ray data show the presence of four propionic acid but only two acetic acid residues, the third acetamide side chain having undergone cyclization at C-8 to yield a new five-membered ring. Chemical and spectral evidence is consistent with a γ-lactam (Figure 1-6).

Treatment of cyanocobalamin in mild alkali (0.1 *N*) at 100° for 10 min in the presence of air gives rise to a neutral product which resembles cyanocobalamin in physical properties but is virtually devoid of biological activity (26). The presence of reducing agents such as ascorbic acid partially protects the vitamin against inactivation by alkali. The inactive vitamin, called "dehydrovitamin B_{12}," has been characterized as cyano[8-amino-α-(5,6-dimethylbenzimidazolyl)cobamic acid- *a,b,d,e,g*-pentaamide-*c*-lactam]. During the reaction cob(II)alamin is formed as an intermediate, suggesting that lactam formation involves a one-electron oxidation which generates a free radical at C-8. This radical then reacts with the *c*-acetamide side chain to form the γ-lactam (Figure 1-6).

A similar cyclization reaction leads to γ-lactones as by-products in the acid hydrolysis of corrinoids (Figure 1-7). Acid hydrolysis of aquocobalamin (0.1 *M* H_2SO_4, 75° for 6 days) in an atmosphere of hydrogen yields the γ-lactones of tri-, tetra-, and pentacarboxylic acids as major products. During this lactone formation the cobalt atom is also reduced, suggesting that the mechanisms of lactam formation in alkali and lactone formation in acid are similar. Lactone formation is hindered when

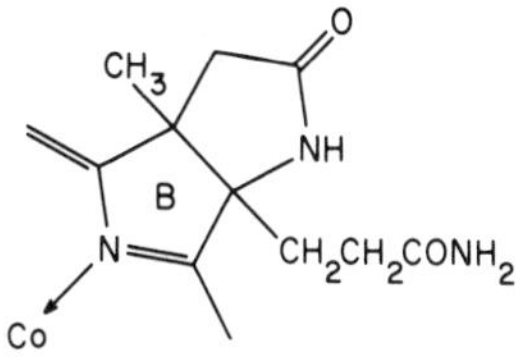

Fig. 1-6. Partial structure of cyano[8-amino-α-(5,6-dimethylbenzimidazolyl)cobamic acid-*a, b, d, e, g*-pentamide-*c*-lactam].

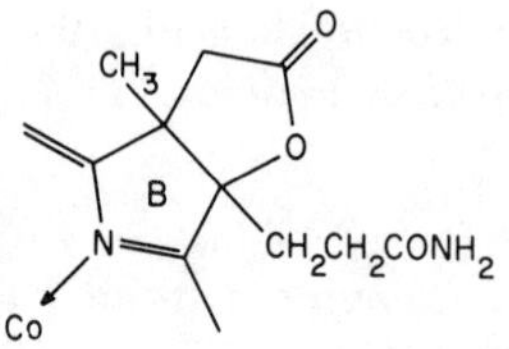

Fig. 1-7. Partial structure of cyano[8-hydroxy-α-(5,6-dimethylbenzimidazolyl)cobamic acid-*a,b,d,e,g*-pentamide-c-lactone]

a strong ligand is coordinated in the β coordination position. For instance, acid hydrolysis of cyanocobalamin or methylcobalamin does not lead to lactone formation. In these cobalamins the cobalt atom is much more resistant to reduction. With these compounds, amide hydrolysis of side chain *c* is faster than lactonization (22).

Treatment of cyanocobalamin with an equimolar amount of chloramine-T or bromine water at pH 4 gives rise to a neutral, halogen-free compound which has been characterized as cyano[8-hydroxy-α-(5,6-dimethylbenzimidazolyl)cobamic acid-*a,b,d,e,g*-pentaamide-*c*-lactone]. This reaction probably involves oxidation at C-8 followed by an attack of the carbonyl oxygen of the acetamide side chain (27). The lactone can be hydrolyzed to the hydroxy acid by gentle treatment, while vigorous alkaline hydrolysis gives rise to a heptacarboxylic acid.

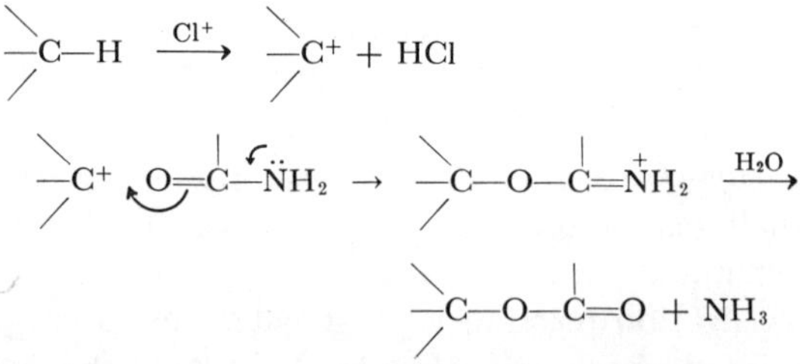

$$\rangle\!-C-H \xrightarrow{Cl^+} \rangle\!-C^+ + HCl$$

$$\rangle\!-C^+ \; O{=}C-\ddot{N}H_2 \rightarrow \rangle\!-C-O-C{=}\overset{+}{N}H_2 \xrightarrow{H_2O}$$

$$\rangle\!-C-O-C{=}O + NH_3$$

The extent of both lactam and lactone formation depends on the nature of the ligand in the β coordination position. Methylcobalamin, adenosylcobalamin, or sulfitocobalamin in 0.1 *N* NaOH at 100° in the presence of oxygen for 10 min yield no lactam. Under the same conditions the γ-lactam is formed from both cyano- and aquocobalamin. Similarly, treatment of cyano- or aquocobalamin with an equimolar amount of chloramine-T gives the lactone, while chloramine-T treatment of methylcobalamin or adenosylcobalamin causes chlorination of the corrin ring at C-10 (28).

Halogenation

When cyanocobalamin or aquocobalamin is treated with an excess of chloramine-T, chlorine-containing lactones are formed. Although these corrinoids have not been fully characterized, the spectral evidence suggests that an electrophilic substitution in the conjugated system has occurred. From a consideration of the electron availability at various positions of the corrin ring, Bonnett and co-workers (9) suggested that halogenation occurs at C-10 to yield cyano[8-hydroxy-10-chloro-α-(5,6-dimethylbenzimidazolyl)cobamic acid-*a,b,d,e,g*-pentaamide-*c*-lactone] Figure 1-8). Treatment of the γ-lactam of cyanocobalamin with 1 equiv of halogenating agent leads to substitution of chlorine at C-10. Similar reactions with *N*-bromosuccinimide leading to the C-10 brominated corrinoids have been described by Renz (29) and Koppenhagen (30): however, the reaction between cyanocobalamin and *N*-iodosuccinimide does not yield the C-10 iodinated cobalamin. Halogenation of cyanoaquocobinamide with chloramine-T, *N*-bromosuccinimide, or *N*-iodosuccinimide gives the C-10 halogenated cobinamides as major products (the cobinamide γ-lactones are minor products). However, C-10 iodocobinamide is very unstable and decomposes to cyanoaquocobinamide. These electrophilic substitutions of halogens at C-10 cause a marked bathochromic shift in the absorption spectra of the corrinoids.

Nitration

Reaction of cyanocobalamin in acetic acid with 2 equiv of nitrosyl chloride gives a new brick-red corrinoid. The change in the absorption spectrum again suggests that electrophilic substitution of the corrin ring has occurred. The same brick-red corrinoid is formed when cyanocobalamin is treated with nitryl chloride, indicating that the nitroso group introduced by the nitrosyl chloride reaction is oxidized to a nitro function. Spectral and chemical evidence is consistent with a nitration at

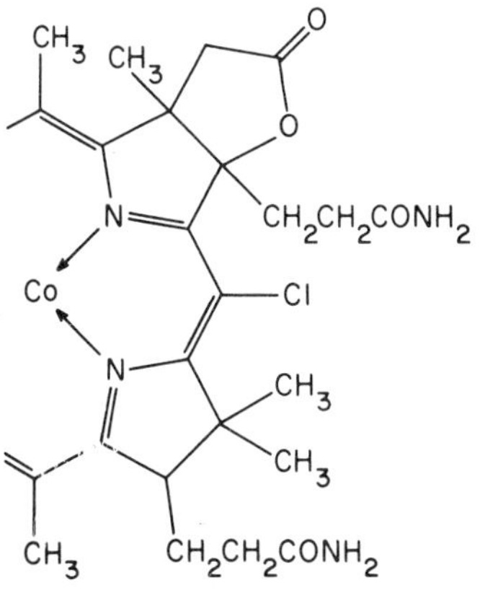

CH_3 CH_3 O O N $CH_2CH_2CONH_2$ Co Cl N CH_3 CH_3 CH_3 $CH_2CH_2CONH_2$

Fig. 1-8. Partial structure of cyano[8-hydroxy-10-chloro-α-(5,6-dimethylbenzimidazolyl)cobamic acid-*a,b,d,e,g*-pentamide-c-lactone].

C-10 of the corrin ring. Sodium borohydride reduction of cyano-10-nitrocobalamin yields cyano-10-aminocobalamin (30).

Isomerization

When cyanocobalamin is treated with trifluoroacetic acid at room temperature for 2 hr, a mixture of cyanocobalamin, cyanocobinamide, and two darker corrinoids are formed. The new corrinoids were named cyanoneocobalamin (neovitamin B_{12}) and cyanoaquoneocobinamide, respectively (31-33). Similar neocorrinoids are formed when corrinoid carboxylic acids such as cobyric acid are dissolved in strong acid. Treatment of corrinoids with strong acid gives an equilibrium mixture of the corrinoids and the neocorrinoids. This equilibration has also been observed in 35% HCl, 42% HBF_4, and 60% $HClO_4$, but anhydrous trifluoroacetic acid is the reagent of choice because hydrolytic side reactions occur in the aqueous acids. The neocorrinoids are virtually indistinguishable from the corrinoids on the basis of electrophoretic behavior or infrared spectra, and the electronic spectra show only small, though significant, differences. The ORD and CD spectra, however, are markedly different. The structure of cyanoneocobalamin has been established by X-ray analysis. The X-ray data show that the propionamide side chain *e* attached to C-13 is projected up instead of down relative to the plane of the corrin ring (Figure 1-9), and thus neocorrinoids should be named 13-epicorrinoids.

The Nucleotide Moiety

Acid hydrolysis of cyanocobalamin in 6 *N* hydrochloric acid at 150° for 20 hr yields 5,6-dimethylbenzimidazole (Figure 2-10) (34). Under these conditions the sugar moiety is destroyed. Such drastic conditions are required because the *N*-glycosidic linkage of the benzimidazole ribosides is very resistant to acid hydrolysis. The purine-containing cobamides such as *Coα*-adenyl-*Coβ*-cyanocobamide [(Ade)CN-Cba] resemble the purine ribonucleotides in acid lability, hydrolysis occurring much more readily (0.05 *N* HCl, 100°, 15 min) (35). Acid hydrolysis of cyanocobalamin in 6 *N* hydrochloric acid at 100° for 8 hr yields 1-α-D-ribo-

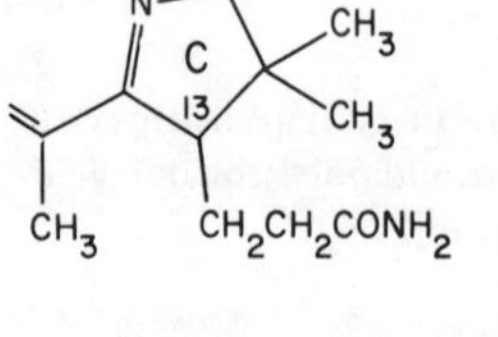

Fig. 1-9. Partial structure of cyano-13-epicobalamin.

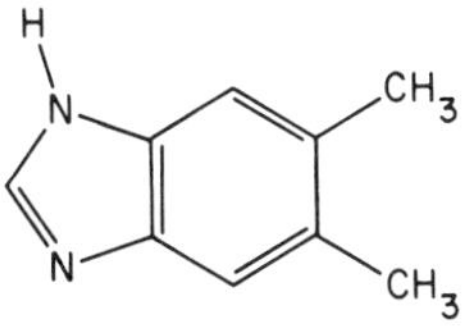

Fig. 1-10. 5,6-Dimethylbenzimidazole.

furanosyl-5,6-dimethylbenzimidazole (Figure 1-11). Final identification of this novel nucleoside was accomplished by synthesis (36). It is noteworthy that the nucleoside has an α configuration at the anomeric carbon while other naturally occurring nucleosides have a β configuration.

A second, more elegant, method for the hydrolysis of the phosphodiester linkages has been described by Friedrich and Bernhauer (37). Treatment of cyanocobalamin in a suspension of cerous hydroxide, pH 8 to 9, at 100° for 2 hr results in the hydrolysis of both phosphodiester linkages, with the release of 1-α-D-ribofuranosyl-5,6-dimethylbenzimidazole, phosphate, and cyanoaquocobinamide. The addition of cyanide ions increases the rate of hydrolysis by cleaving the coordinate bond between cobalt and N-3 of the nucleotide. Using this procedure it is possible to isolate the intact nucleoside, 7-α-D-ribofuranosyladenine, from a hydrolysis reaction of (Ade)CN-Cba (Figure 1-12) (38).

Acid hydrolysis of cyanocobalamin under mild conditions releases the 5,6-dimethylbenzimidazole nucleotide (39, 40). This nucleotide has been characterized as 1-α-D-ribofuranosyl-5,6-dimethylbenzimidazole-3′-phosphate by X-ray analysis (41). Actually the hydrolysis of cyanocobalamin under acid or alkaline condtions gives a mixture of the 2′ and 3′ nucleotides (42).

All the acid hydrolyses under rather severe conditions result in the release of ammonia and of 1-amino-2-propanol (43). The stereochemistry of this amino alcohol has been elucidated by synthesis (44).

$$\begin{array}{c} CH_2NH_2 \\ | \\ H{-}C{-}OH \\ | \\ CH_3 \end{array}$$

(R)-1-Amino-2-propanol

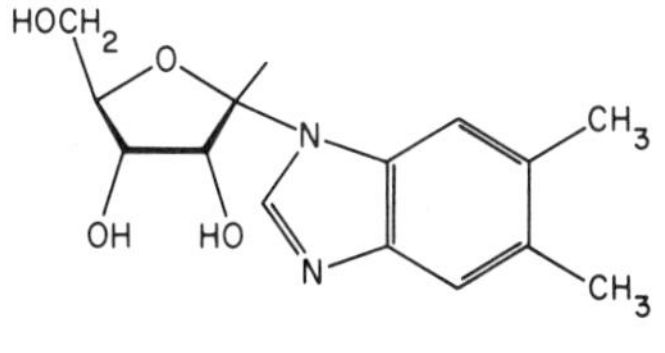

Fig. 1-11. 1-α-D-Ribofuranosyl-5,6-dimethylbenzimidazole.

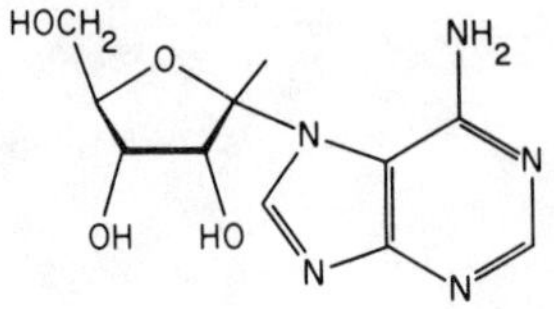

Fig. 1-12. 7-α-D-Ribofuranosyladenine.

The Cobalt Atom

Magnetic susceptibility measurements of cyanocobalamin have shown that the molecule is diamagnetic in the solid state (45, 46) and in solution (47). Similar measurements showed that aquocobalamin (46, 48), nitritocobalamin (49), dicyanocobalamin (48), and adenosylcobalamin (48, 50) are also diamagnetic. X-Ray absorption edge measurements have demonstrated that cyanocobalamin contains trivalent cobalt (51). Cyanocobalamin and the other cobalamins are therefore cobaltic complexes with d^2sp^3 octahedral bonding. In cyanocobalamin four coordinate positions are occupied by the four ring nitrogens while the α and β axial positions are satisfied by N-3 of the 5,6-dimethylbenzimidazole nucleotide and the cyanide ion, respectively. The organocobalamins such as methylcobalamin and adenosylcobalamin can be regarded as hexacoordinate complexes having a carbanion in the β coordination position.

In accord with its diamagnetism, adenosylcobalamin does not give an electron spin resonance (esr) spectrum either as the crystalline solid or in frozen solution (52). Cyanocobalamin is uncharged in neutral solution, the triple charge of the cobalt atom being neutralized by the negative charges of the cyanide ion, one of the corrin ring nitrogens, and the phosphate of the nucleotide. In contrast, aquocobalamin, with a water molecule coordinated to cobalt in the β axial position, has a net positive charge. The organocobalamins such as the alkylcobalamins and adenosylcobalamin are uncharged in neutral solution.

In acid solution the cobalamins acquire a positive charge. Under these conditions the coordinate linkage between cobalt and the nucleotide base is opened while at the same time N-3 of the 5,6-dimethylbenzimidazole moiety is protonated. Water is then coordinated to cobalt in the α coordination position (53, 54).

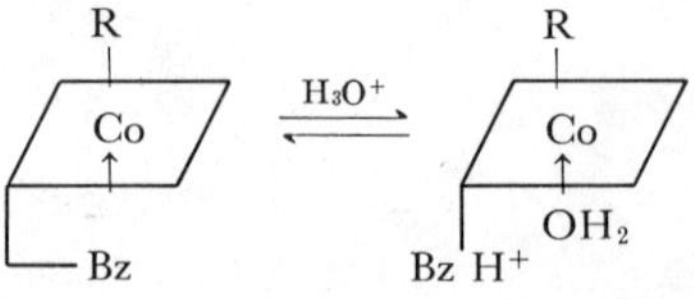

R — Co ← Bz $\underset{}{\overset{H_3O^+}{\rightleftharpoons}}$ R — Co ← OH_2, Bz H^+

The protonated forms (the "base off" forms) have an absorption spectrum very similar to that of the cobinamides. The ease of this reaction depends on the nature of the ligand in the β coordination position, very strong acid being required to protonate aquocobalamin (pK −2.4) while organocobalamins such as methylcobalamin are protonated much more readily (pK 2.7) (55, 56).

Redox Reactions

The cobalt atom of the corrinoids can exist in three possible oxidation states: Co(III), Co(II), and Co(I). Reduction of a solution of cyanocobalamin with hydrogen and a platinum catalyst causes a color change from red to brown (57) (Figure 1-13). The brown compound, cob(II)-alamin, can be oxidized to aquocobalamin with 1 equiv of ferricyanide and thus contains divalent cobalt. Similar conclusions were reached in several laboratories using such methods as controlled potential reduction (58) and polarography (59). Solutions of cob(II)alamin are oxidized to aquocobalamin in the presence of air, but solid dry cob(II)-alamin is quite stable even in the presence of oxygen (60). Cob(II)alamin is also formed when aquocobalamin is treated with carbon monoxide (61) or with monothiols (62, 63). In the presence of oxygen, aquocobalamin catalyzes the oxidation of carbon monoxide to carbon dioxide and the oxidation of monothiols to disulfides. Corrinoids with strong trans ligands such as cyanocobalamin and the alkylcobalamins do not function in

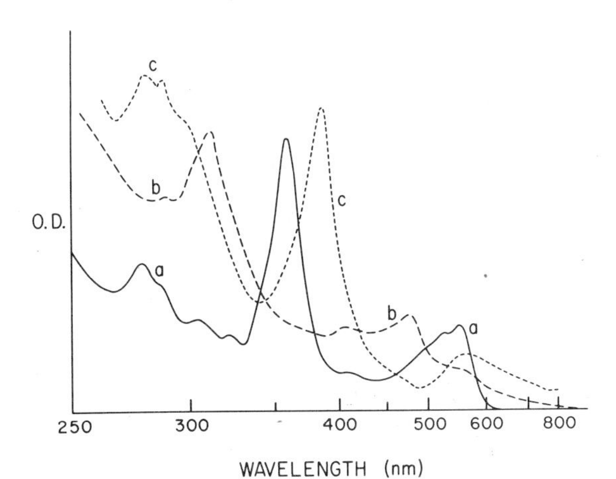

Fig.1-13. Absorption spectra of (*a*) cyanocobalamin, (*b*) cob(II)alamin, and (*c*) cob(I)-alamin [(59)].

these reactions; indeed, diaquocobinamide and diaquocobyric acid are most effective in catalyzing the oxidation of carbon monoxide (61, 64, 65). At low temperatures the Co (II)-containing corrinoids react with oxygen to form paramagnetic complexes. The esr spectra of these complexes suggest that this oxygen adduct of cob(II) alamin is best represented as a superoxide anion coordinated to cob(III)alamin (66, 67).

$$Cbl^{II} + O_2 \rightleftharpoons Cbl^{III}\ O_2^{\cdot -}$$

Controlled potential reduction of aquocobalamin at pH 7 at 0.7 V versus the standard silver-silver chloride reference electrode is a very convenient method for the preparation of cob(II)alamin because, unlike the chemical reductions, no excess of reducing agent is present (68). However, this one-electron reduction is only possible for aquocobalamin, since cobalamins such as cyanocobalamin with a strong nucleophile in the β coordination position undergo a two-electron reduction to cob(I)alamin (69).

Co (II)-containing corrinoids are also formed when organocorrinoids are decomposed by light. Photolysis of an aqueous solution of adenosylcobalamin in the absence of oxygen yields cob(II)alamin (70). In the absence of oxygen, solutions of cob(II)alamin are indefinitely stable, but on exposure to oxygen they are oxidized to aquocobalamin. This oxidation reaction is quite slow ($t_{1/2} = 21$ min) (71) and thus the very fast formation of aquocobalamin upon photolysis of alkylcobalamins in the presence of oxygen suggests that during photolysis cob(II)alamin is oxidized by a more powerful oxidizing agent such as an alkyl peroxide (71, 72).

Reduction of Co(III)- or Co(II)-containing corrinoids with more powerful reducing agents such as sodium borohydride, chromous chloride at pH 9.5, zinc in acetic acid, or zinc in ammonium chloride leads to Co(I) corrinoids. For instance, treatment of a solution of cyanocobalamin with sodium borohydride causes a rapid color change from red to brown [cob(II)alamin] followed by a slower color change to gray-green (Figure 1-13). Addition of cobaltous ions greatly accelerates the rate of the second reduction (68). The gray-green cobalamin [cob(I)alamin] has been shown to contain univalent cobalt by amperometric titration of cyanocobalamin with chromous chloride and by controlled potential reduction of cyanocobalamin and hydroxocobalamin (58, 73, 74). Originally cob(I)alamin was formulated as a hydride and called hydridocobalamin (75). It has been shown (76), however, that in aqueous solution it exists predominantly as the unprotonated species. Cobalt(I) corrinoids are formed by "self-reduction" of cyanocobalamin in strong alkali (77), and cob(I)alamin is produced upon alkaline decomposition of some organocobalamins (78). More recently Schrauzer and Holland (79) have reported that hydridocobalamin can be formed when aquocobalamin is reduced with zinc dust in anhydrous acetic acid under strict exclusion of oxygen. Hydridocobalamin can be considered the conjugate acid of cob(I)alamin and is probably formed at low pH.

$$\mathrm{Cbl^{I} + H^{+} \rightleftharpoons Cbl^{I}\cdot H^{+} \leftrightarrow Cbl^{II}\cdot H \leftrightarrow Cbl^{III}\cdot H^{-}}$$

Aqueous solutions of cob(I)alamin decompose to cob(II)alamin and hydrogen, the rate of this reaction showing dependence on the pH and on the nature of the buffer anion. In aqueous solution cob(I)alamin has a half-life which varies from 355 min at pH 10 to 87 min at pH 8. Cobalt(I) corrinoids are extremely sensitive to oxygen, exposure of an aqueous solution of cob (I)alamin to air causing almost instantaneous oxidation to cob(II)alamin (80).

The reduction of nitrobenzene, nitrosobenzene, and azoxybenzene by potassium borohydride is catalyzed by hydroxocobalamin. This reaction involves the formation of cob(I)alamin which in turn reduces the aromatic compounds (81).

ORGANOCORRINOIDS

Alkyl- and Acylcorrinoids

Synthesis

The organometallic bond of the alkyl- and acylcorrinoids can be formed by three different routes: (a) reaction of Co(I) corrinoids with electrophilic agents, (b) reaction of Co(III) corrinoids with nucleophilic agents, and (c) reaction of Co(II) corrinoids with radicals. The reaction of Co (I) corrinoids with electrophilic agents is the most widely used route to organocorrinoids.

CO(I) CORRINOIDS AND ELECTROPHILIC AGENTS

Almost immediately after the structure of adenosylcobalamin had been elucidated by X-ray crystallography, methods were developed for the synthesis of the carbon-cobalt bond (75, 82). The main corrinoid starting materials for these syntheses are cyanocobalamin or aquocobalamin. Reduction of these two cobalamins with a variety of reducing agents leads to cob(I)alamin which displaces the tosylate ion from 2′,3′-isopropylidene-5′-tosyladenosine to yield 2′,3′-isopropylidene-5′-deoxy-5′-adenosylcobalamin. Removal of the isopropylidene protecting group with mild acid gives adenosylcobalamin. The unambiguous synthesis of the 5′-*O*-*p*-toluenesulfonyl nucleosides requires prior protection of the 2′- and 3′-

hydroxyl function of the ribonucleosides and the 3′-hydroxyl function of the deoxyribonucleosides. The adjacent hydroxyl groups of the ribonucleosides are conveniently protected by reaction with acetone, *p*-dimethylaminobenzaldehyde, *p*-anisaldehyde (83), or diphenylcarbonate (84). The unambiguous synthesis of the 5′-tosyl derivatives of the *arabino*-, *xylo*-, or 2′-deoxyribonucleosides requires a reaction sequence involving tritylation, acetylation, and detritylation. Using such reaction sequences numerous analogs of adenosylcobalamin, differing in the base and/or sugar moiety, have been prepared. Various analogs have also been synthesized without the aid of the isopropylidene protecting group, but generally in lower yield due to the tendency of the unprotected tosyl derivatives to cyclize.

Adenosylcobalamin and some of its analogs can also be prepared by reacting cob(I)alamin with 5′-halogenated nucleosides. A very convenient method for the preparation of these nucleosides has been described by Kikugawa and Ichino (85). Thionyl chloride or thionyl bromide mixed with hexamethylphosphoramide selectively halogenates the 5′ position of the ribonucleoside and thus no prior blocking of the 2′- and 3′-hydroxyl functions is required.

Cob(I)alamin is the most powerful nucleophile known to date. With a nucleophilic reactivity constant $n_{CH_3I} = 14.4$ (80), it undergoes rapid substitution and addition with a variety of electrophiles. The reaction of cob(I)alamin with alkyl halides and acyl halides proceeds by way of a classical S_N2 reaction. Nucleophilic addition to alkenes occurs only if the olefin is activated by electron-withdrawing substituents, while addition to acetvlenes occurs directly (68). Cob(I)alamin also causes ring openings of the cyclic ethers ethylene oxide and tetrahydrofuran to yield hydroxyethyl- and hydroxybutylcobalamin, respectively.

$$
\begin{array}{lll}
[\mathrm{Co^I}] + & \mathrm{RX} & \rightarrow \mathrm{Co{-}R} + \mathrm{X^-} \\
 & \mathrm{RCOX} & \rightarrow \mathrm{Co{-}COR} + \mathrm{X^-} \\
 & \mathrm{RC{\equiv}CX} & \rightarrow \mathrm{Co{-}C{\equiv}CR} + \mathrm{X^-} \\
 & \mathrm{H_2C{=}CH{-}COR} + \mathrm{H^+} & \rightarrow \mathrm{Co{-}CH_2CH_2COR} \\
 & \mathrm{HC{\equiv}CR} + \mathrm{H^+} & \rightarrow \mathrm{Co{-}CH{=}CHR} \\
 & \underbrace{\mathrm{H_2C{-}CH_2}}_{\mathrm{O}} + \mathrm{H^+} & \rightarrow \mathrm{Co{-}CH_2{-}CH_2OH}
\end{array}
$$

The scope of these reactions is only limited by steric hindrance and by the stability of the cobalamin during the synthesis or isolation procedure. No reaction is observed between neopentyl chloride or neopentyltosylate and cob(I)alamin, while the reaction of cob(I)alamin with secondary alkyl halides is fast but the resulting alkylcobalamins are not stable enough to be isolated. Some analogs such as hydroxyethyl- and methoxyethylcobalamin are sensitive to acid while others such as cyanoethylcobalamin and methoxycarbonylethylcobalamin are base labile. For the preparation of these cobalamins neutrality must be maintained during the synthesis and purification. Of course all organocorrinoids are light sensitive and thus their preparation and isolation should be carried out in dim light.

The susceptibility of the cobalt atom to alkylation is markedly affected by the nature of the ligand in the α coordination position. Although cob (I)alamin reacts with secondary alkyl halides, the product of this reaction is not stable enough to be isolated. In contrast, cob(I)inamide reacts with secondary alkyl halides to yield stable alkylcobinamides. For instance, reaction of the cob(I)inamide with cyclohexyl halide yields cyclohexylcobinamide (86). Apparently the absence of a strong ligand in the α coordination position allows the cobalt atom to remain above the plane of the corrin ring and allows alkylation by bulky alkylating agents. In the presence of a strong ligand (dimethylbenzimidazole or cyanide) the cobalt atom is forced into the plane of the corrin ring and as a result steric hindrance precludes the formation of a stable carbon-cobalt bond. Indeed, when coordination of the 5,6-dimethylbenzimidazole is prevented by methylation at N-3, reaction with cyclohexyl halides is possible to give *Coα*-(3,5,6-trimethylbenzimidazolyl)-*Coβ*-cyclohexylcobamide (86).

In aqueous acid cob(I)alamin is unstable and decomposes to cob(II)alamin and molecular hydrogen (76). Under these conditions cob(I)alamin is probably protonated to hydridocobalamin, which is an intermediate in the evolution of hydrogen. However, if aquocobalamin is reduced under conditions which prevent the decomposition of hydridocobalamin, the protonated Co(I) nucleophile can be prepared (79). Reduction of aquocobalamin with zinc dust in anhydrous acetic acid with the exclusion of oxygen yields a green solution. The absorption spectrum of this cobalamin is distinctly different from that of cob(I)alamin generated by reduction of aquocobalamin with sodium borohydride or with zinc dust in aqueous ammonium chloride. Furthermore, the chemical reactivities of hydridocobalamin and cob(I)alamin are strikingly different. Whereas nucleophilic addition of cob(I)alamin to alkenes occurs only if the olefin is activated, hydridocobalamin in anhydrous acetic acid reacts with unactivated olefins such as ethylene and propylene. Reaction with ethylene yields ethylcobalamin and reaction with propylene yields base off isopropylcobalamin, suggesting that the addition of hydridocobalamin to propylene proceeds through a secondary carbonium ion. Like cob(I)inamide, hydridocobalamin reacts with bulky electrophiles to give the corresponding alkylcobalamins in the base off form. Apparently the bulky groups keep the cobalt atom above the plane of the corrin ring system, thus preventing coordination of the base in the lower coordination position.

Co(III) CORRINOIDS AND NUCLEOPHILIC AGENTS

Organocorrinoids can also be prepared by reacting Co(III) corrinoids directly with nucleophilic agents such as alkylmagnesium halides or alkyllithiums. However, the scope of these reactions is very limited because most corrinoids are insoluble in the solvents in which these reactions have to be carried out (28).

Recently a very interesting reaction between a Co(III) corrinoid and electron-rich olefins has been described (87). Reaction of hydroxocobalamin with ethyl vinyl ether in anhydrous ethanol gives 2,2-diethoxyethylcobalamin, while in aqueous ethanol formylmethylcobalamin is also formed. Furthermore, reaction between hydroxocobalamin and 2-hydroxyethyl vinyl ether yields the cyclic acetal which in mild alkali hydrolyzes to formylmethylcobalamin. These reactions have been formulated as proceeding by way of a [Co(III)]π complex which in the presence of nucleophilic agents gives the organocorrinoids.[1]

$$[Co(III)] + H_2C{=}C(H)(OCH_2CH_3) \rightleftharpoons H_2C{-}C(H)(OCH_2CH_3) \cdots^{+} [Co(III)]$$

$$\xrightarrow{CH_3CH_2OH} (CH_3CH_2O)_2CH{-}CH_2{-}[Co(III)] \qquad \xrightarrow{H_2O} OHC{-}CH_2{-}[Co(III)]$$

Co(II) CORRINOIDS AND RADICALS

Cobaloximes are complexes of cobalt with dimethylglyoxime which in certain respects are very similar to cobalamins (89). The cobalt atoms of

[1] Michaely and Schrauzer (88) reported that they were unable to confirm this work. More recent investigations in Schrauzer's laboratory have indicated, however, that under suitable conditions vinyl ethers will react with Co(III) corrinoids as reported by Silverman and Dolphin (88a).

cobaloximes can be alkylated to form organocobalt derivatives in which the carbon-cobalt σ bond is stable to air and water, but is cleaved homolytically by light. Methylcobaloxime, the simplest organocobaloxime, is prepared by reaction of methyl iodide with the Co(I) form of cobaloxime. Photolysis of methylcobaloxime under nitrogen in the presence of cob(II)alamin yields some methylcobalamin, the methyl radical produced in the photolysis reaction combining with cob(II)alamin to give a new carbon-cobalt bond (90). Similarly, heating a solution containing ^{14}C-methylcobalamin and ^{12}C-methylcobinamide in an atmosphere of nitrogen yields ^{14}C-methylcobinamide (91). *Co*-Methylcobyrinic acids are isomerized by heating in an atmosphere of carbon monoxide or by photolysis under anaerobic conditions (92). Both reactions probably involve a homolytic cleavage of the carbon-cobalt bond followed by reaction of the methyl radical with a Co(II) corrinoid.

Reactions

Cleavage of the carbon-cobalt bond of the organocorrinoids can occur by a homolytic or a heterolytic mechanism. Homolytic cleavage leads to cob(II)alamin and an alkyl radical while heterolytic cleavage yields either cob(I)alamin and a carbonium ion or cob(III) alamin and a carbanion.

$$[Co]—R \rightarrow [Co(II)] + R^{\circ}$$
$$[Co(I)] + R^{+}$$
$$[Co(III)] + R^{-}$$

Of course in most cases the reaction products cannot be detected because they undergo further reaction or because the cleavage of the carbon-cobalt bond is concerted with attack of the reagent.

HOMOLYTIC CLEAVAGE OF THE CARBON-COBALT BOND

Photolysis. One of the most striking properties of the organocorrinoids is their instability in light. Exposure of a solution of adenosylcobalamin to light causes loss of coenzyme activity and a change in the absorption spectrum. The first reaction in the photodecomposition is the homolytic cleavage of the organometallic bond to a 5′-deoxyadenosyl radical and cob(II)alamin (52, 70). In the absence of oxygen, cob(II)alamin is indefinitely stable, but the 5′-deoxyadenosyl radical cyclizes to give 8,5′-*cyclic*-adenosine (93) (Figure 1-14). When the photolysis is carried out in the presence of oxygen cob(II)alamin is oxidized to aquocobalamin and the 5-deoxyadenosyl radical is converted to 8,5′-*cyclic*-adenosine and to adenosine-5′-carboxaldehyde (94) (Figure 1-15). The formation of the cyclic nucleoside is suppressed if oxygen is bubbled through the solution during

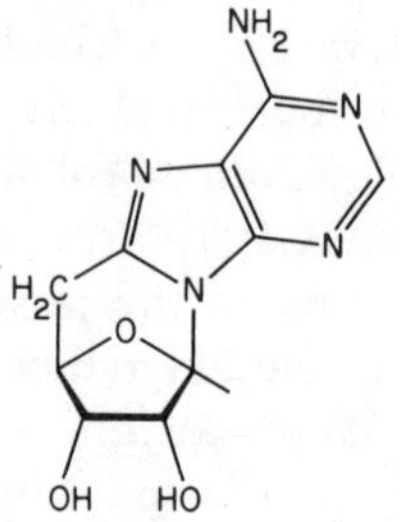

Fig. 1-14. 8,5′-*Cyclic*-adenosine.

photolysis (95). When the anaerobic photolysis is carried out in the presence of a thiol such as homocysteine, 8,5′-*cyclic*-adenosine and a thioether (*S*-adenosylhomocysteine) are formed (96). The photolysis of adenosylcobalamin shows first-order kinetics and the rate is not significantly affected by the oxygen concentration. All analogs of adenosylcobalamin differing from the coenzyme in the nucleoside moiety are photosensitive. For instance, photolysis of inosylcobalamin and 2′-deoxyadenosylcobalamin in the absence of oxygen yields 8,5′-*cyclic*-inosine and 8,5′-*cyclic*-2′-deoxyadenosine, respectively, as well as cob(II)alamin (75, 97). However, the photolytic decomposition of 2′,3′-isopropylideneuridinylcobalamin in the absence of oxygen yields aquocobalamin and 2′,3′-isopropylidenecyclodihydrouridine (Figure 1-16). Evidently the 5′-deoxyuridinyl radical produced as the initial photolysis product first undergoes a cyclization by addition to the 5,6 double bond of the pyrimidine ring and then is reduced by cob(II)alamin to give 2′,3′-isopropylidenecyclodihydrouridine and aquocobalamin (98).

In contrast to the photolysis of adenosylcobalamin, the photolysis of the alkylcorrinoids is very dependent on the reaction conditions. For instance, the photolysis of methylcobalamin in aqueous solution in the presence of oxygen is fast ($k = 1.9 \times 10^{-2}$ sec^{-1}) and yields aquocobalamin and formaldehyde as the major products, with only traces of meth-

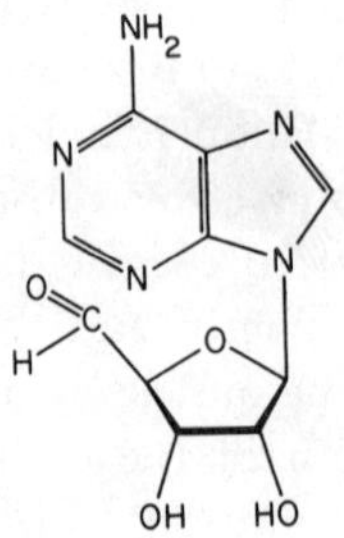

Fig. 1-15. Adenosine-5′-carboxaldehyde (adenine-9-β-D-ribopentofuranosyldialdose).

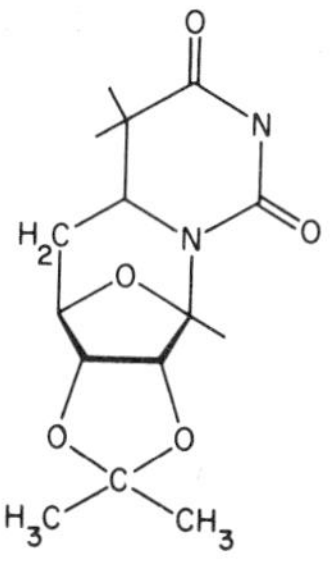

Fig. 1-16. 2′,3′-Isopropylidenecyclodihydrouridine.

anol, methane, and formic acid (71, 72, 99). On the other hand, the photolysis of methylcobalamin under anaerobic condtions is very slow and yields cob(II)alamin, methane, and ethane as the major products (71, 72). Thus the rate of photolysis and the nature of the photolytic products depend on secondary reactions of the Co(II) corrinoid and the alkyl radical. In the absence of oxygen the initial homolytic cleavage of the carbon-cobalt bond can be considered a reversible reaction because cob(II)alamin is a good radical scavenger and recombines with alkyl radical.

$$[Co]—CH_3 \rightleftharpoons [Co(II)] + \cdot CH_3$$

In the presence of oxygen the alkyl radicals are removed by secondary reactions and thus the reverse reaction is eliminated.

The rate of anaerobic photolysis is greatly increased by thiols, quinones, and certain alcohols. Photolysis of methylcobalamin in the presence of cysteine or homocysteine yields S-methylcysteine and methionine, respectively, while 2-methyl-1,4-naphthaquinone is formed when methylcobalamin is photolyzed in the presence of 1,4-naphthaquinone (96, 99). In 1 *M* aqueous alcohol the rate of photolysis is also significantly increased, alcohols with a readily abstractable α-hydrogen accelerating the reaction most effectively. Photolysis of methylcobalamin dissolved in 2-propanol yields cob(II)alamin, pinacol, and presumably methane, suggesting that the methyl radical formed in the primary photolysis reaction has abstracted the α-hydrogen from the alcohol to give methane and a 2-propanol radical. Coupling of 2-propanol radicals gives pinacol (100).

Photolysis of alkylcobalamins with larger alkyl moieties is similar to that of methylcobalamin. In the presence of oxygen, aquocobalamin and aliphatic aldehydes are the major products. However, anaerobic photolysis of these alkylcobalamins is faster than that of methylcobalamin and yields mainly olefins. The anaerobic photolysis of ethylcobalamin yields ethylene, traces of ethane and butane, cob(II)alamin, and cob(I)alamin.

Evidently the conversion of the ethyl radical to ethylene is accompanied by a reduction of cob(II)alamin to cob(I)alamin (90, 99, 101). The rate of photolysis of the carbon-cobalt bond is affected by the trans ligand, but the rate differences are not very striking. Methylcobalamin in acid solution (base off) and methylaquocobinamides are photolyzed at a slower rate than "base on" methylcobalamin, while methylcyanocobinamides are photolyzed more readily than methylaquocobinamides (102).

The photolysis of 2,3-dihydroxypropylcobalamin in air yields aquocobalamin and a mixture of glyceraldehyde and glyceric acid, while photolysis in a nitrogen atmosphere yields cob(II)alamin and glycerol as the main products (103).

Photolysis of carboxymethylcobalamin in air gives aquocobalamin, carbon dioxide, and acetic acid as the major products (104). Photolysis in a nitrogen atmosphere yields acetic acid with smaller amounts of succinic, glycolic, oxalic, and diglycolic acids. Ethyl acrylate is formed upon anaerobic photolysis of 2-ethoxycarbonylethylcobalamin (99).

The products of the photolysis of mono-, di-, and trichloromethylcobalamin in an atmosphere of hydrogen are methyl chloride and cob(II)alamin. The last two cobalamins also yield 1 and 2 equiv of hydrochloric acid, respectively. Studies with ^{14}C-labeled dichloromethylcobalamin have suggested that chlorocarbene may be an intermediate in the photolysis (105, 106).

The quantum yields and action spectra for the aerobic photolysis of several alkylcobalamins were studied by Taylor et al. (106a). In general, the action spectra resembled the absorption spectra. Quantum yields ranged between 0.2 and 0.4 except for adenosylcobalamin, where the yield was of the order of 0.1.

Thermolysis. Solid methylcobalamin under argon decomposes between 215 and 225° to yield approximately equal amounts of methane and ethane. Because these products are very similar to those formed upon photolysis of methylcobalamin, the initial step in the pyrolysis reaction probably involves a homolytic cleavage of the carbon-cobalt bond. At early stages of the pyrolysis more methane is formed, suggesting that abstractable hydrogens are utilized before methyl radical coupling occurs. In the presence of air more ethane is formed. Pyrolysis of ethylcobalamin yields primarily ethylene with traces of ethane and butane (90). The nature of the corrinoid produced has not been established, but since cyanocobalamin crystals darken at 210 to 220° (7) the corrinoid has probably undergone considerable alteration.

Isomerization. The reaction of cob(I)alamin with methyl iodide produces predominantly methylcobalamin with the methyl ligand bound to cobalt in the β coordination position. However, its isomer with the methyl group bound to cobalt in the α coordination position is also formed, though in very low yield. This second form of methylcobalamin, *Co*α-methyl-*Co*β-aquocobalamin, is yellow and its absorption spectrum is very similar to that of *Co*β-methylcobalamin in the base off form (107). The *Co*α-methyl form can also be formed from *Co*β-methylcobalamin by photolysis or thermolysis in an atmosphere of carbon monoxide; under these conditions the two isomers are interconvertible.

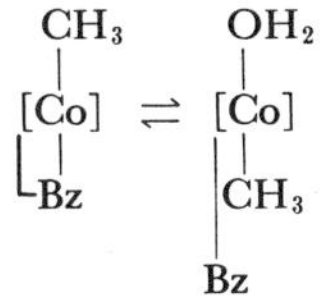

$$\begin{array}{c} CH_3 \\ | \\ [Co] \\ | \\ Bz \end{array} \rightleftharpoons \begin{array}{c} OH_2 \\ | \\ [Co] \\ | \\ CH_3 \\ \\ Bz \end{array}$$

For instance, heating an aqueous solution of either isomer at 95° in an atmosphere of carbon monoxide for several hours yields an equilibrium mixture containing approximately 3% of the Coα-methyl isomer. Corrinoids lacking 5,6-dimethylbenzimidazole in the lower coordination position give a higher yield of this isomer. Thus methylation of cobyric acid [via the Co(I) form] gives approximately 10% of the α isomer, methylation of cobinamide monocarboxylic acid 22%, of cobinamide-phosphoribose 50%, and of cobinamide 50% (91, 92, 108, 109). The isomerization of either isomer of methylcobinamidephosphoribose by photolysis in carbon monoxide yields 30% α isomer and 70% β isomer, while isomerization by thermolysis in carbon monoxide yields 11% α and 89% β. Little or no isomerization occurs in alcoholic solution or in the presence of oxygen. Under these conditions the organocorrinoids are decomposed to the aquo forms. These observations suggest that the isomerization occurs by a radical mechanism; photolysis and thermolysis both cause homolytic cleavage of the carbon-cobalt bond to a Co (II) corrinoid and a methyl radical.

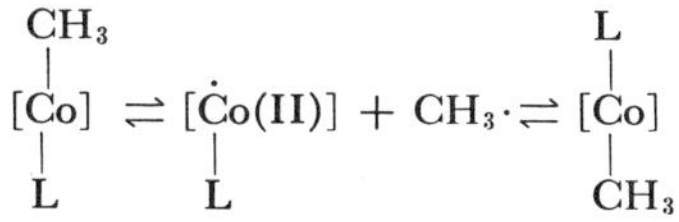

$$\begin{array}{c} CH_3 \\ | \\ [Co] \\ | \\ L \end{array} \rightleftharpoons \begin{array}{c} \\ \\ [\dot{C}o(II)] \\ | \\ L \end{array} + CH_3\cdot \rightleftharpoons \begin{array}{c} L \\ | \\ [Co] \\ | \\ CH_3 \end{array}$$

Recombination of the methyl radicals and the Co(II) corrinoids leads to either Coα or Coβ isomers.

HETEROLYTIC CLEAVAGE OF THE CARBON-COBALT BOND

The carbon-cobalt bond of the adenosylcobamides is stable in neutral aqueous solution in the dark; adenosylcobalamin, *Co*α-benzimidazolyl-

Coβ-adenosylcobamide, and *Coα*-adenyl-*Coβ*-adenosylcobamide are not affected by heating in sodium acetate buffer, pH 6.0 at 100° for 20 min (110, 111). However, these cobamides are decomposed in acid or alkaline solution, particularly at elevated temperatures and in alkaline cyanide. In contrast the simple alkylcobalamins are stable in acid and alkali and are not decomposed by cyanide (55).

Acid Decomposition. Acid treatment of adenosylcobalamin (0.1 N HCl, 100°, 90 min) causes cleavage of the carbon-cobalt bond and yields aquocobalamin, adenine, and D-erythro-2,3-dihydroxy-Δ4-pentenal as the major products (112). Analogs of adenosylcobalamin differing in the

HO OH
H_2O CH_2 O Ade
Co
$\xrightarrow{H^+}$ OH_2 | Co + HC(=CH_2)–CH(OH)–CH(OH)–CH=O + adenine

nucleoside moiety behave similarily (97). The acid decomposition of *Coα*-adenyl-*Coβ*-adenosylcobamide is more complex because the 7-*α*-glycosidic linkage of the purine nucleotide is extremely labile in acid (11). Thus acid hydrolysis of this cobamide in 0.07 *N* HCl at 85° for 90 min releases adenine from the nucleotide moiety and yields *Coα*-aquo-*Coβ*-adenosylcobinamide ribosephosphate. Further treatment with acid causes cleavage of the carbon-cobalt bond with the release of adenine, D-erythro-2,3-dihydroxy-Δ4-pentenal, and diaquocobinamide ribosephosphate. Acid hydrolysis of the organocorrinoids also causes some hydrolysis of the amide side chains of the corrin ring; the *e*-propionamide side chain is particularly susceptible to acid hydrolysis. For instance, mild acid treatment of 2′,3′-O-isopropylideneadenosylcobalamin to remove the protecting group yields not only the desired adenosylcobalamin but also adenosylcobalamin-*e*-carboxylic acid (113).

Hydroxyethyl- and methoxyethylcobalamin are decomposed in mild acid to aquocobalamin and ethylene. This reaction probably involves pro-

$$CH_2{-}CH_2OH\ (\text{Co}) \xrightarrow{H^+} H_2O\ CH_2{-}CH_2{-}\overset{+}{O}H_2\ (\text{Co}) \longrightarrow OH_2\ (\text{Co}) + CH_2{=}CH_2 + H_2O$$

tonation of the oxygen atom followed by a concerted elimination of ethylene and water or methanol (55). Acid treatment of vinylcobalamin also leads to heterolytic cleavage of the carbon-cobalt bond with the formation of aquocobalamin and presumably ethylene. On the other hand, ethynylcobalamin is hydrated in strong acid to acetylcobalamin without cleavage of the carbon-cobalt bond (114). This reaction is similar to the acid-catalyzed hydration of acetylenic ethers and thioethers to their corresponding esters and thioesters (115).

Decomposition by Cyanide. On treatment with cyanide the color of the adenosylcobamides changes from orange to red to purple and the new spectrum is typical of the dicyanocobamides. For instance, treatment of adenosylcobalamin with cyanide yields dicyanocobalamin, adenine, and the two epimeric cyanohydrins of D-erythro-2,3-dihydroxy-Δ4-pentenal, indicating cleavage of both the organometallic bond and glycosidic linkage (95, 111, 116). The susceptibility of the adenosylcobamides to decomposition by cyanide is influenced by the nature of the nucleotide base. The adenosylcobamides containing purines such as adenine and 5,6-diaminopurine as well as adenosylcobinamide are decomposed very rapidly in 0.1 *M* potassium cyanide ($t_{1/2}$ 40–60 sec). On the other hand the adenosylcobamides containing benzimidazoles are decomposed much more slowly under the same conditions ($t_{1/2}$ 6–8 min) (82, 117). In adenosylcobinamide and in the purine-containing adenosylcobamides the ligand trans to the adenosyl moiety is water, a very weak ligand which is readily displaced by cyanide. On the other hand, in the benzimidazole-containing adenosylcobamides one of the imidazole nitrogens is coordinated to cobalt and is much less readily displaced by cyanide. The cyanide decomposition probably proceeds by way of a *Coα*-cyano-*Coβ*-adenosylcobamide (in which the lower ligand is displaced by cyanide). In this intermediate

HO OH
CH₂ O Ade
CN⁻ → Co
CN
→ CN–Co–CN + HC(=CH₂)–CH(OH)–CH(OH)–CH=O + adenine
CN⁻ → HO OH, HC(=CH₂), HC(OH)–CN

the organometallic bond is labilized and the adenosyl ligand is either eliminated by an *E*2 mechanism or displaced by cyanide by way of an S_N2 reaction (118). All analogs of adenosylcobalamin, whether differing in the heterocyclic base or in the sugar moiety, are decomposed by cyanide by a similar mechanism.

Although reaction of the simple *n*-alkylcorrinoids with cyanide gives rise to the *n*-alkylcyanocorrinoids, these intermediates are quite stable because the alkyl group cannot be eliminated as a carbanion. On the other hand, if the alkyl ligand contains a suitable substituent so that an elmination reaction is possible, the carbon-cobalt can be broken. For instance, β-trimethylaminoethylcobalamin is decomposed by cyanide to dicyanocobalamin and probably ethylene and trimethylamine (56).

Decomposition by Alkali. Adenosylcobalamin and the simple *n*-alkylcobalamins are stable in dilute (0.1 *N*) alkali at room temperature. However, at elevated temperatures or in strong alkali adenosylcobalamin is degraded to cob(I)alamin and β elimination products of the adenosyl moiety. 2′,3′-Isopropylideneadenosylcobalamin decomposes in 5% potassium t-butoxide and yields cob(I)alamin and 2′,3′-isopropylidene-4′,5′-didehydro-5′-deoxyadenosine as the major products (30, 119).

HO OH / HO⁻ H— / CH_2 O Ade / Co → Co^I + HO OH / CH_2 O Ade

A similar β elimination of β-cyanoethylcobalamin in alkali yields cob(I)alamin and acrylonitrile. Since β-cyanoethylcobalamin can also be

H ⁻OH / CH_2—CH—CN / Co ⇌ Co^I + CH_2=CH—CN

formed from cob(I)alamin and acrylonitrile this reaction is reversible (78). In contrast the degradation of β-hydroxyethylcobalamin in alkali proceeds by way of a 1,2-hydride shift and yields cob(I)alamin and acetaldehyde (119). *Co*-Acetylcobalamin decomposes in alkali to cob(I)-

H / CH_2—C—O⁻ / H / Co → Co^I + CH_3CHO

alamin and acetate; this reaction probably involves a nucleophilic attack on the carbonyl carbon followed by elimination of cob(I)alamin (120). This type of reaction was first described by Bernhauer and Irion (121), who showed that acylcobalamins react with a nucleophile such as hydroxylamine to yield the corresponding hydroxamic acid and cob(I)alamin.

REDUCTIVE CLEAVAGE

Hydrogenation of methylcobalamin in the presence of platinum yields cob(II)alamin and methane (122). However, while the other simple alkylcorrinoids are undoubtedly also hydrogenated to their corresponding alkanes, no reductive cleavage of adenosylcobalamin could be demonstrated. Carboxymethylcobalamin is cleaved reductively to acetate and probably cob(I)alamin when treated with sodium borohydride (104). Other alkylcobalamins, which are normally stable to sodium borohydride, are also cleaved reductively when treated with borohydride in the presence of copper ions (68).

The electrochemical reduction of adenosylcobalamin, alkylcobalamins, and alkylcobinamides also involves heterolytic cleavage of the carbon-cobalt bond. Adenosylcobalamin and the alkylcobalamins show a two-electron reduction ($E_{1/2} \sim -1.38$ V versus SCE) while the polarograms of the alkylcobinamides show two waves. The two waves probably represent a two-electron reduction of one isomer. (*Co*α-aquo-*Co*β-alkylcobinamide) followed by a two-electron reduction of the other isomer (*Co*α-alkyl-*Co*β-aquocobinamide) (69).

REACTION WITH METAL IONS

The methyl group of methylcobalamin can be displaced by mercuric ion and thallic ion (123–128). The reactions involve an electrophilic attack on the methyl moiety, presumably by an S_E2 mechanism, and yield aquocobalamin and methylmercury cation or methylthallic dication. Methylthallic dication cannot be isolated because it transfers its methyl group to such nucleophiles as chloride and bromide to yield the methylhalide. Methylcobalamin is attacked most readily, the other alkylcobalamins being decomposed at much lower rates. Furthermore, since the reactions involve an electrophilic attack, coordination of 5,6-dimethylbenzimidazole increases the rate of the reaction. Methylcobinamide is dealkylated 10^4 times slower than methylcobalamin. The kinetics of the reaction are complex because mercuric ion also displaces the dimethylbenzimidazole to form methylcobalamin mercuric ion complex (base off).

$$\underset{\underset{Bz}{\uparrow}}{\overset{\overset{CH_3}{|}}{>Co<}} + HgX_2 \rightleftharpoons \overset{\overset{CH_3}{|}}{>Co<} + X^- \quad (\text{Bz}Hg^+X)$$

$$\underset{\underset{Bz}{\uparrow}}{\overset{\overset{CH_3}{|}}{>Co<}} + HgX^+ \rightarrow \underset{\underset{Bz}{\uparrow}}{\overset{\overset{OH_2}{|}}{>Co<}} + CH_3HgX$$

Adenosylcobalamin and the higher alkylcobalamins are dealkylated at a much slower rate than methylcobalamin because the approach of the metal ion to the coordinated carbon atom is sterically hindered. Mercuric ion is not able to remove cyanide from cyanocobalamin, but in contrast silver(I) ion removes cyanide from cyanocobalamin though it does not displace the methyl group of methylcobalamin.

Methylcobalamin can also be dealkylated by platinum and gold salts, but these reactions probably are redox reactions, because they require the presence of both Pt(II) and Pt (IV) or Au(I) and Au(III) (128).

In the presence of palladium salts methylcobalamin is able to alkylate styrene to propenylbenzene (129). The reaction probably involves an electrophilic attack of the metal ion on the methyl moiety, yielding a labile methylpalladium compound which in turn alkylates styrene.

MISCELLANEOUS SUBSTITUTION REACTIONS

Adenosylcobalamin and methylcobalamin are cleaved by iodine to yield iodocobalamin plus 5′-iodo-5′-deoxyadenosine and methyl iodide, respectively. Experiments using iodine monochloride indicate that this heterolytic cleavage of the carbon-cobalt bond is initiated by an electrophilic attack of an iodine cation on the carbon atom coordinated to cobalt (130).

Treatment of acylcobalamins with nucleophilic agents such as ammonia or hydroxylamine causes heterolytic cleavage of the carbon-cobalt bond with the formation of cob(I)alamin. For instance, the reaction between acetylcobalamin and hydroxylamine yields acetylhydroxamic acid and cob(I)alamin (121).

Although adenosylcobalamin and the alkylcobalamins are stable to mercaptans between pH 4 and 6, above pH 6 the carbon-cobalt bond is cleaved heterolytically (131).

$$[Co]—CH_3 + RS^- \rightarrow [Co(I)] + R—S—CH_3$$

Adenosylcobalamin is somewhat less stable to thiols than methylcobalamin (63).

Corrinoids with Cobalt-Sulfur Bonds

In a reaction analogous to the formation of the carbon-cobalt bond, the cobalt-sulfur bond can be formed by reacting Co(I) corrinoids with acid chlorides of sulfur oxoacids. Reaction of cob(I)alamin or cob(I)inamide with sulfuryl chloride yields sulfitocobalamin or the sulfitocobinamides, while reaction with *p*-toluenesulfonyl chloride, benzenesulfonyl chloride, or methylsulfonyl chloride yields the corresponding sulfonylcorrinoids (132, 133).

$$[Co(I)] + SO_2Cl_2 + H_2O \rightarrow \overset{\displaystyle SO_3H \atop |}{[Co]} + 2\ HCl$$

$$[Co(I)] + RSO_2Cl \rightarrow \overset{\displaystyle SO_2R \atop |}{[Co]} + Cl^-$$

Sulfitocorrinoids are also formed when Co(III) corrinoids are treated with sodium bisulfite (134–136).

$$\overset{\displaystyle OH_2}{\underset{|}{}}[Co] + HSO_3^- \rightleftharpoons \overset{\displaystyle SO_3H}{\underset{|}{}}[Co] + H_2O$$

The properties of sulfito- and sulfonylcorrinoids are very similar to those of the alkylcorrinoids. The ultraviolet-visible spectra of sulfitocobalamin and of the sulfonylcobalamins are very similar to that of ethynylcobalamin. In dilute acid solution they change color from red to yellow, indicating that the 5,6-dimethylbenzimidazole moiety is protonated and no longer coordinated to cobalt. In the complete absence of oxygen the sulfito- and sulfonylcorrinoids are stable in light. However, in the presence of oxygen both are photolyzed (136). Photolysis of sulfitocobalamin yields aquocobalamin and sulfate (137). In the presence of cyanide both the sulfito- and sulfonylcorrinoids are decomposed to the dicyanocorrinoids (137). In an analogous reaction to that of the alkylcorrinoids, sulfitocobalamin reacts with chloramine-T or *N*-bromosuccinimide with substitution of halogen into the corrin ring at C-10, but without lactone formation. The products, 10-chloro- or 10-bromosulfitocobalamin, react with β-naphthylamine in the dark and in the absence of oxygen yielding β-naphthylsulfamic acid and 10-halocob(I)alamin (137).

Corrinoids with a cobalt-sulfur bond can also be formed by reacting aquocorrinoids with thiols, hydrogen sulfide, sodium thiosulfate, and potassium thiocyanate (28, 138–142).

Since the Co(III) corrinoids are reduced to the Co(II) forms by thiols in alkaline solution (62), the formation of Co(III) corrinoid-thiol complexes has been studied in slightly acid solution. The glutathione-cobalamin complex is typical of these thiol-cobalamin complexes. It is violet with absorption maxima at 280, 285, 291, 333, 375, 408, 428, 535, and 562 nm. This spectrum does not change with pH between 1 and 10, but in 3 *N* HCl the maxima shift to lower wavelengths. The cobalamin-thiol complexes are relatively stable in the light but are converted to dicyanocobalamin in the presence of cyanide. When the cobalamin-thiol complexes are treated with alkyl halides in the absence of oxygen and light, small amounts of the alkylcobalamins are formed (30, 138). The mechanism of this reaction has not yet been established, but it may involve the Co(III) corrinoids rather than the corrinoid-thiol complex, because Schrauzer (63) has established that the cobalamin-thiol complex slowly decomposes to cob(II)alamin and a sulfhydryl radical.

$$\overset{\displaystyle OH_2}{\underset{|}{}}[Co(III)] + RS^- \rightleftharpoons \overset{\displaystyle SR}{\underset{|}{}}[Co] + H_2O \xrightarrow{\text{slow}} [Co(II)] + RS^{\cdot}$$

A cobalt-sulfur bond is also formed when aquocobalamin is treated with sodium thiosulfate. Although the absorption spectrum of the complex is similar to that of the alkylcobalamins, suggesting a cobalt-sulfur bond, the identity of the complex has not been established (142).

Treatment of aquocobalamin with sodium thiocyanate yields thiocyanatocobalamin (143, 144). The thiocyanate ligand is bound to cobalt through the sulfur atom in crystalline monothiocyanatocobalamin (142), but kinetic data (145) imply that in solution approximately 5 to 10% of the cobalamin is present as [Co]NCS.

$$\overset{\displaystyle OH_2}{\underset{|}{}}\!\!\![Co(III)] + RS^- \rightleftharpoons \overset{\displaystyle SR}{|}\!\!\![Co] + H_2O$$

$$+ HS^- \rightleftharpoons \overset{\displaystyle SH}{|}\!\!\![Co] + H_2O$$

$$+ S{-}SO_3^{2-} \rightleftharpoons \overset{\displaystyle SSO_3^-}{|}\!\!\![Co] + H_2O$$

$$+ SCN^- \rightleftharpoons \underset{\displaystyle \overset{+}{H_2O}}{\overset{\displaystyle SCN}{|}\!\!\![Co]} \rightleftharpoons \overset{\displaystyle NCS}{|}\!\!\![Co]$$

Reaction of cyanocobalamin with sulfurous acid or of aquocobalamin with sulfuric acid yields sulfatocobalamin which has an absorption spectrum similar to that of aquocobalamin, suggesting that the ligand is coordinated through oxygen.

It has been reported that reduced glutathione coordinates to adenosylcobalamin and methylcobalamin by displacing the 5,6-dimethylbenzimidazole moiety from the α coordination position (146). However, in these experiments the cobalamin solutions were inadequately buffered and thus the observed spectral changes probably reflect protonation of the benzimidazole moiety rather than coordination of reduced glutathione.

METAL-FREE CORRINOIDS AND CORRINOIDS CONTAINING METALS OTHER THAN COBALT

Several metal-free corrinoids have been isolated from photosynthetic bacteria (147–149) and from *Streptomyces olivaceous* (150). The corrinoids from *Chromatium* are red and their absorption spectra are similar to those of the cobalt-containing cobinamides (maxima at 524, 497, and 329 nm). The red corrinoids are slowly converted to yellow forms in alkali or acid, or on illumination. This conversion is irreversible if the reactions are carried out in the presence of oxygen. When the red or yellow corrinoids in alkaline solution are treated with cobaltous chloride, cobalt is incorporated into the corrin ring and the spectra of the new corrinoids are identical to that of diaquocobinamide. The neutral and one of the basic cobalt-free corrinoids have been characterized as phenylhydrogenobamide and hydrogenobyric acid, respectively (149, 151). If *Chromatium* is grown in the presence of 5,-6-dimethylbenzimidazole, the base is incorporated in the nucleotide moiety with the result that α-(5,6-dimethylbenzimidazolyl)hydrogenobamide and hydrogenobyric acid are the principal cobalt-free corrinoids formed (151). The cobalt-free corrinoid isolated from *S. olivaceus* grown on a cobalt-free medium has spectral properties similar to those of phenylhydrogenobamide, but attempts to insert cobalt into the corrin ring have been unsuccessful (150). When aqueous solutions of α-(5,6-dimethylbenzimidazolyl)hydrogenobamide, phenylhydrogenobamide, or hydrogenobyric acid are heated with copper sulfate or zinc acetate, these metals are inserted into the corrin ring. The copper-containing corrinoids, α-(5,6-dimethylbenzimidazolyl)cupribamide, phenylcupribamide, and cupribyric acid, are yellow, with absorption maxima at 484, 464, and 330.5 nm. These copper-containing analogs are stable in acid and alkaline solution and are not sensitive to light. However, in contrast with the cobalt-containing corrinoids, the square planar copper-containing corrinoids do not coordinate extra ligands. The zinc-containing corrinoids are red compounds that fluoresce orange-yellow. The absorption spectrum of phenylzincobamide shows maxima at 518, 496, and 335 nm while the spectrum of α-(5,6-dimethylbenzimidazolyl) zincobamide shows maxima at 526, 500, and 338.5 nm, suggesting that N-3 of 5,6-dimethylbenzimidazole is coordinated to the zinc atom. In the presence of oxygen, phenylzincobamide and α-(5,6-dimethylbenzimidazolyl)zincobamide are unstable to light, but the nature of the photochemical reaction is not understood (152).

Treatment of α-(5,6-dimethylbenzimidazolyl)hydrogenobamide in a mixture of ethanol and rhodium carbonyl chloride yields α-(5,6-dimethylbenzimidazolyl)rhodibamide, while reaction with rhodium carbonyl chloride in glacial acetic acid gives rhodibinamide (153a).

SPECTROSCOPY OF THE CORRINOIDS

Absorption Spectroscopy

All corrinoids are beautifully colored compounds, with colors varying from yellow, red, and violet to brown, green, and blue. Consequently each corrinoid has a very distinct ultraviolet and visible spectrum (Figures 1-13 and 1-17). The three most useful bands observed in the visible spectrum of the trivalent cobalt-containing corrinoids (α, β, and γ band) correspond to π–π^* transitions within the corrin ring. The position and intensity of these bands are very sensitive to the nature of the two axial ligands. On the other hand, the absorption spectrum is much less sensitive to changes in the corrin ring or in the side chains.

Since the cobalamins crystallize with large and variable amounts of water of crystallization, the extinction coefficients cannot be determined directly from the weight of the sample and the observed absorbance. The most reliable standard for the determination of the extinction coefficient of a corrinoid is the γ band of dicyanocobalamin or dicyanocobinamide. All cobalamins or cobinamides are readily converted to those two dicyano forms by the action of excess cyanide and light. The spectra of dicyanocobalamin and dicyanocobinamide show a very sharp γ band at 367 to 368 nm with a molar extinction coefficient of $30.8 \times 10^3\ M^{-1}\ cm^{-1}$ (Figure 1-17) (48, 110, 154).

The absorption spectra of a large number of corrinoids have been described by Firth et al. (155). Their studies have shown that the nature of the axial ligands has a pronounced effect on the absorption spectrum and suggest that the major absorption bands and in particular the γ band move to longer wavelengths as the *sum* of the nucleophilicities of the two axial ligands increases. On the other hand, the intensity of the γ band decreases as the *difference* between the nucleophilicities of the two axial ligands increases (118).

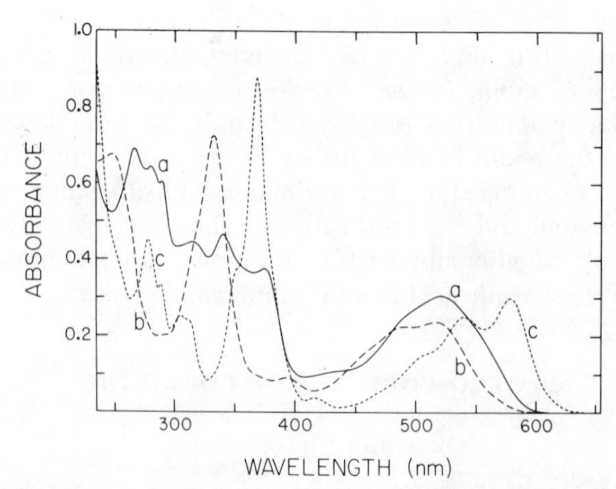

Fig. 1-17. Absorption spectra of *(a)* methylcobalamin, *(b)* aquocobalamin, and *(c)* dicyanocobalamin.

The spectral properties of a few cobalamins are presented in Table 1-1. The spectra of cyano-, aquo-, and hydroxocobalamin are similar and are characterized by the intense γ band in the 350 to 360 nm region, while this γ band is virtually lacking in the spectra of the organocobalamins.

In dilute acid adenosylcobalamin and the alkyl- and acylcobalamins change color from red to yellow; under these conditions the 5,6-dimethylbenzimidazole base is protonated and no longer coordinated to cobalt (Figure 1-18). The pK_a value of the 5,6-dimethylbenzimidazole moiety depends on the nature of the trans ligand. In the organocobalamins this pK_a varies from 2 to 4. However, much stronger acid is required to protonate the nucleotide base of aquo- and cyanocobalamin (Table 1-2) (54).

The spectrum of cob(I)alamin does not change in the pH range from 5 to 14 and is identical to that of cob(I)inamide above 300 nm, suggesting that in cob(I) alamin the 5,6-dimethylbenzimidazole moiety is not coordinated to cobalt (73).

The absorption spectra of the cobinamides are almost identical to the spectra of the corresponding cobalamins in acid solution (base off cobalamins). For instance, the spectra of the alkylaquocobinamides are identical

Table 1-1 Electronic Absorption Spectra of Corrinoids

Corrinoid	Principal Absorption Band (nm) ($\epsilon \times 10^{-3}$ in parentheses)						
Cyanocobalamin	—	278 (16.3)	305 (9.7)	322 (7.9)	361 (28.1)	518 (7.4)	550 (8.7)
Aquocobalamin	—	274 (20.6)	—	317 (6.1)	351 (26.5)	499 (8.1)	525 (8.6)
Hydroxocobalamin	—	278 (19.1)	—	325 (11.4)	358 (20.6)	516 (8.9)	535 (9.3)
Adenosylcobalamin	262 (35.1)	290 (18.2)	—	318 (13.0)	341 (12.8)	376 (11.0)	522 (8.0)
Methylcobalamin	266 (19.1)	280 (18.3)	289 (17.1)	315 (12.5)	340 (13.3)	373 (10.7)	519 (8.7)
Cob(II)alamin	—	288	—	311 (27.5)	—	402 (7.5)	473 (9.2)
Cob(I)alamin	280.5 (29.1)	288 (29.4)	386 (28.0)	455 (2.5)	545 (2.8)	680 (1.7)	800 (1.4)

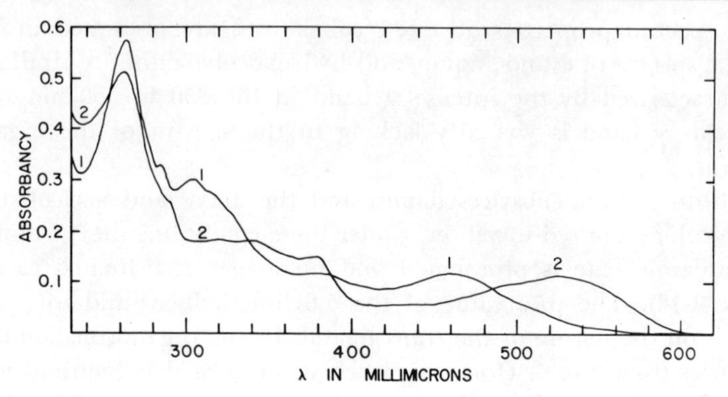

Fig. 1-18. Absorption spectra of adenosylcobalamin. (1) pH 2.0; (2) pH 7.0.

to those of the base off alkylcobalamins in the visible region with only minor differences, primarily in the extinction coefficients, in the ultraviolet region. Displacement of coordinated water in the alkylaquocobinamides by other ligands causes a red shift of the α band of the absorption spectrum. The spectral data show a direct correlation between the wavelength of the α band and the formation constant of the alkylcobinamide-ligand complex (118).

The absorption spectra of the purine-containing adenosylcobamides, such as *Co*α-adenyl-*Co*β-adenosylcobamide, are very similar to the spectrum of adenosylcobalamin in the base off form, suggesting that in these cobamides N-9 of the purine nucleotide is not coordinated to the cobalt atom (156).

Table 1-2 pK_a Values of Some Cobalamins

Cobalamin	Axial Ligand	pK_a
Aquocobalamin	H_2O	−2.4
Cyanocobalamin	CN^-	0.1
Ethynylcobalamin	HC≡C–	0.7
Vinylcobalamin	$H_2C{=}CH$–	2.4
Methylcobalamin	CH_3–	2.72
Adenosylcobalamin	5′-Deoxyadenosyl	3.5
n-Propylcobalamin	$CH_3CH_2CH_2$–	3.8
Cob(II)alamin		2.5

Optical Rotary Dispersion (ORD)

The ORD spectra of a few corrinoids and 13-epicorrinoids have been determined (31, 157). The ORD spectra of dicyanocobinamide and dicyano-13-epicobinamide are quite distinct, suggesting that ORD is a sensitive method for the detection of alterations in the periphery of the corrin ring.

Circular Dichroism (CD)

The CD spectra of a large number of corrinoids have been determined 31, 155, 158, 158a). The CD spectrum of adenosylcobalamin (Figure 1-19) shows negative extremes at 550, 427, 357, and 320 nm and positive extremes at 480, 338, and 330 nm. The CD spectra are more sensitive than the absorption spectra to changes in the axial ligands or in the corrin ring. For instance, the CD spectra of the 13-epicorrinoids are quite distinct from those of the corrinoids (31). Furthermore, changes in solvent or temperature bring about dramatic changes in the CD spectra. In contrast, inversion of the axial ligands causes only minor changes in the circular dichroism. The spectra of the aquocyano- and cyanoaquocorrinoids are only slightly different (159). CD has been used to measure the binding of adenosylcobalamin and hydroxocobalamin to ethanolamine ammonia-lyase. Binding of these two corrinoids to the enzyme leads to significant changes in the CD spectra (160).

Magnetic Resonance Spectroscopy

Electron Spin Resonance

The cobalt atom of the corrinoids can exist in three oxidation states: Co(I), Co(II), and Co(III); of these three forms only the Co(II) corrinoids are paramagnetic. Early experiments showed that adenosylcobalamin photolyzed under anaerobic conditions gave an esr spectrum identical to that of authentic cob(II)alamin (52). Although the early spectra

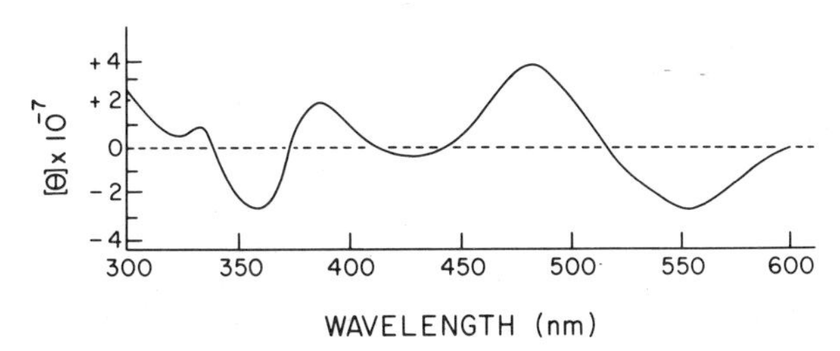

Fig. 1-19. Circular dichroism of adenosylcobalamin [(161)].

showed hyperfine splitting due to the interaction of the unpaired electron with the ^{59}Co nucleus ($I = 7/2$), no superhyperfine splitting could be detected. Spectra with more structure were obtained for solutions of cob(II)alamin containing solutes or solvents such as dimethylglutarate and methanol (161–163). The highly resolved spectra show seven of the eight expected hyperfine lines and also superhyperfine splitting into three lines due to the interaction with the ^{14}N nucleus of the 5,6-dimethylbenzimidazole moiety ($I = 1$). The characteristic features of these spectra (Figure 1-20) are $g_{\perp} = 2.25$; $g_{\parallel} = 2.003$, $A_{\parallel Co} = 196 \pm 2 \times 10^{-4}$ cm^{-1}, and $A_{\parallel N} = 15.8 \pm 0.5 \times 10^{-4}$ cm^{-1} (164, 165).

The hyperfine and superhyperfine splitting parameters are sensitive to the nature of the axial ligand. Thus the spectrum of cob(II)alamin in acid solution, when the 5,6-dimethylbenzimidazole moiety is protonated and no longer coordinated to cobalt, does not show the superhyperfine splitting, while the hyperfine coupling constant ($A_{\parallel Co}$) increases from 103 to 146 × 10^{-4} cm^{-1}. The spectra of the cob(II)inamides are similar to that of base off cob(II)alamin. A study with a series of cob(II)inamides containing different bases in the axial position, showed that the nitrogen coupling constant tends to increase as the cobalt coupling constant decreases (162, 163).

When oxygen is allowed to react with cob(II)alamin the esr spectrum changes from a base on spectrum to a base off spectrum and the superhyperfine structure due to the N-3 of 5,6-dimethylbenzimidazole disappears. Furthermore the hyperfine coupling constant decreases by a factor of about 9, suggesting that the oxygenated complex is a superoxide [superoxocobalamin, Co(III)–O–O$^-$] (66, 166, 167).

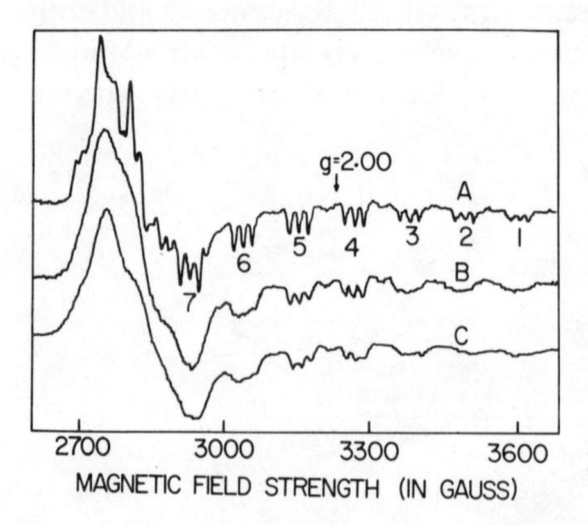

Fig. 1-20. Electron spin resonance spectra of cob(II)alamin formed under various conditions. (*A*) from adenosylcobalamin and ribonucleotide reductase in the presence of sodium dimethylglutarate buffer (pH 7.3), dGTP, and dihydrolipoate. (*B*) From photolysis of adenosylcobalamin in the presence of sodium dimethylglutarate (pH 7.3), dGTP, and dihydrolipoate. (*C*) From reaction of hydroxocobalamin and dihydrolipoate [(164)].

Nuclear Magnetic Resonance

PROTON MAGNETIC RESONANCE

Although most of the early pmr studies have been concerned with the interpretation of the spectra and with the assignment of the resonance positions, more recent work has provided very valuable information concerning the structure of the corrinoids in solution and the effect of the axial ligands on the electronic structure and conformation of the corrin ring (84, 165, 168–170).

The proton magnetic resonance spectrum of methylcobalamin in D_2O at 220 MHz and ambient temperature is shown in Figure 1-21. The low field resonances below $\delta = 6.0$ are predominantly single proton resonances while the strong sharp resonances at high field beyond $\delta = 2$ are mostly methyl groups. When hexadeuterodimethyl sulfoxide is used as solvent instead of D_2O, the low field region of the spectrum shows a complex set of peaks corresponding to the N–H protons of the amide side chains and the O–H protons of the ribose moieties. In the cobinamide spectra the single resonance at low field ($\delta \sim 6.0$) has been assigned to the proton at C-10. This proton exchanges in acidic D_2O and CF_3COOD and is absent in 10-chlorocobinamide. The spectrum of cyanoaquocobinamide shows two resonances in this region ($\delta = 6.53$ and 6.47) which are assigned to the C-10 hydrogens of the two isomers cyanoaquocobinamide and aquocyanocobinamide. The chemical shift of the C-10 vinyl proton is affected by the nature of the axial ligands (168). The hydrogens of the 5,6-dimethylbenzimidazole moiety and also at C-2 and

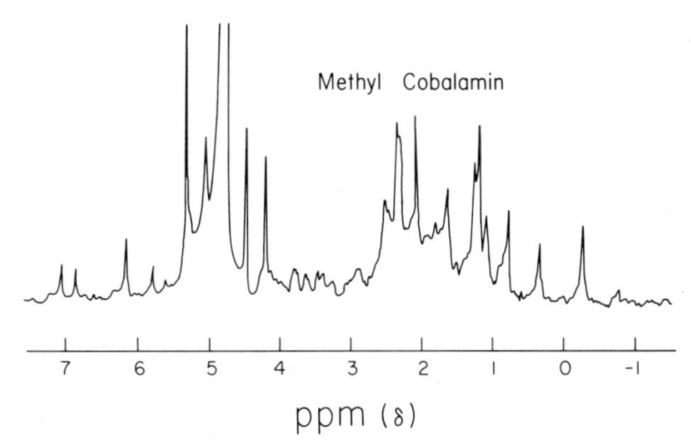

Fig. 1-21. 220-MHz nmr spectrum of methylcobalamin in D_2O at ambient temperature [(173)].

C-8 of the adenosyl group give rise to resonances in the low field region. The chemical shifts of those protons of adenosylcobalamin are tabulated in Table 1-3 (171).

The resonances at very high field around $\delta = 0$ are due to the hydrogens of the alkyl groups coordinated to cobalt. The cobalt-bound methyl group of methylcobalamin exhibits a chemical shift of -0.06 ppm, while the cobalt-bound methyl groups of methylcobinamide and methylcobalamin in the base off form are shifted about 0.3 ppm to higher field. The two prochiral protons on the methylene carbon of the 5′-deoxyadenosyl moiety of adenosylcobalamin exhibit separate proton resonances of 0.34 and 0.93 ppm (171). The large difference in chemical shifts of these two protons suggests that rotation about the carbon-cobalt bond is restricted. A similar nonequivalence has also been found for the two α-CH_2 and β-CH_2 protons of *n*-propylcobalamin and for the two CH_3 groups of *i*-propylcobinamide. In the spectra of the alkylcobinamides the two resonances in the 4.0 and 4.7 ppm region were assigned to the protons of a water molecule coordinated trans to the alkyl group. These two resonances probably do not represent single proton resonances of non-

Table 1-3 Chemical Shift of Protons of Adenosylcobalamin in the Low Field Region, in Parts per Million Downfield from 2,2-Dimethyl-2-silapentane-5-sulfonate (171)

Proton[a]	Chemical Shifts in D_2O (23°)
A2	8.38
A6	8.18
B7	7.34
B4	6.40
B2	7.11
R′1	5.55
R1	6.40
C10	6.10

[a] A2, A6, and R′1 refer to C-8, C-2, and the anomeric carbon of the adenosyl moiety;B2, B4, B7, and R1 refer to C-2, C-4, C-7, and the anomeric carbon of the benzimidazole nucleoside.

equivalent water protons (172) but rather resonances of both water protons in the two alkylcobinamide isomers (alkylaquocobinamide and aquoalkylcobinamide).

Tentative assignments of methyl groups C-20, C-47, C-5, Pr-3, B-10, and B-11 have been made (172). The position of the C-2 resonance of methylcobalamin is strongly pH dependent (173).

CARBON-13 MAGNETIC RESONANCE

Carbon-13 nuclear magnetic resonance (cmr) spectroscopy is one of the most promising tools for the study of the corrinoids. Proton-decoupled natural abundance carbon-13 Fourier transform nmr spectra are much more informative than the proton spectra because nearly all the lines in the ^{13}C spectra are well-resolved single-carbon resonances. Doddrel and Allerhand (174, 175) have assigned many of the resonances of the corrinoids, while some additional assignments have been made by using cyanocobalamin synthesized from [5-^{13}C]δ-aminolevulinic acid (176). Unfortunately in the natural abundance cmr spectrum of adenosylcobalamin the 5′-methylene carbon of the carbon-cobalt bond cannot be detected, probably as a result of line broadening by the ^{59}Co nucleus (174). In contrast the cmr spectrum of methylcobalamin selectively enriched with ^{13}C in the methyl ligand exhibits one relatively sharp resonance only slightly broadened by the ^{59}Co nucleus (109).

The cmr spectrum of the two isomers of monocyanocobyric acid (aquocyanocobyric acid and cyanoaquocobyric acid) shows that a large number of carbon atoms have different chemical shifts in the two isomers. While there are only 38 lines in the spectrum of dicyanocobyric acid, the spectrum of the monocyanocobyric acids exhibits 60 resolved resonances (175). The cmr spectra of methylcobinamide and monocyanocobinamide selectively enriched with ^{13}C in the ligands attached to cobalt show two well-resolved resonances corresponding to the two possible positional isomers. The two ^{13}CN resonances of di[^{13}C]cyanocobinamide can be partially resolved at 10°. The chemical shift of the $^{13}CH_3$ moiety as well as the ^{13}C–H coupling constant are markedly affected by the nature of the trans ligand because the chemical shifts are very sensitive to the electronic environment of the ^{13}C nucleus.

FLUORINE-19 NUCLEAR MAGNETIC RESONANCE

Fluorine-19 nmr has been used to characterize a series of fluoromethylcobalamins (177).

Other Spectroscopic Techniques

Other spectroscopic techniques which have been used to study the corrinoids include resonance Raman spectroscopy (178, 179) and emission Mössbauer spectroscopy (180).

LIGAND EXCHANGE REACTIONS

The great stability of the bonding within the corrin ring restricts ligand exchange reactions to the α and β coordination positions (118).

Ligand Exchange Reactions Involving the Co(III) Cobalamins

The simplest single ligand exchange reactions involve the β coordination position of the cobalamins. Coordinated water in aquocobalamin is readily displaced by a large number of other ligands.

$$\overset{OH_2}{\underset{\llcorner Bz}{[Co]}} + L \underset{k_D}{\overset{k_F}{\rightleftharpoons}} \overset{L}{\underset{\llcorner Bz}{[Co]}} + H_2O$$

In general the equilibrium constant (k_D/k_F) for this ligand substitution reaction depends on the nucleophilicity of the ligand. For instance, for the halide ions the highest stability constant is found with iodide ion (142). Very effective nucleophiles such as the cyanide, azide, thiocyanate, thiosulfate, and sulfite ions yield complexes with very high stability constants (181). Because the stability constant of cyanocobalamin is extremely high ($10^{12}\ M^{-1}$), all cobalamins except the alkylcobalamins are readily converted to cyanocobalamin. Kinetic studies using fast reaction techniques have shown that the rate constants of complex formation (k_F) are relatively independent of the nature of the incoming nucleophile. In contrast the dissociation rate constants (k_D) are very dependent on the nature of the ligand and vary by many orders of magnitude (Table 1-4).

The kinetic data are consistent with a transition state in which both entering and leaving groups are loosely bound to the cobalt atom and not consistent with a mechanism in which the unimolecular release of water from aquocobalamin is the rate-limiting reaction (143, 144, 182).

The nature of the ligand in the β coordination position of the cobalamins has a profound effect on the coordinate linkage between cobalt and N-3 of the 5,6-dimethylbenzimidazole moiety. When solutions of cobalamins are acidified, the coordinated benzimidazole is displaced by water

Table 1-4 Rate Constants for the Formation and Dissociation of a Few Cobalamins (144)

Ligand	k_F (M^{-1} sec^{-1})	k_D (sec^{-1})
SCN^-	2300	1.8
I^-	1400	3.5×10
Br^-	1000	5.9×10^2
N_3^-	1200	2.9×10^{-2}
NCO^-	470	1.1
$S_2O_3^{2-}$	200	3.5×10^{-2}
SO_3^{2-}	$\gtrsim 200$	$\gtrsim 1 \times 10^{-5}$

and becomes protonated. The pK_a of the equilibrium depends on the nature of the trans ligand.

$$\underset{\llcorner Bz}{\overset{L}{[Co]}}\uparrow + H_3O^+ \rightleftharpoons \overset{L}{[Co]}\,\text{—}OH_2 \quad \llcorner BzH^+$$

For the cobalamins the displacement and protonation of the coordinated benzimidazole are accompanied by a color change from red to yellow. The pK_a values range from about 3.9 for *n*-propylcobalamin to 0.1 for cyanocobalamin and -2.4 for aquocobalamin, reflecting the strong nucleophilic character of the alkyl group (Table 1-2). Thus in the alkylcobalamins, including adenosylcobalamin, the 5,6-dimethylbenzimidazole moiety is easily replaced and indeed no ligand is very strongly bound trans to an alkyl group (114).

When potassium cyanide is added to a solution of cyanocobalamin the color of the solution changes from red to purple, indicating the formation of dicyanocobalamin. A similar reaction occurs when some alkylcobalamins are treated with cyanide.

$$\underset{\llcorner Bz}{\overset{L}{[Co]}}\uparrow + CN^- \rightleftharpoons \overset{L}{[Co]}\,\text{—}CN \quad \llcorner Bz$$

The equilibrium constant of this reaction has been determined for a few alkylcobalamins. In the series L = CN^-, $HC{\equiv}C^-$, $H_2C{=}CH^-$, and

H_3C^- the stability constants decrease from $6.3 \times 10^3\ M^{-1}$, $5 \times 10^2\ M^{-1}$, $5\ M^{-1}$ to $1.3\ M^{-1}$, reflecting the very strong nucleophilic character of the methyl carbanion. In the presence of excess cyanide, aquocobalamin is first converted to cyanocobalamin and next to dicyanocobalamin. On acidification the second cyanide is removed, leaving cyanocobalamin. Because the stability constant of cyanocobalamin is extremely high, the first cyanide can only be removed under suitable reducing conditions or by photolysis (57, 183–185).

Ligand Exchange Reactions Involving the Co(III) Cobinamides

Ligand exchange reactions involving the cobinamides are more complex than those involving the cobalamins because the cobinamides lack the nucleotide side chain and thus stereoisomerism is possible. The corrin ring is asymmetric, with all the acetamide side chains projecting toward the β side of the corrin ring and the propionamide side chains projecting toward the α side of the ring.

Diaquocobinamide has a very high affinity for many ligands. For instance, the stability constant of the cyanoaquocobinamide complex is $10^{14}\ M^{-1}$ (187). At equilibrium the two isomers cyanoaquocobinamide and aquocyanocobinamide are present in approximately equal amounts, suggesting that the formation constants for the substitution of coordinated water at the α and β coordination positions by cyanide are almost equal.

$$
\begin{array}{ccccc}
 & & L{-}[Co]{-}OH_2 & & \\
 & \overset{L}{\nearrow}\underset{H_2O}{\swarrow} & & \overset{L}{\searrow}\underset{H_2O}{\nwarrow} & \\
OH_2{-}[Co]{-}OH_2 & & & & L{-}[Co]{-}L \\
 & \overset{L}{\searrow}\underset{H_2O}{\nwarrow} & & \overset{L}{\nearrow}\underset{H_2O}{\swarrow} & \\
 & & OH_2{-}[Co]{-}L & &
\end{array}
$$

At higher concentrations of ligand both the α and β positions are substituted. The stability constant of dicyanocobinamide is $10^8\ M^{-1}$ (187). On acidification one of the cyanides of dicyanocobinamide is lost as HCN and the two isomers of monocyanoaquocobinamide are formed. Diaquocobinamide should be stored in cold with the exclusion of polluted air because in the presence of sulfur dioxide diaquocobinamide is converted to the sulfitaquocobinamides ($K = 10^{11}\ M^{-1}$).

In the pH range from 4 to 14 both coordinated water molecules of diaquocobinamide are titrated, first to hydroxoaquocobinamide (aquohydroxocobinamide; $pK_a = 6.0$), then to dihydroxocobinamide.

The nature of the trans ligand L determines the magnitude of the equilibrium constant for the substitution of water by another ligand.

$$\begin{array}{c} L \\ | \\ [Co] \\ | \\ OH_2 \end{array} + X \rightleftharpoons \begin{array}{c} L \\ | \\ [Co] \\ | \\ X \end{array} + H_2O$$

When L is a very strong nucleophile such as an alkyl carbanion, ligands trans to it are only weakly bound to cobalt. Thus the stability constant of dicyanocobinamide is $10^8\ M^{-1}$ while that of methylcyanocobinamide is only $230\ M^{-1}$ (102). Indeed Firth and co-workers (188, 189) have shown that on varying the temperature solutions of alkylaquocobinamides show reversible changes in the absorption spectrum and in the C-10 proton chemical shift. Since similar changes are observed in the solid state on varying the water content they concluded that these changes represent the reversible removal of coordinated water to form a five-coordinated complex.

$$\begin{array}{c} R \\ | \\ [Co] \\ | \\ OH_2 \end{array} \rightleftharpoons \begin{array}{c} R \\ | \\ [Co] \end{array} + H_2O$$

Ligand Exchange Reactions Involving the Co(II) Corrinoids

When aquocobalamin is reduced to cob(II)alamin the ligand in the β coordination position is detached and the bond to the ligand in the α coordination position is weakened (163). Thus displacement and protonation of the coordinated benzimidazole moiety occur much more readily in cob(II)alamin than in aquocobalamin (Table 1-2). Displacement of the nucleotide by the solvent also occurs in concentrated pyridine solutions. Under these conditions a small percentage of a hexacoordinate complex is formed. At high pH values both cysteine and glutathione coordinate to cob(II)alamin, the nucleotide base becoming detached and replaced by the thiolate ion (190). In the case of aquocob(II)inamide, coordinated water is readily replaced by a large number of ligands. ESR studies have shown that triphenylphosphine, thiocyanate, azide, nitrite, thiourea, dimethyl sulfide, and 2-mercaptoethanol as well as many nitrogen bases coordinate to cob(II)inamide. Cyanide ion, ammonia, hydroxylamine, and a number of amino acids coordinate at high pH but not in neutral solution (163).

Stereoisomerism Involving the Axial Ligands

Friedrich (168, 191, 192) first demonstrated that certain corrinoids carrying cyanide and water as the axial ligands are present in solution as a mixture of the two stereoisomers. These two isomers are interconvertible but the equilibrium between the two isomers is established at a slow rate so that they can be separated by chromatographic techniques at 3°.

$$\begin{array}{c} CN \\ | \\ [Co] \\ | \\ OH_2 \end{array} \rightleftharpoons \begin{array}{c} OH_2 \\ | \\ [Co] \\ | \\ CN \end{array}$$

Cyanoaquocobyric acid has been shown to be present as a mixture of the two isomers by C-13 nuclear magnetic resonance spectroscopy (175). Both isomers of cyanoaquocobinamide were detected by proton and C-13 nuclear magnetic resonance spectroscopy (109, 193). The rate of isomerization is increased by heat, light, carbon monoxide, certain anions, cyanide, and increased concentration of corrinoid, and is inhibited by oxygen and mercury(II) salts (193, 194). The addition of mercury(II) ions suppresses the rate of isomerization by removing free cyanide ion. Since the mercury(II) ion does not decompose aquocyanocobinamide to diaquocobinamide, the mechanism of isomerization does not involve the dissociation of a cyanide ligand from aquocyanocobinamide, but rather probably involves the initial presence of free cyanide ion.

$$\begin{array}{c} CN \\ | \\ [Co] \\ | \\ OH_2 \end{array} + CN^- \rightleftharpoons \begin{array}{c} CN \\ | \\ [Co] \\ | \\ CN \end{array} + H_2O \rightleftharpoons \begin{array}{c} OH_2 \\ | \\ [Co] \\ | \\ CN \end{array} + CN^-$$

The dicyanocorrinoid is also formed when traces of sulfur dioxide and sulfate are present. Addition of very small quantities of sodium sulfite to a solution of aquocyanocobinamide yields sulfitoaquocobinamide and dicyanocobinamide (193).

$$2\begin{array}{c} OH_2 \\ | \\ [Co] \\ | \\ CN \end{array} + SO_3^{2-} \rightleftharpoons \begin{array}{c} SO_3^- \\ | \\ [Co] \\ | \\ OH_2 \end{array} + \begin{array}{c} CN \\ | \\ [Co] \\ | \\ CN \end{array} + H_2O$$

The other anions probably catalyze the isomerization reaction by a similar mechanism. All corrinoids which contain nonidentical axial ligands should exist in solution as a mixture of stereoisomers. However, since most other ligands have much larger dissociation rate constants than cyanide, the free ligand would be formed more readily and thus the isomerization would proceed much faster.

For almost all the aquocyanocorrinoids the two stereoisomers are present in approximately equal amounts, suggesting a lack of preference for the less hindered β coordination position.

Several of the pure isomers of cyanoaquocorrinoids have been isolated in crystalline form. X-ray analysis has shown that in crystalline aquocyanocobyric acid the cyanide ligand is in the α coordination position (12). In general the two isomers have very different crystal habits. However, the absorption spectra of the isomers are virtually identical, and the circular dichroism spectra show small differences (195).

REFERENCES

1. Rickes, E. L., Brink, N. G., Koniuszy, F. R., Wood, T. R., and Folkers, K. (1948). *Science* **107**, 396.
2. Smith, E. L., and Parker, L. F. J. (1948). *Biochem. J.* **43**, viii.
3. Barker, H. A., Weissbach, H., and Smyth, R. D. (1958). *Proc. Natl. Acad. Sci. U.S.A.* **44**, 1093.
4. Weissbach, H., Toohey, J. I., and Barker, H. A. (1959). *Proc. Natl. Acad. Sci. U.S.A.* **45**, 521.
5. IUPAC-IUB CBN (1966). *J. Biol. Chem.* **241**, 2991.
6. IUPAC-IUB CBN. 1973 Recommendations.
7. Smith, E. L. (1965). *Vitamin B_{12}*, 3rd ed., Methuen, London.
8. Wagner, A. F., and Folkers, K. (1964). *Vitamin and Coenzymes,* Interscience, New York, Chapter X.
9. Bonnett, R. (1963). *Chem. Rev.* **63**, 573.
10. Hodgkin, D. C. (1965). *Science* **150**, 979.
11. Weissbach, H., Ladd, J. N., Volcani, B. E., Smyth, R. D., and Barker, H. A. (1960). *J. Biol. Chem.* **235**, 1462.
12. Hodgkin, D. C. (1965). *Proc. Roy. Soc.* **A288**, 294.
13. Bernhauer, K., Wagner, F., and Wahl, D. (1961). *Biochem. Z.* **334**, 279.
14. Lenhert, P. G. (1968). *Proc. Roy. Soc.* **A303**, 45.
15. Kuehl, F. A., Shunk, C. H., Moore, M., and Folkers, K. (1955). *J. Am. Chem. Soc.* **77**, 4418.
16. Garbers, C. F., Schmid, H., and Karrer, P. (1953). *Helv. Chim. Acta* **36**, 65.
17. Brink, N. G., Wolf, D. E., Kaczka, E. Rickes, E. L., Koniuszy, F. R., Wood, T. R., and Folkers, K. (1949). *J. Am. Chem. Soc.* **71**, 1854.
18. Baldwin, R. R., Lowry, J. R., and Harrington, R. V. (1951). *J. Am. Chem. Soc.* **73**, 4968.
19. Boos, R. N., Rosenblum, C., and Woodbury, D. T. (1951). *J. Am. Chem. Soc.* **73**, 5446.
20. Armitage, J. B., Cannon, J. R., Johnson, A. W., Parker, L. F. J., Smith, E. L. Stafford, W. H., and Todd, A. R. (1953). *J. Chem. Soc.* 3849.
21. Bernhauer, K., Wagner, F., Beisbarth, H., Rietz, P., and Vogelmann, H. (1966). *Biochem. Z.* **344**, 289.

22. Bernhauer, K., Vogelmann, H., and Wagner, F. (1968). *Hoppe Seylers Z. Physiol. Chem.* **349**, 1281.
23. Bernhauer, K., Vogelmann, H., and Wagner, F. (1968). *Hoppe Seylers Z. Physiol. Chem.* **349**, 1271.
24. Nockholds, C. K., Waters, T. N. M., Ramaseshan, S., Waters, J. M., and Hodgkin, D. C. (1967). *Nature* **214**, 129.
25. Moore, F. M., Willis, B. T. M., and Hodgkin, D. C. (1967). *Nature* **214**, 130.
26. Bonnett, R., Cannon, J. R., Johnson, A. W., and Todd, A. R. (1957). *J. Chem. Soc.* 1148.
27. Bonnett, R., Cannon, J. R., Clark, V. M., Johnson, A. W., Parker, L. F. J., Smith, E. L., and Todd, A. R. (1957). *J. Chem. Soc.,* 1158.
28. Wagner, F., and Bernhauer, K. (1964). *Ann. N.Y. Acad. Sci.* **112**, 580.
29. Renz, R. (1963). Dissertation, Stuttgart.
30. Koppenhagen, V. (1967). Dissertation. Stuttgart.
31. Bonnett, R., Godfrey, J. M., and Math, V. B. (1971). *J. Chem. Soc. (C),* 3736.
32. Bonnett, R., Godfrey, J. M., and Redman, D. G. (1969). *J. Chem. Soc.,* 1163.
33. Bonnett, R., Godfrey, J. M., Math, V. B., Edmond, E., Evans, H., and Hodder, O. J. R. (1971). *Nature* **229**, 473.
34. Brink, N. G., and Folkers, K. (1950). *J. Am. Chem. Soc.* **72**, 4442.
35. Beaven, G. R., Holiday, E. R., Johnson, E. A., Ellis, B., Mamalis, P., Petrow, V., and Sturgeon, B., (1949). *J. Pharm. Pharmacol.* **1**, 957.
36. Holly, F. W., Shunk, C. H., Peel, E. W., Cahill, J., Lavigne, J. B., and Folkers, K. (1952). *J. Am. Chem. Soc.* **74**, 4521.
37. Friedrich, W., and Bernhauer, K. (1954). *Z. Naturforsch.* **96**, 685.
38. Friedrich, W., and Bernhauer, K. (1956). *Chem. Ber.* **89**, 2507.
39. Buchanan, J. G., Johnson, A. W., Mills, J. A., and Todd, A. R. (1950). *J. Chem. Soc.* 2845.
40. Kaczka, E. A., Heyl, D., Jones, W. H., and Folkers, K. (1952). *J. Am. Chem. Soc.* **74**, 5549.
41. Brink, C., Hodgkin, D. C., Lindsay, J., Pickworth, J., Robertson, J. H., and White, J. G. (1954). *Nature* **174**, 1169.
42. Bonnett, R., Buchanan, J. G., Johnson, A. W., and Todd, A. R. (1957). *J. Chem. Soc.,* 1168.
43. Ellis, B., Petrow, V., and Snook, G. F. (1949). *J. Pharm. Pharmacol.* **1**, 950.
44. Wolf, D. E., Jones, W H., Valiant, J., and Folkers, K. (1950). *J. Am. Chem. Soc.* **72**, 2820.
45. Diehl, H., van der Haar, R. W., and Sealock, R. R. (1950). *J. Am. Chem. Soc.* **72**, 5312.
46. Wallmann, J. C., Cunningham, B. B., and Calvin, M. (1951). *Science,* **113**, 55.
47. Grün, F., and Manassé, R. (1950). *Experientia* **6**, 263.
48. Hill, J. A., Pratt, J. M., and Williams, R. J. P. (1964). *J. Chem. Soc.,* 5149.
49. Smith, E. L., Fantes, K H., Ball, S., Woller, J. G., Emery, W. B., Ambrose, W. K., and Walker, A. D. (1952). *Biochem. J.* **52**, 389.
50. Cunningham, B. B. Unpublished results.

51. Boehm, G., Faessler, A., and Rittmayer (1954). *Z. Naturforsch.* **9b**, 509.
52. Hogenkamp, H. P. C., Barker, H. A., and Mason, H. S. (1963). *Arch. Biochem. Biophys.* **100**, 353.
53. Ladd, J. N., Hogenkamp, H. P. C., and Barker, H. A. (1961). *J. Biol. Chem.* **236**, 2114.
54. Hayward, G. C., Hill, H. A. O., Pratt, J. M., Vanston, N. J., and Williams, R. J. P. (1965). *J. Chem. Soc.,* 6485.
55. Hogenkamp, H. P. C., Rush, J. E., and Swenson, C. A. (1965). *J. Biol. Chem.* **240**, 3641.
56. Hogenkamp, H. P. C. (1966). *Fed. Proc.* **25**, 1623.
57. Beaven, G. H., and Johnson, E. A. (1955). *Nature* **176**, 1264.
58. Tackett, S. L., Collat, J. W., and Abbott, J .C. (1963). *Biochemistry* **2**, 919.
59. Jaselskis B., and Diehl, H. (1954). *J. Am. Chem. Soc.* **76,** 4345.
60. Yamada, R., Shimizu, S., and Fukui, S. (1966). *Arch. Biochem. Biophys.* **117**, 675.
61. Bayston, J. H., and Winfield, M. E. (1967). *J. Catal.* **9**, 217.
62. Peel, J. L. (1963). *Biochem. J.* **88**, 296.
63. Schrauzer, G. N., and Sibert, J. W. (1969). *Arch. Biochem. Biophys.* **130**, 257.
64. Costa, G., Mestroni, G., and Tauzher, G. (1972). *J. Chem. Soc.,* 450.
65. Moskophidis, M., and Friedrich, W. (1972). *Z. Naturforsch.* **27b**, 1175.
66. Bayston, J. H., King, N. K., Looney, F. D., and Winfield, M. E. (1969). *J. Am. Chem. Soc.* **91**, 2775.
67. Zelewsky, von, A. (1972). *Helv. Chim. Acta* **55,** 2941.
68. Dolphin, D. (1971). *Methods Enzymol.* **18C**, 34.
69. Hogenkamp, H. P. C., and Holmes, S. (1970). *Biochemistry* **9**, 1886.
70. Brady, R. O., and Barker, H. A., (1961). *Biochem. Biophys. Res. Commun.* **4**, 373.
71. Pratt, J. M. (1964). *J. Chem. Soc.,* 5154.
72. Hogenkamp, H. P. C. (1966). *Biochemistry* **5**, 417.
73. Das, P. K., Hill, H. A. O., Pratt, J. M., and Williams, R. J. P. (1967). *Biochem. Biophys. Acta* **141**, 644.
74. Das, P. K., Hill, H. A. O., Pratt, J. M., and Williams, R. J. P. (1968). *J. Chem. Soc. (A),* 1261.
75. Johnson, A. W., Mervyn, L., Shaw, N., and Smith, E. L. (1963). *J. Chem. Soc.,* 4146.
76. Collat, J. W., and Abbott, J. C. (1964). *J. Am. Chem. Soc.* **86**, 2308.
77. Yamada, R., Kato, T., Shimizu, S., and Fukui, S. (1966). *Biochem. Biophys. Acta* **117**, 13.
78. Barnett, R., Hogenkamp, H. P. C., and Abeles, R. H. (1966). *J. Biol. Chem.* **241**, 1483.
79. Schrauzer, G. N., and Holland, R. J. (1971). *J. Am. Chem. Soc.* **93**, 4060.
80. Schrauzer, G. N., Deutsch, E., and Windgassen, R. J. (1968). *J. Am. Chem. Soc.* **90**, 2441.
81. Brearly, A. E., Gott, H., Hill, H. A. O., O'Riordan, M., Pratt, J. M., and Williams, R. J. P. (1971). *J. Chem. Soc. (A),* 612.
82. Müller, O., and Müller, G. (1962). *Biochem. Z.* **336**, 299.

83. Hogenkamp, H. P. C., Pailes, W. H., and Brownson, C. (1971). *Methods Enzymol.* **18C**, 57.

84. Law, P. Y., Brown, D. G., Lien, E. L., Babior, B. M., and Wood, J. M. (1971). *Biochem.* **10**, 3428.

85. Kikugawa, K., and Ichino, M. (1971). *Tetrahedron Lett.* **2**, 87.

86. Brodie, J. D. (1969). *Proc. Natl. Acad. Sci. U. S. A.* **62**, 461.

87. Silverman, R. B., and Dolphin, D. (1973). *J. Am. Chem. Soc.* **95**, 1686.

88. Michaely, W. J., and Schrauzer, G. N. (1973). *J. Am. Chem. Soc.* **95**, 5771.

88a. Schrauzer, G. N. Personal communication.

89. Schrauzer, G. N. (1968). *Acc. Chem. Res.* **1**, 97.

90. Schrauzer, G. N., Sibert, J. W., and Windgassen, R. J. (1968). *J. Am. Chem. Soc.* **90**, 6681.

91. Friedrich, W., and Nordmeyer, J. P. (1969). *Z. Naturforsch.* **24b**, 588.

92. Friedrich, W., and Messerschmidt, R. (1970). *Z. Naturforsch.* **25b**, 972.

93. Hogenkamp. H. P. C. (1963). *J. Biol. Chem.* **238**, 477.

94. Hogenkamp, H. P. C., Ladd, J. N., and Barker, H. A. (1962). *J. Biol. Chem.* **237**, 1950.

95. Hogenkamp, H. P. C. (1964). *Ann. N. Y. Acad. Sci.* **112**, 552.

96. Johnson, A. W., Shaw, N., and Wagner, F. (1963). *Biochim. Biophys. Acta* **72**, 107.

97 Hogenkamp, H. P. C., and Oikawa, T. G. (1964). *Biol. Chem.* **239**, 1911.

98. Johnson, A. W., Oldfield, D., Rodrigo, R., and Shaw, N. (1964). *J. Chem. Soc.* 4080.

99. Dolphin, D., Johnson, A. W., and Rodrigo, R. (1964). *J. Chem. Soc.* 3186.

100. Yamada, R., Shimizu, S., and Fukui, S. (1966). *Biochem. Biophys. Acta.* **124**, 195.

101. Yamada, R., Shimizu, S., and Fukui, S. (1966). *Biochem. Biophys. Acta.* **124**, 197.

102. Pailes, W. H., and Hogenkamp, H. P. C. (1968). *Biochemistry* **7**, 4160.

103. Yamada, R., Kato, T., Shimizu, S., and Fukui, S. (1965). *Biochem. Biophys. Acta.* **97**, 353.

104. Ljungdahl, L., and Irion, E. (1966). *Biochemistry* **5**, 1846.

105. Wood, J. M., Kennedy, F. S., and Wolfe, R. S. (1968). *Biochemistry* **7**, 1707.

106. Kennedy, F. S., Buckman, T., and Wood, J. M. (1969). *Biochim. Biophys. Acta* **661**, 177.

106a. Taylor, R. T., Smucker, L., Hanna, M. L., and Gill, J. (1973). *Arch. Biochem. Biophys.* **156**, 521.

107. Friedrich, W., and Nordmeyer, J. P. (1968). *Z. Naturforsch.* **23b**, 1119.

108. Friedrich, W., and Moskophidis, M. (1970). *Z. Naturforsch.* **25b**, 979.

109. Needham, T. E., Matwiyoff, N. A., Walker, T. E., and Hogenkamp, H. P. C. (1973). *J. Am. Chem. Soc.* **95**, 5019.

110. Barker, H. A., Smyth, R. D., Weissbach, H., Munch-Petersen, A., Toohey, J. I., Ladd, J. N., Volcani, B. E., and Wilson, R. M. (1960). *J. Biol. Chem.* **235**, 181.

111. Barker, H. A., Smyth, R. D., Weissbach, J., Toohey, J. I., Ladd, J. N., and Volcani, B. E. (1960). *J. Biol. Chem.* **235**, 480.

112. Hogenkamp, H. P. C., and Barker, H. A. (1961). *J. Biol. Chem.* **236**, 3097.

113. Hogenkamp, H. P. C., and Pailes, W. H. (1967). *Biochem. Prep.* **12**, 124.
114. Hayward, G. C., Hill, H. A. O., Pratt, J. M., Vanston, N. J., and Williams, R. J. P. (1965). *J. Chem. Soc.* 6485.
115. Arens, J. F. (1960). *Adv. Org. Chem.* **2**, 163.
116. Johnson, A. W., and Shaw, N. (1962). *J. Chem. Soc.,* 4608.
117. Toohey, J. I., Perlman, D., and Barker, H. A. (1961). *J. Biol. Chem.* **236**, 2119.
118. Pratt, J. M. (1972). *Inorganic Chemistry of Vitamin B_{12},* Academic Press, New York.
119. Schrauzer, G. N., and Sibert, J. W. (1970). *J. Am. Chem. Soc.* **92**, 1022.
120. Yamada, R., Umetani, T., Shimizu, S., and Fukui, S. (1968). *J. Vitaminol.* **14**, 316.
121. Bernhauer, K., and Irion, E. (1964). *Biochem. Z.* **339**, 530.
122. Dolphin, D. H., Johnson, A. W., and Rodrigo, R. (1964). *Ann. N. Y. Acad. Sci.* **112**, 590.
123. Wood, J. M., Kennedy, F. S., and Rosen, C. G. (1968). *Nature* **220**, 173.
124. Hill, H. A. O., Pratt, J. M., Ridsdale, S., Williams, F. R., and Williams, R. J. P. (1971). *Chem. Commun.,* 341.
125. Imura, N., Sukegawa, E., Pan, S., Nagao, K., Kim, J., Kwan, T., and Ukita, T. (1971). *Science* **172**, 1248.
126. Bertilsson, L., and Neujahr, H. Y. (1971). *Biochemistry* **10**, 2806.
127. Schrauzer, G. N., Weber, J. H., Beckham, T. M., and Ho, R. K. Y. (1971). *Tetrahedron Lett.* 275.
128. Agnes, G., Bendle, S., Hill, H. A. O., Williams, F. R., and Williams, R. J. P. (1971). *Chem. Commun.,* 850.
129. Volpin, M. E., Volkova, L. G., Levitin, I. Y., Boronina, N. N., and Yurkevich, A. M. (1971). *Chem. Commun.,* 849.
130. Bernhauer, K., and Irion, E. (1964). *Biochem. Z.* **339**, 521.
131. Schrauzer, G. N., and Windgassen, R. J. (1967). *J. Am. Chem. Soc.* **89**, 3607.
132. Bernhauer, K., and Wagner, O. (1963). *Biochem. Z.* **337**, 366.
133. Smith, E. L., Mervyn, L., Muggleton, P. W., Johnson, A. W., and Shaw, N. (1964). *Ann. N. Y. Acad. Sci.* **112**, 565.
134. Bernhauer, K., Renz, P., and Wagner, F. (1962). *Biochem. Z.* **335**, 443.
135. Hill, J. A., Pratt, J. M., and Williams, R. J. P. (1962). *J. Theor. Biol.* **3**. 423.
136. Dolphin, D. H., Johnson, A. W., and Shaw, N. (1963). *Nature* **199**, 170.
137. Wagner, F. (1965). *Proc. Roy. Soc.* **A288**, 344.
138. Dolphin, D. H., and Johnson, A. W. (1963). *Proc. Chem. Soc.* 311.
139. Adler, N., Medwick, T., and Poznanski, T. J. (1966). *J. Am. Chem. Soc.* **88**, 5018.
140. Hill, H. A. O., Pratt, J. M., Thorp, R. G., Ward, B., and Williams, R. J. P.(1970). *Biochem. J.* **120**, 263.
141. Kaczka, E. A., Wolf, D. E., Kuehl, A., and Folkers, K. (1951). *J. Am. Chem. Soc.* **73**, 3569.
142. Pratt, J. M., and Thorp, R. G. (1966). *J. Chem. Soc. (A),* 187.
143. Randall, W. C., and Alberty, R. A. (1966). *Biochemistry* **5**, 3189.
144. Thusius, D. (1971). *J. Am. Chem. Soc.* **93**, 2629.

145. Thusius, D. (1968). *Chem. Commun.*, 1183.
146. Law, P. Y., and Wood, J. M. (1973). *J. Am. Chem. Soc.* **95**, 914.
147. Toohey, J. I. (1965). *Proc. Natl. Acad. Sci. U. S. A.* **54**, 934.
148. Toohey, J. I. (1966). *Fed. Proc.* **25**, 1628.
149. Koppenhagen, V. B., and Pfiffner, J. J. (1970). *J. Biol. Chem.* **245**, 5865.
150. Sato, K., Shimizu, S., and Fukui, S. (1970). *Biochem. Biophys. Res. Commun.* **39**, 170.
151. Dinglinger, F., and Braum, I. (1970). *Hoppe Seylers Z. Physiol. Chem.* **351**, 1157.
152. Koppenhagen, V. B., and Pfiffner, J. J. (1971). *J. Biol. Chem.* **246**, 3075.
153. Koppenhagen, V. B., Wagner, F., and Pfiffner, J. J. (1973). *Fed. Proc.* **32**, 589.
153a. Koppenhagen, V. B., Elsenhans, B., Wagner, F., and Pfiffner, J. J. (1974). *J. Biol. Chem.* **249**, 6532.
154. Friedrich, W. (1964). *Biochemisches Taschenbuch,* 2nd ed., H. M. Rauen, Ed., Springer-Verlag, Berlin, p. 708.
155. Firth, R. A., Hill, H. A .O., Pratt, J. M., Williams, R. J. P., and Jackson, W. R. (1967). *Biochemistry* **6**, 2178.
156. Ladd, J. N., Hogenkamp, H. P. C., and Barker, H. A. (1961). *J. Biol. Chem.* **236**, 2114.
157. Eichhorn, G. L. (1961). *Tetrahedron* **13**, 208.
158. Legrand, M., and Viennet, R. (1962). *Bull. Soc. Chim. France* **12**, 1435.
158a. Bonnett, R., Godfrey, J. M., Rath, V. B., Scopes, P. M., and Thomas, R. N. (1973). *J. Chem. Soc., Perkins, I.* 252.
159. Friedrich, W. (1966). *Z. Naturforsch.* **21b**, 595.
160. Babior, B. M., and Li, T. K. (1969). *Biochemistry* **8**, 154.
161. Schrauzer, G. N., and Lee, L. P. (1968). *J. Am. Chem. Soc.* **90**, 6541.
162. Cockle, S. A., Hill, H. A. O., Pratt, J. M., and Williams, R. J .P. (1969). *Biochim. Biophys. Acta* **177**, 686.
163. Bayston, J. H., Looney, F. D., Pilbrow, J. R., and Winfield, M. E. (1970). *Biochemistry* **9**, 2164.
164. Hamilton, J. A., Yamada, R., Blakley, R. L., Hogenkamp, H. P. C., Looney, F. D., and Winfield, M. E. (1971). *Biochemistry* **10**, 347.
165. Hill, H. A. O., Pratt, J. M., and Williams, R. J. P. (1971). *Methods Enzymol.* **18C**, 5.
166. Cockle, S. A., Hill, H. A. O., and Williams, R. J. P. (1970). *Inorg. Nucl. Chem. Lett.* **6**, 131.
167. Schrauzer, G. N., and Lee, L. P. (1970). *J. Am. Chem. Soc.* **92**, 1551.
168. Hill, H. A. O., Pratt, J M., and Williams, R. J. P. (1965). *J. Chem. Soc.*, 2859.
169. Hill, H. A. O., Mann, B. E., Pratt, J. M., and Williams, R. J. P. (1968). *Chem. Soc. (A)*, 564.
170. Cockle, S. A., Hill, H. A. O., Williams, R. J. P., Mann, B. E., and Pratt, J. M. (1970). *Biochim. Biophys. Acta* **215**, 415.
171. Brodie, J. D., and Poe, M. (1972). *Biochemistry* **11**, 2534.
172. Brodie, J. D., and Poe, M. (1971). *Biochemistry* **10**, 914.
173. Wood, J. M., and Brown, D. G. (1972). *Structure and Bonding* **11**, 47.

174. Doddrel, D., and Allerhand, A. (1971). *Proc. Natl. Acad. Sci. U. S. A.* **68**, 1083.
176. Brown, C. E., Kotz, J. J., and Shemin, D. (1972). *Proc. Natl. Acad. Sci. U. S. A.* **69**, 2585.
175. Doddrel, D., and Allerhand, A. (1971). *Chem. Commun.,* 728.
177. Penley, M. W., Brown, D. G., and Wood, J. M. (1970). *Biochemistry* **9**, 4302.
178. Mayer, E., Gardiner, D. J., and Hester, R. E. (1973). *Biochem. Biophys. Acta* **297**, 568.
179. Wozniak, W. T., and Spiro, T. G. (1973). *J. Am. Chem. Soc.* **95**, 3402.
180. Nath, A., Harpold, M., Klein, M. P., and Kündig, W. (1968). *Chem. Phys. Lett.* 471.
181. Firth, R. A., Hill, H. A. O., Pratt, J. M., Thorp, R. G., and Williams, R. J. P. (1969). *J. Chem. Soc. (A),* 381.
182. Randall, W. C., and Alberty, R. A. (1967). *Biochemistry* **6**, 1520.
183. Hogenkamp, H. P. C., and Rush, J. E. (1968). *Biochem. Prep.* **12**, 121.
184. Kaczka, E. A., Denkewalter, R. G., Holland, A., and Folkers, K. (1951). *J. Am. Chem. Soc.* **73**, 335.
185. Veer, W. L. C., Edelhausen, J. H., Wymenga, H. G., and Lens, J. (1950). *Biochim. Biophys. Acta* **6**, 225.
186. Hayward, G. C., Hill, H. A. O., Pratt, J. M., and Williams, R. J. P. (1971). *J. Chem. Soc. (A),* 196.
187. George, P., Irvine, D. H., and Glauser, S. C. (1960). *Ann. N. Y. Acad. Sci.* **88**, 393.
188. Firth, R. A., Hill, H. A. O., Mann, B. E., Pratt, J. M., and Thorp, R. G. (1967). *Chem. Commun.* 1013.
189. Firth, R. A., Hill, H. A. O., Mann, B. E., Pratt, J. M., Thorp, R. G., and Williams, R. J. P. (1968). *J. Chem. Soc. (A),* 2419.
190. Cockle, S., Hill, H. A. O., Ridsdale, S., and Williams, R. J. P. (1972). *J. Chem. Soc. Dalton,* 297.
191. Friedrich, W. (1965). *Biochem. Z.* **342**, 143.
192. Friedrich, W. (1966). *Z. Naturforsch.* **21b**, 138.
193. Firth, R. A., Hill, H. A. O., Pratt, J. M., and Thorp, R. G. (1968). *J. Chem. Soc. (A),* 453.
194. Friedrich, W., and Moskophidis, M. (1968). *Z Naturforsch.* **23b**, 804.
195. Friedrich, W., Ohlons, H., Sandeck, W., and Bieganowski, R. (1967). *Z. Naturforsch.* **22b**, 839.

CHAPTER TWO

BIOSYNTHESIS OF CORRINOIDS

HERBERT C. FRIEDMANN

Department of Biochemistry
University of Chicago
Chicago, Illinois

CONTENTS

Child, if you have a rummy kind of name,
Remember to be thankful for the same.

Hillaire Belloc
Cautionary Verses

INTRODUCTION

Emphasis on Comparisons with Biosyntheses of Simpler Substances

Corrinoids are the most complex nonpolymeric natural products known. They can be dissected into distinct parts, each of which tells a separate biosynthetic story. The structural similarities of these parts to other substances invite biosynthetic comparisons. Thus the corrin ring strongly resembles the porphyrin ring, and, in fact, these two groups of substances have common biosynthetic features. Similarly, the formation and assembly of the components of the "nucleotide loop" may be compared with the formation of the more common nucleotides. Again, one may compare the mode of origin and attachment of the *Coβ*-adenosyl group of corrinoids and of the adenosyl group of *S*-adenosylmethionine. In each case the differences from the better known pathways rather than the similarities to them provide the prime challenge to present-day research in this field. Many facets in the natural assembly of these complex molecules have come to light since this subject was surveyed in 1970 (1) so that the time for review [see also Alworth in Plaut et al. (2)] is opportune.

CORRIN RING FORMATION

δ-Aminolevulinic Acid as Precursor

The structure of the corrins was announced (3, 4) shortly after it had been shown (5, 6) that δ-aminolevulinic acid was the precursor of the tetrapyrrolic porphyrin ring. The similarities between the corrin and the porphyrin ring systems suggested that δ-aminolevulinic acid was the precursor of the former ring system as well. This was verified by experiments in Shemin's laboratory in which δ-aminolevulinic acid variously labeled with ^{14}C was fed to a streptomycete. Partial chemical degradation of the cobalamin formed revealed labeling patterns in accord with the expected pathway by way of porphobilinogen (7–9), the monopyrrole produced by the condensation of two δ-aminolevulinic acid molecules. Other workers (10) showed that porphobilinogen itself was incorporated into corrin.

Uroporphyrinogen III as Precursor, and the Origin of the C_1-Methyl Group

One of the most striking structural differences between the corrin and the porphyrin ring system (Figure 2-1) is the absence from the former of a methene bridge between one of the pairs of pyrroline-type rings (*A* and *D*). A methyl group is attached to C_1, that is, to the ring *A* α-carbon which is joined directly to ring *D*. The position of this methyl would appear to correspond to the missing methene bridge. Moreover, the sequence of amide side-chain substituents at the β position of the pyrroline-type rings in corrinoids corresponds to that of the carboxylic acid groups (AP–AP–AP–PA) in uroporphyrinogen (urogen) III (Figure 2.1). Hence a corrin precursor function for urogen III seemed plausible. Burnham and Plane (11) reported that a number of clostridia, bacteria thought to make corrinoids but not heme proteins, produced uroporphyrin (type undertermined) upon incubation with δ-aminolevulinic acid and fumarate. Cell-free extracts of *Clostridium tetanomorphum* were shown by Porra (12) to convert δ-aminolevulinic acid to uroporphyrin III. Furthermore, it was tentatively concluded by Shemin and Bray that the unique methyl group at C_1 was derived from the aminomethyl end of δ-aminolevulinic acid (8, 9). It seemed likely, therefore, that urogen III was converted to the corrin ring, most probably by a pathway that preserved the methene bridge between rings *A* and *D* in the form of the C_1 methyl group of ring *A*.

It was shown, however, that the methyl at C_1 is not derived from δ-aminolevulinic acid (13, 14). Experiments with *Propionibacterium shermanii*, using ^{13}C nuclear magnetic resonance (15, 16, 16a) demon-

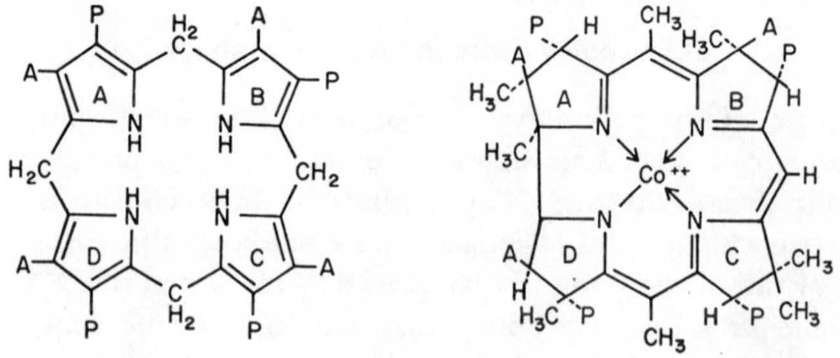

A = CH_2COOH

P = CH_2CH_2COOH

Fig. 2-1. Structures of uroporphyrinogen III (left) and of cobyrinic acid (right).

strated what chemical analysis had left uncertain: that this methyl group is in fact derived from the methyl group of methionine (14, 17). This finding was unexpected, not merely because of the earlier interpretation of the chemical degradation studies, but because nobody would have guessed from direct inspection of the corrin ring structure that this very methyl group did not correspond to the "missing" methene bridge carbon.

Paradoxically, the candidacy of urogen III as corrin precursor was strengthened in separate experiments from Scott's laboratory at the very time that δ-aminolevulinic acid was eliminated as source of the C_1 methyl group. When [^{14}C]urogen III was administered to *P. shermanii,* cyano-[^{14}C]cobalamin could be isolated (18). Degradation of the labeled cobalamin indicated that the label had not been incorporated into the carbons of the nucleotide loop or into the "extra" methyl groups. No incorporation occurred with [^{14}C]urogen I, the uroporphyrinogen whose carboxylic acid groups show the regularly alternating sequence AP–AP–AP–AP. It was concluded that urogen III was incorporated specifically and intact into the corrin ring system. These results were confirmed by the nuclear magnetic resonance spectra of the cyano[^{13}C]cobalamin obtained from *P. shermanii* after feeding δ-aminolevulinic acid, porphobolinogen, and urogen isomers, all labeled with ^{13}C in equivalent positions. It follows that in the conversion of urogen III to cobalamin, the methene bridge carbon between rings *A* and *D* of urogen III is expelled in what has been called (18) a reductive contraction, or a hydrolytic ring contraction.

Müller and Dieterle (19) and Franck et al. (20) have concluded that urogen III is essentially not converted to corrinoids by *P. shermanii.* These negative results were probably due to the incubation conditions and to the concentrations of urogen III administered (18).

Origin of the Six Other "Extra" Methyl Groups

In addition to the "extra" methyl group attached to C_1, corrinoids contain six other "extra" methyl groups. One "extra" methyl group is found in the β position of each of the four rings *A, B, C,* and *D* (Figure 2-1). The remaining two "extra" methyl groups are attached to two of the three methene bridges (the bridge between rings *B* and *C* is unmethylated). Early work from Shemin's laboratory showed that all of these methyl groups are derived from methionine (8, 9).

The pattern by which the aminomethyl group of δ-aminolevulinic acid and the methyl group of methionine are distributed into the corrinoid molecule was fully confirmed in ^{13}C spectroscopy experiments with *P. shermanii* by Battersby et al. (21). A heptamethyl ester derived from cyanocobalamin was used. Related experiments with [Me-^{13}C]methionine were reported from Scott's (22) and from Shemin's (17) laboratories. A new refinement was the demonstration that of the two methyl groups attached to C_{12} of ring *C*, one of which was known to originate from δ-aminolevulinic acid, it was the pro-*R* methyl group that was derived from methionine (21, 22), a result that, although it appeared to be contradicted by some other data (17), has been firmly established by later work with both ^{13}C and ^{2}H labeling (17a).

The sequence of methylations in the course of biosynthesis of the corrin ring remains unknown.

Corrinoid Synthesis by a Cell-Free System

A report has recently appeared [Scott et al. (23), see also Scott (24)] on the formation of cobyrinic acid from δ-aminolevulinic acid, as well as from a urogen mixture containing urogen III, by a crude cell-free system from *P. shermanii* incubated with a complex mixture that includes *S*-adenosylmethionine. This system therefore contains not only the enzymes for converting δ-aminolevulinic acid to urogen III but also those required for the seven methylations, for the net reduction of two of the eight double bonds in urogen (the corrin ring has six double bonds), for ring *C* acetate decarboxylation, and for cobalt insertion (not necessarily in this order).[1] The rate of conversion (about 2 pmole of cobyrinic acid formed per milligram of protein per hour at 37°) was very much lower than the lowest rates calculated for the crude *P. shermanii* extracts in some of the later biosynthetic steps (1). However, the conversion of about 0.113 μmole of δ-aminolevulinic acid to cobyrinic acid in 16 hr at 37° in a 10-ml incubation mixture containing 400 mg of protein is gratifying indeed. Future results will undoubtedly elucidate the pathway and the mechanisms of these reactions, and might indicate the rate-limiting step.

UROPORPHYRINOGEN III FORMATION: MECHANISM OF INVERSION OF RING *D*

General

If, as discussed above, urogen III is an obligatory precursor of the corrin ring, then the biosynthesis of urogen III from porphobilinogen is related to that of corrinoids.

[1] A hypothetical sequence for these transformations (24a), including the crucial reaction for the formation of the corrin ring structure, was published after the completion of this manuscript.

The central problem in urogen III formation is the mechanism of inversion of the side chains in ring *D*. It is known from the work of Bogorad (25, 26) that two enzymes are required for the *in vitro* enzymatic condensation of four molecules of porphobilinogen to urogen III: urogen I synthase, also known as porphobilinogen deaminase, and urogen III co-synthase. Urogen I synthase by itself catalyzes the condensation of porphobilinogen to urogen I, the cyclic tetrapyrrole with regularly alternating side chains. However, simultaneous incubation of urogen I synthase and urogen III co-synthase with porphobilinogen yields urogen III. Urogen III co-synthase by itself causes no condensation of porphobilinogen, nor does it change urogen I to urogen III.

An immense amount of experimentation and model building has attempted to explain these findings. To quote a recent article, "Few biochemical mechanisms have aroused as much interest as the enzymatic conversion of porphobilinogen into uroporphyrinogen III, the basic skeleton of all the natural porphyrins and chlorins. More than 20 hypotheses have been put forward to explain this reaction, but none has yet been proved" (27).

To form the one inverted set of side chains in urogen III it is not enough simply to visualize a "flip" (28) of one porphobilinogen, followed by direct condensation of the most adjacent carbon atoms. Such an inversion would lead to a structure with no carbon atom on one side of a pyrrole ring and two carbons on the other (Figure 2-2). This problem has stimulated the 20-odd number of hypotheses (27–31) for urogen III formation. This surfeit of models is an object lesson of the fact that the availability of an *in vitro* enzyme system does not guarantee that it will be easy to determine a reaction sequence, let alone a mechanism.

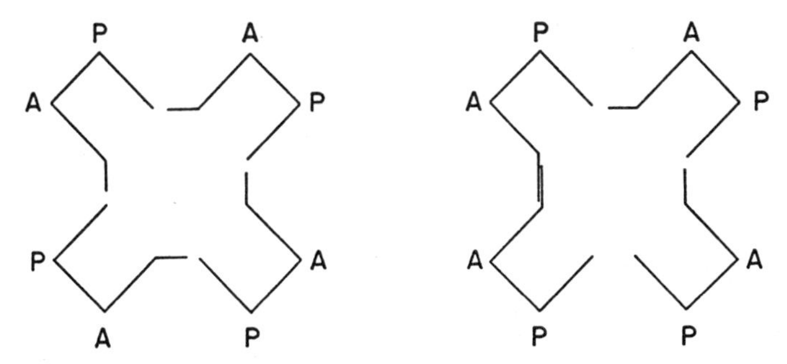

Fig. 2-2. Schematic illustration of the fact that for the condensation of four porphobilinogen (PEG) molecules to form uroporphyrinogen III it does not suffice to visualize a simple "flip" of one of the porphobilinogens.

Intermediates

Recently the problem of urogen III formation has come nearer to a solution. Two experimental approaches have opened the field. In one approach a partially inhibited system was used. Intermediates were isolated, identified, and added to the uninhibited system. In the other approach the fate of various synthetic dipyrrylmethanes was studied.

The first approach, initiated in Bogorad's laboratory (32, 33), demonstrated that di-, tri-, and tetrapyrrylmethanes accumulate when urogen I synthase from spinach leaves is incubated with porphobilinogen in the presence of ammonium ions or of hydroxylamine. The di- and tetrapyrrylmethanes could be isolated. The dipyrrylmethane was shown to be identical to the synthetic (34) dipyrrylmethane A (Figure 2-3) by nuclear magnetic resonance and by enzymatic measurements (32). The tetrapyrrylmethane was concluded to be a linear tetrapyrrole with the regularly alternating side-chain pattern found in urogen I and with the aminomethyl group at the side corresponding to an acetic acid side chain (Figure 2-3). Accumulation of pyrryl intermediates was also shown by the group of Llambías and Batlle with enzyme from soybean callus tissues [see Stella et al. (35) and references cited therein], and by Davies and Neuberger with the photosynthetic microorganism *Rhodopseudomonas spheroides* (36). In Bogorad's laboratory it was shown that urogen I was not formed by the condensation of two molecules of the dipyrrylmethane A (Figure 2-3) since A was incorporated into urogen I only in the presence of porphobilinogen (32). It was concluded, furthermore, that the formation of urogen III probably does not go by way of the rearrangement of a preformed tetrapyrrole (33).

The conclusion that the committed step in urogen III synthesis occurs

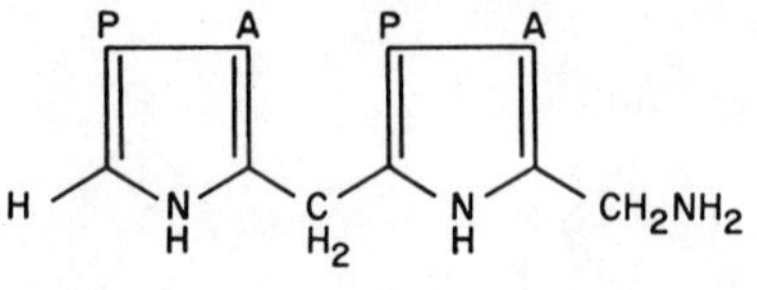

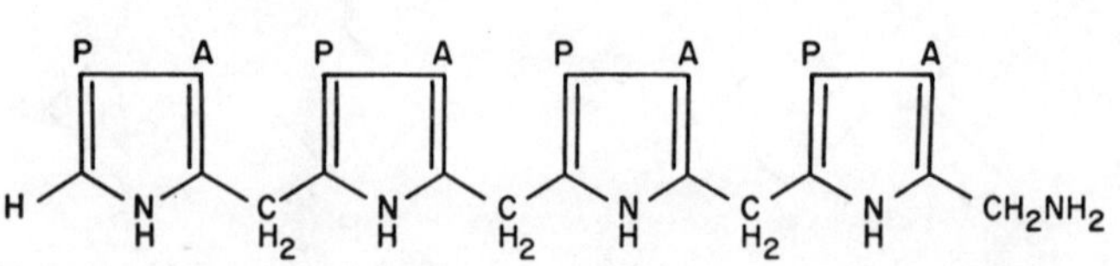

Fig. 2-3. Di- and tetrapyrrylmethanes with regularly alternating side chains. Methylamine group in each case is on the acetic acid side, as in porphobilinogen.

early in the biosynthetic sequence is also based on work done in the laboratories of Frydman and of Rapoport (27, 37) with a number or synthetic dipyrrylmethanes. They found, again, that the dipyrrylmethane A in the presence of porphobilinogen leads to urogen I only, even when urogen III cosynthase was added with urogen I synthase. On the other hand, the dipyrrylmethane B (Figure 2-4), in which the side chains are inverted relative to each other and which still carries the aminomethyl group on the original acetic acid side of one of the rings, was not utilized enzymatically. It followed, in contrast to most models and most discussions of the problem, which had assumed the very opposite, that the pathways for the formation of urogen I and of urogen III diverge at the very start of the polymerization. This conclusion has been confirmed in more recent work with a synthetic tripyrrylmethane (38).

In accord with the above-mentioned conclusion of an early committed step for urogen III biosynthesis, Frydman et al. (27) postulated the existence of a dipyrrylmethane that can give rise to urogen III upon repeated reaction with porphobilinogen. In addition to the inversion of the side chain this dipyrrylmethane must carry its aminomethyl group on the propionic acid side chain of one of the rings. The required compound, the dipyrrylmethane C (Figure 2-5) would have to be formed by a head-to-head condensation of two porphobilinogen molecules, followed by an intramolecular migration. A repetition of this process would lead to urogen III. This pathway, first suggested by Sir Robert Robinson (39), has been demonstrated (40) with the synthetic dipyrrylmethane C. Isotopic data indicate that when this substance is incubated with urogen I synthase and urogen III co-synthase in the presence of porphobilinogen, urogen III is formed. At long last, therefore, the mechanism by which porphobilinogen units condense to form urogen III is beginning to be understood. This scheme, which requires the aminomethyl group to migrate after rather than before condensation, is in accord with the finding that iso-porphobilinogen (Figure 2-6) is not an intermediate in porphyrin biosynthesis (41). The contrasting pathways of urogen I and urogen III biosynthesis by way of the dipyrrylmethanes A and C have

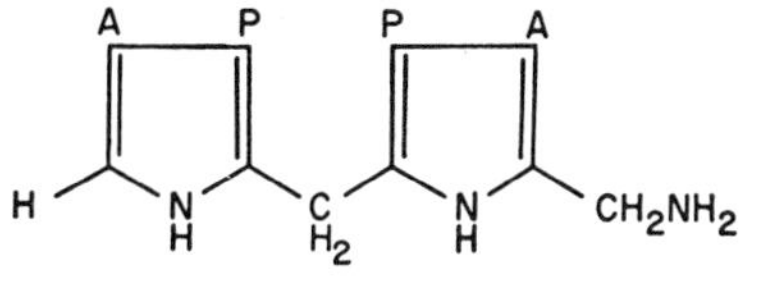

B

Fig. 2-4. Dipyrrylmethane with inverted side chains. Methylamine group is on the acetic acid side, as in porphobilinogen.

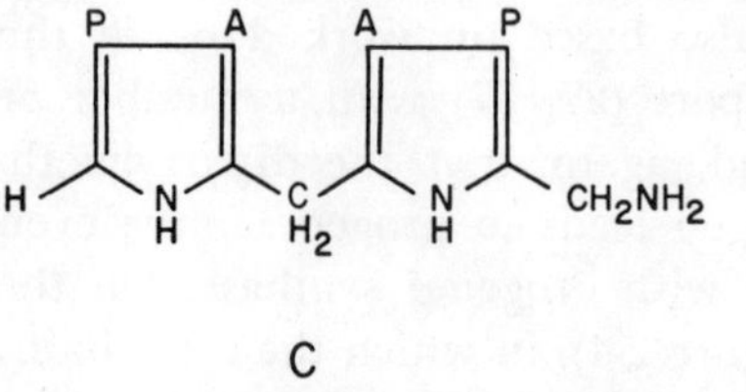

P A A P

H N C N CH_2NH_2
H H_2 H

C

Fig. 2-5. Dipyrrylmethane with inverted side chains. Methylamine group is on the propionic acid side.

been recently reviewed by Frydman et al. (41a). Recent experiments from Battersby's laboratory (42) establish an intramolecular rearrangement of the porphobilinogen unit forming ring *D,* but not of the other three porphobilinogen units.

These results suggested that urogen III co-synthase is a "specifier protein" which changes the way in which urogen I synthase brings about porphobilinogen polymerization (27). Since the rate of reaction of the dipyrrylmethane A (Figure 2-3) with porphobilinogen was much slower than the rate of polymerization of porphobilinogen alone, it was concluded that the whole process takes place on the enzyme surface without liberation of free pyrrylmethanes.

The wide distribution and fundamental importance of cyclic tetrapyrroles and the ease with which they are formed by the condensation of monopyrroles have stimulated experiments and discussions about their role in the origin of life; for recent articles see Krasnovskii and Umrikhina (43) and Petryka and Watson (44). At least one paper has postulated that the corrin ring may actually predate the porphyrin ring in molecular evolution (45).

Labeling Patterns

Labeling patterns of cobalamin formed *in vivo* from [5-^{14}C]δ-aminolevulinic acid clearly indicate a methylene group migration relative to the porphobilinogen side chains, since the only three adjacent carbons labeled under these condtions comprise the bridge carbon between rings *C* and *D* and a carbon of each of these two rings (Figure 2-7). It is not known whether this bridge carbon is the one lost in corrin ring formation from the porphobilinogen corresponding to ring *A,* or whether it is

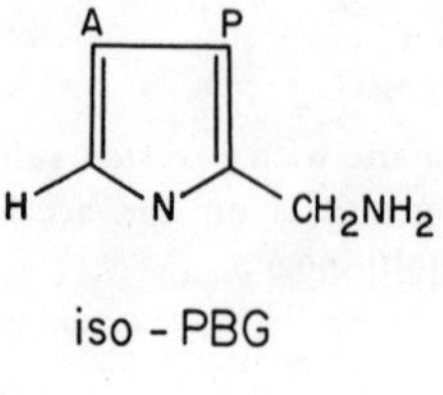

A P

H N CH_2NH_2

iso - PBG

Fig. 2-6. Iso-porphobilinogen (iso-PBG). The methylamine group is on the propionic acid side, rather than, as in porphobilinogen, on the acetic acid side.

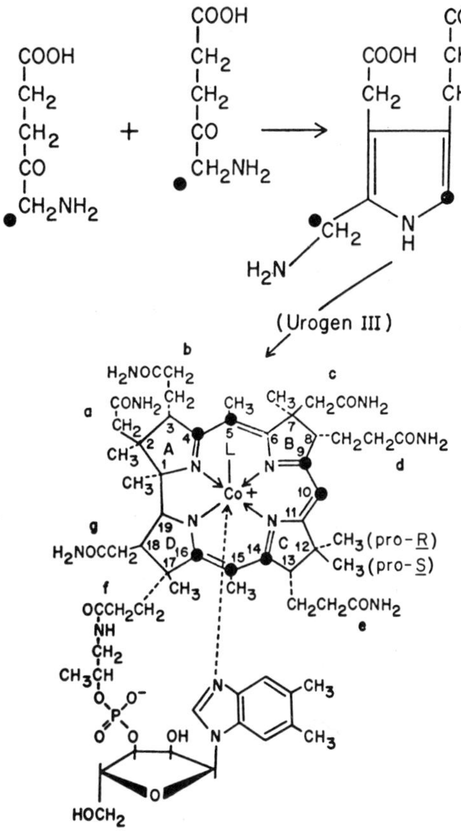

Fig. 2-7. Labeling pattern of cobalamin formed, by way of the correspondingly labeled porphobilinogen, from two molecules of δ-carbon-labeled δ-aminolevulinic acid. In rings *B* and *C*, and the corresponding bridge-carbons 5 and 10, the label distribution corresponds to that in the porphobilinogen; in ring *A* the methyl group at carbon-1 is not labeled, while the derivation of the label in the bridge-carbon 15 is not established. L = Coβ ligand. [After Brown et al. (13), Scott et al. (14), and Battersby et al. (21).]

the carbon that was originally on the acetic acid side of the prophobilinogen corresponding to ring *D* (13). In the scheme for urogen III biosynthesis advanced by Bullock et al. (46) "the methylene group originally attached to ring D migrates successively to rings A, B, and C and finally forms the link between rings C and D."

Mechanism of Urogen I Synthase: Models

The question of the binding of porphobilinogen and its polymerization by urogen I synthase has elicited far less discussion than the question of "flipping" for urogen III formation. Observations consistent with the idea that urogen I synthase has a polypyrrole cyclizing site distinct from a porphobilinogen polymerizing site have been advanced by Frydman and Frydman (47). A model for porphobilinogen polymerization and tetrapyrrole cyclization that invokes two binding sites was offered by Jordan and Shemin (48). This model is based on the finding that the urogen I synthase from *R. spheroides* exists as a single polypeptide chain. It proposes an intramolecular transfer after the condensation of the first two porphobilinogen molecules. Thus one of the binding sites is vacated for a third porphobilinogen molecule. Further sequences of polymerization and translocation follow, with final cyclization. This model is reminiscent of the two-site model of ribosomes for the elongation of a peptide chain.

It is likely that further progress in this field will be facilitated by the possibility of making the hitherto scarce porphobilinogen in large amounts by means of the enzymatic conversion of δ-aminolevulinic acid to porphobilinogen on a Sepharose column containing bound δ-aminolevulinic dehydratase (49).

SIDE-CHAIN MODIFICATIONS: ONE DECARBOXYLATION, SIX AMIDATIONS, ONE SUBSTITUTED AMIDATION

Comparisons with Side-Chain Modifications of Porphyrinogens

Side-chain modification is quite common in the tetrapyrrole field. The conversions of uroporphyrinogens to coproporphyrinogens to protoporphyrinogens are accomplished by decarboxylations and desaturations (50, 51). The side-chain modifications in corrinoid biosynthesis are unique. There is but one decarboxylation, that of the acetate side chain of ring *C* to a methyl group. The remaining seven carboxylic acid groups are all present as amides, either unsubstituted or, for the propionic acid side chain of ring *D* (position *f*), as a substituted amide formed with 1-amino-2-propanol.

These modifications pose two questions: (a) In what sequence do they occur? (b) How are they effected?

Sequences of Side-Chain Modifications

Cobyrinic acid, the most primitive known corrinoid, has already undergone the decarboxylation of the ring *C* acetate side chain. This, then, is the first side-chain modification. Nothing is known about the stage in corrinoid biosynthesis at which this step occurs, or about the enzymology of the process, although an *in vitro* system that includes this conversion has been obtained (23, 24). It has been suggested that it takes place before addition of the "extra" methyl to ring *C* (14; see also 24a).

The unraveling of the sequence of amidations represents a formidable problem; no less than 190 different structures can be written between cobyrinic acid and cobamide (52). Extensive studies on the sequence of attachments of unsubstituted amino groups and of 1-amino-2-propanol have been performed in several laboratories. This topic can be summarized only briefly here. More detail can be found in an earlier review (1). Either the various corrinoids from intact *P. shermanii* were isolated and identified (52–57) or the conversions of partially amidated corrinoids added to intact *P. shermanii* (58–60) were investigated.

The principal conclusion from the elaborate studies by the Bernhauer group (56) is that the amidations follow a preferred but not a unique sequence. The following generalizations can be made: (1) The acetic acid side chain *c* (Figure 2-7) is always the first to be amidated. The amidation of the three acetic acid side chains occurs before that of the propionic acid groups, but one of the latter (*d*) can be amidated before all of the acetic acid groups are amidated. (2) The 1-amino-2-propanol group can be added to the propionic acid side chain *f* at any stage after the amidation of group *c*. Thus one obtains a dynamic picture in which attachment of 1-amino-2-propanol to the *f* side chain is concurrent with the step-by-step amidations of the other carboxylic acid groups. The stage at which 1-amino-2-propanol is added depends on the *P. shermanii* growth conditions: In a poor medium this addition follows all the other amidations, that is, it takes place preferentially at the stage of cobyric acid, while in a rich medium it takes place early, that is, on one of the cobyrinic acid amides. Clearly a control is exerted by the relative amounts of L-threonine, the precursor of 1-amino-2-propanol (see the following section), and of the donors of unsubstituted amino groups.

This picture of the sequence of amidations, particularly concerning the stage of 1-amino-2-propanol attachment, is somewhat at variance with the conclusions of other workers. Friedrich and Sandeck, for example, basing their results on the addition of incomplete corrinoids to *P. shermanii,* found only one sequence of amidations, the 1-amino-2-propanol attaching last (58–60). However, added corrinoids that already contained 1-amino-2-propanol could be fully amidated. Possibly the deamidation of cobinamide to cobyric acid (61, 62–64) plays a role here. Pawełkiewicz's group (57) also found evidence for late 1-amino-2-propanol attachment, since in late *P. shermanii* cultures a high proportion of the monocarboxylic corrinoids was in fact cobyric acid. On the other hand Rapp, by *in vitro* studies of amino group and 1-amino-2-propanol attachment to corrinoids that were added to crude *P. shermanii* extracts, generally confirmed the *in vitro* findings from Bernhauer's laboratory, although no cobyric acid was found in these experiments (65). In light of the nutritional results mentioned above, at least some of these discrepancies between different workers may be the result of differences in growth conditions of *P. shermanii.* Certain conclusions regarding the sequence of amidations, particularly the conclusions that the propionic acid of ring *C* is the last to be amidated, were based on X-ray and neutron diffraction identification (66, 67) of a monocarboxylic corrinoid (cMS_1) which had to be revised (68, see 1) after the Bernhauer studies were published. Finally, there are species differences in corrinoid composition and in overall amidating capacity that may possibly reflect differences in the sequence of side-chain modifications (52, 69–76).

In *P. shermanii,* cobyrinic acid is never found as the *Coβ*-adenosyl derivative, but all other corrinoids isolated from this organism are in this form. It is not known whether adenosylation is a prerequisite for the subsequent amidations, or whether the rate of adenosylation far exceeds that of the amidations. Small amounts of incompetely amidated corrinoids that already contain the whole nucleotide loop have also been observed (52, 77–79). This indicates either that completely amidated corrinoids are not required for attachment of the nucleotide loop (78) or that partial deamidations of corrinoids are possible (56, 61–64).

Source and Mechanism of Attachment of the 1-Amino-2-Propanol Group

Much work has been done on the source and mode of attachment of the (R)-1-amino-2-propanol group. It is established that this group is derived from L-threonine. The initial evidence for this conclusion, from Sprinson's laboratory, was the observation that upon administration of L-[^{15}N]-threonine to *Streptomyces griseus,* the (R)-1-amino-2-propanol of the cobalamin formed contained ^{15}N (80). The suggestion that the carbon skeleton of this group is also derived from L-threonine (80) has been amply proved in several laboratories since radioactivity from L-[U-^{14}C]-threonine was incorporated by *P. shermanii* (81) and by *Streptomyces olivaceus* (82) specifically into the (R)-1-amino-2-propanol moiety of cobalamin.

(R)-1-Amino-2-propanol is simply L-threonine minus the carboxyl group. However, to date no direct L-threonine decarboxylase has been observed. One way around this difficulty, suggested by Neuberger and Tait (83), invokes the spontaneous decarboxylation (84) of the β-keto-acid, α-amino-β-ketobutyrate, formed from L-threonine by dehydrogenation. Stereospecific reduction of the resulting aminoacetone would yield

(R)-1-amino-2-propanol. The net result of such a process is decarboxylation of the threonine. L-threonine dehydrogenase was known from *Staphylococcus aureus* (85) and from *R. spheroides* (83, 86) at the time that this suggestion was made. (R)-1-Amino-2-propanol dehydrogenase (or aminoacetone reductase) was discovered later in *E. coli* by Turner (87) and was studied in Turner's laboratory with a number of microorganisms (88–90) and by Dekker and Swain with *E. coli* (91). The separated *E. coli* L-threonine and (R)-1-amino-2-propanol dehydrogenases can be used in a coupled system to yield (R)-1-amino-2-propanol (92).

In spite of the fact that L-threonine can be converted enzymatically to (R)-1-amino-2-propanol, there is no evidence that this pathway operates in corrinoid biosynthesis. Rather, the opposite may obtain. Thus Lowe and Turner made the highly intriguing observation that extracts of anaerobically grown *P. shermanii* and *P. freudenreichii,* both excellent producers of corrinoids, have no demonstrable dehydrogenase activity for either (R)-1-amino-2-propanol or L-threonine, while extracts from a number of other bacteria, including aerobically grown *P. shermanii,* do contain these enzymes. In fact *E. coli,* which cannot make corrinoids from any precursor simpler than cobinamide (52, 70–74)—the corrinoid that already contains (R)-1-amino-2-propanol—had among the highest concentrations of (R)-1-amino-2-propanol dehydrogenase. It was concluded that "a relationship between L-threonine and D-1-amino-2-propanol dehydrogenases and B_{12} biosynthesis does not appear likely" (90). This conclusion was strengthened by the observation that, while growing *S. olivaceus* cultures do indeed incorporate L-[U-^{14}C]threonine specifically into the (R)-1-amino-2-propanol of cobalamin, added (R)-[U-^{14}C]1-amino-2-propanol is not incorporated into this moiety of cobalamin but instead into other parts of the cobalamin molecule. It was suggested that the free amino alcohol does not participate in cobalamin formation (82). On the other hand, Müller et al. (93) found that *P. shermanii* suspensions incorporated both L-[U-^{14}C]threonine and (R)-[U-^{14}C]1-amino-2-propanol specifically into the (R)-1-amino-2-propanol moiety of cobalamin. It is not known whether species differences may account for these conflicting observations.

Despite the unresolved situation with regard to the incorporation of free (R)-1-amino-2-propanol into corrinoids, the possibilities of the initial attachment of L-threonine, L-threonine phosphate, or aminoacetone have been eliminated by work with numerous synthetic corrinoid analogs. The corrinoids containing L-threonine or L-threonine phosphate are not decarboxylated by *P. shermanii,* nor do they undergo exchange with free (R)-1-amino-2-propanol (94). Furthermore, the cobinamide analog containing aminoacetone instead of (R)-1-amino-2-propanol is converted by *P. shermanii* to the (S) rather than the (R) stereoisomer of cobalamin (95). In conclusion, it appears that the route of incorporation of L-threonine into corrinoids remains to be discovered.

CONVERSION OF COBINAMIDE TO COMPLETED CORRINOID: INTRODUCTION OF *N*-α-GLYCOSIDIC NUCLEOSIDE 3′-PHOSPHATE

Mode of Linkage of *N*-α-Glycosidic 3′-Nucleotide to the Rest of the Corrinoid Molecule

The so-called complete corrinoids, such as adenosylcobalamin, adenyl-*Coβ*-adenosylcobamide (adenosyl-"pseudoB_{12}"), and 2-methyladenyl-*Coβ*-adenosylcobamide (adenosyl-"factor A"), can be regarded as consisting of adenosylcobinamide plus an unusual *N*-α-glycosidic 3′-nucleotide. This nucleotide is linked by way of its 3′-phosphate to the hydroxyl of (R)-1-amino-2-propanol of the cobinamide, and is coordinated by way of a certain nitrogen of the given nucleotide base to the central cobalt. The strength of this latter linkage depends on the type of base, nature of Coβ ligand, pH, and temperature (see Chapter 1).

Role of Cobinamide Phosphate and of Guanosine Diphosphate Cobinamide

The 3′-nucleotide is not preassembled and then incorporated. The cobinamide first receives the phosphate moiety. A corrinoid, later recognized as cobinamide phosphate, has been observed in a variety of bacterial sources (71–73, 96–98). The phosphate is attached by an ester bond to the hydroxyl group of the 1-amino-2-propanol at corrinoid position *f*. A very weak cobinamide kinase activity has been demonstrated, using a large excess of ATP, in *P. shermanii* extracts (99, 100).

Another corrinoid, now known to lie on the pathway of nucleotide incorporation, and occurring in somewhat larger amounts than cobinamide phosphate (72, 73, 98), was isolated from *Nocardia rugosa* extracts and demonstrated by DiMarco's group (101–103) to be guanosine diphosphate cobinamide. This substance has also been obtained from *P. shermanii* (104), and in the *Coβ*-adenosyl form from *Clostridium thermoaceticum* (105). GDP-cobinamide contains a conventional N-β-glycosidic ribonucleoside 5′-phosphate attached in anhydride linkage through its phosphate to the phosphate of cobinamide phosphate (Figure 2-8). It is formed by reaction between *Coβ*-adenosylcobinamide phosphate and GTP (106). The decrease in the level of cobinamide and GDP-cobinamide in *N. rugosa* under conditions where complete corrinoids are made

(103, 107, 108) and analogy to the biosynthetic function of numerous nucleoside diphosphates (109) suggested that GDP-cobinamide was a biosynthetic intermediate. This conclusion was supported by the observation that in *P. shermanii* the level of cobalamin rises while that of GDP-cobinamide falls upon administration of 5,6-dimethylbenzimidazole (110). Earlier experiments with ^{32}P had shown that the phosphate of cobinamide phosphate and the corresponding inner phosphate of GDP-cobinamide are retained in cobamide formation (111).

Interlude: Formation of *N*-α-Glycosidic 5′-Nucleotide; Mechanism and Enzymology

The incorporation of the ribose and of the 5,6-dimethylbenzimidazole parts of the *N*-α-glycosidic nucleotide remains to be discussed. *In vivo* experiments with ^{14}C-labeled ribose established that the nucleoside was incorporated as a unit (112). It was therefore postulated that the last step in corrinoid biosynthesis consisted in the formation of a 3′-phosphate ester bond between the inner phosphate of GDP-cobinamide and the ribose of an *N*-α-glycosidic nucleoside, to yield GMP and completed corrinoid. When this proposal was offered, no evidence was at hand for the occurrence or formation of this nucleoside. Shortly thereafter, however, Friedmann and Harris (113) showed that an *N*-α-glycosidic nucleoside is in fact formed by *P. shermanii* from free 5,6-dimenthylbenzimidazole, and that this arises by dephosphorylation of the corresponding *N*-α-glycosidic 5′-nucleotide (114, 115). This type of nucleotide arises by a single-displacement reaction between free base and the nicotinate moiety of nicotinate mononucleotide (114–117):

$$\text{nicotinate-R-5}'\text{-P}^{+} + \text{base} \rightarrow \text{base-}\alpha\text{-R-5}'\text{-P} + \text{nicotinate} + \text{H}^{+}$$

The driving force for this reaction is given by the conversion of the quaternary nicotinate-ribose bond to the uncharged base-ribose bond in the product, with the concomitant liberation of a proton. In this sense, then, the nicotinate may be regarded as activating the C_1' of the ribose 5′-phosphate of nicotinate mononucleotide. In contrast, the formation of the common *N*-β-glycosidic nucleotides always involves a single-displacement reaction at the C_1 of 5-phosphoribosyl-1-pyrophosphate, that is, of ribose 5-phosphate activated at C_1 by pyrophosphate (109). This *trans*-*N*-glycosidase or phosphoribosyltransferase (nicotinatenucleotide-dimethylbenzimidazole phosphoribosyltransferase, EC 2.4.2.21) is the only enzyme known thus far to produce *N*-α-glycosidic bonds.

The formation of this type of nucleotide suggested that it had a biosynthetic function. This conclusion was strengthened by the analysis of the adenine-containing 5′-nucleotide formed by the phosphoribosyltransferase. Friedmann and Fyfe showed that in this compound the ribose was linked to the adenine by an N-α-glycosidic bond to nitrogen-7 and not, as in conventional AMP, by an N-β-glycosidic bond to nitrogen-9 of the base (118, 119). The nucleotide moiety of *Co*α-adenylcobamides also contains this 7-α-D-ribofuranosyladenine bond (120, 121). This structural agreement was judged unlikely to be purely fortuitous. Recently it has been shown by Dinglinger and Renz that the Coα ligand in factor C_x, a corrinoid obtained from two strains of *P. shermanii,* is a variant of the Coα ligand in *Co*α-adenylcobamide. In factor C_x the Coα ligand is unexpectedly 9-α-D-ribofuranosyladenine rather than 7-α-D-ribofuranosyladenine (122). The biosynthetic origin of this linkage is unknown.

The phosphoribosyltransferase does not react with nicotinate ribonucleoside or with nicotin*amide* mononucleotide (NMN). A slow reaction, between free base and NMN, seen in crude bacterial extracts, was shown (116) to be due to the initial deamidation of NMN to nicotinate mononucleotide by a hitherto unrecognized NMN deamidase.

Variability in Incorporated Bases: Guided Biosynthesis

Although the phosphoribosyltransferase is highly specific for the ribose 5-phosphate donor, it is remarkably unselective regarding the type of base it can use. The properties of this enzyme hence explain not only the mode of formation of the N-α-glycosidic bond, but also the mechanism by which many bases other than 5-6-dimethylbenzimidazole, either endogenously derived or supplied to the bacteria from outside ("guided biosynthesis"), are incorporated into N-α-glycosidic nucleotides and transferred in cobalamin analogs. Specificity restrictions correlate with *in vivo* observations on the types of cobalamin analogs made. Thus 2-methylbenzimidazole, which is not incorporated into a corrinoid (123), does not react with the enzyme (115). Similarily, although the *P. shermanii* enzyme reacts with various benzimidazoles, it does not appear to react with adenine (115), a base not usually incorporated by this organism into pseudo B_{12}. However, the purified enzyme from *C. sticklandii,* an organism that incorporates both benzimidazoles and adenine into corrinoids, reacts with benzimidazoles as well as with adenine (116–119). Clearly the enzyme that catalyzes the reaction between GDP-cobinamide and the N-α-glycosidic nucleotides (see the following section) must also be unspecific for the base. This prediction from *in vivo* studies has not been tested *in vitro*. A remarkable variety of bases, including among others benzimidazoles, naphthimidazole, 4-5-dimethylimidazole, bentriazoles, quinazolines, phenazines, and purines, have been shown to be incorporated into corrinoids (64, 124–128).

Incorporation of *N*-α-Glycosidic 5′-Nucleotide: Formation of Cobalamin 5′-Phosphate

We have already seen that the phosphate of the 3′-nucleotide of complete corrinoids is derived from ATP by way of cobinamide phosphate and GDP-cobinamide. The phosphate of the *N*-α-glycosidic 5′-nucleotide formed by the *trans-N*-glycosidase does not appear in the finished product. There are two possible pathways: (1) the free 5′-nucleotide is dephosphorylated, and the resulting nucleoside reacts with GDP-cobinamide to form a cobamide directly; or (2) the 5′-nucleotide is incorporated intact, forming yet another intermediate, a cobamide 5′-phosphate[2] which is subsequently dephosphorylated.

It appears that the latter pathway obtains. Friedmann showed that [2-^{14}C]5,6-dimethylbenzimidazole administered to *P. shermanii* was incorporated into a corrinoid more acidic than cobalamin (129). Time-course and pulse-chase experiments indicated that this substance was a cobalamin precursor. By chemical evidence and by comparison with the synthetic substance (130, 131) this precursor was identified as cobalamin 5′-phosphate (129). The identification was confirmed by the X-ray diffraction analysis of the isolated, crystalline material (132, 133). The formation of cobalamin 5′-phosphate was demonstrated independently by Renz in *P. shermanii* extracts (134–136). Although the evidence indicates that cobalamin formation by way of cobalamin 5′-phosphate is the major route *in vivo,* the specificity of the enzyme catalyzing the reaction between GDP-cobinamide and the 5′-nucleotide does not appear to be absolute, since with *in vitro* systems the nucleoside itself gave rise to cobalamin, though in somewhat lower amounts (135). Evidence for this latter reaction was first obtained by Ronzio and Barker (106).

Dephosphorylation of Cobalamin 5′-Phosphate

The last step in cobalamin biosynthesis is the dephosphorylation of the cobalamin 5′-phosphate. Although the X-ray crystallographic data indicated that the 5′-phosphate group in the cobalamin 5′-phosphate is not

[2] The use of 5′ is in a sense ambiguous, since in the corrinoid field 5′ is primarily used to denote the important 5′-carbon of the adenosyl group. Since, however, during its linkage to cobalt this adenosyl carbon is saturated; since in the case of the Coβ-ligands cyano, hydroxo, methyl, and so on, there is no ambiguity; and since biosynthetically the phosphate in question is derived from a 5′-nucleotide, it was felt that the use of a notation such as 5″ or $5'^{\text{low}}$ would be more cumbersome than helpful.

buried, dephosphorylation of cyanocobalamin 5′-phosphate by conventional alkaline and acidic phosphatases took place only with very great difficulty (129). It was therefore of interest that the dephosphorylation by *E. coli* alkaline phosphatase was facilitated markedly when the Coβ ligand of cobalamin 5′-phosphate was adenosyl, ethyl, or methyl, rather than cyano. Similarily, crude *P. shermanii* extract dephosphorylated adenosylcobalamin 5′-phosphate more readily than cyanocobalamin 5′-phosphate (137). The rate of dephosphorylation by the *E. coli* enzyme of the various cobalamin 5′-phosphates correlated well with the degree of dissociation of the *Co*α-5,6-dimethylbenzimidazole bond; for references, see Schneider and Friedmann (137), and Pratt (138). It appeared that size or shape of the Coβ ligand had little influence on the binding of the cobalamin 5′-phosphate to the phosphatase. The determining factor seemed to be an increase in the proportion of the "base off" form over the "base on" form mediated by the *trans*-effect exerted by these Coβ ligands. It appeared that the nucleotide had to swing free of the cobalt prior to enzymatic 5′-phosphate dephosphorylation (137). During investigation of cobalamin 5′-phosphate formation by *P. shermanii* it was observed that heat treatment of this organism under carefully controlled conditions led to a more than 40-fold increase in the concentration of the compound (139). Although a marked decrease of adenosylcobalamin 5′-phosphate dephosphorylation by heated *P. shermanii* extracts (137) correlates well with the heat-induced accumulation of this precursor in these bacteria (129, 139), the existence of a specific cobalamin 5′-phosphate phosphatase remains undemonstrated.

Summary of Pathway from Cobinamide to Cobalamin

The pathway of incorporation of the constituents of the *N*-α-glycosidic 3′-ribonucleotide into cobalamin is summarized in Figure 2-8. It is seen that, starting with cobinamide and free base, at least three corrinoid intermediates, one nucleotide intermediate, and at least five different enzymes are required. *In vivo* experiments with *P. shermanii* have established that such transformations take place even if the Coβ ligand is not adenosyl (140). This circumstance may have physiological importance for the completion of methylcorrinoids in organisms such as *Clostridium thermoaceticum* (69, 105, 141).

Base Exchange Reaction

It was seen above how bases are attached to incomplete corrinoids in the final stages of *de novo* corrinoid biosynthesis. There also exists a

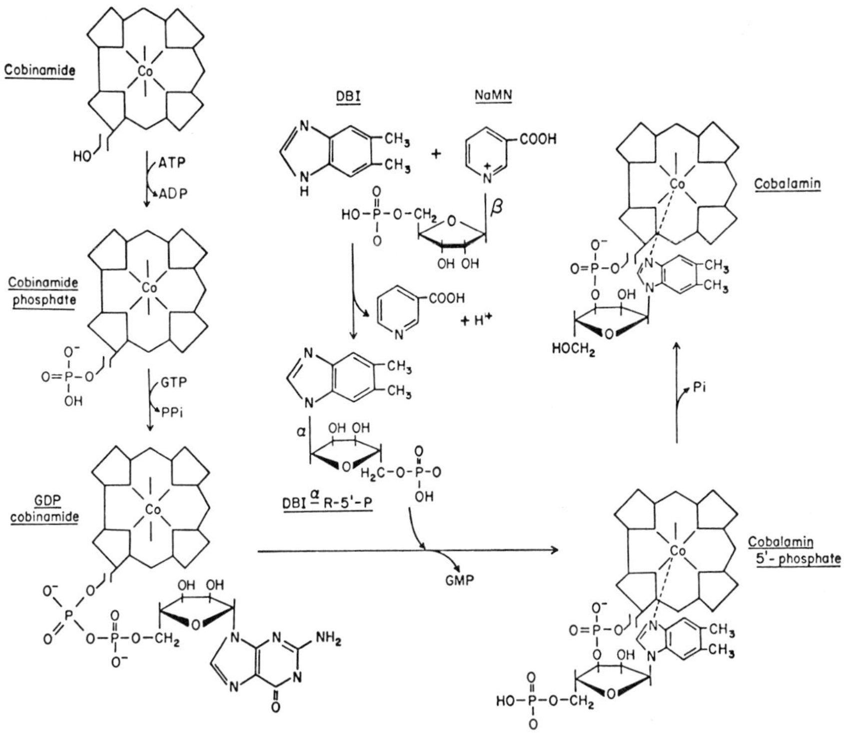

Fig. 2-8. The last steps in cobalamin biosynthesis. The pathway on the left prepares cobinamide in the form of GDP-cobinamide for reaction with the *N*-α-glycosidic nucleotide formed by the reaction in the center. The dephosphorylation of the resulting cobalamin phosphate is shown in the reaction on the right. For clarity details have been omitted from the corrin rings. DBI = 5,6-dimethylbenzimidazole, NaMN = nicotinate mononucleotide.

pathway, less well understood, for the conversion of one complete corrinoid into another by a net exchange of free and bound base (142, 143). This transformation does not entail a simple exchange of bases, since the bond between ribose and the replaced base remains intact during the exchange (144). Furthermore, cobalamin 5′-phosphate is formed as an intermediate when *Co*α-2-methyladenyl-*Co*β-cyanocobamide (factor A) and 5,6-dimethylbenzimidazole are given to *P. shermanii* (145). As in the case of the *de novo* incorporation of bases, the exchange of bases does not require the *Co*β ligand to be adenosyl (140). Although these are obvious analogies to the *de novo* formation of complete corrinoids, the detailed pathway of this exchange reaction is unknown.

In all of the cases just mentioned, free 5,6-dimethylbenzimidazole is exchanged for bound purine or bound benzimidazole. Apparently, exchange of a free purine for a bound benzimidazole has not been observed.

Corrinoids that Contain Phenols

A quite novel type of corrinoid, which contains an *O*-α-glycosidic 3′-ribonucleotide, has been isolated from sewage sludge. Instead of a base, this substance, formerly known as factor 1b (146), contains an aglycone identified as *p*-cresol (147). The bacterial source of this corrinoid, a *p*-cresylcobamide, and the mechanism of incorporation of the phenol remain unknown.

Descobaltocorrinoids containing endogenous phenol (148) or added p-cresol (149) as aglycones have been obtained from the purple sulfur bacterium *Chromatium* strain D.

INCORPORATION OF COBALT, AND RELATED MATTERS

It is not known how and exactly when cobalt is incorporated into the corrin ring. By analogy to the insertion of iron and of magnesium in the course of heme and of chlorophyll formation [see Burnham (150) for review] one might expect cobalt to be inserted into a cyclic tetrapyrrole. However, there is no evidence on this point. The relevance of a bacterial cobalt porphyrin synthase (12, 151, discussed in 1), is not clear. One certainty is that cobalt is inserted before amidations of the side chains since cobyrinic acid, the simplest corrinoid isolated from *P. shermanii* (56), contains cobalt. The recently observed cell-free *P. shermanii* extract (22, 23) that catalyzes the conversion of δ-aminolevulinic acid to cobyrinic acid clearly contains the cobalt-inserting system, but details are not yet available. A recent hypothetical biosynthetic corrin sequence (24a) proposes late cobalt insertion.

Highly intriguing questions about cobalt insertion are posed by the discovery of descobaltocorrinoids, first isolated from the purple sulfur bacterium *Chromatium* strain D (152, 153). The simplest explanation for their formation is that *Chromatium* merely lacks a cobalt-inserting enzyme, and that descobaltocorrinoids are formed through the normal operation of the rest of the pathway. Again, there is no evidence for or against this hypothesis. However, a number of related observations can be explained by postulating variations in specificities and amounts of enzymes in different organisms, rather than a completely different pathway for descobaltocorrinoids. Thus corrinoids and descobaltocorrinoids are both formed in varying proportions by certain purple nonsulfur bacteria grown on a cobalt-containing medium (153, 154). Possibly here the cobalt-inserting enzyme is present in limiting amounts and is specific for one particular precursor. *S. olivaceus* forms small amounts of a descobaltocorrinoid containing unamidated side chains when grown without cobalt (155). In the presence of cobalt normal corrinoids are formed (156, 157). By contrast *P. shermanii* apparently makes no descobaltocorrinoids when grown without cobalt (153). Possibly the *P. shermanii* enzymes are more specific for the presence of cobalt than the *S. olivaceus enzymes, P. shermanii,* again, does not incorporate cobalt into descobaltocorrinoids (149), while *Clostridium perfringens* has been reported to do so (158). On the other hand, *P. shermanii* as well as *Chromatium* strain D can incorporate 5,6-dimethylbenzimidazole into descobaltocobalamin (159) so that at this stage of biosynthesis neither of these organisms requires that the precursor contain cobalt.

BIOSYNTHESIS OF 5,6-DIMETHYLBENZIMIDAZOLE: RELATION TO RIBOFLAVIN

Structural Similarities to Riboflavin; Formation of Riboflavin

Our knowledge of the origin of the unique 5,6-dimethylbenzimidazole moiety of the cobalamin molecule has increased enormously in the last few years as a result of work in the laboratories of Alworth and of Renz. The 1,2-diamino-4,5-dimethylbenzene structure is found only in two relatively rare types of natural products, namely in riboflavin and its derivatives, and in the 5,6-dimethylbenzimidazole moiety of cobalamin. Shortly after the discovery of cobalamin it was suggested by Wooley (160, 161) that riboflavin and 5,6-dimethylbenzimidazole have a common biosynthetic pathway.

Formation of 5,6-Dimethylbenzimidazole from 6,7-Dimethyl-8-Ribityllumazine and from Riboflavin

The very first evidence for a metabolic relationship between 5,6-dimethylbenzimidazole and riboflavin is contained in early papers by Ford and Holdsworth (162) and by Ford et al. (71). These authors showed that when cobinamide is given to *E. coli* in the presence of riboflavin, small amounts of cobalamin are formed. The authors concluded that either riboflavin was directly converted to 5,6-dimethylbenzimidazole or that a limiting amount of a common precursor could be channeled to 5,6-dimethylbenzimidazole in the presence of excess riboflavin. More than 12 years later studies with isotopically labeled molecules began to appear that supported a relationship between riboflavin and 5-6-dimethylbenzimidazole biosynthesis. Renz and Reinhold (163) noted that the labeling pattern of 5,6-dimethylbenzimidazole derived from [2-^{14}C]lactate and [2,3-^{14}C] lactate (presumably incorporated into 5,6-dimethylbenzimidazole after conversion to acetate) was analogous to the labeling pattern known (164) for riboflavin biosynthesis from labeled acetate. Work from Alworth's group showed that although ^{14}C-formate was not preferentially utilized as precursor of C_2 of 5,6-dimethylbenzimidazole (165), the label of [1-^{14}C]ribose was incorporated very efficiently into the C_2 position (166, 167). Renz demonstrated (168) that randomly labeled riboflavin gave rise to labeled 5,6-dimethylbenzimidazole. All of the work just mentioned, done with intact or with broken cells of *P. shermanii,* suggested a close relationship between the biosynthetic routes leading to riboflavin and to 5,6-dimethylbenzimidazole. More detailed information on the mode of formation of 5,6-dimethylbenzimidazole was obtained by the independent demonstration in the two laboratories that *P. shermanii* utilizes 6,7-dimethyl-8-ribityllumazine (Figure 2-9) for the formation of

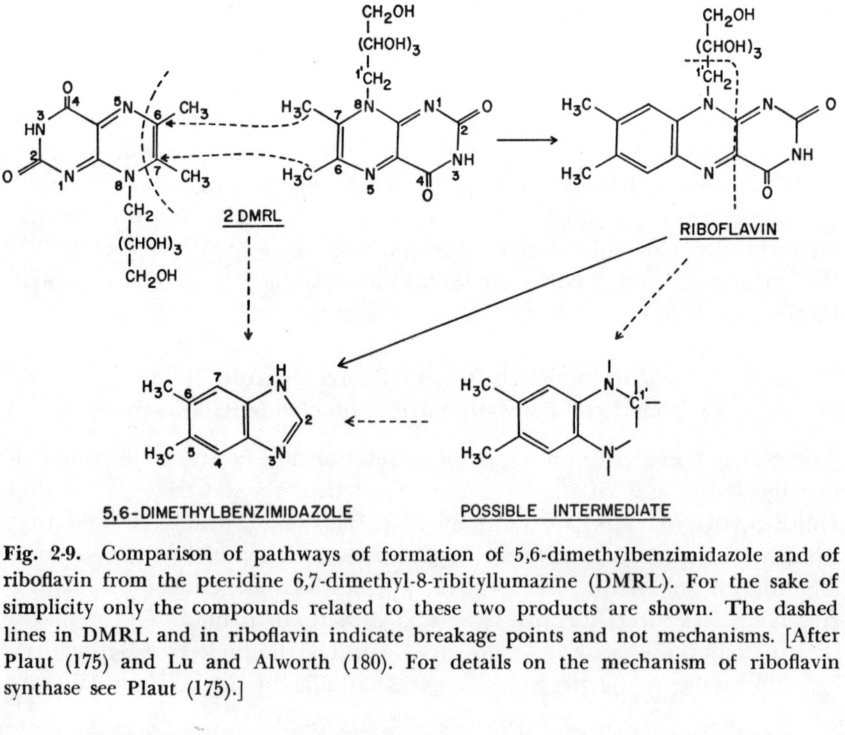

Fig. 2-9. Comparison of pathways of formation of 5,6-dimethylbenzimidazole and of riboflavin from the pteridine 6,7-dimethyl-8-ribityllumazine (DMRL). For the sake of simplicity only the compounds related to these two products are shown. The dashed lines in DMRL and in riboflavin indicate breakage points and not mechanisms. [After Plaut (175) and Lu and Alworth (180). For details on the mechanism of riboflavin synthase see Plaut (175).]

5,6-dimethylbenzimidazole (169–171). Dimethylribityllumazine, a green fluorescent pteridine (172, 173) formed from guanosine or a related compound by a series of reactions the details of which are not completely understood (174–176), is the key intermediate in riboflavin biosynthesis. Riboflavin is made from dimethylribityllumazine in a fascinating bimolecular condensation, catalyzed by riboflavin synthase. This enzyme has been purified several thousand-fold from an extract of bakers' yeast (177, 178). The distribution of ^{14}C in 5,6-dimethylbenzimidazole formed from dimethylribityllumazine labeled with ^{14}C in its two methyl groups was completely analogous to the labeling pattern observed in riboflavin formed from similarly labeled dimethylribityllumazine (179). It was concluded that the pair of methyl groups from one dimethylribityllumazine molecule formed the "central" carbons of the benzene ring of 5,6-dimethylbenzimidazole, while the two methyl groups from a second molecule of dimethylribityllumazine provided the two methyl groups of 5,6-dimethylbenzimidazole. These observations and the preferential incorporation of C_1 of ribose into C_2 of 5,6-dimethylbenzimidazole (166, 167) were extended by the demonstration that the 5,6-dimethylbenzimidazole formed from dimethylribityllumazine labeled with ^{14}C at C_1 of the ribityl moiety and with ^{15}N at N_5 contained ^{14}C and ^{15}N roughly in the ratio present in the starting material. Moreover, ^{14}C was found exclusively in C_2. These findings indicated that the C_1 and N_5 atoms of dimethylribityllumazine are incorporated into 5,6-dimethylbenzimidazole as a unit (180).

It will be seen from the equation depicting the overall conversion of two molecules of dimethylribityllumazine to one molecule of 5,6-dimethylbenzimidazole (Figure 2-9) that the formation of the relatively simple 5,6-dimethylbenzimidazole molecule represents an even more complex process than the formation of the more elaborate riboflavin molecule.

The reactions required to produce the dimethylphenylene ring can be written in a number of different sequences, in one of which riboflavin is an intermediate. It is therefore of interest that, in addition to the demonstration that uniformly labeled riboflavin can act as a precursor of 5,6-dimethylbenzimidazole (168), it was further shown that the conversion of [1'-^{14}C]riboflavin by intact *P. shermanii* into the 5,6-dimethylbenzimidazole of cobalamin is associated with ^{14}C-incorporation exclusively into the C_2 position of the 5,6-dimethylbenzimidazole (181). These observations were consistent with the labeling patterns observed in 5,6-dimethylbenzimidazole following the incorporation of specifically labeled dimethylribityllumazine. Furthermore, since the conversion of riboflavin back to dimethylribityllumazine has not been observed, these results strongly suggest that riboflavin, known to be made by *P. shermanii* (182), may be an obligatory intermediate in 5,6-dimethylbenzimidazole biosynthesis. Further labeling experiments, the identification of intermediates, and work with isolated enzymes will be able to answer these questions. For recent reviews of this topic, see Plaut et al. (2) and Schlee (183).

P. shermanii, with which all of the work just discussed was done, requires aeration to form 5,6-dimethylbenzimidazole for incorporation into cobalamin (184, 185). However, in the case of *Butyribacterium rettgeri,* cobalamin, and hence 5,6-dimethylbenzimidazole, is formed under anaerobic conditions (186).

If riboflavin will indeed turn out to be an obligatory precursor of 5,6-dimethylbenzimidazole, then we have the unique situation wherein the formation of cobalamin requires the direct participation of two other vitamins: nicotinate for the formation of the nucleotide, and riboflavin both as a source of 5,6-dimethylbenzimidazole and, in the form of flavoproteins, to serve as reductants of cob(III)- to cob(I)alamin prior to *Coβ*-adenosylation (see the section on corrinoid *Coβ*-adenosylation) and possibly methylations (Chapter 3).

Formation of Other Benzimidazoles

The mode of formation of other benzimidazole derivatives such as 5-hydroxybenzimidazole (187–189) and 5-methoxybenzimidazole (69, 141, 190, 191), found in the corrinoids of some strictly anaerobic bacteria, remains obscure. Possible schemes have been discussed by Bernhauer et al. (192). It is not known whether the corresponding *o*-phenylenediamines, which can be incorporated as the respective benzimidazoles into corrinoids by various bacteria (107, 193–195), serve as normal biosynthetic precursors.

ORIGIN OF THE *Coβ* LIGAND

Types of *Coβ* Ligand

By and large one of three different groups, the hydroxo group, or one of the two alkyl groups, namely adenosyl or methyl, constitutes the *Coβ* ligand in naturally occurring corrinoids. Of these, the adenosyl ligand is by far the most prevalent. Many other groups can be attached chemically (see Chapter 1). Among these the strongly bound cyano ligand is the most important. This ligand is regarded as an artifact of isolation, fortuitously introduced in early experiments by way of cyanide present in charcoal used as adsorbent (196) or by cyanide added as papain activator in a proteolytic preparative step (197). Hence this group appears to be of no biosynthetic interest in the corrinoid field, in spite of sporadic reports on the physiological occurrence of small amounts of cyanocobalamin in some organisms (198, 199). It suffices to repeat the statement (1) that the activity of enzymes in man that can convert cyanocobalamin to the physiologically active *Coβ*-adenosyl form "both accelerated the discovery of vitamin B_{12} and delayed the discovery of the light-sensitive coenzyme form."

Deficiencies among enzymes involved in the *Coβ*-alkylation of cobalamin are the bases of some human genetic diseases, as is discussed in Chapter 8. The understanding of these genetic lesions was helped enormously by the prior study of the pertinent enzymes obtained from bacteria.

Corrinoid *Coβ*-Adenosylation

The introduction of the alkyl ligand adenosyl, in distinction to the introduction of ligands such as cyano or sulfito, requires that the cobalt be in the Co(I) form. This reduction must proceed by way of the Co(II) form. Enzymatic studies are complicated by the spontaneous disproportionation of corrinoid at the Co(II) stage to corrinoid at the Co(III) and Co(I) stages of reduction (200).

In *C. tetanomorphum* it has been shown that the two reductions are catalyzed by two distinct flavoproteins, one bringing about the first, and the other the second reduction (201). The physiological electron donor for both flavoproteins is NADH. A small dithioprotein may function as electron carrier between flavoproteins and corrinoids (202). The adenosyl moiety is derived from ATP; see Brady et al. (203) and earlier references cited therein, and Peterkofsky, and Weissbach (204). Here ATP has the somewhat unusual functions of a biological alkylating agent. In *C. tetanomorphum* the three phosphates are eliminated as inorganic tripolyphosphate (204, 205), while in *P. shermanii* they are liberated as inorganic pyrophosphate plus orthophosphate, possibly formed from enzyme-bound inorganic tripolyphosphate (203, 206).

The analogy to the alkylation of methionine by ATP to form *S*-adenosylmethionine has often been pointed out (205. 207). In the case of S-adenosylmethionine formation there is no reduction, and a sulfonium ion is formed. Persuasive evidence for enzyme-bound tripolyphosphate formed from ATP in the course of *S*-adenosylmethionine formation has been presented (207).

Essentially all of the work on corrinoid *Coβ*-alkylation has been done with cobalamins. However, while the cobalamins are clearly participants in the enzymatic *Coβ*-adenosylation of added cyanocobalamin or hydroxocobalamin, they are not normally the substrates for *Coβ*-adenosylation in corrinoid biosynthesis (see the section on sequences of side-chain modifications). It is of interest that the physiological substrates are converted only with some difficulty (56, 65) to the *Coβ*-adenosyl form by *P. shermanii* extracts. It is also interesting that the alkylating enzyme is not specific for ATP (208). The 300-fold purified *Coβ*-adenosylating enzyme from *C. tetanomorphum* also utilizes CTP, and somewhat less well ITP, UTP, and GTP (208). Clearly, *in vivo* controls exist that assure that only the adenosyl group is attached to corrinoids.

Corrinoid *Coβ*-Methylations

The *Coβ*-methyl ligand functions in at least three different types of reactions: the formation of methionine, of methane, and of acetic acid (206, 209, 210). In each case the methylation of a protein-bound corrinoid is part of the metabolic system for the formation of one of these three products. The *Coβ*-methyl group is thus essentially a bird of passage. Here, then, biosynthesis and function are closely connected, and each of these three cases must be considered separately. The various *Coβ*-methylations and other aspects related to these reactions are discussed in Chapter 3.

SUMMARY

The number and variety of quite distinct pathways covered in this chapter are probably unique in the study of the formation of but a single nonpolymeric substance. One is reminded of Gulliver, surrounded on all sides by hard-working Lilliputians who wonder how this monster came to be. Since corrinoids are relative newcomers on the biochemical horizon, the study of its biosynthesis has benefited from comparisons with the formation and metabolism of simpler substances structurally related to their various components. Corrinoid biosynthesis represents a coalescence of variations on biosynthetic themes from the fields of amino acid, vitamin, pyrrole, nucleotide, and metal biochemistry. Its study will occupy many a Lilliputian for some time to come.

ACKNOWLEDGMENTS

This work was supported by Research Grant AM-09134 from the National Institutes of Health, United States Public Health Service.

The author wishes to express his appreciation to the many colleagues who supplied results before publication and who critically read the manuscript.

REFERENCES

1. Friedmann, H. C., and Cagen, L. M. (1970). *Ann. Rev. Microbiol.* **24**, 159–208.
2. Plaut, G. W. E., Smith, C. M., and Alworth, W. L. (1974). *Ann. Rev. Biochem.* **43**, 899–908 and 916–922.
3. Hodgkin, D. C., Pickworth, J., Robertson, J. H., Trueblood, K. N., Prosen, R. J., and White, J. G. (1955). *Nature* **176**, 325–328.
4. Bonnett, R., Cannon, J. R., Johnson, A. W., Sutherland, I., Todd, A. R., and Smith, E. L. (1955). *Nature* **176**, 328–330.
5. Shemin, D., and Russell, C. S. (1953). *J. Am. Chem. Soc.* **75**, 4873–4874.
6. Neuberger, A., and Scott, J. J. (1953). *Nature* **172**, 1093–1094.
7. Corcoran, J. W., and Shemin, D. (1957). *Biochim. Biophys. Acta* **25**, 661–662.
8. Bray, R. C., and Shemin, D. (1963). *J. Biol. Chem.* **238**, 1501–1508.
9. Shemin, D., and Bray, R. C. (1964). *Ann. N. Y. Acad. Sci.* **112**, 615–621.
10. Schwartz, S., Ikeda, K., Miller, I. M., and Watson, C. J. (1959). *Science* **129**, 40–41.
11. Burnham, B. F., and Plane, R. A. (1966). *Biochem. J.* **98**, 13C–15C.
12. Porra, R. J. (1965). *Biochim. Biophys. Acta.* **107**, 176–179.
13. Brown, C. E., Katz, J. J., and Shemin, D. (1972). *Proc. Natl. Acad. Sci. U. S. A.* **69**, 2585–2588.
14. Scott, A. I., Townsend, C. A., Okada, K., Kajiwara, M., Whitman, P. J., and Cushley, R. J. (1972). *J. Am. Chem. Soc.* **94**, 8267–8269.
15. Doddrell, D., and Allerhand, A. (1971). *Proc. Natl. Acad. Sci. U. S. A.* **68**, 1083–1088.
16. Anet, F. A. L., and Levy, G. C. (1973). *Science* **180**, 141–148.

16a. Séquin, U., and Scott, A. I. (1974). **186**, 101–107.

17. Brown, C. E., Shemin, D., and Katz, J. J. (1973). *J. Biol. Chem.* **248**, 8015–8021.

17a. Battersby, A. R., Ihara, M., McDonald, E., Stephenson, J. R., and Golding, B. T. (1974). *J. Chem. Soc. Chem. Comm.* 458–459.

18. Scott, A. I., Townsend, C. A., Okada, K., Kajiwara, M., and Cushley, R. J. (1972). *J. Am. Chem. Soc.* **94**, 8269–8271.
19. Müller, G., and Dieterle, W. (1971). *Z. Physiol. Chem.* **352**, 143–150.
20. Franck, B., Gantz, D., Montforts, F.-P., and Schmidtchen, F. (1972). *Angew. Chem. (Int. Ed.)* **11**, 421–422.
21. Battersby, A. R., Ihara, M., McDonald, E., and Stephenson, J. R. (1973). *J. Chem. Soc. Chem. Commun.* 404–405.
22. Scott, A. I., Townsend, C. A., and Cushley, R. J. (1973). *J. Am. Chem. Soc.* **95**, 5759–5761.
23. Scott, A. I., Yagen, B., and Lee, E. (1973). *J. Am. Chem. Soc.* **95**, 5761–5762.
24. Scott, A. I. (1974). *Science* **184**, 760–764.

24a. Scott, A. I., Lee, E., and Townsend, C. A. (1974). *Bioorg. Chem.* **3**, 229–237.

25. Bogorad, L. (1958). *J. Biol. Chem.* **233**, 510–515.
26. Bogorad, L. (1962). *Methods Enzymol.* **5**, 885–895.

27. Frydman, B., Reil, S. Valasinas, A., Frydman, R. B., and Rapoport, H. (1971), *J. Am. Chem. Soc.* **93**, 2738–2745.
28. Burnham, B. F. (1969). In *Metabolic Pathways,* 3rd ed., Vol. III, Greenberg, D. M., ed., Academic Press, New York, pp. 403–537.
29. Marks, G. S. (1962). *Ann. Rep. Prog. Chem.* **59**, 385–399.
30. Marks, G. S. (1966). *Bot. Rev.* **32**, 56–94.
31. Bogorad, L., and Troxler, R. F. (1967). In *Biogenesis of Natural Compounds,* Bernfeld, P., Ed., Pergamon Press, Oxford, pp. 247–313.
32. Pluscec, J., and Bogorad, L. (1970). *Biochemistry* **9**, 4736–4743.
33. Radmer, R., and Bogorad, L. (1972). *Biochemistry* **11**, 904–910.
34. Osgerby, J. M., Pluscec, J., Kim, Y. C., Boyer, F., Stojanac ,N., Mah, H. D., and MacDonald, S. F. (1972). *Can. J. Chem.* **50**, 2652–2660.
35. Stella, A. M., Parera, V. E., Llambías, E. B. C., and Batlle, A .M. del C. (1971). *Biochim. Biophys. Acta* **252**, 481–488.
36. Davies, R. C., and Neuberger, A. (1973). *Biochem. J.* **133**, 471–492.
37. Frydman, R. B., Valasinas, A., and Frydman, B. (1973). *Biochemistry* **12**, 80–85.
38. Frydman, R. B., Valasinas, A., Levy, S., and Frydman, B .(1974). *FEBS Lett.* **38**, 134–138.
39. Robinson, R. (1955). *The Structural Relations of Natural Products,* Clarendon Press, Oxford.
40. Frydman, R. B., Valasinas, A., Rapoport, H., and Frydman, B. (1972). *FEBS Lett.* **25**, 309–312.
41. Carpenter, A. T., and Scott, J. J. (1961). *Biochim. Biophys. Acta* **52,** 195–198.
41a. Frydman, R. B., Valasinas, A., and Frydman, B. (1973). *Enzyme* **16**, 151–159.
42. Battersby, A. R., Hunt, E., and McDonald, E. (1973). *J. Chem. Soc. Chem. Commun.,* 442–443.
43. Krasnovskii, A. A., and Umrikhina, A. V. (1972). *Dokl. Nauk SSSR Biochem. Sec. (Engl. trans.)* **202**, 4–7.
44. Petryka, Z. J., and Watson, C. J. (1972). *Perspect. Biol. Med.* **15**, 443–449.
45. Decker, K., Jungermann, K., and Thauer, R. K. (1970). *Angew. Chem. (Int. Ed.)* **9**, 138–158.
46. Bullock, E., Johnson, A. W., Markham, E., and Shaw, K. B. (1958). *J. Chem. Soc.,* 1430–1440.
47. Frydman, R. B., and Frydman, B. (1970). *Arch. Biochem. Biophys.* **136**, 193–202.
48. Jordan, P. M., and Shemin, D. (1973). *J. Biol. Chem.* **248**, 1019–1024.
49. Gurne, D., and Shemin, D. (1973). *Science* **180**, 1188–1190.
50. Lascelles, J. (1964). *Tetrapyrrole Biosynthesis and Its Regulation,* Benjamin, New York, pp. 49–51.
51. Burnham, B. F. (1969). In *Metabolic Pathways,* 3rd ed., Vol. III, Greenberg, D. M., Ed., Academic Press, New York, pp. 435-439.
52. Bernhauer, K., Becher, E., Gross, G., and Wilharm, G. (1960). *Biochem. Z.* **332**, 562–572.
53. Bernhauer, K., Wagner, F., Beisbarth, H., Rietz, P., and Vogelmann, H. (1966). *Biochem. Z.* **344**, 289–309.

54. Bernhauer, K., Vogelmann, H., and Wagner, F. (1968). *Z. Physiol. Chem.* **349**, 1271–1280.
55. Bernhauer, K., Vogelmann, H., and Wagner, F. (1968). *Z. Physiol. Chem.* **349**, 1281–1296.
56. Bernhauer, K., Wagner, F., Michna, H., Rapp, P., and Vogelmann, H. (1968). *Z. Physiol. Chem.* **349**, 1297–1309.
57. Bartosiński, B., Zagalak, B., and Pawełkiewicz, J. (1967). *Biochim. Biophys. Acta* **136**, 581–584.
58. Friedrich, W., and Sandeck, W. (1964). *Biochem. Z.* **340**, 465-470.
59. Friedrich, W. (1965). *Biochem. Z.* **342**, 143–160.
60. Friedrich, W., and Sandeck, W. (1965). *Z. Naturforsch.* **20b**, 79–80.
61. Friedrich, W., and Sandeck, W. (1964). *Z. Naturforsch.* **19b**, 538–539.
62. Bernhauer, K., Wagner, F., and Zahn, W. (1965). *Biochem. Z.* **342**, 207–216.
63. Friedrich, W. Quoted in Bernhauer et al. (1965), ref. 62.
64. Friedrich, W. (1966). *Nutr. Dieta (Swiss)* **8**, 178–225, esp. p. 188 et seq.
65. Rapp, P. (1973). *Z. Physiol. Chem.* **354**, 136–140.
66. Nockolds, C. K., Waters, T. N. M., Ramaseshan, S., Waters, J. M., and Hodgkin, D. C. (1967). *Nature* **214**, 129–130.
67. Moore, F. M., Willis, B. T. M., and Hodgkin, D. C. (1967). *Nature* **214**, 130–133.
68. Hodgkin, D. C. Personal communication.
69. Irion, E., and Ljungdahl, L. G. (1965). *Biochemistry* **4**, 2780–2790.
70. Bernhauer, K., and Wagner, F. (1960). *Z. Physiol. Chem.* **322**, 184–189.
71. Ford, J. E., Holdsworth, E. S., and Kon, S. K. (1955). *Biochem. J.* **59**, 86–93.
72. Dellweg, H., Becher, E., and Bernhauer, K. (1956). *Biochem. Z.* **328**, 81–87.
73. DiMarco, A., Boretti, G., Ghione, M., Migliacci, A., and Sanfilippo, A. (1957). *Ital. J. Biochem. (in English)* **6**, 259–269.
74. Volcani, B. E., Toohey, J. I., and Barker, H. A. (1961). *Arch. Biochem. Biophys.* **92**, 381–391.
75. Rapp, P., and Hildebrand, R. (1972). *Z. Physiol. Chem.* **353**, 1141–1152.
76. Bernhauer, K., Dellweg, H., Friedrich, W., Gross, G., Wagner, F., and Zeller, P. (1960). *Helv. Chim. Acta* **43**, 693–696.
77. Kelemen, A. M., and Simon, A. (1961). *Acta Microbiol. Acad. Sci. Hung.* **8**, 223–230.
78. Keleman, A. M., and Simon, A. (1961). *Acta Microbiol. Acad. Sci. Hung.* **8**, 231–235.
79. Kelemen, A. M., Csanyi, E., Simon, A., and Eckhardt, S. (1962). In *Vitamin B_{12} und Intrinsic Factor, 2nd European Symposium, Hamburg 1961,* Heinrich, H. C., Ed., Enke Stuttgart, pp. 241–244.
80. Krasna, A. J., Rosenblum, C., and Sprinson, D. B. (1957). *J. Biol. Chem.* **225**, 745–750.
81. Müller, G., and Müller, O. (1966). *Z. Naturforsch.* **21b**, 1159–1164.
82. Lowe, D. A., and Turner, J. M. (1970). *J. Gen. Microbiol.* **64**, 119–122.
83. Neuberger, A., and Tait, G. H. (1960). *Biochim. Biophys. Acta* **41**, 164–165.
84. Laver, W. G., Neuberger, A., and Scott, J. J. (1959). *J. Chem. Soc.,* 1483–1491.
85. Elliott, W. H. (1960). *Biochem. J.* **74**, 478–485.

86. Neuberger, A., and Tait, G. H. (1962). *Biochem. J.* **84**, 317–328.
87. Turner, J. M. (1966). *Biochem. J.* **99**, 427–433.
88. Turner, J. M. (1967). *Biochem. J.* **104**, 112–121.
89. Lowe, D. A., and Turner, J. M. (1968). *Biochim. Biophys. Acta* **170**, 455–456.
90. Lowe, D. A., and Turner, J. M. (1970). *J. Gen. Microbiol.* **63**, 49–61.
91. Dekker, E. E., and Swain, R. R. (1968). *Biochim. Biophys. Acta* **158**, 306–307.
92. Campbell, R. L., and Dekker, E. E. (1973). *Biochem. Biophys. Res. Commun.* **53**, 432–438.
93. Müller, G., Gross, R., and Siebke, G. (1971). *Z. Physiol. Chem.* **352**, 1720–1722.
94. Bernhauer, K., and Wagner, F. (1962). *Biochem. Z.* **335**, 325–339.
95. Bernhauer, K., Wagner, F., Wahl, D., and Glatzle, D. (1964). *Biochem. Z.* **340**, 171–185.
96. Ford, J. E., and Porter, J. W. G. (1952). *Biochem. J.* **51**, v.
97. Ford, J. E., Holdsworth, E. S., Kon, S. K. ,and Porter, J. W. G. (1953). *Nature* **171**, 150–151.
98. Pawełkiewicz, J. (1956). *Acta Biochim. Polon.* **3**, 581–590.
99. Renz, P. (1968). *Biochem. Biophys. Res. Commun.* **30**, 373–378.
100. Renz, P. (1971). *Methods Enzymol.* **18C**, 86–88.
101. Barchielli, R., Boretti, G., Julita, P., Migliacci, A., and Minghetti, A. (1957). *Biochim. Biophys. Acta.* **25**, 452.
102. DiMarco, A., Boretti, G., Migliacci, A., Julita, P., and Minghetti, A. (1957). *Boll. Soc. Ital. Biol. Sper.* **33**, 1513–1516.
103. Barchielli, R., Boretti, G., DiMarco, A., Julita, A., Migliacci, A., Minghetti, A., and Spalla, C. (1960). *Biochem. J.* **74**, 382–387.
104. Pawełkiewicz, J., Walerych, W., and Bartosiński, B. (1959). *Acta Biochim. Polon.* **6**, 431–440.
105. Irion, E., and Ljungdahl, L. G. (1968). *Biochemistry* **7**, 2350–2355.
106. Ronzio, R. A., and Barker, H. A. (1967). *Biochemistry* **6**, 2344–2354.
107. DiMarco, A., Alberti, C. G., Boretti, G., Ghione, M., Migliacci, A., and Spalla, C. (1957). In *Vitamin B_{12} und Intrinsic Factor, 1st European Symposium, Hamburg 1956,* Heinrich, H. C., Ed., Enke, Stuttgart, pp. 55–59.
108. DiMarco, A., Boretti, G., and Spalla, C. (1961). *Sci. Rep. 1st Super. Sanità* (in English) **1**, 355–367.
109. Kornberg, A. (1957). *Adv. Enzymol.* **18**, 191–240.
110. Bartosiński, B. (1966). *Bull. Acad. Polon. Sci., Ser. Sci. Biol.* **14**, 143–147.
111. Boretti, G., DiMarco, A., Fuoco, L., Marnatti, M. P., Migliacci, A., and Spalla, C. (1960). *Biochim. Biophys. Acta* **37**, 379–380.
112. Barbieri, P., Boretti, G., DiMarco, A., Migliacci, A., and Spalla, C. (1962). *Biochim. Biophys. Acta.* **57**, 599–600.
113. Friedmann, H. C., and Harris, D. L. (1962). *Biochem. Biophys. Res. Commun.* **8**, 164–168.
114. Friedmann, H. C., and Harris, D. L. (1965). *J. Biol. Chem.* **240**, 406–412.
115. Friedmann, H. C. (1965). *J. Biol. Chem.* **240**, 413–418.
116. Fyfe, J. A., and Friedmann, H. C. (1969). *J. Biol. Chem.* **244**, 1659–1666.

117. Fyfe, J. A., and Friedmann, H. C. (1971). *Methods Enzymol.* **18B**, 197–204.
118. Friedmann, H. C., and Fyfe, J. A. (1969). *J. Biol. Chem.* **244**, 1667–1671.
119. Friedmann, H. C. (1971). *Methods Enzymol.* **18C**, 96–98.
120. Hodgkin, D. C. (1955). *Biochem. Soc. Symp.* **13**, 28.
121. Friedrich, W., and Bernhauer, K. (1956). *Chem. Ber.* **89**, 2507–2512.
122. Dinglinger, F., and Renz, P. (1971). *Z. Physiol. Chem.* **352**, 1157–1161.
123. Dellweg, H., Becher, E., and Bernhauer, K. (1956). *Biochem. Z.* **327**, 422–449.
124. Kon, S. K., Pawełkiewicz, J. (1960). In *Vitamin Metabolism. Symp. 11, Proc. 4th Int. Congr. Biochem., Vienna 1958*, pp. 115–149.
125. Perlman, D., Barrett, J. M., and Jackson, P. W. (1962). *In Vitamin B_{12} und Intrinsic Factor, 2nd European Symposium, Hamburg 1961*, Heinrich, H. C., Ed., Enke, Stuttgart, pp. 58–69.
126. Mervyn, L., and Smith, E. L. (1964). *Prog. Ind. Microbiol.* **5**, 152–201.
127. Kamikubo, T., Narahara, H., Murai, K., and Kumamoto, N. (1966). *Biochem Z.* **346**, 159–166.
128. Hayashi, M., and Kamikubo, T. (1971). *FEBS Lett.* **15**, 213–216.
129. Friedmann, H. C. (1968). *J. Biol. Chem.* **243**, 2065–2075.
130. Wagner, F. (1962). *Biochem. Z.* **336**, 99–101.
131. Friedmann, H. C. (1971). *Methods Enzymol.* **18C**, 54–57.
132. Coulter, C. L., Hawkinson, S. W., and Friedmann, H. C. (1969). *Biochim. Biophys. Acta* **177**, 293–302.
133. Hawkinson, S. W., Coulter, C. L., and Greaves, M. L. (1970). *Proc. Roy. Soc. London Ser. A.* **318**, 143–167.
134. Renz, P. (1967). *Angew. Chem. (Int. Ed.)* **6**, 368–369.
135. Renz, P. (1968). *Z. Physiol. Chem.* **349**, 979–981.
136. Renz, P. (1971). *Methods Enzymol.* **18C**, 82–92.
137. Schneider, Z., and Friedmann, H. C. (1972). *Arch. Biochem. Biophys.* **152**, 488–495.
138. Pratt, J. M. (1972). *Inorganic Chemistry of Vitamin B_{12}*, Academic Press, New York.
139. Friedmann, H. C. (1971). *Methods Enzymol.* **18C**, 92–95.
140. Müller, O., and Müller, G. (1963). *Biochem. Z.* **337**, 179–194.
141. Ljungdahl, L. G., Irion, E., and Wood, H. G. (1965). *Biochemistry* **4**, 2771–2780.
142. Ford, J. E., and Porter, J. W. G. (1953). *Brit. J. Nutr.* **7**, 326–337.
143. Bernhauer, K., Becher, E., and Wilharm, G., (1959). *Arch. Biochem. Biophys.* **83**, 248–258.
144. Renz, P. (1965). *Angew. Chem. (Int. Ed.)* **4**, 527.
145. Ohlenroth, K., and Friedmann, H. C. (1968). *Biochim. Biophys. Acta* **170**, 465–467.
146. Dellweg, H. W., and Bernhauer, K. (1957). *Arch. Biochem. Biophys.* **69**, 74–80.
147. Dinglinger, F., and Braun, I. (1970). *Z. Physiol. Chem.* **351**, 1157–1160.
148. Koppenhagen, V. B., and Pfiffner, J. J. (1970). *J. Biol. Chem.* **245**, 5865–5867.
149. Koppenhagen, V. B. (1973). Personal communication.

150. Burnham, B. F. (1969). In *Metabolic Pathways,* 3rd ed., Vol. III, Greenberg, D. M., Ed., Academic Press, New York, pp. 439–456.
151. Porra, R. J., and Ross, B. D. (1965). *Biochem. J.* **94**, 557–562.
152. Toohey, J. I. (1965). *Proc. Natl. Acad. Sci. U. S. A.* **54**, 934–942.
153. Toohey, J. I. (1966). *Fed. Proc.* **25**, 1628–1632.
154. Toohey, J. I. (1971). *Methods Enzymol.* **18C**, 71–75.
155. Sato, K., Shimizu, S., and Fukui, S. (1970). *Biochem. Biophys. Res. Commun.* **39**, 170–174.
156. Maitra, P. K., and Roy, S. C. (1960). *Biochem. J.* **75**, 483–487.
157. Shaposhnikov, V. N., Konova, I. V., and Rybakova, R. K. (1966). *Dokl. Akad. Nauk SSSR* **171**, 1439–42.
158. Burnham, B. F. (1969). In *Metabolic Pathways,* 3rd ed., Vol. III, Greenberg, D. M., Ed., Academic Press, New York, p. 469.
159. Koppenhagen, V. B., and Pfiffner, J. J. (1971). *Fed. Proc.* **30**, 1088Abs.
160. Wooley, D. W. (1950). *Proc. Soc. Exp. Biol. Med.* **75**, 745–746.
161. Wooley, D. W. (1951). *J. Exp. Med.* **93**, 13–24.
162. Ford, J. E., and Holdsworth, E. S. (1954). *Biochem. J.* **56**, xxxv.
163. Renz, P., and Reinhold, K. (1967). *Angew. Chem. (Int. Ed.)* **6**, 1083.
164. Plaut, G. W. E. (1954). *J. Biol. Chem.* **211**, 111–116.
165. Alworth, W. L., and Baker, H. N. (1968). *Biochem. Biophys. Res. Commun.* **30**, 496–501.
166. Alworth, W. L., Baker, H. N., Lee, D. A., and Martin B. A. (1969). *J. Am. Chem. Soc.* **91**, 5662–5663.
167. Alworth, W. L., Baker, N., Winkler, M. F., and Keenan, A. M. (1970). *Biochem. Biophys. Res. Commun.* **40**, 1026–1031.
168. Renz, P. (1970). *FEBS Lett.* **6**, 187–189.
169. Lu, S.-H., Winkler, M. F., and Alworth, W. L. (1971). *J. Chem. Soc., Chem. Commun.,* 191–192.
170. Alworth, W. L., Lu, S.-H., and Winkler, M. F. (1971). *Biochemistry* **10**, 1421–1424.
171. Kühnle, H. F., and Renz, P. (1971). *Z. Naturforsch.* **26b**, 1017–1020.
172. Masuda, T. (1956). *Pharm. Bull. (Tokyo)* **4**, 71–72; 72–74; 375–381.
173. Masuda, T. (1957). *Pharm. Bull. (Tokyo)* **5**, 28–30; 136–141.
174. Goodwin, T. W. (1970). In *Metabolic Pathways,* 3rd ed., Vol. IV, Greenberg, D. M., Ed., Academic Press, New York, pp. 353–368.
175. Plaut, G. W. E. (1970). In *Comprehensive Biochemistry,* Florkin, M., and Stotz, E. H., Eds., Elsevier, New York, pp. 11–45.
176. Demain, A. L. (1972). *Ann. Rev. Microbiol.* **26**, 369–388.
177. Harvey, R. A., and Plaut, G. W. E. (1966). *J. Biol. Chem.* **241**, 2120–2136.
178. Plaut, G. W. E., Beach, R. L., and Aogaichi, T. (1970). *Biochemistry* **9**, 771–785.
179. Plaut, G. W. E. (1963). *J. Biol. Chem.* **238**, 2225–2243.
180. Lu, S.-H., and Alworth, W. L. (1972). *Biochemistry* **11**, 608–611.
181. Renz, P., and Weyhenmeyer, R. (1972). *FEBS Lett.* **22**, 124–126.

182. Janicki, J., Chelkowski, J., and Nowakowska, K. (1966). *Acta Microbiol.. Polon.* **15**, 249–253.

183. Schlee, D. (1973). *Die Pharmazie* **28**, 284–287.

184. Speedie, J. D., and Hull, G. W. (1960). U.S. Patent No. 2,951,017.

185. Riley, P. B., Jackson, P. W., Ross, D., and Savage, P. A. (1961). *Soc. Chem. Ind. (London) Monogr.* **12**, 127–139.

186. Perlman, D., and Semar, J. B. (1963). *Biotechnol. Bioeng.* **5**, 21–25.

187. Friedrich, W., and Bernhauer, K. (1953). *Angew. Chem.* **65**, 627–628.

188. Friedrich, W., and Bernhauer, K. (1954). *Angew. Chem.* **66**, 776–780.

189. Lezius, A. G., and Barker, H. A. (1965). *Biochemistry* **4**, 510–518.

190. Friedrich, W., and Bernhauer, K. (1956). *Chem. Ber.* **89**, 2030–2044.

191. Ljungdahl, L. G., LeGall, J., and Lee, J.-P. (1973). *Biochemistry* **12**, 1802–1808.

192. Bernhauer, K., Müller, O., and Wagner, F. (1962). In *Vitamin B_{12} und Intrinsic Factor, 2nd European Symposium, Hamburg 1961,* Heinrich, H. C., Ed., Enke, Stuttgart, pp. 53–56.

193. Dulaney, E. L., and Williams, P. L. (1953). *Mycologia* **45**, 345–358.

194. Fantes, K. H., and O'Callaghan, C. H. (1955). *Biochem. J.* **59**, 79–82.

195. Perlman, D., and Barrett, J. M. (1958). *Can. J. Microbiol.* **4**, 9–15.

196. Smith, E. L. (1965). *Vitamin B_{12},* 3rd ed., Methuen, London, p. 26.

197. Wijmenga, H. G. (1951). Doctorate Thesis, University of Utrecht, quoted in Smith (196, p. 26).

198. Gebgardt, A. G., Kucheras, R. V., Laska, D. V., and Vogrin, A. G. (1970). *Mikrobiologiya* **39**, 447–452.

199. Kreuzig, F. (1971). *Int. J. Vitam. Nutr. Res.* **41**, 424–428.

200. Yamada, R., Shimizu, S., and Fukui, S. (1968). *Biochemistry* **7**, 1713–1719.

201. Walker, G. A., Murphy, S., and Huennekens, F. M. (1969). *Arch. Biochem. Biophys.* **134**, 95–102.

202. Walker, G. A., Murphy, S., Schmidt, R. R., and Huennekens, F. M. (1967). *Fed. Proc.* **26**, 343.

203. Brady, R. O., Castanera, E. G., and Barker, H. A. (1962). *J. Biol. Chem.* **237**, 2325–2332.

204. Peterkofsky, A., and Weissbach, H. (1963). *Ann. N. Y. Acad. Sci.* **112**, 622–637.

205. Peterkofsky, A., and Weissbach, H. (1964). *J. Biol. Chem.* **238**, 1491–1497.

206. Barker, H. A. (1967). *Biochem. J.* **105**, 1–15.

207. Mudd, S. H. (1965). In *Transmethylation and Methionine Biosynthesis,* Shapiro, S. K., and Schlenk, F., Eds. University of Chicago Press, Chicago, pp. 33–47.

208. Vitols, E., Walker, G. A., and Huennekens, F. M. (1966). *J. Biol. Chem.* **241**, 1455–1461.

209. Barker, H. A. (1972). *Ann. Rev. Biochem.* **41**, 75–85.

210. Taylor, R. T., and Weissbach, H. (1973). In *The Enzymes,* 3rd ed., Vol. 9, Boyer, P. D., Ed., Academic Press, New York, Part B, pp. 121–165.

CHAPTER THREE

COBAMIDES AS COFACTORS

METHYLCOBAMIDES AND THE SYNTHESIS OF METHIONINE, METHANE AND ACETATE

J. MICHAEL POSTON AND THRESSA C. STADTMAN

National Institutes of Health
Bethesda, Maryland

TABLE OF CONTENTS

INTRODUCTION

The biological origin of the methyl group of methionine and the mechanism of its synthesis from various one-carbon compound precursors and transfer to homocysteine have occupied the attention of numerous investigators and research groups in several countries for many years. Studies in the 1950s and early 1960s established that in certain microorganisms and in mammalian liver the biosynthesis of methionine from homocysteine and a methyl donor (e.g., N^5-methyltetrahydrofolate) depended on cobamides. During this same period of time a cobamide coenzyme discovered by Barker and co-workers (1, 2) was shown to contain a light- and acid-labile deoxyadenosyl moiety linked to the vitamin. The X-ray analyses of Lenhert and Hodgkin (3) demonstrated that this deoxyadenosyl moiety was covalently linked through its 5′-carbon atom to the cobalt atom of the vitamin. The natural occurrence of an alkylcobalt compound of this type prompted D. D. Woods to suggest that by analogy a methyl group covalently bonded to the cobalt atom of a cobamide would be an attractive intermediate to explain the need for the cobamide in methionine synthesis. Indeed when synthetic methylcobalamin was prepared and tested in the cobalamin-dependent methyltransferase system of *Escherichia coli* (4) it was found to be an efficient methyl donor for synthesis of methionine from homocysteine. Although homocysteine also was methylated by methylcobalamin in a nonenzymic reaction, addition of the enzyme increased the rate of synthesis of methionine many-fold. This focused attention on the possibility that the true intermediate in methionine synthesis might be an enzyme-bound *Co*-methylcobalamin and suggested that this bound methyl group could be donated by free methylcobalamin as well as by the normal methyl donor, methyltetrahydrofolate. The correctness of the assumption has been borne out by the subsequent investigations in many laboratories on the mechanism of methionine biosynthesis (see later). Moreover, it was the success of these imaginative early experiments of D. D. Woods and associates that prompted methylcobalamin to be considered as a likely chemical model of an intermediate that could be reductively cleaved to yield methane by the methane bacteria or be converted to the methyl group of acetate by certain anaerobic acetate-synthesizing bacteria. Methylcobalamin proved to be an efficient precursor of methane and also of the methyl moiety of acetate. In both of these processes there now is substantial experimental evidence to support the view that an enzyme-bound methylcobamide is the true intermediate. However, final proof of the generality of a methylcobalt type of intermediate in the biosynthesis of methane and in the synthesis of acetate from carbon dioxide awaits more detailed investigation. Both of these biosynthetic processes are catalyzed by complex oxygen-labile enzyme systems that, to date, have been studied by only a few investigators.

METHIONINE SYNTHESIS

The final step in the biosynthesis of methionine in bacteria, fungi, higher plants, and animals involves the transfer of a methyl group from N^5-methyltetrahydrofolate to the sulfhydryl group of homocysteine to form *S*-methylhomocysteine or methionine (Scheme 3-1). There are two types

Scheme 3-1. Metabolic pathways for the generation of the methyl of N^5-methyltetrahydrofolate and homocysteine and their reaction to form methionine.

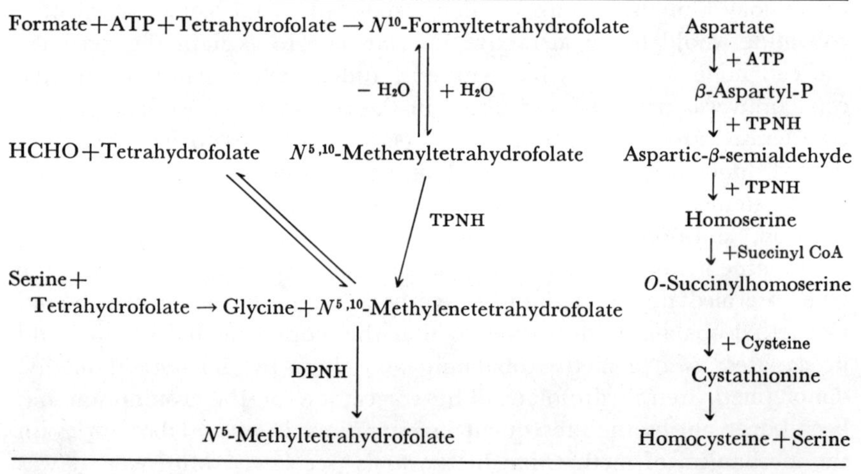

N^5-Methyltetrahydrofolate+$HSCH_2CH_2CH(NH_2)COOH$ → CH_3-S-$CH_2CH_2CH(NH_2)$-COOH+Tetrahydrofolate

of methyltransferases that catalyze the methylation of homocysteine. One, a cobalamin-dependent N^5-methyltetrahydrofolate homocysteine methyltransferase that can utilize either the monoglutamate or the polyglutamate forms of methyltetrahydrofolate as methyl donor, is found in many bacteria and in animals but not in organisms such as yeast and higher plants that lack cobamides (Table 3-1). The other type does not possess a cobamide chromophore, requires polyglutamate forms of methyltetrahydrofolates as methyl donor[1] and occurs in yeasts, higher plants, and also in *E. coli* and several other bacteria (Table 3-1).

Table 3-1 Distribution of N^5-Methyltetrahydrofolate Homocysteine Methyltransferases

Organism	Cobalamin Dependent	Cobamide Independent	Ref.
Escherichia coli	+	+	5
Aerobacter aerogenes	+	+	6
Salmonella typhimurium	+	+	7
Corynebacterium simplex	+	±	8
Rhodopseudomonas spheroides	+		9
Pseudomonas denitrificans	+		10
Streptomyces olivaceus	+		11
Ochromonas malhamensis	+		12
Mammalian liver	+		13, 14
Mammalian cells in tissue culture	+		15
Avian liver	+		13
Higher plants		+	16
Chlorella pyrenoidosa		+	17
Coprinus lagopus		+	18, 19
Neurospora crassa		+	17
Saccharomyces cerevisiae		+	17, 20
Bacillus subtilis		+	21

In view of the many excellent reviews in the literature on methionine biosynthesis and, in particular, on the cobalamin-dependent methyltransferase reaction, only a general summary of information concerning these methylation processes is given in this chapter. The emphasis is on the reactions catalyzed by bacteria; discussions of methionine synthesis in animals are to be found elsewhere in this volume. For other details the reader is referred to recent reviews by Taylor and Weissbach (23, 24) and by Rüdiger and Jaenicke (25).

Studies initiated in 1951 in the laboratory of D. D. Woods at Oxford on the biosynthesis of methionine in wild type and mutant stains of

[1] Germinating pea seeds (*Pisum sativum*) contain a methyltransferase that uses the monoglutamate form of N^5-methyltetrahydrofolate as methyl donor for methionine synthesis (22). It is markedly sensitive to the metal chelator, ethylenediaminetetraacetic acid (EDTA).

E. coli provided the groundwork for subsequent studies of other investigators and established, at least in principle, many of the details of the process that elsewhere have been merely confirmed. It is therefore unfortunate that, because of Dr. Woods' ill health in the middle 1950s and his untimely death in November of 1964, there was a great delay in the publication of accounts of many of the discoveries of his research group regarding the pathways of methionine biosynthesis. As a result, the truly original contributions of Woods and his collaborators are not generally appreciated, because by the time many of the papers describing details of the earlier phases of the work finally appeared in 1960 (5, 26–33), research groups elsewhere also were engaged in similar studies.

The particular strength of Woods and his co-workers in their studies on methionine biosynthesis was due in large part to their extensive use of mutant strains of *E. coli,* an approach which proved to be exceedingly useful in unraveling this complex biological process. These included mutants unable to form the one-carbon compound donor, serine; strains defective in the ability to form the folate intermediates involved in the reduction and transfer of the one-carbon moiety; and an auxotrophic strain (121/176) that required either cobalamin or methionine for growth. Using these microorganisms, it was established that there are two independent pathways by which methionine is formed (34–36). In one, the triglutamate (but not the monoglutamate) derivative of methyltetrahydrofolate serves as substrate for the methylation of homocysteine catalyzed by an enzyme that requires magnesium as the only added cofactor (35, 36). The other, a more complex pathway, is catalyzed by a cobalamin-containing enzyme that requires *S*-adenosylmethionine and a strong reducing system for activity but is less exacting in its requirement for the methyl donor substrate, utilizing either the monoglutamate or triglutamate form of methyltetrahydrofolate (35–37). It was established that wild-type *E. coli* and certain of the mutant strains produce both the cobalamin-dependent and the cobamide-independent methyltransferases, whereas strain 121/176 (the methionine/cobalamin auxotroph) lacks the cobamide-independent pathway entirely and produces only the cobalamin-containing enzyme when cobalamin is provided in the growth medium. Measurement of levels of these enzymes in the various strains of *E. coli* as a function of several different growth condtions also provided considerable information concerning the mechanism of control of their activities in the cell (38–40).

The important thing is that the deductions made on the basis of the earlier studies have stood the test of time and, with few exceptions, have been confirmed by later investigators.

Cobalamin-Dependent N^5-Methyltetrahydrofolate Homocysteine Methyltransferase

The most highly purified cobalamin-dependent methyltransferase that has been obtained was prepared from pig kidney by Mangum and North (41) but the yield was very low and the amount of pure enzyme was insufficient for detailed characterization. A large number of the studies on mechanism of the transmethylase reaction and its cofactor requirements have been made on partially purified preparations of the enzyme isolated from various mutant strains of *E. coli* (36), from *E. coli* B (23, 24, 42), or from unspecified strains of *E. coli* (25, 43). As a result there are points of disagreement among various research groups concerning some aspects of the transmethylase reaction, but this may be due to different impurities in the enzyme preparations or difficulties in methodology rather than to inherent differences in properties of the catalysts themselves. In particular, failure to maintain the strictly anaerobic conditions required for the initial activation and continued turnover of the cobalamin-dependent methyltransferase has led to many erroneous conclusions concerning cofactor requirements of the reaction.

Physical Properties

The molecular weight of the holoenzyme form of the cobalamin-dependent methyltransferase from *E. coli* is 140,000 to 150,000 (42–44). However, Rüdiger and Jaenicke (45) report that if an isolation procedure is employed that avoids all types of purification steps that cause marked changes in ionic strength, a more active preparation of greater stability is obtained. The molecular weight of this preparation is 255,000. Also, there is a report by Galivan and Huennekens (46) that in addition to the large cobalamin-containing subunit a small sulfhydryl protein subunit of about 3000 molecular weight is essential for maximal catalytic activity of the enzyme. There is no information concerning the precise role of this accessory sulhydryl protein, but analogy with the cobamide coenzyme-dependent lysine mutases (47) suggests that it may maintain the catalytic subunit in a more active conformation and prevent inactivation of the cobamide cofactor during continued turnover of the enzyme.

The colorless apoprotein form of the methyltransferase can be obtained from extracts of *E. coli* grown in the absence of cobalamin (4, 48, 49) or prepared by resolution of the holoenzyme with 6 *M* urea and 1,4-dithiothreitol (50). Conversion to the catalytically active holoenzyme form occurs upon incubation either with methylcobalamin itself (4, 50) or with a mixture of components that lead to the production of methylcobalamin on the enzyme. For this purpose, hydroxocobalamin together with a reducing system (e.g., FAD or FMN, hydrogen, and platinic oxide) and *S*-adenosylmethionine has frequently been employed (4, 37, 48, 49).

Mechanism of the Cobalamin-Dependent Methyl Transfer Reaction

It is now well established that enzyme-bound methylcobalamin is formed as an intermediate in the transfer of the methyl moiety of N^5-methyltetrahydrofolate to homocysteine by the cobalamin-dependent methyltransferase. Free methylcobalamin itself can be used as an alternative substrate for the methylation of homocysteine in many instances, although not all cobalamin-dependent methyltransferases appear to exhibit activity with this donor (25). In fact, as pointed out earlier, it was the demonstration by D. D. Woods and his colleagues (4) that the methyl group of methylcobalamin could be transferred to homocysteine by a methyltransferase preparation from *E. coli* that provided the initial clue to the role of cobalamin in methionine biosynthesis.

Evidence that the enzyme-bound cobamide participated directly in the methyl transfer reaction was provided by the observation that incubation of the enzyme-cobamide complex with propyl iodide led to inactivation of the enzyme (51). This inactivation, like the normal catalytic reaction, occurred only in the presence of a reducing system and, moreover, was inhibited by *S*-adenosylmethionine, an activator of the methyltransferase activity. Restoration of activity of the propyl iodide-treated enzyme was achieved by irradiation with visible light. From this it was concluded that an inactive *Co*-propylcobalamin derivative of the enzyme had been formed and only upon its destruction by photolytic cleavage of the cobalt-propyl bond could the enzyme again catalyze the normal methyl transfer process. This implied that the enzyme-bound cobalamin must participate directly in the mechanism of the methyl transfer.

The nature of this participation was clarified by experiments with isotopically labeled substrates showing that the methyl group of N^5-methyltetrahydrofolate could be transferred to the enzyme-bound cofactor to form methylcobalamin (43, 52–56). Incubation of the methylated enzyme with homocysteine resulted in the loss of the methyl group from methylcobalamin concomitant with the production of methionine. While methylation of cobalamin by N^5-methyltetrahydrofolate required *S*-adenosylmethionine and a reducing system, the transfer of the methyl group to homocysteine occurred in the absence of either of these components (54, 55). These experiments showed that, in essence, the methyltransferase reaction can be viewed as the sum of the two half-reactions, 1 and 2, consisting of the successive methylation and demethylation of cobalamin bound to the enzyme. Under the usual circumstances the methyl donor is N^5-methyltetrahydrofolate, and the final acceptor is homocysteine.

$$N^5\text{-Methyltetrahydrofolate} + \text{enzyme} \cdot \text{cobalamin} \rightarrow \text{tetrahydrofolate} + \text{enzyme} \cdot \text{methylcobalamin} \quad (1)$$

$$\text{enzyme} \cdot \text{methylcobalamin} + \text{homocysteine} \rightarrow \text{enzyme} \cdot \text{cobalamin} + \text{methionine} \quad (2)$$

Reversal of reaction 1, that is, transfer of a methyl group from the methylated enzyme to tetrahydrofolate to form methyltetrahydrofolate, has been demonstrated with several enzyme preparations (37, 43, 56), but the equilibrium for the overall reaction lies far to the right, so that only a trace amount of methyltetrahydrofolate was formed from methionine and tetrahydrofolate (23).

Activation of the Cobalamin-Dependent Methyltransferase by S-Adenosylmethionine

The part played by *S*-adenosylmethionine in the methyltransferase reaction has until recently been the subject of controversy, even though it was known for many years to be required for methionine formation with N^5-methyltetrahydrofolate as methyl donor. However, it is now generally accepted that the effect of *S*-adenosylmethionine on the methyltransferase reaction is mediated through methylation of the enzyme-bound cobalamin. In contrast to methylation by N^5-methyltetrahydrofolate, methylation by *S*-adenosylmethionine is not involved in catalysis *per se* in the sense that *S*-adenosylmethionine is able to serve as an efficient donor of methyl groups for the overall biosynthesis of methionine. Rather it appears that methylation by *S*-adenosylmethionine is necessary to activate the enzyme, that is, to render it capable of catalyzing the transfer of methyl groups from N^5-methyltetrahydrofolate to homocysteine.

Direct experimental evidence that *S*-adenosylmethionine activates the methyltransferase by alkylating the enzyme-bound cobalamin was furnished by a number of research groups (43, 51–53, 56, 57). Thus, incubation of the transferase in the presence of a reducing system and [^{14}C-methyl]*S*-adenosylmethionine resulted in the formation of ^{14}C-labeled enzyme-bound methylcobalamin (42, 56). Both the normal activation process and the transfer of the labeled methyl group to the enzyme were inhibited by propylation of the enzyme-bound cofactor (58).

A number of other alkylating agents, including *S*-adenosylethionine, *S*-methylmethionine, and the highly unphysiological compounds methyl and ethyl iodide (57, 59), were shown to be able to substitute for *S*-adenosylmethionine as activators of the methyltransferase. Moreover, reaction of the enzyme with [^{14}C]methyl iodide in the presence of a reducing system resulted in the formation of enzyme-bound [^{14}C]methylcobalamin (55). Finally, it was shown (50) that apomethyltransferase which had been resolved of its bound cofactors by treatment with urea could be reconverted to the active holoenzyme form simply by incubation with methylcobalamin. The fact that this reconstituted enzyme was able to form methionine from N^5-methyltetrahydrofolate and homocysteine in the absence of *S*-adenosylmethionine and a reducing system showed that these latter components were required solely to alkylate the inactive form of the holoenzyme, that is, the form which contains bound cobalamin but no attached methyl group.

In a series of elegant experiments, Taylor and Weissbach (56) showed that the initial alkylation of the enzyme-bound cobalamin by *S*-adenosylmethionine was followed by transfer of the methyl group to homocysteine, after which the activated cobalamin enzyme participated in the normal cyclic methylation and demethylation reaction with N^5-methyltetrahydrofolate serving as the methyl donor. In contrast to the activating effect of the methyl group initially donated to the enzyme-bound cobalamin by *S*-adenosylmethionine, introduction of a propyl group by treatment with propyl iodide resulted in the formation of an inhibited enzyme (51) which was unable to accept a methyl group from N^5-methyltetrahydrofolate. However, irradiation of the inhibited, propylated enzyme with visible light resulted in cleavage of the cobalt-propyl bond and allowed the enzyme to be reactivated by *S*-adenosylmethionine in the normal fashion. Although these experiments indicate that the methyltransferase is unable to catalyze the transfer of an enzyme-bound propyl group to homocysteine, Rüdiger and Jaenicke (25) report that their methyltransferase preparations utilize both the ethyl and propyl analogs of N^5-methyltetrahydrofolate as substrates. However, the rates of the dealkylation steps with the ethyl and propyl derivatives were stated to be slower than the transfer of the methyl group.

Since the cobalamin bound to the methyltransferase can be methylated both by N^5-methyltetrahydrofolate and by *S*-adenosylmethionine, the question arises as to why *S*-adenosylmethionine is required at all. The answer appears to be related to the susceptibility to oxidation of the species of cobalamin generated by transfer of the methyl group from enzyme-bound methylcobalamin to homocysteine. This susceptibility is shown in experiments in which a preactivated methylcobalamin-containing methyltransferase preparation was permitted to catalyze the transfer of a methyl group from N^5-methyltetrahydrofolate to homocysteine in the absence of a reducing system and *S*-adenosylmethionine. When the reaction was carried out in vessels open to the air, the methyltransferase preparation was inactivated after catalyzing the formation of only 8 to 10 molecules of methionine per active site of enzyme. The same enzyme preparation, when incubated anaerobically under otherwise identical conditions with N^5-methyltetrahydrofolate and homocysteine, was able to catalyze approximately 100 turnovers before inactivation occurred (58).

In a formal sense, the methyl group is transferred as a cation, while cob(I)alamin, a compound that is readily oxidized by agents as weak as H^+ (see Chapter 1), remains bound to the enzyme. In fact, optical spectroscopy (60) has indicated that transfer of the methyl group from enzyme-bound methylcobalamin to homocysteine results in the formation of enzyme-bound cob(I)alamin. Ordinarily (58, 61, 62), the cob(I)alamin is realkylated by N^5-methyltetrahydrofolate and the catalytic cycle continues (Scheme 3-2). Occasionally, however, the cob(I) alamin is intercepted by an oxidizing agent before it can be realkylated, thereby being converted to a compound no longer susceptible to methylation by N^5-methyltetrahydrofolate. The precise chemical nature of the cobalamin bound to the inactivated methyl transferase is uncertain, but its spectrum resembles that of cob(II)alamin and base off alkylcobamides (42).

Scheme 3-2. Activation of the cobalamin-dependent methyltransferase by *S*-adenosylmethionine.

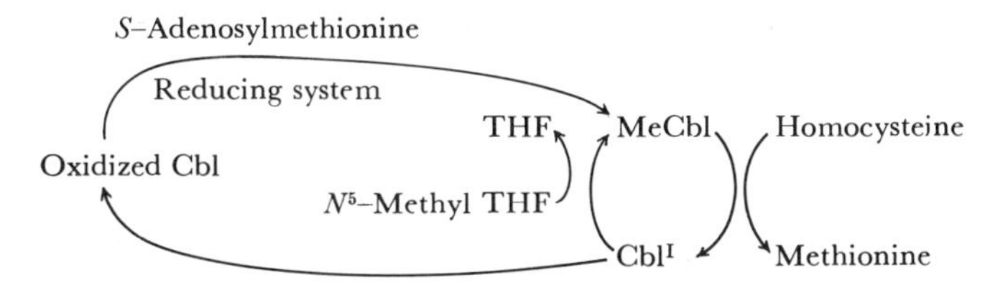

For reactivation, this enzyme-bound cobalamin must be reduced and methylated with a methylating agent more powerful than N^5-methyltetrahydrofolate—for example, *S*-adenosylmethionine or methyl iodide. Whether this is related to the relative inability of the weaker alkylating agent, N^5-methyltetrahydrofolate, to react with the very low concentration of cob(I)alamin generated by the disproportion of cob(II)alamin (25) or is the result of a conformational change that is difficult to reverse is not known. Evidence that there are two separate methylcobalamin binding sites on the methyltransferase (63) suggests that the latter possibility may also be an important factor.

Regulation of Methyltransferase Activities

Repression of several enzymes involved in methionine biosynthesis (Scheme 3-1) is observed in many *E. coli* strains when they are grown in the presence of methionine (39, 40, 64, 65). Formation of the cobamide-independent N^5-methyltetrahydrofolate homocysteine methyltransferase and the $N^{5,10}$-methylenetetrahydrofolate reductase is repressed either by cobalamin (1 nM) or by methionine (10 mM). A series of mutants with lesions at different loci prepared by Kung et al. (66) have been particularly valuable in studying these effects. Detailed analysis of the mechanism of repression of the cobamide-independent methyltransferase by cobalamin revealed that a functional holoenzyme form of the cobalamin-dependent transmethylase is essential if repression by cobalamin is to be observed (66, 67).

ROLE OF COBAMIDES IN METHANE BIOSYNTHESIS

The physiological group of strictly anaerobic bacteria that reduce carbon dioxide to methane has been known since the beginning of this century, but information concerning the nature of the individual reactions and the intermediates involved in the overall four-step reduction process is still fragmentary. It was clear from the early experiments (68) that simple one-carbon compounds (e.g., formate, formaldehyde, and methanol) were not free intermediates in the process. Furthermore, pulse-type experiments with highly radioactive carbon dioxide and other one-carbon compounds failed to reveal any derivatives that might participate in a cyclic process (69, 70). In particular, there was no information concerning the terminal reduction step, and for a long time there was not even a plausible chemical model of an intermediate that might be reductively cleaved under mild conditions by a bacterial enzyme to give the hydrocarbon, methane. However, with the discovery of the alkylcobalamins and the demonstration of the importance of methylcobalamin as methyl donor for methionine biosynthesis (4), an organometallic compound of this sort containing a methyl group covalently bonded to a cobalt atom seemed attractive from the chemical point of view as an immediate precursor of methane.

There were already several suggestive bits of evidence to indicate that cobamides might somehow be involved in methane biosynthesis. The bacterial cell mass that remains in the final methane-producing sewage digestion tanks and that is known as sewage sludge or "Faulschlamm" was the source of the 5-hydroxybenzimidazolylcobamide isolated by

Bernhauer and Friedrich (71). Neujahr (72, 73) showed that sewage sludge is a rich source of corrinoid compounds and that enrichment cultures of methane bacteria obtained from sewage, particularly those that fermented methanol or acetate, produced exceptionally high levels of corrinoids. Cobalamin was produced only when the methane-producing cultures were supplied with 5,6-dimethylbenzimidazole. In the absence of this supplement, other cobamides, one of which resembled 5-hydroxybenzimidazolylcobamide in its chromatographic behavior, were synthesized.

Microbiological assays[2] on dried cells of *Methanobacterium omelianskii*[3] and *Methanococcus vannielii* which had been grown in pure culture showed both of these organisms to be rich sources of cobamides. In particular, *M. vannielii*, whose entire energy metabolism is geared to C_1-compound metabolism, was found to contain twice as much cobamide on a dry weight basis as cells of the *M. omelianskii* culture. Appreciable amounts of the cobamide compounds present in both of these microorganisms were later found to be *Coβ*-adenosylcobamide derivatives (74). Careful analysis of the cobamides formed by the *M. omelianskii* culture (75) showed that the *Coβ*-aquo and adenosyl derivatives of 5-hydroxybenzimidazolylcobamide were the principal compounds present. *Methanosarcina barkeri,* which produces such high levels of cobamides that the cells are colored pink to red, also synthesizes almost exclusively the forms containing 5-hydroxybenzimidazole (76).

In *Clostridium sticklandii,* an organism which has the ability to synthesize acetate from carbon dioxide, a direct correlation was noted between the availability of one-carbon compound precursors provided in the growth medium and the level of corrinoid compounds that were isolated from the various batches of cells (74). By analogy with many specialized microorganisms in which exaggeration of a particular metabolic pathway for energy-yielding purposes is accompanied by the presence in the cells of exceptionally high levels of the coenzymes that participate in the overall process, this observation suggested a role of cobamide compounds in the anaerobic pathway involving reduction of C_1 compounds.

This type of reasoning, together with the aforementioned discovery of the conversion of methylcobalamin to methionine (4), prompted us to test methylcobalamin directly as a substrate for methane biosynthesis

[2] Assays kindly performed at the Lederle Laboratories in 1950 by Dr. W. L. Williams.
[3] Later shown to be a culture composed of two microorganisms, one that oxidizes ethanol to acetate and hydrogen, and another, Methanobacterium strain M. O. H., that utilizes the hydrogen to reduce carbon dioxide to methane.

by enzyme preparations of methane bacteria. The first experiments, using broken-cell preparations of *M. barkeri* and methylcobalamin labeled with ^{14}C in the methyl group, showed (77, 78) that the methyl moiety was indeed reduced to methane in good yield when pyruvate was provided as a source of reducing equivalents (Table 3-2). In subsequent experiments it was demonstrated that the particulate fraction of the extract of *M. barkeri* was not required. The soluble enzyme preparations, if protected from exposure to oxygen, efficiently reduced the methyl moiety of methylcobalamin to methane with either pyruvate or molecular hydrogen as reducing agent. At the same symposium where these results with the biological system were presented, Dolphin et al. (79) reported that, in a strictly nonenzymic system, methylcobalamin is readily cleaved by hydrogen and a platinum catalyst to methane and cob(II)-alamin according to equation 1:

$$\text{Methylcobalamin} + \tfrac{1}{2}\,H_2 \rightarrow CH_4 + \text{Cob(II)alamin} \qquad (1)$$

This provided sound chemical evidence that the methyl-cobalt bond of methylcobalamin is susceptible to direct reductive cleavage under mild conditions and that, from the chemical point of view, it can be considered a satisfactory model of the biological intermediate.

When methylcobalamin was tested as a substrate for methane production by extracts of the *M. omelianskii* organisms which are unable to ferment methanol (80), it was found to be used even more efficiently than the natural one-carbon substrate, carbon dioxide. This showed that the reduction of the methyl moiety of methylcobalamin to methane by *M. barkeri* was not uniquely related to the ability of this organism to utilize methanol as a substrate for growth and suggested a general role

Table 3-2 $^{14}CH_4$ Formation from [^{14}C] Methylcobalamin by Broken-Cell Preparations of *Methanosarcina barkeri*

[^{14}C] Methylcobalamin Added (cpm)	Total $^{14}CH_4$ Formed (cpm)	Total $^{14}CO_2$ Formed (cpm)
600,000	228,000	3870
312,000	92,700	1050

The added labeled substrate, 600,000 cpm/μmole, was incubated in a 2-ml reaction mixture containing 50 mM potassium phosphate buffer, pH 6.8, 30 μmoles of potassium pyruvate, and 0.7 ml of *M. barkeri* cells that had been disrupted in a Branson sonifier. The gas phase was helium. From Blaylock and Stadtman (77,78).

of a methylcobamide type of intermediate in methane biosynthesis. The failure to detect any methylated derivatives among the cobamides isolated from *M. omelianskii* (75) could be due to the fact that, under normal reducing conditions, the enzyme-bound intermediate is rapidly cleaved to methane and hence has only a transient existence. However, it has been reported (81, 82) that methylcobalamin was isolated from cells of *Methanobacillus kuzneceovii,* a thermophilic organism that synthesizes both methane and acetate from methanol according to equation 2:

$$4CH_3OH \rightarrow 2CH_4 + CH_3COOH + 2H_2O \qquad (2)$$

Types of Experimental Data That Support an Intermediary Role of a Methylcobamide in Methane Biosynthesis

Experimental support for the postulated role of a cobamide in the final reduction step of methane biosynthesis was obtained in studies with *M. barkeri* at the National Institutes of Health (77, 78, 83–86) and with the *M. omelianskii* culture at the University of Illinois, (80, 87–90). These studies involved (a) use of inhibitors known to react specifically with cobamides, (b) characterization of the enzyme systems that utilize free methylcobalamin as methyl donor for methane biosynthesis, (c) purification and characterization of methyltransferase system that forms methylcobalamin from methanol and reduced cobalamin, and (d) purification of a red component of the methane-forming enzyme system that contains a cobamide as its prosthetic group.

The methane-forming system of *M. barkeri* is sensitive to intrinsic factor, the glycoprotein that specifically binds cobamide compounds. As shown in Table 3-3, both the reduction of methanol to methane and the transfer of the methyl group of methanol to reduced cobalamin [cob(I)alamin] to form methylcobalamin are markedly inhibited by intrinsic factor. Likewise, when methylcobalamin is added as substrate, its conversion to methane is inhibited, suggesting that intrinsic factor is tightly bound to an essential cobamide on the enzyme and is not displaced by the added free methylcobalamin.

A procedure involving alkylation with propyl iodide that was used by Taylor and Weissbach (51) to inactivate the cobalamin-dependent methionine synthetase also proved to be effective with the methane-forming system of *M. omelianskii* (89, 90). Treatment of reduced enzyme preparations with propyl iodide prior to introduction of the substrate completely inhibited methane formation; this inhibition could be overcome by irradiation of the treated extracts with visible light (Table 3-4). These results indicate that a cobamide enzyme that normally undergoes methylation to form a *Co*-methylcobamide intermediate instead is alkylated

Table 3-3 Effect of Intrinsic Factor on Reactions Catalyzed by *Methanosarcina barkeri*

Reaction	Inhibition (%)
$^{14}CH_3OH$ reduction to $^{14}CH_4$	63
[1–^{14}C] Pyruvate reduction to $^{14}CH_4$	88
N^5-[^{14}C] Methyltetrahydrofolate reduction to $^{14}CH_4$	80
[$^{14}CH_3$] Methylcobalamin reduction to $^{14}CH_4$	84
[$^{14}CH_3$] Methylcobalamin synthesis from $^{14}CH_3OH$ + Cob(I)alamin	100

From Blaylock and Stadtman (84).

by propyl iodide and the resulting *Co*-propylcobamide derivative cannot be cleaved by the reductase system. Only after the cobalt-propyl bond is cleaved by irradiation with light can the cobamide again participate in the methane-forming reaction.

Inhibition of methane formation in the rumen of sheep by the Freon 12 (CF_2Cl_2) propellant of Antifoam A (91) eventually was attributed to the formation of a difluorochloromethylcobamide (92). Cob(I)alamin readily reacts with CF_2Cl_2 and yields difluorochloromethylcobalamin; this fluoroalkyl derivative competitively inhibits methane formation from methylcobalamin by extracts of Methanobacterium strain M. O. H. (92). These observations suggest that a cobamide is an essential cofactor for methane biosynthesis in the rumen, one of the natural environments of the methane bacteria.

Table 3-4 Reversible Inhibition of Methane Biosynthesis by Treatment with Propyl Iodide

	Methane Produced by Crude Bacterial Extract		
Incubation Time (min)	Untreated (μmoles)	+ Propyl Iodide (μmoles)	+ Propyl Iodide then Irradiated (μmoles)
20	2.5	0	1.5
40	3.4	0	1.95
60	3.75	0	2.0

Data from Wood and Wolfe (89).

Using standard protein fraction procedures, a red protein containing a bound cobamide as its chromophore was isolated in highly purified form from extracts of *M. barkeri* (86) and from the *M. omelianskii* culture (90). To facilitate isolation of this protein from the *M. omelianskii* extracts it was first converted to a radioactive propyl derivative by treatment of the crude extract with ^{14}C-labeled propyl iodide. During the purification procedure the radioactive protein fractions were protected from light. The final red preparation was irradiated to cleave the labeled propyl group and regenerate the biologically active protein. Addition of this cobamide protein to crude preparations of the methane-forming system allowed active formation of methane from either methylcobalamin or N^5-methyltetrahydrofolate. Similarly, the red protein, isolated in its native form from *M. barkeri* (86), was required for maximal synthesis of methane from methanol and from carbon dioxide by numerous enzyme preparations that were deficient in this component.

A more precise assay for the cobamide protein from *M. barkeri* was provided by a methyltransferase reaction wherein the ^{14}C-labeled methyl group of methanol is transferred to reduced cobalamin to form radioactive methylcobalamin (84–86). Ferredoxin and an additional protein fraction required for the formation of methylcobalamin also were partially purified from *M. barkeri* extracts and, when these were combined with the appropriate cofactors (Table 3-5), the system exhibited an absolute dependency on the red cobamide protein for catalytic activity. The same components, plus a small amount of crude extract to supply the oxygen-labile methyl reductase, were capable of catalyzing the reduction of methanol to methane (Table 3-5). The red protein isolated from *M. barkeri* is acidic and has a molecular weight of 150,000 to 200,000.

It is likely that the cobamide moiety of the red protein from *M. barkeri* and also that from the *M. omelianskii* culture undergo methylation to form a *Co*-methylcobamide intermediate in the terminal reduction step that leads to the formation of methane, but, as yet, neither protein has been prepared in amounts sufficient to establish its precise role in the reaction. If such a methylated cobamide protein can be shown to be a product of a methyltransferase reaction and a substrate for cleavage to methane by a purified methyl reductase enzyme system, it would provide strong evidence that a protein-bound methylcorrinoid is the intermediate that is reductively cleaved to methane.

Methane Formation by Cobamide-Independent Pathways

By analogy with methionine biosynthesis, which occurs by a cobalamin-dependent and also by a cobamide-independent pathway, it is possible

Table 3-5 Components of Methyltransferase and Methyl Reductase Systems of *Methanosarcina barkeri*

	CH_3OH + Cob(I)alamin → Methylcobalamin	$CH_3OH + H_2 \rightarrow CH_4$
Red cobamide protein	+	+
Protein A_2 fraction	+	+
Ferredoxin	+	+
Methyl reductase	−	+
H_2	+	+
Acid-stable cofactor[a]	+	+
ATP	+	+
Mg^{2+}	+	+
Methanol	+	+
Cob(I)alamin	+	−

[a] This acidic cofactor may contain 2-thiolethanesulfonic acid, the methyl transfer cofactor isolated from Methanobacterium strain M.O.H. (93,94), as part of its structure.

that there may be alternative mechanisms of methane biosynthesis. In fact, it already has been demonstrated that nitrogenase can reduce cyanide to equimolar amounts of methane and ammonia (95, 96). Although cyanide and several other unusual substrates with which nitrogenase reacts are reduced at very low rates compared with the normal substrate, N_2, these observations are of interest in that they indicate that a catalytic system containing iron, molybdenum, and sulfur can reductively cleave the C–N bond of cyanide and form methane. Numerous chemical models of nitrogenase have been prepared in efforts to elucidate the mechanism of action of the enzyme. One model, a crystalline complex of glutathione and molybdenum (97), catalyzed the reduction by borohydride of some of the unnatural nitrogenase substrates (e.g., hydrazine and acetylene were reduced), but activity on cyanide was not reported. Although the distribution of nitrogenase among the methane-producing bacteria is not known, the mixed culture, *M. omelianskii,* fixes atmospheric nitrogen (98), and other organisms of this physiologic group also may be able to reduce nitrogen. A catalyst of the nitrogenase type that exhibited greater reactivity with C–N compounds might be able to produce methane rapidly and thus allow the process to take place by a cobamide-independent pathway.

A new methyl-group carrier isolated from Methanobacterium strain M. O. H. has been identified (94) as 2-thiolethanesulfonic acid ($HSCH_2$-CH_2SO_3H). The methylated form of this compound is *S*-methyl-2-thioethanesulfonic acid. The free thiol is methylated enzymically by a purified methyltransferase from Methanobacterium strain M. O. H. that uses methylcobalamin as methyl donor. It is also methylated in crude extracts by a methyl group generated by the reduction of carbon dioxide with hydrogen (93). When *S*-methyl-2-thioethanesulfonic acid is added in substrate amounts to crude extracts of the methane organism, the methyl group is reduced quantitatively to methane in the presence of molecular hydrogen. Neither the mechanism of the methyl transfer reaction from methylcobalamin to 2-thiolethanesulfonic acid nor that of the reduction of the *S*-methyl group to methane is known. It will be interesting to know whether methyl transfer and methane biosynthesis occur in Methanobacterium strain M. O. H. by a cobamide-independent pathway or if the series of reactions involves the red cobamide protein characterized from the mixed *M. omelianskii* culture (90). In *M. barkeri,* both a cobamide protein and an acidic, heat- and acid-stable cofactor are required for the methyl transfer reaction that forms methylcobalamin and for reduction of a methyl group to methane (Table 3-5).

Energetics of Methane Biosynthesis from Carbon Dioxide and Hydrogen

The growth of several species of methane bacteria is supported by the strictly anaerobic reduction of carbon dioxide to methane in the presence of molecular hydrogen as the sole oxidizable substrate. A consideration of the individual steps involved in this overall process leads one to the conclusion that the various microorganisms must derive their energy for growth by catalysis of the final reduction step. Although, as already pointed out, the precise intermediates of the first three reduction steps are unknown, carbon dioxide must undergo successive reductions to compounds at the level of oxidation of formate, formaldehyde, and methanol and the free energy changes calculated for the formation of these free one-carbon compounds should serve as a rough estimate of the energetics of reduction of carbon dioxide to the level of a methyl group. That catalysis of the first three reduction steps occurs with little if any net free energy change is suggested by the values shown in Table 3-6. The calculated free energy change associated with the reduction of methanol to methane indicates that this is a highly exergonic reaction and accounts for most of the energy that should be made available by catalysis of the overall reduction process.

ACETATE BIOSYNTHESIS

Clostridium thermoaceticum, isolated by Fontaine et al. in 1942 (100), carries out a homoacetate fermentation of glucose in which 3 moles of

Table 3-6 Energetics[a] of Methane Biosynthesis from Carbon Dioxide and Hydrogen

$$CO_2 + 4H_2 \rightarrow CH_4 + 2H_2O$$

$$HCO_3^- + H^+ + 4H_2 \rightarrow CH_4 + 3H_2O$$

Overall free energy change = −32 kcal/mole CH_4

$$CO_2 \xrightarrow{+2H} [HCOOH] \xrightarrow{+2H} [HCHO] \xrightarrow{+2H} [CH_3OH] \xrightarrow{+2H} CH_4$$

$CO_2 + H_2 \rightarrow HCOOH$	+ 2 kcal
$HCOOH + H_2 \rightarrow HCHO + H_2O$	+ 8 kcal (1 ATP required)
$HCHO + H_2 \rightarrow CH_3OH$	− 10 kcal
$CH_3OH + H_2 \rightarrow CH_4 + H_2O$	
Free energy change =	−26 kcal

[a] Values calculated from data of Krebs et al. (99).

acetate are produced for each mole of hexose utilized. In an early tracer experiment, Barker and Kamen (101) showed that when glucose was fermented in the presence of $^{14}CO_2$, both carbons of the acetate produced were labeled. To explain these results it was suggested that glucose is catabolized to pyruvate and then the latter is oxidized to acetate, carbon dioxide, and reducing equivalents. Utilization of the reducing equivalents to form additional acetate from 2 moles of carbon dioxide would provide the electron sink for the fermentation process.

The question as to whether the labeled acetate was a doubly labeled molecule or a mixture of singly labeled molecules, some containing ^{14}C in the methyl group and some in the carboxyl group, was answered in an elegant study by Wood (102) . He showed by mass analysis that the acetate formed from ^{13}C-labeled carbon dioxide by fermenting cells had a molecular weight 2 mass units greater than normal acetate and therefore must be $^{13}CH_3{}^{13}COOH$. Continued studies by Wood and his collaborators (103–107) on the mechanism of the overall anaerobic process indicated that none of the known pathways accounted for the *de novo* total synthesis of acetate from carbon dioxide by *C. thermoaceticum*.

Corrinoid Involvement in Acetate Biosynthesis

The report from Oxford by Guest et al. (4) that methylcobalamin serves as a methyl donor in the biosynthesis of methionine prompted investigation of the possibility that methylcobalamin might also be an efficient precursor of the methyl group of acetate (108). Three types of experimental evidence were obtained in these studies that supported the intermediary role of a methylcobamide in acetate synthesis: (1) Synthesis of doubly labeled acetate from $^{14}CO_2$ by cell-free extracts of *C. thermoaceticum* was inhibited by addition of the specific cobamide-binding protein, intrinsic factor. Of greater significance was the fact that there was a proportionately greater decrease in incorporation of label in the methyl carbon than in the carboxyl carbon. (2) Extracts of *C. thermoaceticum* selectively labeled the methyl carbon of acetate with ^{14}C when [^{14}C]methylcobalamin was included as a substrate in reaction mixtures that contained the components essential for acetate synthesis (108, 109). (3) When unlabeled methylcobalamin was incubated with acetate-forming extracts in the presence of $^{14}CO_2$ the methyl moiety of the methylcobalamin became labeled; alternatively, addition of reduced cobalamin to the complete reaction mixtures resulted in methylation of the reduced cobalt atom of the vitamin and a net synthesis of [^{14}C]methylcobalamin was observed (110).

Further evidence in support of the participation of a cobamide in acetate synthesis was provided by the demonstration that alkylation with propyl iodide inhibited the reaction in a light reversible fashion (111). It was established earlier by Weissbach and associates (51) in their studies on the cobalamin-dependent methionine synthetase of *E. coli* that a propyl group covalently bound to the cobalt atom of the cobamide cannot be transferred by the enzyme, so that unless this carbon-cobalt bond is cleaved by exposure to visible light, the enzyme is prevented from participation in its normal cyclic acceptance and transfer of a methyl group. When *C. thermoaceticum* extracts were treated with a reducing agent and then with propyl iodide prior to the addition of the labeled methyl donor ([^{14}C]methylcobalamin or N^5-[^{14}C]methyltetrahydrofolate) there was a marked decrease in synthesis of ^{14}C-labeled acetate (Table 3-7). Restoration of activity of the treated extracts by irradiation with visible light indicated that inhibition had been caused by propylation of an essential cobamide catalyst.

Although cobalamin itself is not among the corrin compounds normally synthesized by *C. thermoaceticum,* a closely related compound, 5-methoxybenzimidazolylcobamide, is produced (112, 113). This complete corrin compound, together with cobyric acid and lesser amounts of several other incomplete factors, accounts for the corrinoid compounds present in cells of the organism. Of especial interest is the finding that to a small extent, both cobyric acid and 5-methoxybenzimidazolylcobamide exist in *C. thermoaceticum* as their Coβ-methyl derivatives. Furthermore, when cells actively fermenting glucose were exposed briefly to $^{14}CO_2$,

Table 3-7 Inhibition of Acetate Biosynthesis by Propyl Iodide Treatment and Reversal by Visible Light Irradiation

Enzyme Treatment		^{14}C-Acetate Formed (μmole) from	
		N^5-[^{14}C]Methyltetrahydrofolate	[^{14}C]Methylcobalamin
Control	Dark	0.128	0.294
Control	Light	0.104	0.346
$CH_3CH_2CH_2I$, 1.6 mM	Dark	0.066	0.170
$CH_3CH_2CH_2I$, 1.6 mM	Light	0.130	0.330

Ghambeer et al (111).

the *Co*-methyl groups of both of these compounds became highly radioactive.

These observations that *Co*-methylcorrinoids occur naturally and that a close chemical analog, *Co*-methylcobalamin, can both serve as a methyl donor for acetate synthesis and be formed from reduced cobalamin and acetate precursors support the theory that a *Co*-methylcorrinoid is an intermediate in the biosynthesis of acetate. The available data indicate that the naturally occurring cobamide on the enzyme can be methylated either by N^5-methyltetrahydrofolate or by free *Co*-methylcobalamin (114, 115).

There is some experimental evidence to suggest that the immediate precursor of acetate may be a *Co*-carboxymethylcorrinoid. Trace amounts of labeled *Co*-carboxymethylcorrinoids were detected in cells of *C. thermoaceticum* which were pulsed with highly radioactive carbon dioxide (116). Also, the carboxymethyl moiety of *Co*-carboxymethylcobalamin is rapidly reduced to acetate by cell-free extracts of *C. thermoaceticum* and the reaction is stimulated by pyruvate and TPNH (112). However, because of the marked liability of *Co*-carboxymethylcorrinoids to light and their tendency to undergo a variety of cleavage reactions under mild conditions, the role of an intermediate of this type in acetate synthesis is not yet established.

Protein Components and Coenzymes Required for Acetate Synthesis

Extracts of *C. thermoaceticum* have been separated into three protein fractions that are required to convert *Co*-methylcobalamin to acetate (109, 117, 118). Two of these (A and B) are relatively crude fractions while the third has been purified and identified as ferredoxin; none has activity alone. Pyruvate, also required for this reaction, can be replaced by α-ketobutyrate but not by reduced pyridine nucleotides plus sources of high energy. Coenzyme A is also required, and the reaction is stimulated by the addition of glutathione, Mg^{2+}, and Fe^{2+}.

Ljungdahl et al. (119) demonstrated that fractions A and B were labeled with ^{14}C when cell-free extracts were incubated with pyruvate and $^{14}CO_2$. Exposure to light caused loss of radioactivity from fraction A, a finding consistent with the presence of a *Co*-[^{14}C]alkylcorrinoid. When the labeled fraction A was incubated with whole-cell-free extract, part of the label was converted to methyl-labeled acetate. Fraction B catalyzed the TPNH-dependent cleavage of synthetic *Co*-carboxymethylcobalamin. Neither DPNH nor pyruvate is able to replace TPNH in this reaction (105), although in unfractionated extracts, either pyruvate or TPNH was effective in the cleavage (112). From fraction B, Ljungdahl and co-workers (120) have purified a small protein (about 27,000 daltons) that contains 1 mole of 5-methoxybenzimidazolecobamide [existing as the Co(II) corrinoid in the native state] tightly bound to each mole of protein.

Distribution of the Ability to Catalyze *de novo* Synthesis of Acetate

In addition to the homoacetate fermentation of glucose by *C. thermoaceticum,* several other clostridia have been shown to synthesize acetate *de novo* from one-carbon fragments. One of the early demonstrations of total synthesis of acetate from CO_2 and H_2 was that of Wieringa (121) in *C. aceticum*. A closely related organism, *C. formicoaceticum,* has been shown to carry out many of the same reactions of *C. thermoaceticum* (122–124). *C. sticklandii,* an organism that ferments amino acids by way of the Stickland reaction, has also been shown to convert formate to acetate (125) and can also convert *Co*-methylcobalamin to acetate if supplemented with fraction B of *C. thermoaceticum* (109). All of these organisms are mesophiles except for *C. thermoaceticum,* and *C. aceticum* is the single autotroph. Thus the ability to catalyze *de novo* synthesis of acetate is not limited to a single nutritional type or temperature of cultivation.

Pathway for the Generation of the Methyl Moiety of Acetate

The reduction of CO_2 to the level of a methyl group probably proceeds by way of a tetrahydrofolate intermediate. Ljungdahl et al. (111, 114, 115) have shown that N^5-methyltetrahydrofolate is converted to [^{14}C]acetate by extracts of *C. thermoaceticum* in the presence of pyruvate. It was shown that the conversion of N^5-[^{14}C]methyltetrahydrofolate or of [^{14}C]-

methylcobalamin to acetate was inhibited by treatment of the enzyme with propyl iodide, an inhibition relieved by exposure to light (111). This indicates that the corrinoid component of the synthetic pathway was alkylated by the propyl group, and that since the alkylation inhibited conversion of N^5-[^{14}C]methyltetrahydrofolate as well as [^{14}C]methylcobalamin, it is probable that the methyl group is transferred from the tetrahydrofolate derivative to the corrinoid (Scheme 3-3).

Scheme 3-3. Metabolic pathways of the homoacetate fermentation of glucose by *Clostridium thermoaceticum*.

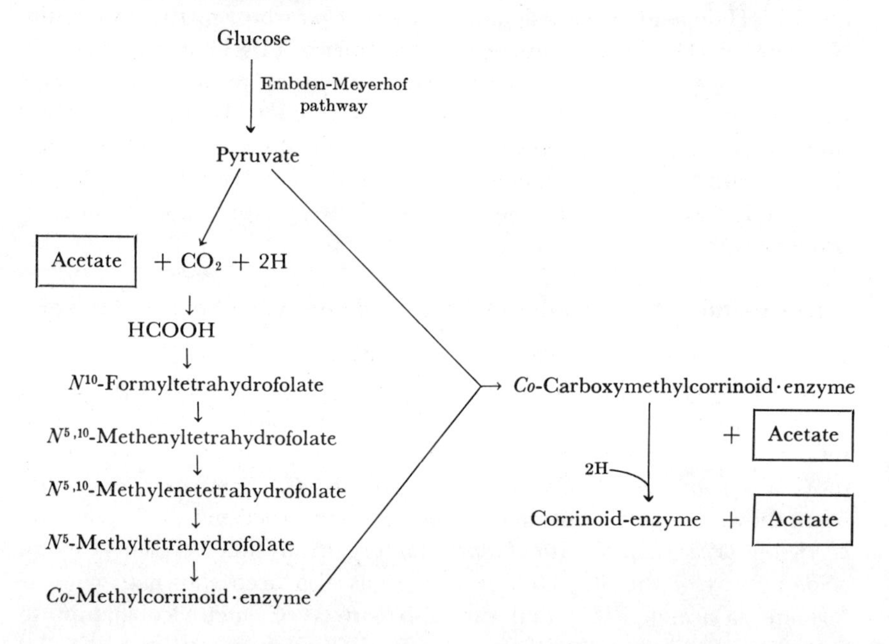

C. thermoaceticum readily catalyzes a direct reduction of CO_2 to formate by formate dehydrogenase (126, 127). Ljungdahl et al. (128–130), reported the purification of an ATP-dependent enzyme that converts formate to N^{10}-formyltetrahydrofolate. Subsequent steps that lead, in turn, to the formation of $N^{5,10}$-methenyltetrahydrofolate, $N^{5,10}$-methylenetetrahydrofolate and N^5-methyltetrahydrofolate have been demonstrated in whole-cell preparations (132) and in extracts (131).

Carboxylation of the Methyl Group

Three pathways have been proposed to account for the conversion of a *Co*-methylcorrinoid to acetate. One invokes the carboxylation of the

Co-methylcorrinoid followed by reductive cleavage of the *Co*-carboxymethylcorrinoid to yield acetate. A second considers Co-methylcorrinoid as a biological Grignard reagent after the proposal of Ingraham (133). The third is the dithiol-facilitated reductive production from *Co*-methylcorrinoid of a methyl carbanion which then reacts with CO_2 to form acetate (134). Parker et al. (135) attempted to distinguish among these pathways by following the migration of deuterium from N^5-trideuteriomethyltetrahydrofolate or trideuteriomethylcobalamin. They found that more than half of the acetate formed was trideuterioacetate. This supports the idea of a Grignard reaction in which *Co*-acetylcobalamin is an intermediate. A reasonable case might be made, however, for carboxylation of the methyl group with displacement of hydrogen to form a carboxymethyl group followed by return of the displaced hydrogen without exchange with solvent to form the acetate. Since their results do not exclude a carboxymethylcorrinoid as an intermediate, any mechanism proposed must account for the predominant formation of trideuterioacetate.

Source of the Carboxyl Carbon

Although Barker and Kamen (101) originally proposed that the carboxyl carbon of acetate was derived from CO_2, a collaboration between Wood's group at Case Western Reserve and Ljungdahl's at the University of Georgia (136, 137) led to the conclusion that the carboxyl carbon of acetate arises from the carboxyl carbon of α-ketoacids and not directly from CO_2. This conclusion was based on three lines of evidence: the α-ketoacid could not be replaced by CO_2, a reducing system, and a supply of ATP; the rate of conversion of N^5-methyltetrahydrofolate was independent of CO_2 between 50 mM and 1μM; and when using either N^5-methyltetrahydrofolate or *Co*-methylcobalamin plus $^{14}CO_2$ and unlabeled α-ketobutyrate, the specific activity of acetate was initially low and increased with time, whereas when unlabeled CO_2 and CH_3CH_2-$CO^{14}COOH$ were used, the initial specific activity was high and decreased with time. Thus the radioactivity of acetate paralleled that of the carboxyl carbon of α-ketobutyrate. From this is was concluded that the carboxyl group of acetate is donated from an α-ketoacid in a transcarboxylation reaction. Since CO_2 has been shown to exchange rapidly with both formate and the carboxyl group of pyruvate, the ultimate donor is carbon dioxide.

The fermentation of glucose by *C. thermoaceticum* generally is believed to proceed through the production of 2 moles of pyruvate, one of which is catabolized to yield CO_2, acetate, and reducing equivalents

(Scheme 3-3), the other donating its carboxyl group in the *de novo* synthesis of acetate by transcarboxylation. The end result of such a scheme easily accords with the long-recognized homoacetate fermentation of glucose by *C. thermoaceticum*.

REFERENCES

1. Barker, H. A., Weissbach, H., and Smyth, R. D. (1958). *Proc. Natl. Acad. Sci. U. S. A.* **44**, 1093.
2. Weissbach, H., Toohey, J., and Barker, H. A. (1959). *Proc. Natl. Acad. Sci. U. S. A.* **45**, 521.
3. Lenhert, P. G., and Hodgkin, D. C. (1961). *Nature* **192**, 937.
4. Guest, J. R., Friedman, S., Woods, D. D., and Smith, E. L. (1962). *Nature* **195**, 340.
5. Guest, J. R., Helleiner, C. W., Cross, M. J., and Woods, D. D. (1960). *Biochem. J.* **76**, 396.
6. Morningstar, J. F., and Kisliuk, R. L. (1965). *J. Gen. Microbiol.* **39**, 43.
7. Cauthen, S. E., Foster, M. A., and Woods, D. D. (1966). *Biochem. J.* **98**, 630.
8. Fujii, K., Takeuchi, H., Shimizu, S., and Fukui, S., (1972). *Agric. Biol. Chem.* **36**, 2323.
9. Cauthen, S. E., Pattison, J. R., and Lascelles, J. (1967). *Biochem. J.* **102**, 774.
10. Lago, B. D., and Demain, A. L. (1969). *J. Bacteriol.* **99**, 347.
11. Ohmori, H., Sato, K., Shimizu, S., and Fukui, S. (1971). *Agric. Biol. Chem.* **35**, 338.
12. Griffiths, J. M., and Daniel, L. J. (1969). *Arch. Biochem. Biophys.* **134**, 463.
13. Dickerman, H., Redfield, B. G., Bieri, J., and Weissbach, H. (1964). *J. Biol. Chem.* **239**, 2545.
14. Laughlin, R. E., Elford, H. L., and Buchanan, J. M. (1964). *J. Biol. Chem.* **239**, 2888.
15. Mangum, J. H., Murray, B. K., and North, J. A. (1969). *Biochemistry* **8**, 3496.
16. Burton, E., and Sakami, W. (1969). *Biochem. Biophys. Res. Commun.* **36**, 228.
17. Burton, E., Selhub, J., and Sakami, W. (1969). *Biochem. J.* **111**, 793.
18. Salem, A. R., Foster, M. A., and Wilson, R. H. (1970). *Biochem. J.* **118**, 16p.
19. Wilson, R. H. (1970). *Biochem. J.* **118**, 16p.
20. Botsford, J. D., and Parks, L. W. (1967). *J. Bacteriol.* **94**, 966.
21. Salem, A. R., Pattison, J. R., and Foster, M. A. (1972). *Biochem. J.* **126**, 993.
22. Dodd, W. A., and Cossins, E. A. (1970). *Biochim. Biophys. Acta* **201**, 461.
23. Taylor, R. T., and Weissbach, H. (1971). In *Methods in Enzymology*, Vol. 17, Tabor, H., and Tabor, C. W., Eds., Academic Press, New York, Part B, p. 379.
24. Taylor, R. T., and Weissbach, H. (1973). In *The Enzymes*, 3rd ed., Boyer, P.D., Ed., Academic Press, New York, Chapter IX, Part B, p. 121.
25. Rüdiger, H., and Jaenicke, L. (1973). *Mol. Cell. Biochem.* **1**, 157.
26. Gibson, F., and Woods, D. D. (1952). *Biochem. J.* **51**, v.

27. Gibson, F., and Woods, D. D. (1960). *Biochem. J.* **74**, 160.
28. Szulmajster, J., and Woods, D. D. (1955). *Abstr. 3rd Int. Congr. Biochem. Brussels,* p. 44.
29. Szulmajster, J., and Woods, D. D. (1960). *Biochem. J.* **75**, 3.
30. Cross, M. J., and Woods, D. D. (1954). *Biochem. J.* **58**, xvi.
31. Helleiner, C. W., and Woods, D. D. (1956). *Biochem. J.* **63**, 26p.
32. Kisliuk, R. L., and Woods, D. D. (1957). *J. Gen. Microbiol.* **18**, xv.
33. Kisliuk, R. L., and Woods, D. D. (1960). *Biochem J.* **75**, 467.
34. Foster, M. A., Tejerina, G., and Woods, D. D. (1961). *Biochem. J.* **81**, 1p.
35. Foster, M. A., Tejerina, G., Guest, J. R., and Woods, D. D. (1964). *Biochem. J.* **92**, 476.
36. Guest, J. R., Friedman, S., Foster, M. A., Tejerina, G., and Woods, D. D. (1964). *Biochem. J.* **92**, 497.
37. Foster, M. A., Dilworth, M. J., and Woods, D. D. (1964). *Nature* **201**, 39.
38. Foster, M. A., Rowbury, R. J., and Woods, D. D. (1963). *J. Gen. Microbiol.* **31**, xix.
39. Rowbury, R. J., and Woods, D. D. (1961). *J. Gen. Microbiol.* **24**, 129.
40. Rowbury, R. J., and Woods, D. D. (1964). *J. Gen. Microbiol.* **35**, 145.
41. Mangum, J. H., and North, J. A. (1971). *Biochemistry* **10**, 3765.
42. Taylor, R. T., and Weissbach, H. (1967). *J. Biol. Chem.* **242**, 1502.
43. Stavrianopoulos, J., and Jaenicke, L. (1967). *Eur. J. Biochem.* **3**, 95.
44. Taylor, R. T. (1971). *Biochim. Biophys. Acta* **242**, 355.
45. Rüdiger, H., and Jaenicke, L. (1970). *Eur. J. Biochem.* **16**, 92.
46. Galivan, J., and Huennekens, F. M. (1970). *Biochem. Biophys. Res. Commun.* **38**, 46.
47. Baker, J. J., van der Drift, C., and Stadtman, T. C. (1973). *Biochemistry* **12**, 1054.
48. Weissbach, H., Redfield, B. G., Dickerman, H., and Brot, N. (1965). *J. Biol. Chem.* **240**, 856.
49. Brot, N., and Weissbach, H. (1966). *J. Biol. Chem.* **241**, 2024.
50. Taylor, R. T. (1970). *Arch. Biochem. Biophys.* **137**, 529.
51. Taylor, R. T., and Weissbach, H. (1967). *J. Biol. Chem.* **242**, 1509.
52. Brodie, J. D. (1967). *Biochem. Biophys. Res. Commun.* **26**, 261.
53. Taylor, R. T., and Weissbach, H. (1967). *Biochem. Biophys. Res. Commun.* **27**, 398.
54. Taylor, R. T., and Weissbach, H. (1967). *Arch. Biochem. Biophys.* **119**, 572.
55. Taylor, R. T., and Weissbach, H. (1968). *Arch. Biochem. Biophys.* **123**, 109.
56. Taylor, R. T., and Weissbach, H. (1969). *Arch. Biochem. Biophys.* **129**, 728.
57. Taylor, R. T., and Weissbach, H. (1966). *J. Biol. Chem.* **241**, 3641.
58. Taylor, R. T., and Weissbach, H. (1969). *Arch. Biochem. Biophys.* **129**, 745.
59. Taylor, R. T., and Weissbach, H (1967). *J. Biol. Chem.* **242**, 1517.
60. Taylor, R. T., and Hanna, M. L. (1970). *Biochem. Biophys. Res. Commun.* **38**, 758.
61. Rüdiger, H. (1971). *Eur. J. Biochem.* **21**, 264.

62. Burke, G. T., Mangum, J. H., and Brodie, J. D. (1971). *Biochemistry* **10**, 3079.
63. Taylor, R. T. (1971). *Arch. Biochem. Biophys.* **144**, 352.
64. Katzen, H. M., and Buchanan, J. M. (1965). *J. Biol. Chem.* **240**, 825.
65. Milner, L., Whitfield, C., and Weissbach, H. (1969). *Arch. Biochem. Biophys.* **133**, 413.
66. Kung, H.-F., Spears, C., Greene, R. C., and Weissbach, H. (1972). *Arch. Biochem. Biophys.* **150**, 23.
67. Greene, R. C., Williams, R., Kung, H.-F., Spears, C., and Weissbach, H. (1973). *Arch. Biochem. Biophys.* **158**, 249.
68. Barker, H. A. (1956). *Bacterial Fermentations,* Wiley, New York.
69. Stadtman, T. C. Unpublished data.
70. Lezius, A., Stadtman, T. C., and Lynen, F. (1960). Unpublished data.
71. Bernhauer, K., and Friedrich, W. (1953). *Angew. Chem.* **65**, 627.
72. Neujahr, H. Y. (1955). *Acta Chem. Scand.* **9**, 622.
73. Neujahr, H. Y. (1959). *Acta Chem. Scand.* **13**, 1453.
74. Stadtman, T. C. (1960). *J. Bacteriol.* **79**, 904.
75. Lezius, A., and Barker, H. A. (1965). *Biochemistry* **4**, 510.
76. Lezius, A., Blaylock, B. A., and Stadtman, T. C. Unpublished data .
77. Blaylock, B. A., and Stadtman, T. C. (1963). *Biochem. Biophys. Res. Commun.* **11**, 34.
78. Blaylock, B. A., and Stadtman, T. C. (1964). *Ann. N. Y. Acad. Sci.* **112**, 799.
79. Dolphin, D., Johnson, A. W., and Rodrigo, R. (1964). *Ann. N. Y. Acad. Sci.* **112**, 590.
80. Wolin, M. J., Wolin, E. A., and Wolfe, R. S. (1963). *Biochem. Biophys. Res. Commun.* **12**, 464.
81. Pantskhava, E. S. (1970). *Dokl. Akad. Nauk SSSR* **193**, 1419.
82. Pantskhava, E. S., and Plechkina, V. V. (1969). *Prikl. Biokhim. Mikrobiol.* **5**, 416.
83. Blaylock, B. A., and Stadtman, T. C. (1964). *Biochem. Biophys. Res. Commun.* **17**, 475.
84. Blaylock, B. A., and Stadtman, T. C. (1966). *Arch. Biochem. Biophys.* **116**, 138.
85. Stadtman, T. C., and Blaylock, B. A. (1966). *Fed. Proc.* **25**, 1657.
86. Blaylock, B. A. (1968). *Arch. Biochem. Biophys.* **124**, 314.
87. Wolin, M. J., Wolin, E. A., and Wolfe, R. S. (1964). *Biochem. Biophys. Res. Commun.* **15**, 420.
88. Wood, J. M., and Wolfe, R. S. (1966). *J. Bacteriol.* **92**, 696.
89. Wood, J. M., and Wolfe, R. S. (1966). *Biochem. Biophys. Res. Commun.* **22**, 119.
90. Wood, J. M., and Wolfe, R. S. (1966). *Biochemistry* **5**, 3598
91. Bauchop, T. (1967). *J. Bacteriol.* **94**, 171.
92. Penley, M. W., Brown, D. G., and Wood, J. M. (1970). *Biochemistry* **9**, 4302.
93. McBride, B. C., and Wolfe, R. S. (1971). *Biochemistry* **10**, 2317.
94. Taylor, C. D., and Wolfe, R. S. (1973). *Fed. Proc.* **32**, Abstr. 589, 2102.
95. Hardy, R. W. F., and Knight, E., Jr. (1967). *Biochim. Biophys. Acta* **139**, 69.
96. Hardy, R. W. F., and Burns, R. C. (1968). *Ann. Rev. Biochem.* **37**, 331.

97. Werner, D., Russell, S. A., and Evans, H. J. (1973). *Proc. Natl. Acad. Sci. U. S. A.* **70**, 339.

98. Pine, M. J., and Barker, H. A. (1954). *J. Bacteriol.* **68**, 589.

99. Krebs, H. A., Kornberg, H. L., and Burton, K. (1957). *Energy Transformations in Living Matter,* Springer-Verlag, Berlin and New York.

100. Fontaine, F. E., Peterson, W. H., McCoy, E., and Johnson, M. J. (1942). *J. Bacteriol.* **43**, 701.

101. Barker, H. A., and Kamen, M. D. (1945). *Proc. Natl. Acad. Sci. U. S. A.* **31**, 219.

102. Wood, H. G. (1952). *J. Biol. Chem.* **194**, 905.

103. Wood, H. G. (1952). *J. Biol. Chem.* **199**, 579.

104. Lentz, K. E., and Wood, H. G. (1955). *J. Biol. Chem.* **215**, 645.

105. Lentz, K. E. (1956). Studies on the Synthesis of Acetate by *Clostridium thermoaceticum.* Ph.D. Thesis, Western Reserve University, Cleveland, Ohio.

106. Ljungdahl, L., and Wood, H. G. (1963). *Bacteriol. Proc. Abstr.* 109.

107. Ljungdahl, L., and Wood, H. G. (1965). *J. Bacteriol.* **89**, 1055.

108. Poston, J. M., Kuratomi, K., and Stadtman, E. R. (1964). *Ann. N. Y. Acad. Sci.* **112**, 804.

109. Poston, J. M., Kuratomi, K., and Stadtman, E. R. (1966). *J. Biol. Chem.* **241**, 4209.

110. Kuratomi, K., Poston, J. M., and Stadtman, E. R. (1966). *Biochem. Biophys. Res. Commun.* **23**, 691.

111. Ghambeer, R. K., Wood, H. G., Schulman, M., and Ljungdahl, L. (1971). *Arch. Biochem. Biophys.* **143**, 471.

112. Ljungdahl, L., Irion, E., and Wood, H. G. (1965). *Biochemistry* **4**, 2771.

113. Irion, E., and Ljungdahl, L. (1965). *Biochemistry* **4**, 2780.

114. Ljungdahl, L., Irion, E., and Wood, H. G. (1966). *Fed. Proc.* **25**, 1642.

115. Ljungdahl, L., and Wood, H. G. (1969). *Ann. Rev. Microbiol.* **23**, 515.

116. Ljungdahl, L., and Irion, E. (1966). *Biochemistry* **5**, 1846.

117. Poston, J. M., and Stadtman, E. R. (1967). *Biochem. Biophys. Res. Commun.* **26**, 550.

118. Poston, J. M. (1971). *Fed. Proc.* **30**, Pt. 2, 1202.

119. Ljungdahl, L., Glatzle, D., Goodyear, J., and Wood, H. G. (1967). *Bacteriol. Proc. Abstr.,* 128.

120. Ljungdahl, L., LeGall, J., and Lee, J.-P. (1973). *Biochemistry* **12**, 1802.

121. Wieringa, K. T. (1936). *Antonie van Leewenhoek J. Microbiol. Serol.* **3**, 263.

122. el Ghazzawi, E. (1967). *Arch. Mikrobiol.* **57**, 1.

123. Andreesen, J. R., Gottschalk, G., and Schlegel, H. G. (1970). *Arch. Mikrobiol.* **72**, 154.

124. O'Brien, W. E., and Ljungdahl, L. G. (1972). *J. Bacteriol.* **109**, 626.

125. Stadtman, T. C., and White, F. H., Jr. (1954). *J. Bacteriol.* **67**, 651.

126. Thauer, R. K. (1972). *FEBS Lett.* **27**, 111.

127. Li, L. F., Ljungdahl, L., and Wood, H. G. (1966). *J. Bacteriol.* **92**, 405.

128. Sun, A. Y., Ljungdahl, L., and Wood, H.G. (1969). *J. Bacteriol,* **98**, 842.

129. Ljungdahl, L., Brewer, J. M., Neece, S. H., and Fairwell, T. (1970). *J. Biol. Chem.* **245**, 4791.

130. Brewer, J. M., Ljungdahl, L., Spencer, T. E., and Neece, S. H. (1970). *J. Biol. Chem.* **245**, 4798.

131. Andreesen, J. R., Schaupp, A., Neurauter, C., Brown, A., and Ljungdahl, L. G. (1973). *J. Bacteriol.* **114**, 743.

132. Parker, D. J., Wu, T.-F., and Wood, H. G. (1971). *J. Bacteriol.* **108**, 770.

133. Ingraham, L. L. (1964). *Ann. N. Y. Acad. Sci.* **112**, 713.

134. Schrauzer, G. N., and Sibert, J. W. (1970). *J. Am. Chem. Soc.* **92**, 3509.

135. Parker, D. J., Wood, H. G., Ghambeer, R. K., and Ljungdahl, L. (1972). *Biochemistry* **11**, 3074.

136. Schulman, M., Ghambeer, R. K., and Ljungdahl, L. G. (1973). *Fed. Proc.* **32**, 627.

137. Schulman, M., Ghambeer, R. K., Ljungdahl, L. G., and Wood, H. G. (1973). *J. Biol. Chem.* **248**, 6255.

CHAPTER FOUR

COBAMIDES AS COFACTORS

ADENOSYLCOBAMIDE-DEPENDENT REACTIONS

BERNARD M. BABIOR

Department of Medicine
Tufts-New England Medical Center
Boston, Massachusetts

CONTENTS

INTRODUCTION

Two classes of cobamides serve as cofactor in enzyme-catalyzed reactions: the adenosylcobamides and the methylcobamides. There are several members of each class, differing in the nature of the heterocyclic base coordinated to the α position of the cobalt. Each class, however, possesses a characteristic β ligand. In methylcobamides the β ligand is a CH_3– group covalently attached to the central cobalt atom. With adenosylcobamides, the β ligand is the 5′-deoxyadenosyl group, attached to the rest of the molecule by a covalent bond between the metal and the terminal carbon of the 5′-deoxyribose (Figure 4-1).

In this chapter, enzymatic reactions requiring adenosylcobamides are discussed. These reactions fall into two categories. The first category contains a number of very characteristic *rearrangements* brought about with the aid of an adenosylcobamide. The second category is represented by only a single reaction—the adenosylcobamide-dependent *reduction* of ribonucleotide triphosphates to deoxyribonucleotide triphosphates. The reactions are primarily of importance to microorganisms, each of them having been demonstrated to occur in one or more species of bacteria. Adenosylcobamide-dependent ribonucleotide reduction, however, has also been demonstrated in the eukaryotic flagellate *Euglena gracilis*,[1] the cofactor requirement for this reaction forming the basis for a widely used bioassay of cobamides, while the interconversion between methylmalonyl CoA and succinyl CoA, a rearrangement dependent on adenosylcobalamin, has been found in many higher animals, including man. There is evidence implicating a failure of this rearrangement in the pathogenesis of the neurological lesions associated with pernicious anemia and certain other cobalamin deficiency states (see Chapter 9).

THE REARRANGEMENTS

The adenosylcobamide-dependent rearrangements are a series of reactions having in common the feature that each of them involves the transfer of a hydrogen atom from one carbon atom to an adjacent carbon atom in exchange for another group (X, 4-1) which migrates in the opposite

$$\begin{array}{c} \text{X} \\ | \\ \text{C} \end{array}\!\!-\!\!\begin{array}{c} \\ \\ \text{C} \\ | \\ \text{H} \end{array} \rightleftharpoons \begin{array}{c} \\ \\ \text{C} \\ | \\ \text{H} \end{array}\!\!-\!\!\begin{array}{c} \text{X} \\ | \\ \text{C} \end{array} \qquad (4\text{–}1)$$

[1] In most organisms, including those multicellular species hitherto examined, ribonucleotide reduction appears to be independent of cobamides, being catalyzed by an iron-containing enzyme with no other cofactor requirements (1).

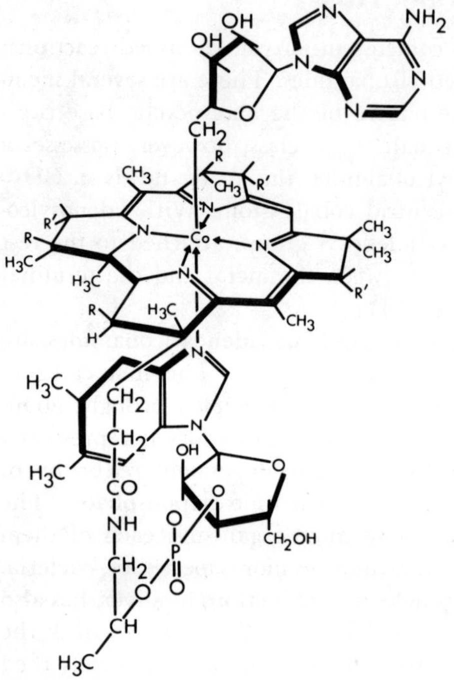

Fig. 4-1. Adenosylcobalamin. (Reprinted from *Fed. Proc.* **25**, 1623–1627, 1966.)

direction. These reactions are of great interest mechanistically, because for a number of them there is no obvious analogy in organic chemistry, while for others the similarity between the adenosylcobamide-dependent reaction and the apparently analogous model reaction has turned out to be deceptive. The enzymes catalyzing the reactions can be classified in terms of the group (X) which trades places with a hydrogen atom during the course of the rearrangement. In the first class of reactions, an interconversion between a branched-chain and straight-chain compound is accomplished by the adenosylcobamide-dependent migration of a bulky alkyl or acyl group.

$$\mathrm{H_2\overset{\overset{\displaystyle R}{|}}{C}\!-\!CH_2\!-} \rightleftharpoons \mathrm{H_3C\!-\!\overset{\overset{\displaystyle R}{|}}{C}H\!-}$$

In the second class, X is a nucleophilic group (OH– or NH2–) which migrates to a carbon atom already bearing a hydroxyl group. The resulting *gem*-diol or amino alcohol then eliminates to form an aldehyde.

$$\mathrm{\overset{\overset{\displaystyle X}{|}}{C}\!-\!\underset{H}{\overset{H}{C}}\!-\!OH} \rightarrow \mathrm{HC\!-\!\underset{H}{\overset{\overset{\displaystyle X}{|}}{C}}\!-\!OH} \xrightarrow{\ XH\ } \mathrm{HC\!-\!\overset{\overset{\displaystyle O}{\|}}{C}H}$$

The third class of reactions involves the migration of NH_2– between a terminal carbon atom and an adjacent methylene group, a rearrangement resulting in the conversion of a primary amine to a secondary amine.

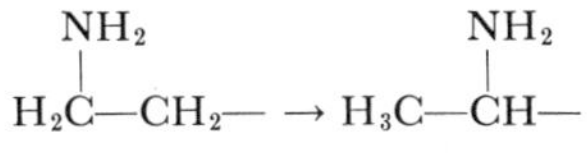

$$\underset{H_2C}{\overset{NH_2}{|}}\!\!-\!CH_2\!- \rightarrow H_3C\!-\!\underset{CH}{\overset{NH_2}{|}}\!-$$

Properties of the Enzymes

Enzymes Catalyzing Branched-Chain ⇌ Straight-Chain Rearrangements

GLUTAMATE MUTASE

Glutamate mutase, discovered in *Clostridium tetanomorphum* during an investigation of an unusual pathway for the degradation of glutamate (2) and subsequently shown to occur in certain photosynthetic organisms (3), was the first enzyme for which an adenosylcobamide requirement was demonstrated. The enzyme catalyzes the reversible isomerization of L-glutamate to L-threo-β-methylaspartate (4), a reaction in which a glycine moiety is the migrating alkyl group (see 4-2) (5, 6). The

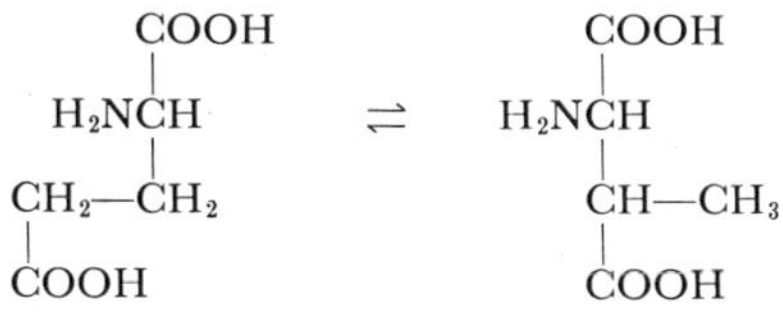

$$\begin{array}{ccc} \text{COOH} & & \text{COOH} \\ | & & | \\ \text{H}_2\text{NCH} & \rightleftharpoons & \text{H}_2\text{NCH} \\ | & & | \\ \text{CH}_2\text{—CH}_2 & & \text{CH—CH}_3 \\ | & & | \\ \text{COOH} & & \text{COOH} \end{array} \qquad (4\text{–}2)$$

β-methylaspartate produced in this reaction is subsequently deaminated to form mesaconic acid (methylfumaric acid) which is then metabolized to yield acetate, butyrate, CO_2, and hydrogen gas (7).

In addition to adensylcobamide, a sulfhydryl compound is necessary for the reaction to take place, but no requirement for monovalent or divalent cation has been found (5). The substrate specificity of glutamate mutase is quite high, no reaction being observed with α-methyl-DL-glutamate, L-glutamine, L-glutamate γ-methyl ester, or DL-α-aminoadipate, all analogs of glutamic acid, nor with β-ethylaspartic acid, a β-methylaspartate analog (4, 5). Stereospecificity of the reaction is exhibited in both directions: Isomerization of glutamate gives rise exclusively to the L-threo isomer of β-methylaspartate (4, 8), while rearrangement of β-methylaspartate to glutamate takes place with a stereospecific migration of one of the hydrogen atoms of the substrate methyl group such that the carbon which receives the migrating hydrogen atom undergoes a net inversion of configuration during the course of the reaction (4-3) (9, 10).

$$\begin{array}{c} R \\ | \\ C \\ H_3C \diagup \quad \diagdown COOH \\ H \end{array} \rightleftharpoons \begin{array}{c} H \\ | \\ C \\ H \diagup \quad \diagdown COOH \\ RH_2C \end{array} \tag{4-3}$$

Early attempts to purify the enzyme indicated that it consisted of more than one protein component (5). Further purification led to the isolation of two components, neither of which was active alone but which together were able to catalyze the glutamate mutase reaction. These were called component S and component E, because the former was found in the supernatant and the latter in the eluate from a calcium phosphate gel adsorption step used in the purification of the enzyme. It happens that the terminology is also appropriate to the function of the two components, since component S is a protein whose function depends on reduced –SH groups, while component E binds the substrates and cofactor and therefore can in a certain sense be regarded as the "enzymatic" component of glutamate mutase.

The first component to be isolated was component E, purified to about 75% homogeneity by Suzuki and Barker (11). Sedimentation velocity measurements indicated a molecular weight of about 128,000. A proposal that the glutamate mutase reaction might require the participation of pyridoxal led to a measurement of this cofactor in the enzyme preparation; none was detected. The binding to component E of L-glutamate and L-threo-β-methylaspartate, as well as of L-glutamine and L-aspartic acid, was indicated by the observation that these amino acids protected the component against inactivation by Tris • HCl, a buffer in which the enzyme is highly unstable. No protection was afforded by other substrate analogs, by EDTA, by saturating amounts of component S or by (Bza)AdoCba. Nonetheless, direct binding experiments showed that component E took up cobamide. By itself, component E was found to bind approximately 1 mole of AdoCbl per mole of enzyme. In the presence of a seven-fold molar excess of component S, the binding of cofactor rose to 2 moles/mole of component E.

Component S, purified to homogeneity by Switzer and Barker (12), is a much smaller protein, sedimentation velocity indicating a molecular weight of 17,000. By its spectrum, that of a simple protein without bound cofactors, as well as by direct analysis, it was concluded that this component too was free of pyridoxal. On exposure to air, component S was converted to an inactive dimer, but it was reactivated and converted back to the monomer by exposure to sulfhydryl compounds. This finding explains the requirement for a sulfhydryl compound in the glutamate mutase reaction. Further study revealed that component S was very sensitive to sulfhydryl reagents such as mercuribenzoate and iodoacetate, as well as to AsO_2^-, a reagent which affects enzymes possessing a vicinal dithiol group at the active site. Sulfhydryl titrations under various conditions showed that component S possessed five –SH groups. Inhibition by AsO_2^- indicated that two of these groups are adjacent and that at least one of these adjacent groups is required for activity. Production of the inactive dimer in air was proposed to take place by the formation of a disulfide bridge between two molecules of component S. Reduction of this bridge by thiols led to the regeneration of the active monomer (Figure 4-2). Though neither component E nor component S is active alone, glutamate mutase activity appears when the two components are mixed together. When glutamate mutase activity is determined as a function of the concentration of one component with the other component held constant, saturation behavior is observed with both component S and component E (Figure 4-3) (12). Half-maximal activity is observed at a component S/component E ratio of 1 : 1. It is virtually certain from these findings that an enzymatically active complex is formed from these components. This complex, however, must be in rapid equilibrium with the two separate components, since it was not detected in gel filtration experiments. Although component S does not bind cobamides by itself, the formation of the active complex between

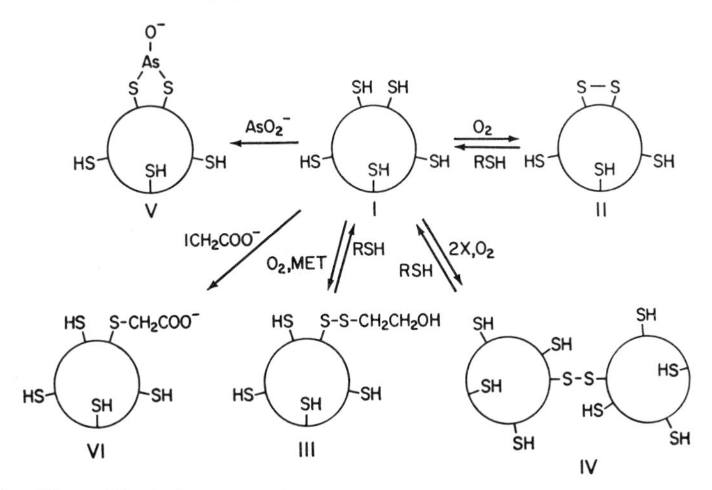

Fig. 4-2. The sulfhydryl groups of component S of glutamate mutase (12).

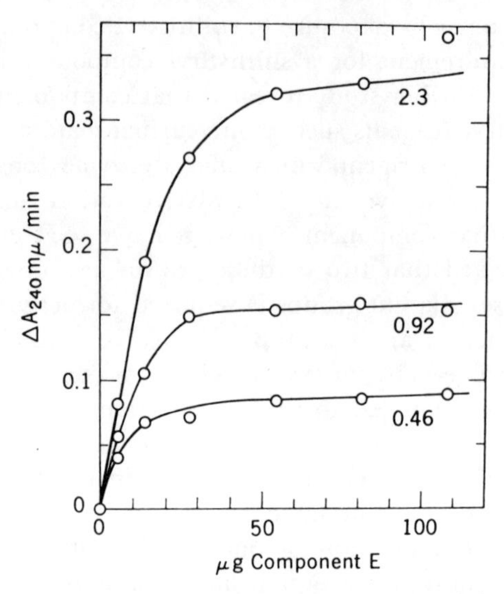

Fig. 4-3. The effect of the concentration of component E (left) and component S (right) on the activity of glutamate mutase at fixed concentrations of the other com-

component S and component E greatly alters the properties of cobamide binding by component E. Reciprocal plots of reaction velocity versus cofactor [in this case, (Bza)AdoCba] concentration at varying ratios of component S to component E indicate that the affinity of component E for the cobamide increases markedly as the component S/component E ratio rises (Figure 4-4). Conversely, complex formation between component S and component E is greatly augmented by an increase in the concentration of (Bza)AdoCba (Figure 4-5). It appears that cobamide and component S each facilitate the binding of the other to component E of the mutase.

METHYLMALONYL COA MUTASE

Like all other adenosylcobamide-requiring enzymes so far discovered, methylmalonyl CoA mutase is found in a microorganism. Unique among adenosylcobamide-requiring enzymes, however, it is also present in mammalian tissues. The reaction catalyzed by this enzyme is the reversible conversion of methylmalonyl CoA to succinyl CoA, a reaction involving an intramolecular migration of the bulky –COSCoA group in exchange for a hydrogen atom (4-4) (13–19). The conversion takes place

$$\begin{array}{c} \text{COOH} \\ | \\ \text{CH}_3\text{—CH} \\ | \\ \text{COSCoA} \end{array} \rightleftharpoons \begin{array}{c} \text{COOH} \\ | \\ \text{CH}_2\text{—CH}_2 \\ | \\ \text{COSCoA} \end{array} \qquad (4\text{–}4)$$

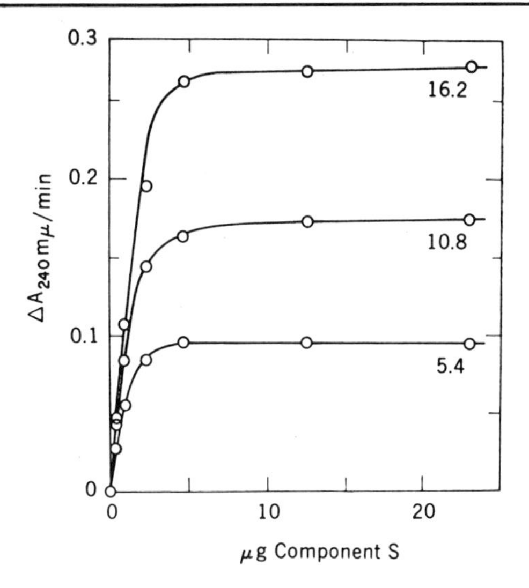

ponent (12). The *number* on each line indicates the amount of fixed component (in micrograms) present in the assay mixture.

with retention of the configuration of the carbon atom receiving the migrating hydrogen atom (20, 21), a stereochemical outcome opposite to that observed in the glutamate mutase reaction. The configuration of the carbon atom receiving the migrating acyl group is also retained, at least with ethylmalonyl CoA as substrate (21a).

In mammals, this reaction is on the pathway for the degradation of propionyl CoA (see 4-5) (18) produced by the catabolism of isoleucine

$$CH_3CH_2COSCoA \xrightarrow[\text{propionyl CoA carboxylase}]{ATP,\ CO_2} CH_3\underset{\displaystyle COOH}{\overset{\displaystyle COSCoA}{CH}} \underset{\text{racemase}}{\rightleftharpoons}$$

$$CH_3\underset{\displaystyle COSCoA}{\overset{\displaystyle COOH}{CH}} \underset{\text{mutase}}{\overset{\text{AdoCbl}}{\rightleftharpoons}} \text{succinyl CoA} \qquad (4\text{–}5)$$

and valine. In the overall pathway, propionyl CoA is carboxylated to D-methylmalonyl CoA in a biotin-dependent reaction. D-Methylmalonyl CoA then undergoes an enzyme-catalyzed racemization, a step which is necessary because the mutase is specific for the L-enantiomer (20, 22, 23). L-Methylmalonyl CoA is then converted to succinyl CoA by the mutase.

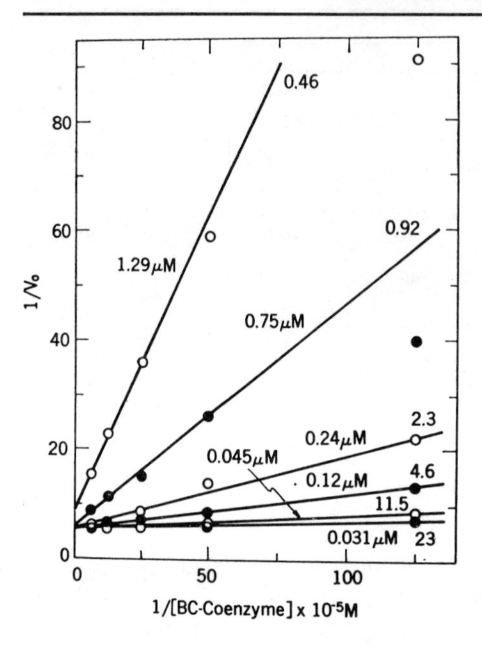

Fig. 4-4. Effect of the component S-to-component E ratio on the K_m for (Bza)AdoCba (12). The *number* on each line indicates the amount of component E. The *concentration* on each line gives the apparent K_m derived from the slope of that line.

The microorganism in which the mutase has been found is *Propionibacterium shermanii.* In this organism, the reaction is involved in energy production (18). *Propionibacterium shermanii* is an anaerobe which uses pyruvic acid as an electron sink, reducing this compound to propionic acid in the course of its energy-producing reactions. The pathway shown above, operating in reverse, represents part of the sequence of reactions by which propionic acid is synthesized by this microorganism.

Both the bacterial and mammalian enzymes have been purified to homogeneity. From *P. shermanii* two different forms of the enzyme have been isolated (18, 24). One form, a pink preparation with the optical

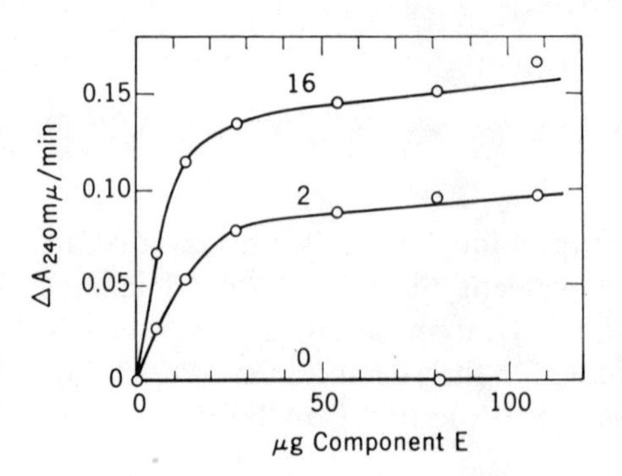

Fig. 4-5. Effect of (Bza)AdoCba concentration on the mutase activity of component E at various concentrations of component S (12). The *numbers* indicate the concentrations of (Bza)-AdoCbl (micromolar).

spectrum of OH-Cbl and a specific activity of 9.7 units/mg, was obtained from a portion of crude extract which was carried through the entire purification procedure without interruption. The remainder of this crude extract was carried through part of the procedure, then stored at –20° for 3 months before completing the purification. The enzyme from this portion of the crude extract was colorless, with a specific activity of 14.4 units/mg. The difference in specific activity between the two enzyme preparations was attributed to partial inhibition of the pink preparation by OH-Cbl bound to the active sites of the enzyme.

Recently, a pink enzyme preparation, obtained by a modification of the method previously used (18, 24), was chromatographically resolved into two proteins, one colorless and the other red (24a). The colorless protein catalyzed the adenosylcobalamin-dependent interconversion of methylmalonyl CoA and succinyl CoA with a specific activity of 14 units/mg. The red protein, which contained bound hydroxocobalamin, was inactive.

The molecular weight of methylmalonyl CoA mutase from *P. shermanii* was originally reported to be 56,000, as measured by ultracentrifugation (24). A subsequent study, however, indicates a molecular weight of 124,000, the enzyme being composed of two nonidentical subunits of molecular weights 61,000 and 63,000 (24a). The OH-Cbl associated with the pink form of the enzyme is bound very tightly, since neither charcoal treatment nor acid-ammonium sulfate (25) was able to resolve the enzyme of its cobamide. The enzyme requires no cofactors other than an appropriate adenosylcobamide. Neither univalent nor divalent cations are necessary for the activity of the mutase (24). The enzyme is not inhibited by EDTA (24), and is quite insensitive to the sulfhydryl reagents *p*-hydroxymercuribenzoate and N-ethylmaleimide (24).

The mammalian mutase has been partially purified from sheep kidney (26, 27), and purified to homogeneity from sheep liver (22, 28). Both enzymes carried adenosylcobamide with them during the purification, and prior to resolution were almost fully active in the absence of added AdoCbl (27, 28). It was possible to resolve the sheep liver enzyme of bound cofactor by an acid-ammonium sulfate step (22). After this step, the liver enzyme showed an absolute requirement for adenosylcobalamin. Completeness of resolution was indicated by the observation that the specific activity of the apoenzyme in the presence of excess cofactor was the same as the specific activity of the holoenzyme prior to acid-ammonium sulfate treatment, a finding which implies that there is no inhibition of the resolved mutase by cofactor analogs (e.g., OH-Cbl) bound to the active site. The cobamide, however, remained associated with the enzyme in the form of the hydroxo derivative, bound presumably at locations remote from the active site. The sheep kidney mutase was resolved by treatment with charcoal following exposure of the enzyme to low pH; charcoal treatment without prior acidification was ineffective (26).

The purified sheep liver enzyme (22) was an orange protein of molecular weight 165,000 (ultracentrifugation). The enzyme possessed one cobamide-binding active site per 75,000 daltons. There are no data concerning the cation requirements of the liver enzyme, but no requirement for monovalent or divalent metal ions was seen with the partially purified kidney mutase (27). There was little inhibition of the apoenzyme by free OH-Cbl or CN-Cbl at analog/AdoCbl ratios of 20 : 1 (22). The apoenzyme was very sensitive to inhibition by *p*-hydroxymercuribenzoate and *N*-ethylmaleimide, while the holoenzyme was rather resistant (22). The binding of cobamides to the active site of sheep liver mutase thus appears to protect the enzyme against sulfhydryl reagents. A similar situation is seen with several other adenosylcobamide-dependent enzymes (see below). In none of these cases is it known whether cobamide blocks access of sulfhydryl reagents to an –SH group at the cobamide binding site, or whether the protection is the result of a cobamide-induced conformational change leading to burial of sulfhydryl groups exposed in the apoenzyme.

α-METHYLENEGLUTARATE MUTASE

Methyleneglutarate mutase catalyzes the reversible interconversion between α-methyleneglutarate and methylitaconate (29–31). The reaction presumably involves the migration of an acrylyl residue from one carbon to an adjacent one (4-6), although there is at present no direct evidence

$$H_2C{=}C(COOH)CH_2CH_2COOH \rightleftharpoons H_2C{=}C(COOH){-}CH(CH_3){-}COOH \qquad (4\text{–}6)$$

pertaining to this point. The mutase appears to require no cofactor other than an adenosylcobamide; in particular, neither a monovalent nor a divalent cation is necessary for catalysis. The enzyme is found in a species of Clostridium which grows on nicotinic acid. Both α-methyleneglutarate and methylitaconate are intermediates in the pathway by which the organism converts nicotinic acid to propionic acid, acetic acid, and CO_2, the end products of the fermentation (32).

A proposal that the rearrangement might involve an intermediate cyclopropyl compound led to an investigation of the effects of the enzyme on the postulated cyclic intermediates. These investigations showed that neither 1-methyl-1,2-*cis*- (I), nor 1-methyl-1,2-*trans*-cyclopropanedicarboxylate (II), nor 1,2-cyclobutanedicarboxylate (III) (stereochemistry not specified; see 4-7) was a substrate of the enzyme. The cyclopropane acids

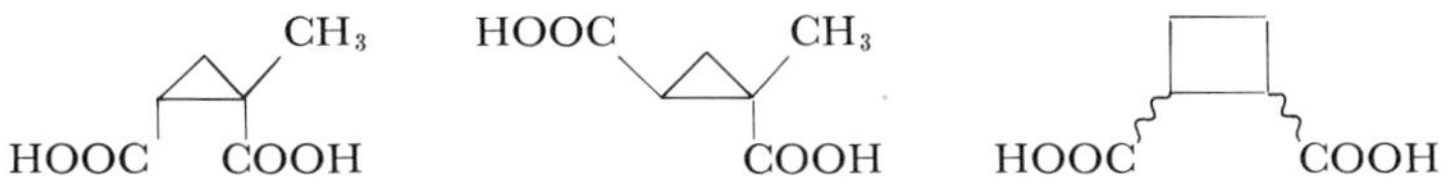

CH_3 HOOC CH_3

HOOC COOH COOH HOOC COOH (4–7)

I II III

were inhibitors, however, as were several other substrate analogs, including L-malate, succinate, itaconate, mesaconate, and glutaconate (32).

Methyleneglutarate mutase was partially purified from crude extracts (32), but it was not possible to separate the enzyme completely from methylitaconate isomerase. The latter enzyme catalyzes the reversible isomerization of methylitaconate to dimethylmaleic acid, the next step in the pathway by which nicotinic acid is degraded. Gel filtration showed that the molecular weight of the mutase was 170,000. Cobamides seem to bind less avidly to the mutase than to most of the other enzymes of this class, since passage of the enzyme over a DEAE column during the course of purification resolved it completely of the bound cobamide which was present in less pure preparations. This enzyme too appears to possess an essential sulfhydryl group, since it was inhibited by iodoacetic acid. Arsenite, however, had no effect, suggesting that vicinal sulfhydryl groups were not required for activity.

Enzymes Catalyzing the Production of Aldehydes from Diols or Amino Alcohols

DIOL DEHYDRASE

Diol dehydrase was one of the first adenosylcobalamin-dependent enzymes to be discovered, and the first shown to catalyze the formation of an aldehyde. The first indication of the existence of such an enzyme was the finding that crude extracts from a strain of *Aerobacter aerogenes* were able to catalyze the adenosylcobamide-dependent conversion of ethylene glycol, propylene glycol, and glycerol to acetaldehyde, propionaldehyde, and β-hydroxypropionaldehyde, respectively (33). Purification subsequently showed that the first two of these subtrates—ethylene glycol and propylene glycol—were acted on by a single enzyme, diol dehydrase (34). Since its discovery, diol dehydrase has come to be one of the most extensively studied of all the adenosylcobamide-dependent enzymes.

Early observations made with the purified enzyme concerned the time course of the reaction (34). With both ethylene glycol and propylene glycol, an initial lag was noted at low concentrations of AdoCbl. The lag was abolished by preincubation of the enzyme with the cofactor. Beyond the lag period the reaction rate with propylene glycol was constant. With ethylene glycol, however, the reaction rate decreased with time, until by 20 min the conversion of ethylene glycol to acetaldehyde had ceased. Cessation of catalysis was accompanied by destruction of the enzyme-bound cofactor, a process which has been found to be quite characteristic of adenosylcobamide-dependent enzymes, and which is discussed in detail below.

The stereochemistry of the diol dehydrase-catalyzed dehydration of propylene glycol has been completely defined. In these stereochemical studies, the remarkable observation was made that diol dehydrase will accept either (R)- or (S)-propanediol as substrate (35–37). With stereospecifically deuterated substrates it was shown that the conversion of propylene glycol to propionaldehyde is associated with the migration of hydrogen from C-1 of substrate to C-2 of product (35–37). With (S)-propane-1,2-diol the *pro*-(S)-C-1 hydrogen migrates, while in the case of the (R) substrate the migrating atom is the *pro*-(R) hydrogen. In both cases, the configuration of the carbon which receives the migrating hydrogen atom undergoes inversion during the course of the reaction (4-8).

$$\begin{array}{c} OH \\ | \\ HC\mathbf{H} \\ | \\ HC{-}OH \\ | \\ CH_3 \end{array} \rightarrow \begin{array}{c} CHO \\ | \\ \mathbf{H}CH \\ | \\ CH_3 \end{array} \qquad (4\text{–}8)$$

Thus the reactions with each of the two enantiomeric substrates are, in a sense, themselves enantiomeric.

The adenosylcobamide-dependent reactions hitherto discussed are all rearrangements in which a hydrogen atom changes places with another group on the substrate molecule. Even though the conversion of a glycol to an aldehyde is formally a dehydration reaction, the peculiar stereochemical indifference of diol dehydrase permitted the demonstration that the propanediol → propionaldehyde reaction involved an exchange of places between a C-1 hydrogen atom and the C-2 hydroxyl group. The experimental observations were as follows: (1) propionaldehyde produced from (S)-1-[^{18}O]propane-1,2-diol retained all the isotopically labeled oxygen originally present in the substrate, while the propionaldehyde produced from (R)-1-[^{18}O]propane-1,2-diol contained only unlabeled oxygen (4-9) (38); (2) propionaldehyde from (RS)-2-[^{18}O] propane-1,2-diol retained half the ^{18}O initially present. These findings were interpreted to indicate a reaction mechanism whereby the migration of oxygen from C-2 to C-1 to form propionaldehyde hydrate is followed by a stereospecific (i.e., enzyme-catalyzed) elimination reaction yielding propionaldehyde and water.

$$
\begin{array}{lcccc}
 & \begin{array}{c} {}^{18}OH \\ | \\ HCH \\ | \\ HC{-}OH \\ | \\ CH_3 \end{array} & \rightarrow & \begin{array}{c} H \\ HO{-}C{-}{}^{18}OH \\ | \\ CH_2 \\ | \\ CH_3 \end{array} & \xrightarrow{H_2{}^{18}O \uparrow} \begin{array}{c} CHO \\ | \\ CH_2 \\ | \\ CH_3 \end{array} \\
(R) & & & & \\
 & \begin{array}{c} {}^{18}OH \\ | \\ HCH \\ | \\ HO{-}CH \\ | \\ CH_3 \end{array} & \rightarrow & \begin{array}{c} H \\ H^{18}O{-}C{-}OH \\ | \\ CH_2 \\ | \\ CH_3 \end{array} & \xrightarrow{H_2O \uparrow} \begin{array}{c} CH^{18}O \\ | \\ CH_2 \\ | \\ CH_3 \end{array} \\
(S) & & & &
\end{array}
\qquad (4\text{-}9)
$$

In addition to an adenosylcobamide, diol dehydrase requires a monovalent cation for activity (34, 39, 40). K^+ is the ion usually employed, but other cations, including Na^+, NH_4^+, Rb^+, Cs^+, and Tl^+, are also suitable. Li^+ is inactive. The effectiveness of a given ion in the activation of the diol dehydrase reaction has been correlated with its radius, maximal effectiveness occurring with ions whose radii are in the vicinity of 1.4Å (K^+, Rb^+, and NH_4^+) (Figure 4-6).

The requirement for monovalent cations in the diol dehydrase reaction has recently been explained by observations showing that the complex between the enzyme and the cobamide dissociates in the absence

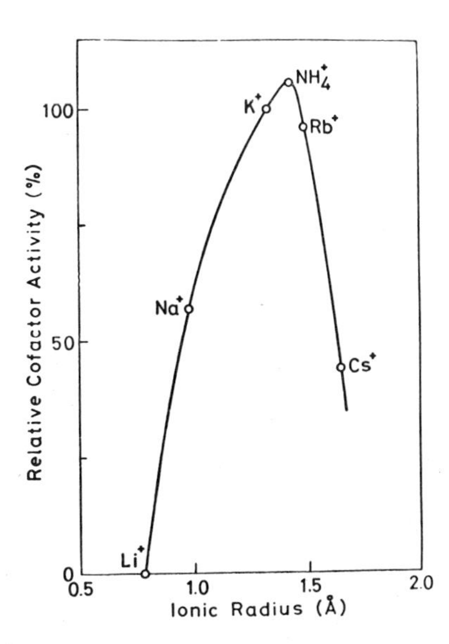

Fig. 4-6. Relation between ionic radii and relative cofactor activities of monovalent cations in the diol dehydrase system. [Reprinted with permission from Toraya, T., Sugimoto, Y., Tamao, Y., Shimizu, S., and Fukui, S. (1971). *Biochemistry* **10**, 3475–3484. Copyright by the American Chemical Society.]

of an appropriate cation (40–42). Gel filtration of the enzyme-adenosylcobalamin complex in the absence of K+ or other suitable ion led to the complete resolution of the complex into apoenzyme and cofactor, while in the presence of K+ resolution proceeded to only a limited extent. If substrate (ethylene glycol) was present in the eluting buffer in addition to K+, the complex passed through the column intact. Other cations besides K+ also stabilized the enzyme-adenosylcobalamin complex, the degree of stabilization correlating with the effectiveness of the ion as a replacement for K+ in the catalytic reaction. K+ also stabilized the complex between enzyme and cobalamin derivatives which are not coenzymatically active (40, 41, 43). These derivatives, including CN-Cbl, MeCbl, and OH-Cbl, are powerful inhibitors of the enzyme (34). Gel filtration of the enzyme • CN-Cbl and enzyme • MeCbl complexes in the absence of a suitable cation yielded apoenzyme free of cobalamin. The enzyme • OH-Cbl complex was not dissociated by this procedure, but the binding of OH-Cbl to diol dehydrase appears to differ from that of the other analogs inasmuch as there is a great deal of nonspecific binding of OH-Cbl to diol dehydrase, a situation not seen with other cobalamin derivatives, which generally bind to the enzyme in a 1 : 1 molar ratio. Diol dehydrase was successfully resolved of OH-Cbl by techniques involving *inter alia* the prior conversion of the enzyme-bound OH-Cbl to other cobalamin derivatives (34).

Diol dehydrase is a protein of molecular weight 250,000. The active enzyme possesses one active site per molecule (44). By chromatography on DEAE-cellulose, it can be dissociated into two types of nonidentical subunits which, though inactive by themselves, can be combined to reconstitute active diol dehydrase (45).

Sulfhydryl reagents inhibit diol dehydrase in a manner similar to that seen with sheep liver methylmalonyl CoA mutase (34, 42, 43). The apoenzyme is very sensitive to inhibition by organic mercurials such as *p*-chloromercuribenzoate and *p*-hydroxymercuribenzoate, but protection against these agents is conferred by the binding of certain cobamides to the active site. Cobamides affording protection include CN-Cbl, MeCbl, and AdoCbl, but not OH-Cbl. This is another illustration of the difference between the binding of OH-Cbl and other cobamides to diol dehydrase. Protection by AdoCbl is greatly augmented in the presence of propanediol, but propanediol alone has no effect on the inhibition of enzyme by organic mercurials. The combination of AdoCbl and propanediol also protects the enzyme against inactivation by *N*-ethylmaleimide and iodoacetate, compounds which alkylate sulfhydryl groups. In contrast to its failure to protect against organic mercurials, propanediol alone does provide some protection against these alkylating agents. The essential sulfhydryl group appears to reside specifically on one of the two types of subunit, since incubation with iodoacetamide, a sulfhydryl reagent, destroys the ability of this subunit to form active diol dehydrase upon combination with the complementary subunit (45).

ETHANOLAMINE AMMONIA-LYASE

Ethanolamine ammonia-lyase, discovered in an unclassified Clostridium by Bradbeer (46) and later purified to homogeneity by Kaplan and Stadtman (47, 48), catalyzes the conversion of vicinal amino alcohols to aldehydes and ammonia. Originally it was thought that ethanolamine was the only substrate for the enzyme, the reaction products being acetaldehyde and ammonia.

$$H_2NCH_2—CH_2OH \rightarrow CH_3CHO + NH_3$$

More recently it has been found that the enzyme also catalyzes the deamination of L-2-aminopropanol to propionaldehyde and ammonia (49, 50). The latter reaction, however, is accompanied by a side reaction leading to irreversible destruction of the cofactor. Whether the enzyme can also act on the D isomer is not known.

In the enzyme-catalyzed deamination of ethanolamine, the carbinol carbon of the substrate becomes the carbonyl carbon of the product (48, 51). Experiments in $[^{18}O]H_2O$ showed that alcoholic oxygen atom is retained in the product, excluding mechanisms involving the formation of an intermediate imine which hydrolyzes to product (51). As usual, migration of a hydrogen atom takes place, the atom moving from the carbinol carbon of the substrate to the methyl carbon of acetaldehyde. In the transfer of hydrogen from the carbinol carbon, the enzyme distinguishes between the *pro*-(R) and *pro*-(S) atom, stereospecifically selecting one of the two to migrate and leaving the other behind (it is not known which atom migrates) (51).

Stereoselectivity in the formation of the methyl group was recently investigated (52). The hydrogens of this group may be distinguished from each other in structural formulas by assigning a label to each atom (4-10). Similarly, they can be distinguished experimentally by means of

```
     R
     |
H—C—T          (4–10)
     |
     D
```

isotopic labeling. A methyl group which carries one hydrogen, one deuterium, and one tritium exists as two experimentally distinguishable enantiomers, and experiments based on this principle can be conducted to determine changes in the configuration of a methyl carbon atom during the course of a reaction. Methyl groups carrying the three different hydrogen isotopes were formed in the ethanolamine ammonia-lyase reaction by incubating enzyme and adenosylcobalamin with each of the two enantiomers of the doubly labeled substrate 2-amino[2-D,2-T]ethanol. It was expected that since enzymatic reactions are highly stereospecific and only one of the enantiomers of substrate was present in each incubation mixture, only one of the two possible enantiomeric methyl groups would be produced. Analysis revealed, however, that both enantiomeric methyl groups appeared in about equal amounts. These results indicate that the amino carbon atom of the substrate racemizes during the course of the reaction.

There is little evidence concerning the route of ammonia loss. Consideration of the mechanisms of other adenosylcobamide-dependent rearrangements supports the notion that the reaction involves the migration of ammonia to the carbinol carbon atom to produce 1-aminoethanol, followed by its elimination to form the final products.
The ammonia requirement in an ethanolamine ammonia-lyase-catalyzed exchange of hydrogen between acetaldehyde and adenosylcobalamin (53) is consistent with such a mechanism, but constitutes only weak evidence in its favor. The major evidence supporting ammonia migration is analogy, since whenever studied, it has been found that adenosylcobalamin-dependent rearrangements occur by group migration.

The course of the ethanolamine ammonia-lyase reaction is shown in formula 4-11. The evidence to date indicated that it is very similar to the diol dehydrase reaction discussed above.

$$\underset{\substack{|\\ NH_2}}{CH_2}{-}\overset{H}{\underset{|\\ H}{COH}} \rightarrow \underset{|\\ H}{CH_2}{-}\overset{\substack{NH_2\\ |}}{\underset{H}{COH}} \xrightarrow{NH_3} CH_3CHO \qquad (4\text{–}11)$$

Ethanolamine ammonia-lyase is a very large enzyme, the molecular weight being estimated by ultracentrifugation as 520,000, but it dissociates into subunits of about 50,000 molecular weight in 5 *M* guanidine (48). Titration and kinetic experiments have shown two independent active sites per 520,000 daltons (54–56). Like diol dehydrase, ethanolamine ammonia-lyase requires a monovalent cation for activity (47). K^+ and NH_4^+ are most effective, but activity is also seen with Rb^+. Na^+ and Li^+, on the other hand, show no activity of their own, but inhibit the enzyme by competing with the required cations. The enzyme is inhibited by certain sulfhydryl reagents, including pCPMS, but appears to be relatively resistant to others (47).

Ethanolamine ammonia-lyase is also inhibited by cobalamin derivatives which lack cofactor activity. The characteristics of inhibition by cobalamin derivatives were studied in some detail (55). The deamination of ethanolamine in a reaction mixture containing AdoCbl together with an inhibitory analog such as MeCbl was initially rapid, but the rate of deamination decreased over a minute or two to a final reaction velocity much lower than the rate initially observed (Figure 4-7). Measurements of the initial velocity at several concentrations of analog showed that at the start of the reaction the analog behaved like a competitive inhibitor of AdoCbl. Similar measurements made after the final velocity had been attained indicated that at this time too, the analog was a competitive inhibitor of the cofactor. Determinations of K_i for the analog during the two stages of the reaction (initially, and after the final velocity had been attained) indicated that the analog was much more tightly bound to the enzyme in the final stage of the reaction than in the initial stage. Moreover, the V_{max} in the final stage was only one-quarter the V_{max} based on measurements of initial velocities. The findings were interpreted in terms of a mechanism whereby the binding of the inhibitory cobalamin induced a conformational change which produced an enzyme form with a lower V_{max} and a much higher affinity for the cobalamin analog than the original enzyme (4-12). The possibility cannot be ruled out,

$$\begin{array}{ccc} E\cdot Cbl & \longrightarrow & E'\cdot Cbl \\ \rightleftharpoons & & \rightleftharpoons \\ E & & E' \\ \rightleftharpoons & & \rightleftharpoons \\ E\cdot AdoCbl & & E'\cdot AdoCbl \end{array} \qquad (4\text{–}12)$$

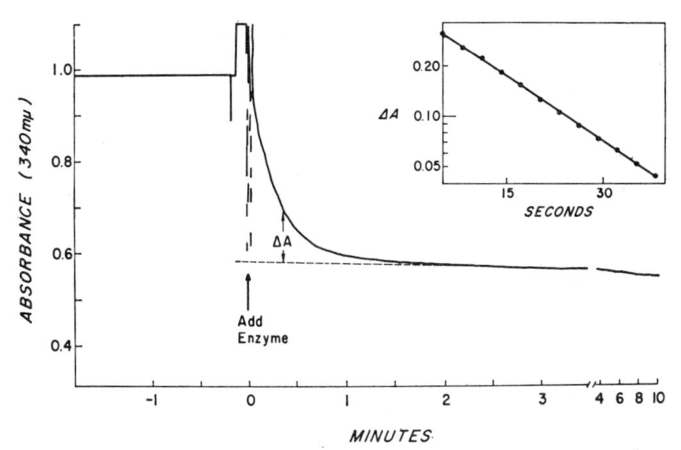

Fig. 4-7. Time course of the ethanolamine ammonia-lyase reaction in the presence of MeCbl (55).

however, that the enzyme preparation contains two forms of ethanolamine ammonia-lyase which differ in their susceptibilities to inhibition by cobalamin derivatives.

GLYCEROL DEHYDRASE

Glycerol dehydrase, an enzyme present in a strain of *Aerobacter aerogenes* and in a Lactobacillus, catalyzes the conversion of glycerol to β-hydroxypropionaldehyde (4-13). The presence of the enzyme in Lactobacillus

$$\begin{array}{c} CH_2OH \\ | \\ CHOH \\ | \\ CH_2OH \end{array} \xrightarrow{\nearrow H_2O} \begin{array}{c} CHO \\ | \\ CH_2 \\ | \\ CH_2OH \end{array} \qquad (4\text{–}13)$$

was of considerable interest to the brewing industry, since during distillation the hydroxyaldehyde produced in a fermentation broth containing this organism underwent dehydration to acrolein, an irritant which imparted characteristically unpleasant qualities to the otherwise highly desirable distillate (57).

As with diol dehydrase and ethanolamine ammonia-lyase, the activity of glycerol dehydrase depends on a monovalent cation as well as an adenosylcobamide (58, 59). The physiological ion is probably K^+, but the enzyme is also active with Li^+, NH_4^+, and Rb^+.

Purification of glycerol dehydrase from *A. aerogenes* has revealed the enzyme to be a 1 : 1 complex of two nonidentical subunits (60, 61). The smaller of the two subunits (MW 22,000) is itself composed of two subunits of approximately 12,000 molecular weight. The molecular weight of the entire complex is 188,000. Association of the subunits is promoted by glycerol, by cobamides, and by cations such as K^+ which are required for enzymatic activity, but is inhibited by Na^+.

Although neither subunit is capable of binding cobamides by itself, cobamides are bound to the complete enzyme at a stoichiometry of 1 mole/mole of enzyme (61, 62). The enzyme will take up corrins of varying structure, with the consequence that those not possessing cofactor activity (for example, MeCbl and OH-Cbl) act as inhibitors (57, 63). The enzyme • AdoCbl and enzyme • MeCbl complexes are resolved by gel filtration when eluted with water, but not when potassium phosphate buffer is used to elute the complexes from the column (61, 64); this finding was the first indication that monovalent cations are involved in the binding of cofactor by adenosylcobamide-dependent enzymes. The enzyme • OH-Cbl complex, on the other hand, is stable to elution with water. It is possible, however, to remove OH-Cbl from the enzyme by exchange with AdoCbl, provided Mg^{2+} and SO_3^{2-} are present during the exchange reaction (62). This procedure resembles those used to remove OH-Cbl from diol dehydrase, in that the exchange reaction probably takes place after the conversion of enzyme-bound OH-Cbl to another cobalamin derivative (in this case, sulfitocobalamin). It is not likely that SO_3^{2-} merely reduces OH-Cbl to cob (II)alamin, since BH_4^-, a substance which converts OH-Cbl to cob(II)alamin, could not replace SO_3^{2-} in the exchange reaction.

The binding of OH-Cbl to glycerol dehydrase is accompanied by a marked change in the cobalamin spectrum (Figure 4-8) (62). This is the only enzyme for which such a spectral change has been reported, although spectral changes following the addition of substrate or a substrate analog to enzyme • *AdoCbl* complexes are well known (see below). It is likely that this spectral change reflects a change in the coordination sphere of the cobalt atom whereby one or both of the axial ligands are replaced by groups on the protein.

The complex between enzyme and adenosylcobalamin undergoes rapid inactivation, a process accompanied by the cleavage of the coenzyme at the carbon-cobalt bond (61, 64). Inactivation takes place both in the presence and absence of substrate. Enzyme-bound OH-Cbl has been identified spectroscopically as one of the cleavage products; what becomes of the adenosyl fragment is not known.

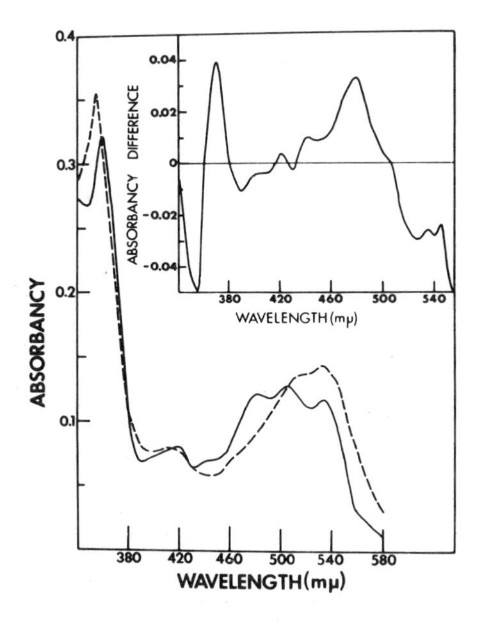

Fig. 4-8. Comparison of the spectrum of free OH-Cbl (- - - -) to that of OH-Cbl bound to glycerol dehydrase (—) (62).

Glycerol dehydrase is inhibited by the sulfhydryl reagents pCMB, phenylmercuric acetate, and *N*-ethylmaleimide, the apoenzyme being much more sensitive than either the holoenzyme (i.e., the enzyme • AdoCbl complex) or the enzyme • OH-Cbl complex (57–59, 61, 62). The subunits are also affected by sulfhydryl reagents, but to very different extents. While the small subunit still displays 25% of its original activity after exposure to *N*-ethylmaleimide, the activity of the large subunit is totally abolished by the same treatment (61).

Aminomutases

The aminomutases are a group of enzymes found in certain microorganisms which grow on lysine. The enzymes catalyze the reversible migration of an amino group from a terminal to an adjacent carbon atom—that is, they interconvert a primary and a secondary amine. Three of these enzyme have so far been described: L-β-lysine mutase, D-α-lysine mutase, and D-ornithine mutase. They catalyze the following reactions:

$$H_2NCH_2CH_2CH_2-\underset{H}{\overset{NH_2}{\overset{|}{C}}}-CH_2COOH \xrightarrow[\text{L-}\beta\text{-lysine mutase}]{} CH_3-\underset{H}{\overset{NH_2}{\overset{|}{C}}}-CH_2-\underset{H}{\overset{NH_2}{\overset{|}{C}}}-CH_2COOH$$

L-β-Lysine → L-Erythro-3,5-diaminohexanoic acid (65–68)

$$H_2NCH_2CH_2CH_2CH_2-\underset{NH_2}{\underset{|}{\overset{H}{C}}}-COOH \xrightarrow[\text{D-}\alpha\text{-lysine mutase}]{} CH_3-CHNH_2-CH_2CH_2-\underset{NH_2}{\underset{|}{\overset{H}{C}}}-COOH \qquad (4\text{–}14)$$

D-α-lysine → D-2,5-Diaminohexanoic acid (69,70) (The configuration of the 4-amino group is not established.)

$$H_2NCH_2CH_2CH_2-\underset{NH_2}{\underset{|}{\overset{H}{C}}}-COOH \xrightarrow[\text{ornithine mutase}]{} CH_3-\underset{H}{\overset{NH_2}{\overset{|}{C}}}-CH_2-\underset{NH_2}{\underset{|}{\overset{H}{C}}}-COOH$$

D-Ornithine → D-Threo-2,4-diaminovaleric acid (71–74)

The enzymes are highly specific for their own substrates, and are inhibited by closely related substrate analogs, including the substrates of the other amino mutases (74, 75).

All three enzymes are present in lysine-grown *Clostridium sticklandii* (65, 68, 70–73, 76–79). In addition, L-β-lysine mutase has been found in a strain of Clostridium denoted SB-4 (65, 67, 80), while another strain of Clostridium, a soil-dwelling organism called Clostridium M-E (68, 76, 79, 81), presumably contains one or more of these amino mutases, since crude extracts from this organism carry out an adenosylcobalamin-requiring fermentation of lysine. In all these organisms, the amino mutase reaction, an early step in the fermentation of α-ω-diamino acids [the first step, with D-lysine and ornithine; the second with L-lysine, the mutase reaction being preceded by a pyridoxal-dependent reaction which converts L-lysine into L-3,6-diaminohexanoic acid—i.e., L-β-lysine (80)], is followed by a pyridine nucleotide-dependent oxidation of one of the amino groups to produce an aminoketoacid (65, 68, 72, 76,79–81). It is presumed that the subsequent reactions involve thiolytic cleavage of the aminoketoacid by coenzyme A followed by appropriate oxidation-reduction reactions (65, 68, 76, 79, 81) to yield acetate and butyrate (or in the case of ornithine, acetate and CO_2), the final products of the fermentation.

All three amino mutases have been purified from *C. sticklandii*. D-α-Lysine mutase is a cobamide-containing protein of molecular weight 250,000 (75), while L-β-lysine mutase, also isolated with bound cobamide, is somewhat smaller, its molecular weight having been determined as 160,000 (82). L-β-Lysine mutase has been further separated into subunits of molecular weight 32,000 and 52,000 (78), suggesting that this enzyme is a tetramer composed of two pairs of nonidentical subunits. For full activity, each mutase requires a colorless protein of molecular weight about 60,000 (75, 82). While the mutases themselves are specific for each of the substrates, there is evidence that the small proteins isolated with each of the two mutases are interchangeable (75). Destruction of the activity of the small proteins by incubation with iodoacetamide indicates that each of them contains an essential sulfhydryl group (75, 82). The mutases themselves, however, are relatively resistant to sulfhydryl reagents.

Ornithine mutase also carries bound cobamide with it during purification. The purified ornithine mutase (74), an enzyme of molecular weight 170,000, can be dissociated into two subunits of molecular weight about 90,000. This enzyme requires no additional sulfhydryl protein for activity, but appears itself to contain an essential sulfhydryl group, since it is sensitive to pCMB, *N*-ethylmaleimide, and iodoacetate.

The binding of AdoCbl by all three enzymes is very firm. Like several other AdoCbl-dependent enzymes, the lysine amino mutases are partially active without added cofactor (75, 82), indicating the presence of adenosylcobamide at the active site of the purified enzyme. All three enzymes are strongly inhibited by intrinsic factor (67, 71, 72, 79, 81, 82), confirming the presence of cofactor which is not separated from the enzymes during purification.

In addition to adenosylcobamide, two of the enzymes require metal ions for activity (75, 78). Both Mg^{2+} and a monovalent cation are required by each of the lysine mutases. D-α-Lysine mutase is active in the presence of K^+, Rb^+, and NH_4^+, but is inhibited by Na^+ and Li^+. L-β-Lysine mutase, on the other hand, is active in the presence of Li^+, Na^+, K^+, Rb^+, Cs^+, and NH_4^+, although the level of activity varies from ion to ion. Ornithine mutase appears to require neither a monovalent nor a divalent cation (74).

D-α-Lysine mutase also requires ATP for full activity (75). This requirement is not shared by either of the other purified amino mutases, although ATP is required for the fermentation of each of the three substrates by crude extracts of *C. sticklandii.* ATP is not consumed during the rearrangement of D-α-lysine, and in fact can be replaced by either AMPPCP or AMPCPP. These observations show that ATP acts as an allosteric activator of D-α-lysine mutase. The presence of the nucleotide causes an increase in both V_{max} and the affinity of the enzyme for the substrate.

Finally, each of the mutases requires pyridoxal phosphate for activity (66, 74, 75, 78). This finding is of great interest with respect to the mechanism of action of these enzymes, since it invites the speculation that the amino group migrates not as ammonia but as pyridoxamine or pyridoxaldimine (4-15). Supporting this speculation is the observation

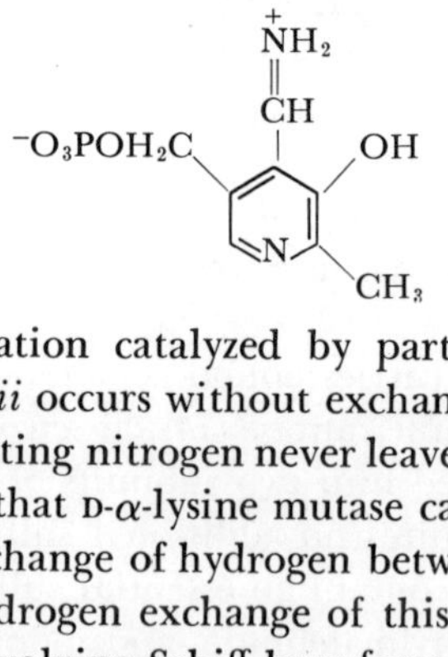

$\overset{+}{N}H_2$ ‖ CH; $^-O_3POH_2C$, OH, N, CH_3 (4–15)

that the ammonia migration catalyzed by partially purified L-β-lysine mutase from *C. sticklandii* occurs without exchange with free NH_4^+ (83), indicating that the migrating nitrogen never leaves the enzyme. Even more suggestive is the finding that D-α-lysine mutase catalyzes a slow pyridoxal phosphate-dependent exchange of hydrogen between the solvent and C-6 of the substrate (69). Hydrogen exchange of this sort is very characteristic of certain reactions involving Schiff base formation between pyridoxal and an amine (see 4-16).

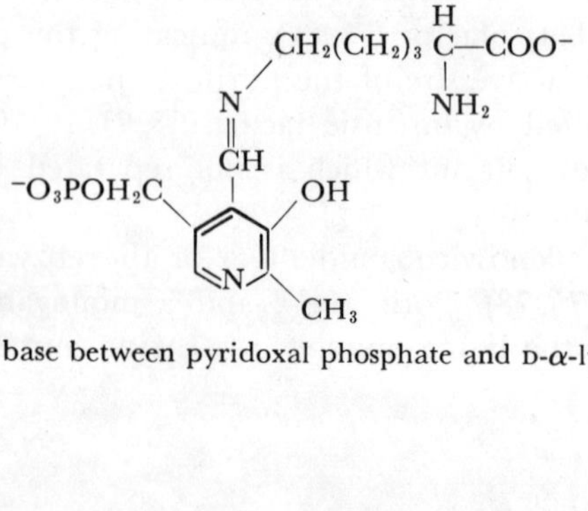

H; $CH_2(CH_2)_3$ C—COO^-; NH_2; N ‖ CH; $^-O_3POH_2C$, OH, N, CH_3 (4–16)

Schiff base between pyridoxal phosphate and D-α-lysine.

Purified ornithine mutase is not active unless pyridoxal phosphate is added to the reaction mixture (74). With D-α-lysine mutase, however, the requirement for pyridoxal phosphate only appears after treatment of the enzyme with hydroxylamine (66). These findings suggest that pyridoxal phosphate is an integral part of the active site of D-α-lysine mutase. It has been proposed that pyridoxal phosphate is attached to the enzyme through a Schiff base linkage involving the ϵ-amino group of a lysine residue on the protein. This hypothesis could account for the appearance of a pyridoxal requirement following hydroxylamine treatment, since hydroxylamine would be expected to disrupt the Schiff base linkage and inactivate the enzyme-bound pyridoxal.

Pyridoxal phosphate is not necessary for L-β-lysine mutase activity in crude extracts or partially purified preparations (65, 67, 76, 79–81). With these systems, though, the conversion of L-β-lysine to 3,5-diaminohexanoic acid requires the presence of a ketoacid. When the enzyme from *C. sticklandii* was purified to homogeneity, the requirement for a ketoacid vanished, but a pyridoxal phosphate requirement appeared (78). Further investigation revealed that pyridoxal phosphate took part, together with AdoCbl, Mg^{2+}, and dithiothreitol, in a process whereby the enzyme, inactive in the form in which it was obtained after purification, was converted to a catalytically active species. Once formed, the catalytically active species was able to convert L-β-lysine to 3,5-diaminohexanoate in the absence of free AdoCbl, Mg^{2+}, and dithiothreitol. The only (free) cofactor which was necessary was monovalent cation. During the course of catalysis, however, the enzyme rapidly lost activity. The inactivation of the enzyme, a process which required the presence of 3,5-diaminohexanoate and was stimulated by oxygen, was associated with the cleavage of the enzyme-bound AdoCbl at the carbon-cobalt bond. The adenosyl group was converted to 5′-deoxyadenosine; the corrinoid product was not identified.

The *initial rate* of the reaction catalyzed by activated L-β-lysine mutase was the same whether or not the sulfhydryl protein was present in the reaction mixture (78). However, the rate of inactivation of the mutase was greatly accelerated in the absence of the sulfhydryl protein. It was proposed that this protein acts by potentiating the binding of AdoCbl to the mutase. The mechanism of action of adenosylcobamide-dependent enzymes is thought to involve cleavage of the carbon-cobalt bond as one of the catalytic steps (see below). Inactivation of mutase was postulated to result from dissociation of one of the fragments produced by the cleavage of the cofactor. By retarding this dissociation, the sulfhydryl protein would reduce the rate of inactivation of L-β-lysine mutase.

Summary of the Properties of the Enzymes

Some of the properties of the enzymes catalyzing adenosylcobamide-dependent rearrangements are summarized in Table 4-1.

The Binding of Cobamides to Rearrangement-Catalyzing Enzymes: Structural Requirements for Cofactor Activity

Because a large number of corrinoids of diverse structure are available for experimental use, it has been possible to carry out rather extensive studies of the relationship between corrinoid structure and the ability of the corrinoid to bind and to serve as cofactor with various of the enzymes discussed above (18, 22, 24–26, 32–34, 41, 43, 54, 55, 57–59, 63, 72, 82, 84–102). Corrinoids which vary in the nature of the Coα and Coβ ligands, in the number and identity of the carboxyl substituents at the periphery of the ring, and in the nature of the groups on the corrin ring itself, have all been investigated. From the data gathered in these studies it has been possible to come to certain conclusions regarding the forces which hold the corrinoid to the active site of the enzyme, and to specify certain structural features which are necessary for a corrinoid to act as a cofactor.

Corrinoids which bind to the active sites of these enzymes are attached very firmly, and (under the proper conditions) are released very slowly, if at all. As a result, several of the cobamide-dependent enzymes carry cobamides with them through all stages of the purification (22, 47, 62, 74, 75, 78, 82). For this reason the pure enzymes may show only a partial requirement for AdoCbl (22, 75, 82) even after charcoal treatment to remove bound cofactor (22, 82). With such enzymes, absolute dependence on AdoCbl may be shown by total inhibition of the enzyme by intrinsic factor which is reversed by the addition of AdoCbl, or by the appearance of an absolute requirement for AdoCbl after the enzyme is subjected to a specific and sometimes rather rigorous procedure to resolve it of its bound cofactor.

Table 4-2 lists the corrinoids which have been tested in enzymatic systems, classified according to whether they are active as cofactors, active as inhibitors, or inactive. The K_m for the cofactors and (where investigated) the K_i for the inhibitors have all been found to be of the order of 1 μM or less. These figures indicate that both cofactors and inhibitors bind tightly to the enzymes, the inhibitors binding somewhat more tightly than the cofactors in those systems in which the comparison has been made.

Evidence for the low rate of dissociation of cofactor from the active site of the enzyme has been obtained using AdoCbl labeled with tritium in the C-5′ of the *Coβ*-adenosyl group. Tritium in this location is transferred to the product during the course of adenosylcobalamin-dependent rearrangements. (The mechanistic implications of this finding are discussed below.) With diol dehydrase, ethanolamine ammonia-lyase, and D-α-lysine mutase (39, 69, 103), it was found that when the concentration of [^{3}H]AdoCbl exceeded the concentration of active sites, only that fraction of cofactor which had become attached to the enzyme at the beginning of the incubation participated in tritium transfer. If exchange between free and bound cofactor were rapid, it would be expected that all the AdoCbl would participate in tritium transfer, regardless of the molar ratio of AdoCbl to enzyme. These results show, therefore, that with the enzymes mentioned, enzyme-bound AdoCbl does not exchange freely with AdoCbl in solution. With methyleneglutarate mutase, on the other hand, similar experiments suggested that all the AdoCbl takes part in tritium transfer, even when the cofactor is in molar excess over enzyme (30). With this enzyme, then, exchange between free and bound cofactor is facile. This is in accord with the finding that methyleneglutarate mutase is separated from bound cofactor during purification. Even so, the low K_m (0.073 μM for AdoCbl in the methyleneglutarate mutase reaction) indicates that the association between this enzyme and its cofactor is firm in a thermodynamic sense.

A survey of the compounds listed in Table 4-2 shows that the binding of a corrinoid to an adenosylcobalamin-dependent enzyme, as indicated by catalytic or inhibitory activity of the compound in question, is relatively insensitive to the identity of the Coα and Coβ substituents. While the affinity between the corrinoid and the enzyme varies as these substituents are altered, every compound tested was found to bind to enzyme, regardless of the nature of the Coα or Coβ substituent (provided the peripheral carboxyl groups were in the form of amide derivatives; see below). The double-bond system of the ring itself appears to be relatively uninvolved in the binding process, at least with ethanolamine ammonia-lyase, since little spectral change occurs when a corrinoid binds to this enzyme (major distortion of the conformation of the macrocycle upon binding, or significant participation of the conjugated double-bond system of the ring in the binding process, should lead to significant spectral changes upon attachment of the corrinoid to its binding site) (104). With diol dehydrase, similar conclusions can be drawn from experiments with a ring-substituted adenosylcobamide. The H atom in the 10 position of the ring of adenosylcobalamin can be replaced by the bulkier Cl atom with retention of cofactor activity in the diol dehydrase reaction, suggesting only limited interaction of the enzyme with the conjugated double-bond system, at least in the vicinity of C-10 (94, 96). By contrast, substitutions involving the amide groups at the periphery of the ring

Table 4-1 Some properties of enzymes catalyzing

Enzyme	Source	Structural Properties: Subunit Properties[a]	Active Sites	Molecular Weight
Glutamate mutase	*C. tetanomorphum*	α, σ	?	α: 128,000 σ: 17,000
Methylmalonyl CoA mutase	*Propionibacterium shermanii*	α, β	1	α: 61,000 β: 63,000
Methylmalonyl CoA mutase	Sheep liver	?	2	165,000
α-Methylene-glutarate mutase	*Clostridium* sp.	?	?	170,000
Glycerol dehydrase	*Aerobacter aerogenes*	$\alpha_2\beta$	1	α: 12,000 β: 170,000
Diol dehydrase	*A. aerogenes*	α, β	1	α: ? β: ? Total: 250,000
Ethanolamine ammonia-lyase	*Clostridium* sp.	?	2	520,000
L-β-Lysine mutase	*C. sticklandii*	$\alpha_2\beta_2\sigma$	?	α: 32,000 β: 52,000 σ: 60,000
D-α-Lysine mutase	*C. sticklandii*	α, σ	?	Complete, 250,000
Ornithine mutase	*C. sticklandii*	α_2		α: 90,000

[a] σ refers to a sulfhydryl protein which separates from the cobamide binding protein upon purification but which is required for catalytic activity.
[b] Enzyme shows only partial AdoCbl requirement.

Catalytic Properties			
K_m	pH Optimum	K_{eq} for Reaction	Other Cofactor Requirements
(Bza)AdoCba: 0.031 μ*M* L-Glutamate: 1–2 m*M* β-Methylaspartate: 0.5 m*M*	8.5	10.7 favoring glutamate	None
AdoCbl: 0.035 μ*M* Succinyl CoA: 35 μ*M* Methylmalonyl CoA: 80 μ*M*	6.0–8.2, maximum 7.4	10.5 favoring succinyl (with natural isomer)	None
AdoCbl: 0.021 μ*M* Succinyl CoA: 62 μ*M* Methylmalonyl CoA: 240 μ*M*	7.0–8.6	10.5 favoring succinyl (with natural isomer)	None
AdoCbl: 0.073 μ*M* α-Methyleneglutarate: 7.1 m*M*	~7.7	0.23 in α-methyleneglutarate → methylitaconate direction	None
Glycerol: 6 m*M* AdoCbl: 0.036 μ*M*	8.0	Irreversible	K^+
AdoCbl: 0.6 μ*M* Propylene glycol: 0.3 m*M*	6.0–10.0	Irreversible	K^+
AdoCbl: 1.5 μ*M* Ethanolamine: 10 μ*M*	6.5–8.0	Irreversible	K^+
AdoCbl: ca. 2 μ*M* [for activation; see Baker et al. (78)] β-Lysine: 0.3 m*M*	?	Reversible	Pyridoxal phosphate K^+ Mg^{2+}
AdoCbl: ca. 5 μ*M*[b] α-Lysine: 0.2–1.5 m*M*[c]	9.0	1.9 favoring D-α-lysine	Pyridoxal phosphate Mg^{2+} ATP K^+
AdoCbl: 3 μ*M* D-ornithine: 6.7 or 0.44 μ*M*[d]	9.0	~1.0	Pyridoxal phosphate

[c] Varies with ATP concentration.
[d] Depending on assay conditions.

Table 4-2

Cofactor[a]			
AdoCbl	(2,3-OH-4-Ade-Cyclopentyl)methylCbl		(5(6)-CF_3Bza)AdoCba
(Bza)AdoCba	(N^6-MeAdo)Cbl		Iso-AdoCbl
(Ade)AdoCba	(2′MeAde)AdoCba	(2-NH_2Ade)AdoCba	AdoCbl (*e*-ϕNH)
(6-SHPur)AdoCba	(Pur)AdoCba	(5(6)-NO_2Bza)AdoCba	AdoCbl (*e*-MeNH)
2′dAdoCbl	(OH)AdoCbi	(5(6)-NH_2Bza)AdoCba	AdoCbl (*e*-EtNH)
AdoCbl(10-Cl)	3′-dAdoCbl	(5(6)-MeBza)AdoCba	AdoCbl (*e*-2,4-$(NO_2)_2\phi$NH)
Inhibitor[b]			
ThdCbl	4-(9-Ade)BuCbl		(OH)(Bu-2)Cbi
OHCbl	InoCbl		(OH)PeCbi
MeCbl	β-(2-Tetrahydropyryloxy)EtCbl		(OH)OctylCbi
CNCbl	(OH)PrCbi		(OH)DecylCbi
(1-MeAdo)Cbl	(OH)BuCbi		(OH)CarbethoxyCbi
β-OHEtCbl	CydCbl		UrdCbl
(OH)MeCbi	(1-O-Methyl-5-deoxyribos-5-yl)Cbl		(3,5,6-Me_3Bza)AdoCba
(OH)EtCbi	(SO_3^{2-})Cbl		Cbl^{II}
UrdCbl	(OH,SO_3^{2-})Cbi		2′,3′-IsopropylideneAdoCbl
No Activity			
(CN,Aq)Cbi	(OH)AdoCby		(3,4,6-Me_3Bza)CydCba
Cbl$(CH_2)_4$Cbl			AdoCbl (*e*-OH)

[a] Reported to be active with one or more enzymes. Not all are active with every enzyme.
[b] See *Ann. N.Y. Acad. Sci.* **112**, 565 and 703.

have profound effects on the binding of corrins to adenosylcobalamin-specific enzymes. While replacement of one of the amide groups by –NHR or –NHAr does not alter binding to a great extent, the attachment of a corrinoid to the active site of an adenosylcobalamin-dependent enzyme is abolished by conversion of an amide to a carboxylic acid. The evidence available thus points to the peripheral amide groups as the important, or at least among the important, points of attachment of the corrinoid to the enzyme.

To exhibit cofactor activity, a corrinoid must not only be capable of binding to the active site of the enzyme, but in addition must possess an alkyl group of defined structure in the Co*β* position. The alkyl group must be adenosine, attached to the cobalt by the 5′-carbon of the sugar, or a very closely related compound (4-17).

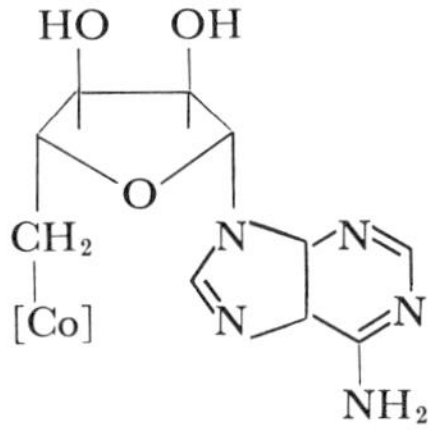

HO OH

O

CH_2 N N

[Co] N N

NH_2

(4–17)

A *Coβ*-adenosyl cobamide.

Replacement of the *Coβ*-ribosyl group by other groups with five-member rings is consistent with the retention of cofactor activity. Thus, 2′-deoxyadenosylcobalamin and 3′-deoxyadenosylcobalamin are both active with glycerol dehydrase (85), and 2′-deoxyadenosylcobalamin is active as well with diol dehydrase and ethanolamine ammonia-lyase (55, 89). With diol dehydrase, activity is seen with an adenosylcobalamin analog in which the oxygen of the furanose ring of the *Coβ*-ribose is replaced by $-CH_2-$ (95). These results, showing that none of the three oxygen atoms of the *Coβ*-ribosyl group is absolutely essential for cofactor activity, suggest that the ribosyl group may function by virtue of its size and its configurational rigidity rather than because of any specific chemical property. It is, for example, conceivable that *Coβ*-ribosyl could be replaced by *Coβ*-cyclopentylmethyl, substituted in the 3 position of the ring by adenine cis to the methyl group (see 4-18), with retention of cofactor activity.

Although a certain amount of structural variability in the ribosyl portion of the alkyl group is consistent with cofactor activity, the heterocycle must be adenine. Substitution on the exocyclic nitrogen atom is permitted, and the adenine may be attached to the sugar through ring

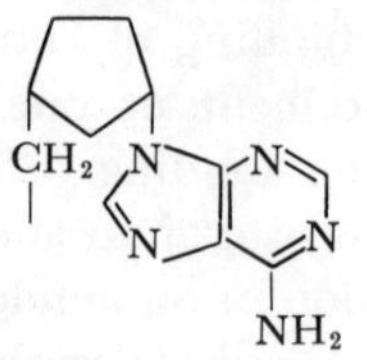

CH_2 N N N N NH_2 (4–18)

Will cobalamin with this Coβ ligand be catalytically active?

nitrogen atoms other than N-9 with retention of cofactor activity (84, 85), but to date no cobamide displaying cofactor activity in the above-listed rearangement reactions has been described which did not possess an adenyl nucleoside in the Coβ position.

From these observations, it can be inferred that enzymes catalyzing adenosylcobamide-dependent rearrangements possess a binding site for the adenyl group. This conclusion is supported by the finding that adenosine is an inhibitor of the diol dehydrase reaction (92). Further evidence favoring this conclusion is the observation that 4-(adenyl-9)butylcobalamin (4-19) is extremely resistant to photolysis when bound to ethanol-

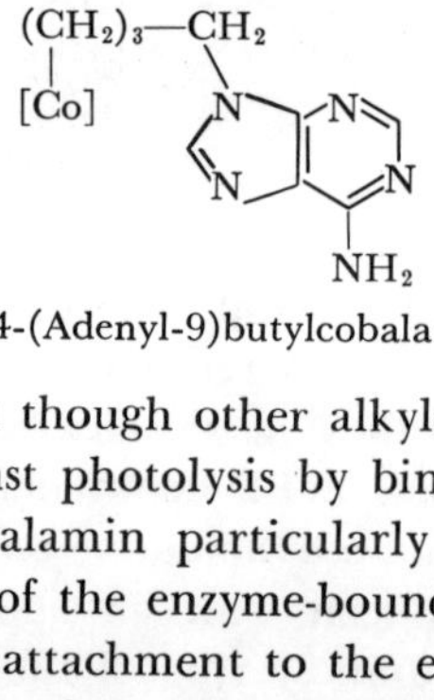

$(CH_2)_3$—CH_2 [Co] N N N N NH_2 (4–19)

Coβ-4-(Adenyl-9)butylcobalamin

amine ammonia-lyase, even though other alkylcobalamin derivatives are not greatly stabilized against photolysis by binding to this enzyme, nor is free 4-(adenyl-9)butylcobalamin particularly resistant to cleavage by light (104). The resistance of the enzyme-bound derivative to photolysis has been attributed to the attachment to the enzyme of both the corrin ring and the adenyl residue of the *Coβ*-alkyl group. A similar stabilization against photolysis is conferred upon adenosylcobalamin by diol dehydrase, provided substrate is present while the reaction mixture is being illuminated (44, 92). Finally, the activities observed with methylmalonyl CoA mutase (22), ornithine mutase, and D-α-lysine mutase (71, 72, 75) in the absence of added cofactor are resistant to photodestruction, indicating that these enzymes too protect adenosylcobamides against photolysis.

It thus appears that enzymes which catalyze adenosylcobalamin-dependent rearrangements bind the cofactor by attachment both to the corrin ring and to the *Coβ*-alkyl group, the latter probably through the adenyl residue. As will be discussed subsequently, the mechanism of catalysis appears to involve cleavage of the cofactor at the carbon-cobalt bond. It is likely that the binding process leads to distortions of the cofactor which facilitate the catalytically essential carbon-cobalt cleavage.

Mechanism of the Rearrangements

The Transfer of Hydrogen

Of the various steps comprising the mechanism of the adenosylcobamide-dependent rearrangements, the best understood are those which involve hydrogen transfer. It has been shown that the cofactor is an intermediary hydrogen carrier, accepting the migrating hydrogen atom from the substrate and transferring it back to the product during the course of the rearrangement. Details of the steps by which hydrogen is transferred between substrate and cofactor have also been elucidated, so that it is now possible to formulate a reasonable sequence of reactions describing this process.

Investigations of the mechanism of hydrogen transfer began with the discovery that in adenosylcobamide-dependent rearrangements, a hydrogen atom is lost from one carbon atom and gained by an adjacent one. It was found that when the rearrangement was carried out in isotopically labeled water, the hydrogen atom participating in the rearrangement did not exchange with solvent protons (18, 23, 30, 51, 74, 105–107). This observation was widely interpreted to indicate an intramolecular shift of hydrogen, the atom moving with both its electrons as a hydride ion. While consistent with the data, this is not the only conclusion which can be drawn from these results. As pointed out by Iodice and Barker (106), failure of exchange with solvent protons would also be seen if hydrogen were transferred with only one electron (a free radical mechanism). For that matter, there are many instances of enzyme-catalyzed reactions in which a hydrogen transferred *as a proton* from one carbon atom of substrate to another fails to exchange with solvent protons, or exchanges only to a limited extent (108). It is therefore difficult to draw firm conclusions regarding the nature of an enzyme-catalyzed hydrogen transfer step on the basis of an observation that the migrating hydrogen does not exchange with the protons of the solvent.

The first unequivocal evidence concerning the mechanism of hydrogen transfer was provided by Zagalak and Abeles (109) working with diol dehydrase. As discussed previously, this enzyme catalyzes the conversion of both ethylene glycol and propylene glycol to the respective aldehydes. In experiments in which unlabeled ethylene glycol and L-[1,1-D_2]propylene glycol were present in the same reaction mixture, these investigators showed that deuterium appeared not only in propionaldehyde, as expected if hydrogen transfer were intramolecular, but in acetaldehyde as well (Table 4-3). Hydrogen transfer was thus shown to be at least in part *inter*molecular. From this observation, it was concluded that the enzyme-cofactor complex served as an intermediate hydrogen carrier, accepting a deuterium atom from a molecule of propylene glycol during the course of one catalytic turnover and then transferring it to another molecule, either acetaldehyde or propionaldehyde, during a subsequent turnover.

Table 4-3 Intermolecular Hydrogen Transfer in the Reaction Catalyzed by Diol Dehydrase

	^3H (% of total) in Product	
Starting Material	Acetaldehyde	Propionaldehyde
1,2-Propanediol-1-T plus ethylene glycol	38	62
1,2-Propanediol-1-T plus acetaldehyde	0	100

Shortly thereafter, Abeles and his associates identified the cofactor as the intermediate hydrogen carrier (110–112). Hydrogen was found to be transferred from the substrate to the carbon atom attached directly to the cobalt (hereafter termed the 5′-carbon), and from there to the product (Table 4-4; Formula 4-20). Subsequently, similar hydrogen transfer

$$CH_3-\overset{OH}{\underset{H}{C}}-\overset{OH}{\underset{\mathbf{H}}{C}}H \rightarrow \overset{R}{\underset{[Co]}{H\mathbf{C}\mathbf{H}}} \rightarrow CH_3-\overset{H}{\underset{\mathbf{H}}{C}}-\overset{O}{\overset{\|}{C}}H \qquad (4\text{-}20)$$

processes were shown to occur with several other adenosylcobalamin-dependent rearrangements (30, 69, 113–119). In those rearrangements which are reversible, hydrogen transfer from cofactor takes place in both directions. (For example, in the methylmalonyl CoA mutase reaction, hydrogen is transferred from adenosylcobalamin to both methylmalonyl CoA and succinyl CoA.) By contrast, in the irreversible diol dehydrase and ethanolamine ammonia-lyase reactions, it has not been possible to show transfer of hydrogen *from cofactor back to substrate,* but exchange of hydrogen *from cofactor into product* has been found to take place (53, 111) (with ethanolamine ammonia-lyase, this exchange is absolutely dependent on ammonia). Thus, with the irreversible systems there is direct hydrogen transfer from cofactor to product as well as hydrogen

Table 4-4 Adenosylcobalamin as Intermediate H Carrier in the Diol Dehydrase Reaction

	Specific Activity of AdoCbl (cpm/μmole)	
Tritiated Constituent	Before Reaction	After Reaction
1,2-Propanediol-1-T[a]	0	2.4×10^5
AdoCbl[b]	4.3×10^6	2.2×10^6

[a] Specific activity 1.1×10^5 cpm/μg-at H.
[b] Prepared enzymatically.

transfer occurring during the course of the catalytic rearrangement of substate.

Study of the diol dehydrase-catalyzed transfer of hydrogen to and from adenosylcobalamin revealed that the reaction was kinetically competent—that is, that its rate was as rapid as the overall reaction rate (44). Of equal interest with regard to the mechanism of the reaction, both of the 5′-hydrogen atoms of adenosylcobalamin were found to participate in the reaction (110, 112). This was established by the observation that when adenosylcobalamin tritiated chemically in both 5′ positions was used in the reaction, all the tritium was transferred to the product (Table 4-4). Since the two 5′-hydrogen atoms are distinguishable in principle in an enzymatic system, the fact that they both participated in the hydrogen transfer reaction raised the possibility that at one point in the catalytic sequence, these two atoms became equivalent. An alternative mechanism in which the enzyme distinguished between the hydrogen atoms, transferring to the product a hydrogen atom different from the one which was donated to the cofactor by the substrate, was ruled out by experiments in which substrate mixtures containing propylene glycol labeled in the migrating hydrogen atom together with unlabeled ethylene glycol were incubated with diol dehydrase and adenosylcobalamin (44, 110). Extrapolating from determinations of the amount of tritium in propionaldehyde as a function of the ethylene glycol/propylene glycol ratio, it was estimated that even with infinite amounts of ethylene glycol, tritium would still appear in propionaldehyde. This situation could only occur if the opportunity existed for a molecule of product to receive a hydrogen atom originally donated to the cofactor by the corresponding substrate molecule. In the same study, experiments were conducted showing that the hydrogen atom from the substrate became equivalent to at least two other hydrogen atoms—presumably, the two on the 5′ position of the coenzyme. Using a different approach, other groups came to similar conclusions with regard to the reactions catalyzed by glutamate mutase and methylmalonyl CoA mutase (118, 119). In their experiments, reactions containing mixtures of unlabeled and trideuterated (with methylmalonyl CoA mutase) or tetradeuterated (with glutamate mutase) substrate were allowed to proceed part way toward equilibrium, after which the distribution of various deuterated species (e.g., undeuterated, monodeuterated, etc.) in the starting material and product was determined. A comparison of these figures with those calculated from various models of the hydrogen transfer reaction supported the proposal that the hydrogen atom from substrate became equivalent to at least two other hydrogen atoms during the course of the rearrangement.

The most immediately obvious way for the migrating substrate hydrogen to become equivalent to the two 5′-hydrogen atoms of the cofactor would be through the formation of 5′-deoxyadenosine during the catalytic sequence. This compound could be formed by rupture of the carbon-cobalt bond of the cofactor, followed by the abstraction of the migrating hydrogen atom of the substrate by the resulting adenosyl fragment (4-21).

$$\begin{array}{cccccccccc} & & & & & R & & & & \\ & & R & & & | & & & R & \\ | & & | & & | & CH_2 & & | & | & \\ -C-H & & CH_2 & \rightarrow & -C-H & | & \rightarrow & -C- & CH_3 & \\ | & & | & & | & | & & | & | & \\ & & [Co] & & & [Co] & & & [Co] & \end{array} \qquad (4\text{–}21)$$

Rotation of the methyl group about the C-4′–C-5′ bond would ensure that its hydrogen atoms, one from substrate and two from the cofactor, would become equivalent.

The participation of 5′-deoxyadenosine formed by cleavage of adenosylcobalamin in the mechanism of adenosylcobalamin-dependent rearrangements was first proposed by Ingraham (120). Its production from the cofactor by an adenosyl cobamide-requiring enzyme was first demonstrated by Wagner et al. (121), who found that incubation of adenosylcobalamin with diol dehydrase and glycolaldehyde, an analog of the substrate ethylene glycol, led to irreversible cleavage of the carbon-cobalt bond with the production of 5′-deoxyadenosine from the adenosyl fragment. The glycolaldehyde, oxidized to glyoxal during the course of the reaction, was the source of the third hydrogen atom of the methyl group of 5′-deoxyadenosine. Ethanolamine ammonia-lyase was subsequently shown to participate in a similar reaction (122). Incubation of this enzyme with adenosylcobalamin and ethylene glycol resulted in the rapid formation ($k = 0.2\ \text{sec}^{-1}$) of stoichiometric amounts of acetaldehyde, 5′-deoxyadenosine, and an unidentified cobalamin derivative which reverted to hydroxocobalamin upon denaturation of the enzyme with urea. In this reaction too, the extra hydrogen atom on 5′-deoxyadenosine was shown to come from the substrate analog. Since then, it has been found that the irreversible formation of 5′-deoxyadenosine by adenosylcobamide-requiring enzymes is a rather general reaction, having been demonstrated with four such enzymes (50, 53, 78, 123–125) using a variety of substrates and substrate analogs (Table 4-5). In addition, it has recently been reported that 5′-deoxyinosine is produced when diol dehydrase is incubated with ethylene glycol and inosylcobalamin (98).

Table 4-5 Formation of 5′-Deoxyadenosine from AdoCbl by Enzymes Which Catalyze AdoCbl-Dependent Rearrangements

Enzyme	Substrate or Analog with Which 5′-Deoxyadenosine Formation Has Been Demonstrated
Diol dehydrase	Glycolaldehyde Chloroacetaldehyde
Methylmalonyl CoA mutase	Malonyl CoA Succinyl CoA
L-β-Lysine mutase	L-Erythro-3,5-diaminohexanoate
Ethanolamine ammonia-lyase	Ethylene glycol Acetaldehyde Ethanolamine 2-Aminopropanol

All of the reactions just mentioned, however, are processes in which 5′-deoxyadenosine formation accompanies the irreversible destruction of the cofactor. If it could be shown that under some conditions the formation of 5′-deoxyadenosine from cofactor is a reversible process, the evidence that it is a true intermediate rather than merely the product of an abortive side reaction would be greatly strengthened. Many attempts to demonstrate reversibility by experiments involving enzyme-catalyzed exchange of *free* 5′-deoxyadenosine into adenosylcobalamin were unsuccessful. It was possible, however, to obtain evidence that *enzyme-bound* 5′-deoxyadenosine can be converted to adenosylcobalamin. 5′-Deoxyadenosine formed by the cleavage of adenosylcobalamin was isolated when reaction mixtures containing ethanolamine ammonia-lyase, adenosylcobalamin, and ethanolamine were denatured while the enzyme was in the act of catalysis (123). Reversibility of 5′-deoxyadenosine formation was inferred from the observation that the amount of nucleoside isolated from the reaction mixture—that is, the extent of cleavage of the cofactor—depended on the agent used to denature the enzyme. The difference between the cleavage observed under different conditions of denaturation can be explained by the hypothesis that the enzyme-coenzyme complex is a mixture containing a species with an intact carbon-cobalt bond in equilibrium with a certain amount of a species in which this bond is broken, and that the relative resistance of the two species to denaturation is a function of the denaturing agent. Under conditions in which the species with the broken bond is substantially more stable than the species with the intact coenzyme, inactivation of the former would only occur after the adenosyl residue had been rejoined to the cobalamin in the reverse reaction, and little or no cleavage would be seen; under conditions in which the stability of the species is reversed, however, fragments released from the species with the broken carbon-cobalt bond would be observed (Figure 4-9). Though consistent, the differences were small (4% cleaved when the enzyme was denatured with hot propanol, compared with 1% with tetrahydrofuran), so that the evidence for reversibility, while suggestive, was inconclusive.

Much stronger evidence of a similar nature has been obtained with the same system, using L-2-aminopropanol as substrate (49, 50). With this substrate, about 80% cleavage was seen when the enzyme was denatured with heat or trichloroacetic acid, but only about 15% when ethanol was used to denature the enzyme. Addition of the natural substrate ethanolamine to the propanolamine-containing reaction mixture prior to denaturation with trichloroacetic acid reduced the extent of cleavage from 80 to 11%. Apparently, the displacement of propanolamine by ethanolamine at the active site of the enzyme caused the steady-stage level of

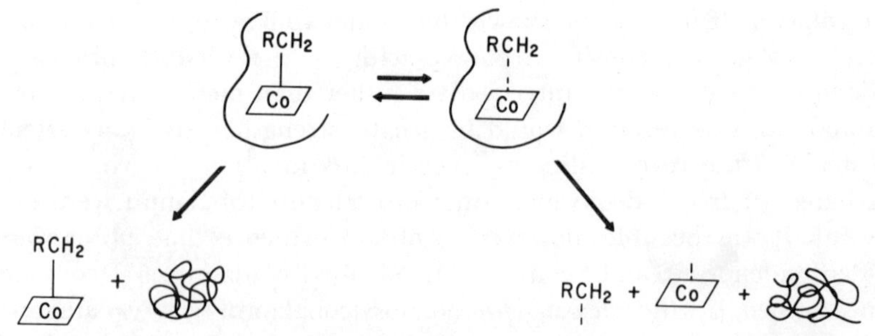

Fig. 4-9. Differences in the extent of cleavage of cofactor under varying conditions of denaturation as evidence for reversibility of cleavage.

dissociated adenosylcobalamin to fall from the very high level characteristic of the enzyme • cofactor • propanolamine complex to the much lower level seen with the ethanolamine-containing complex (see the previous paragraph). Finally, it was found that hydrogen present on the 5′-deoxyadenosine isolated from a propanolamine-containing reaction mixture could be transferred into product. This was shown as follows: Two identical reaction mixtures were prepared, each containing enzyme, adenosylcobalamin, and [1-^{3}H]propanolamine. In these mixtures, the tritium from the propanolamine was rapidly exchanged into the cofactor. One reaction mixture was denatured, and the distribution of tritium was determined. Analysis showed that 90% of the tritium in the cofactor was present in 5′-deoxyadenosine, the remainder in intact adenosylcobalamin (Table 4-6). At the same time, a large excess of unlabeled ethanolamine was added to the other reaction mixture, which was incubated for a short while longer. When this portion of the reaction mixture was denatured, it was found that most of the cofactor was once again in its original form, and that virtually all of the tritium from both 5′-deoxyadenosine and adenosylcobalamin had been transferred to acetaldehyde, the product of deamination of ethanolamine. These experiments showed that enzyme-bound 5′-deoxyadenosine formed in the presence of propanolamine could be converted back to a species with the catalytic activity of the cofactor—that is, that the formation of 5′-deoxyadenosine from adenosylcobalamin was reversible.

Table 4-6 Reversibility of Carbon-Cobalt Bond Cleavage during the Deamination of 2-Aminopropanol by Ethanolamine Ammonia-lyase

	^{3}H (cpm × 10^{-4})	
Compound	TCA only	Ethanolamine, then TCA
5′-Deoxyadenosine	19.4	0.7
Adenosylcobalamin	2.0	0.6
Acetaldehyde	—	19.1

Free Radicals and Hydrogen Transfer

The results presented in the preceding section provide support for a mechanism involving cleavage of the carbon-cobalt bond followed by transfer of hydrogen from the substrate to the 5′-deoxyadenosyl fragment produced by the cleavage reaction. Evidence based primarily on electron spin resonance (esr) spectroscopy, but supplemented to a certain extent by radiochemical data, has indicated that both carbon-cobalt cleavage and the subsequent abstraction of hydrogen from the substrate are processes involving species with unpaired electrons.

Experiments with ethanolamine ammonia-lyase provided the first indication that species with unpaired electrons might be involved in adenosylcobalamin-dependent rearrangements. The evidence consisted of an esr signal detected in a reaction mixture in which the adenosylcobalamin-dependent deamination of ethanolamine was taking place (126). The appearance of the signal depended on the presence of enzyme, cofactor, and substrate; omission of enzyme or cofactor abolished the signal, while in the absence of substrate the esr signal was replaced by a smaller signal with different characteristics. Previous experiments showing substrate-induced changes in the optical spectra of complexes between adenosylcobalamin and both diol dehydrase (121, 127) and ethanolamine ammonia-lyase (128) were consistent with the production of cob(II)alamin during catalysis; inasmuch as other processes involving alkylcobamides (e.g., displacement of the heterocyclic base from the Coα position) may lead to similar spectral changes, however, interpretations based solely on optical spectroscopy are somewhat ambiguous. The appearance of an esr signal constituted unequivocal evidence for the presence of paramagnetic species in the reaction mixture.

Experiments with radioactive adenosylcobalamin provided further support for the idea that ethanolamine ammonia-lyase was able to catalyze the homolysis of the carbon-cobalt bond of the cofactor (123). Denaturation of [^{14}C]AdoCbl-containing reaction mixtures during the act of catalysis suggested that under steady-state conditions, 3 to 4% of the cofactor was in a form in which the carbon-cobalt bond was broken. This figure agreed with the estimate of the concentration of species with unpaired electrons based on the size of the esr signal observed under similar conditions (127, 129). In the absence of substrate, steady-state cleavage of the carbon-cobalt bond in the enzyme-coenzyme complex amounted to about 1%. Moreover, when isolated from a reaction mixture containing only the enzyme-coenzyme complex, the adenosyl derivatives corresponding to the small fraction of cofactor with a broken carbon-cobalt bond were identified as adenosine-5′-aldehyde and 5′,8-cycloadenosine, compounds which had previously been shown to be produced by the aerobic photolysis of adenosylcobalamin, a homolytic process.

A spin-labeled adenosylcobinamide with cofactor activity, *Coα*-(4-hydroxy-2,2,6,6-tetramethylpiperidine-*N*-oxyl)-*Coβ*-(5′-deoxyadenosyl) cobinamide, was prepared by Wood and his co-workers (130) (4-22). In this compound, the paramagnetic nitroxyl group is in the Coα position of the metal coordination sphere. Addition of ethanolamine to the com-

Ado
|
[Co]
↑
O
|
N
OH

(4–22)

plex between ethanolamine ammonia-lyase and the spin-labeled cofactor led to a marked decrease in the amplitude of the spin-label esr signal. The signal returned almost to its original size when the substrate was used up (Figure 4-10). This observation provided additional support for homolytic cleavage of the carbon-cobalt bond during catalysis by ethanolamine ammonia-lyase.

Electron spin resonance spectra have now been observed with two additional enzymes catalyzing adenosylcobalamin-dependent rearrangements, namely, diol dehydrase (131) and glycerol dehydrase (132). In addition, spectra with a much improved signal-to-noise ratio have been obtained with the ethanolamine ammonia-lyase system (129). Representative spectra with these enzyme systems are shown in Figure 4-11. Each of these spectra consists of two components: a broad signal, in some instances displaying superimposed hyperfine structure, assigned to cob(II)alamin; and a narrower signal, attributed to a free radical. The concentration of unpaired electrons represented by these signals corresponds to an extent of carbon-cobalt bond cleavage of 25, 38, and 4% for glycerol dehydrase, diol dehydrase, and ethanolamine ammonia-lyase, respectively. Electron spin resonance spectra have also been observed upon incubation of the

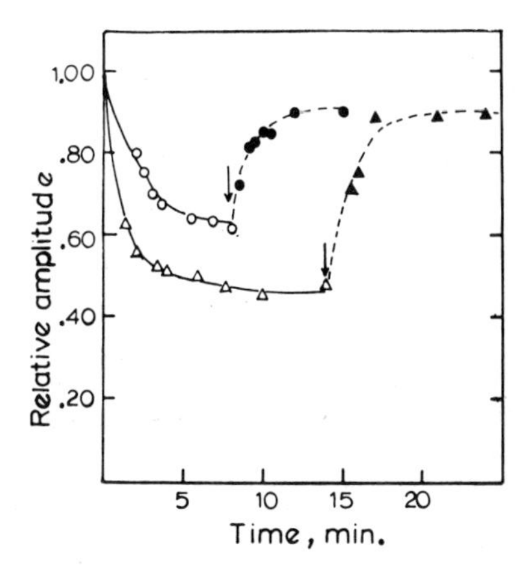

Fig. 4-10. Changes in the intensity of the nitroxide esr signal upon incubation of a spin-labeled *Coβ*-adenosylcobamide with ethanolamine ammonia-lyase and substrate. The arrows indicate the points at which substrate was exhausted. [Reprinted with permission from Law, P. Y., Brown, D. G., Lien, E. L., Babior, B. M., and Wood, J. M. (1971). *Biochemistry* **10,** 3428–3435. Copyright by the American Chemical Society.]

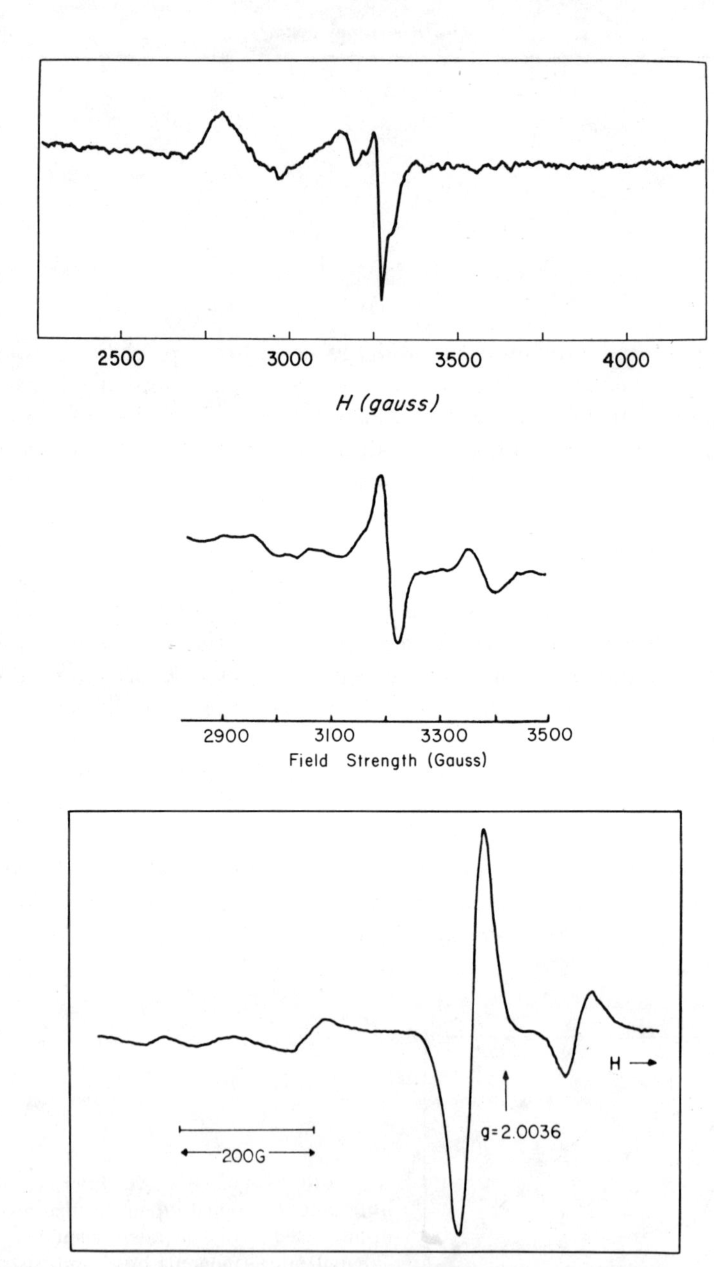

Fig. 4-11. Esr signals observed upon incubation of ethanolamine ammonia-lyase [top (129)], glycerol dehydrase (middle), and diol dehydrase [bottom (131)] with adenosylcobalamin and their respective substrates. [Middle reprinted with permission from Cockle, S. A., Hill, H. A. O., Williams, R. J. P., Davies, S. P., and Foster, M. A. (1972). *J. Am. Chem. Soc.* **94**, 275–277. Copyright by the American Chemical Society.]

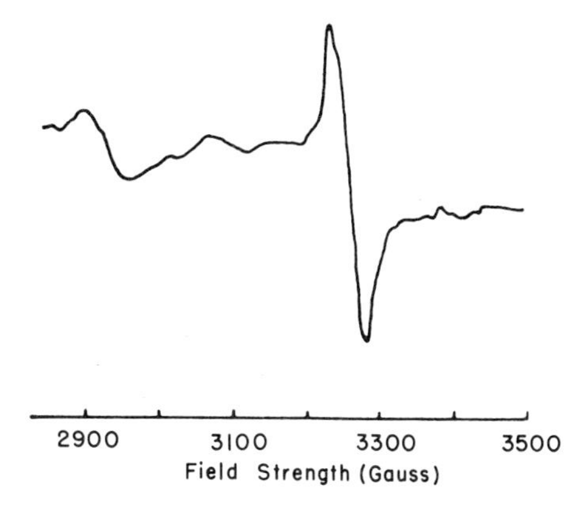

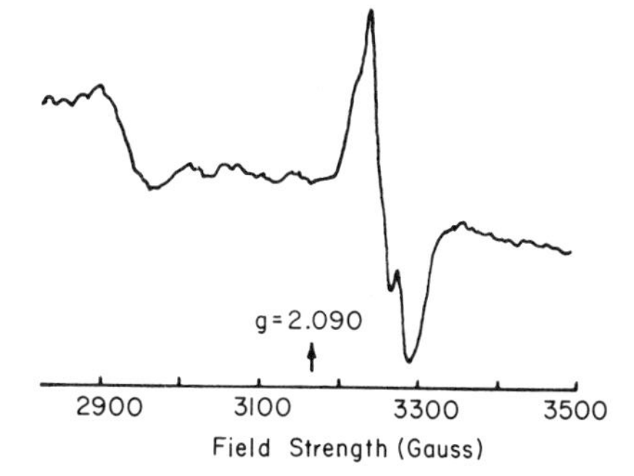

Fig. 4-12. Esr signals observed upon incubation of diol dehydrase and adenosylcobalamin with glycolaldehyde (above) and chloroacetaldehyde (below) (131).

diol dehydrase • adenosylcobalamin complex with glycolaldehyde and with chloroacetaldehyde (131), substrate analogs which cause the carbon-cobalt bond to be split by diol dehydrase (Figure 4-12). The general features of these spectra [i.e., two signals, one corresponding to cob(II)-alamin and one to a free radical] are similar to those of spectra generated in the presence of substrate.

The strongest evidence for free radical participation in hydrogen transfer has been obtained with the ethanolamine ammonia-lyase system, using propanolamine as substrate (133). As discussed above, deamination of this substrate is associated with extensive reversible cleavage of the carbon-cobalt bond of the cofactor. Electron spin resonance spectroscopy of a reaction mixture frozen during catalysis of propanolamine deamination (Figure 4-13) revealed a large signal, consisting as expected of a broad component representing cob(II)alamin and a narrower component, an asymmetrical doublet with overlapping peaks, attributable to a free radical. Spectra obtained with reaction mixtures prepared with isotopi-

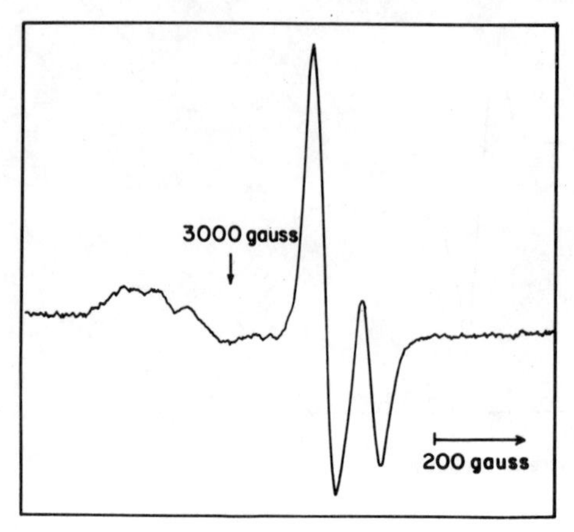

Fig. 4-13 Esr spectrum of the ethanolamine ammonia-lyase • AdoCbl complex in the presence of propanolamine (133).

cally labeled substrates (Figure 4-14) identified the free radical as the 2-aminopropanol-1-yl radical (see 4-23). The concentration of unpaired

$$CH_3-\overset{\overset{NH_2}{|}}{\underset{H}{C}}-\overset{\overset{OH}{|}}{\underset{H}{C}}\cdot \qquad (4\text{–}23)$$

electrons corresponded to homolysis of 65% of the adenosylcobalamin bound to the enzyme. The reaction in which the unpaired electrons were produced was found to be first order, with a rate of 7 sec^{-1}. Comparison of this value with the overall rate of the enzyme-catalyzed deamination of propanolamine, 1 to 2 sec^{-1}, shows that the reaction (or sequence of reactions) responsible for the production of the unpaired electrons is kinetically competent—that is, is rapid enough to participate in the overall reaction. Both the concentration of the species with unpaired electrons and the rate at which they appear make their participation in the deamination of propanolamine extremely likely.

Based on the experimental evidence discussed above, the view currently held most widely is that hydrogen transfer in adenosylcobalamin-dependent reactions is a two-step process (4-24). In the first step, the

$$\underset{[Co]}{\overset{}{CH_2R}} \rightleftharpoons \underset{[\dot{C}o]}{} \quad H-\overset{\overset{CH_2R}{|}}{\underset{|}{C}}- \rightleftharpoons \underset{[\dot{C}o]}{CH_3R} \quad \cdot\overset{|}{\underset{|}{C}}- \qquad (4\text{–}24)$$

carbon-cobalt bond of the cofactor is cleaved homolytically, with the production of a cob(II)alamin and the 5′-deoxyadenos-5′-yl radical. This

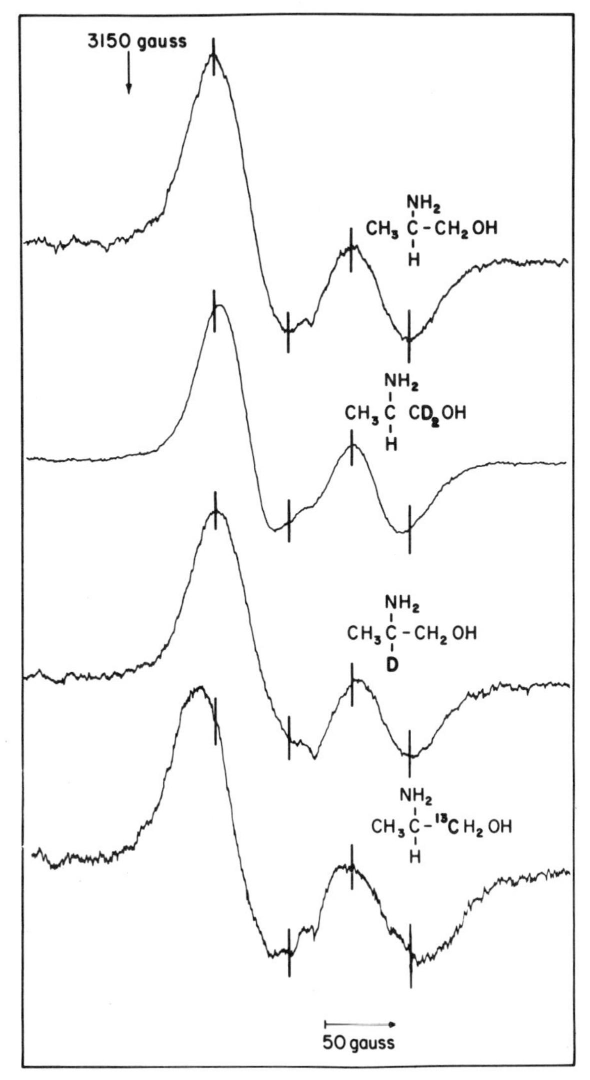

Fig. 4-14 Effect of isotopic labeling of propanolamine on the free radical component of the signal shown in Figure 4-13 (133).

radical then abstracts a hydrogen atom from the substrate to produce 5′-deoxyadenosine and the substrate radical.

An alternative hypothesis, based largely on experiments with model compounds, holds that carbon-cobalt bond cleavage is heterolytic, and that the unsaturated nucleoside 4′,5′-anhydroadenosine is involved in catalysis together with cob(I)alamin, rather than 5′-deoxyadenosine and cob(II)-

alamin. The evidence for this hypothesis is discussed in detail by Schrauzer and his co-workers (134–136, 136a, and references cited therein). Arguments against this hypothesis, and against the experimental interpretations on which it is based, are to be found in several papers from Abeles' laboratory (44, 125, 137).

Hydrogen Isotope Effects in Adenosylcobalamin-Dependent Rearrangements

Through the use of isotopically labeled substrates, it has been possible to show that hydrogen transfer is the rate-limiting step in several adenosylcobamide-dependent rearrangements. Deuterium isotope effects for these reactions are shown in Table 4-7 (35, 36, 44, 51, 95, 103, 118, 119). In each case, the difference between the rates with unlabeled and deuterated substrates indicates a primary isotope effect. For these reactions, then, the rupture of a carbon-hydrogen bond is the slow step in the sequence comprising the mechanism of catalysis of the rearrangement.

Adenosylcobalamin-dependent rearrangements, however, involve two hydrogen transfer steps—transfer of hydrogen from substrate to cofactor and transfer from cofactor to the reaction product. Either of these can be the slow step in the rearrangement. With glutamate mutase and methylmalonyl CoA mutase, the isotope effect for the transfer of hydrogen from substrate to cofactor has been calculated to be approximately

Table 4-7 Overall Deuterium Isotope Effects for Some Adenosylcobalamin-Dependent Rearrangements

Enzyme	Deuterated Substrate	k_H/k_D
Diol dehydrase	$HOCD_2CD_2OH$	6.4
	$CH_3CHOHCD_2OH$	10
Ethanolamine ammonia-lyase	$NH_2CH_2CD_2OH$	7.4
Methylmalonyl CoA mutase	$CD_3—CH(COSCOA)(COOH)$	3.5
Glutamate mutase	$HOOC—C(NH_2)(H)—C(CD_3)(D)—COOH$	6.6

equal to the overall isotope effect, implying that the abstraction of hydrogen from the substrate is the rate-limiting step in these reactions (118, 119). No information was obtained regarding the isotope effect for the transfer of hydrogen from cofactor to product.

With diol dehydrase and ethanolamine ammonia-lyase, on the other hand, it was possible to measure the isotope effects for each of the hydrogen transfer steps. These isotope effects (tritium rather than deuterium) were determined by comparing the rate of transfer of label from tritiated cofactor into product derived from *unlabeled substrate* with the rate of transfer of label from *tritiated substrate* into both cofactor and product (44, 103). The isotope effects measured by these experiments are shown in Table 4-8. While k_H/k_T for the transfer of hydrogen from substrate to cofactor is in the conventional range for a primary isotope effect, k_H/k_T for the subsequent hydrogen transfer is remarkably high in both reactions. The reason for the unprecedented magnitude of this isotope effect is obscure. The deuterium isotope effect for the transfer of hydrogen from cofactor to product was also determined, this being accomplished by measuring the rate of washout of tritium from labeled cofactor into product derived from deuterated substrate. The rate ratios obtained in these experiments, also shown in Table 4-8, actually represented k_D/k_T, but from these numbers k_H/k_D was easily calculated. Comparison of this figure with the deuterium isotope effect for the overall reaction showed that the rate-limiting step in the diol dehydrase reaction was the first hydrogen transfer step, as observed with glutamate mutase and methylmalonyl CoA mutase, but that the rate-limiting step in the ethanolamine ammonia-lyase reaction was the transfer of hydrogen from the cofactor to the product.

Table 4-8 Isotope Effects for the Individual Hydrogen Transfer Steps in Certain Adenosylcobalamin-Dependent Rearrangements

	Isotope Effect		
	Substrate → Cofactor	Cofactor → Product	
Enzyme	k_H/k_T	k_H/k_D	k_H/k_T
Diol dehydrase	20	28	125
Ethanolamine ammonia-lyase	4.7	7.3	160

The Migration of Group X

Although the mechanism of the hydrogen transfer reaction is reasonably well understood, there is little information from biochemical studies regarding the mechanism by which the migration of group X occurs. The only biochemical experiments which bear on this question have been the repeated failures to demonstrate an enzyme-catalyzed exchange between the products of an adenosylcobamide-dependent rearrangement and compounds postulated to represent the form in which group X migrates. The failure of L-β-lysine mutase to exchange free NH_4^+ into either L-β-lysine or 3,5-diaminohexanoate has been discussed previously. Glutamate mutase was unable to exchange glycine, propionic acid, acrylic acid, or α-ketoglutaric acid into either glutamate or β-methylaspartate (8). Neither propionic acid (13) nor acrylic acid (20) was exchanged into methylmalonyl CoA or succinyl CoA by methylmalonyl CoA mutase. No radioactivity was incorporated into methylitaconate when the α-methyleneglutarate mutase reaction was carried out in the presence of [^{14}C]-acrylic acid (32).

Lacking direct enzymological evidence, biochemists have fallen back on model reactions to explain migration of group X. Many chemical reactions have been proposed as models for this migration. The variety of mechanisms represented by these model reactions leaves the imaginative armchair biochemist a wide scope for speculation concerning pathways for the enzyme-catalyzed migration of group X, each of which can be supported by reference to the appropriate model.

The model reactions fall into two categories: those in which the cobamide participates directly as a reactant, and those in which it does not. In the latter category are processes in which a reactive intermediate generated by the removal of a hydrogen from the substrate undergoes spontaneous rearrangement. Examples have been provided in which the hydrogen has been abstracted as a proton, as an atom, and as a hydride ion, the intermediates undergoing rearrangement being a carbanion, free radical, and carbonium ion, respectively. The rearrangement of the 1-carbethoxy-2-ketocyclopentylmethyl carbanion to ethyl-3-ketocyclohexanecarboxylate has been proposed as one model for the methylmalonyl CoA mutase reaction (138) (see 4-25). A free radical rearrangement, the

$$\text{(cyclopentanone with } \bar{\ddot{C}}H_2 \text{ and COOEt; =O)} \xrightarrow{H^+} \text{(cyclohexanone with COOEt; =O)} \qquad (4\text{–}25)$$

conversion of neophyl chloride to *t*-butylbenzene (and other hydrocarbons), has also been proposed as a model of the methylmalonyl CoA mutase reaction (16):

$$\phi\text{—}\underset{CH_3}{\overset{CH_3}{C}}\text{—}CH_2Cl \rightarrow \phi\text{—}\underset{CH_3}{\overset{CH_3}{C}}\text{—}CH_2\cdot \rightarrow \cdot\underset{CH_3}{\overset{CH_3}{C}}\text{—}CH_2\phi \rightarrow H\underset{CH_3}{\overset{CH_3}{C}}\text{—}CH_2\phi \qquad (4\text{–}26)$$

In the realm of carbonium ion rearrangements, the conversion of propylene glycol to acetaldehyde by way of an oxirane intermediate (4-27) has

$$CH_3\text{—}\underset{H}{\overset{OH}{C}}\text{—}CH_2OH \rightarrow CH_3\text{—}\underset{H}{C}\overset{O}{\frown}CHOH \rightarrow CH_3CH_2CHO \qquad (4\text{–}27)$$

been put forth as a plausible mechanism for the diol dehydrase reaction (38), based on the well-known acid-catalyzed conversion of vicinal diols to carbonyl compounds (see 4-28), a reaction postulated to involve carbonium ion intermediates.

$$\text{—}\underset{HO}{\overset{R}{C}}\text{—}\underset{OH}{C}\text{—} \xrightarrow{H^+} \text{—}\underset{\|O}{C}\text{—}\overset{R}{C}\text{—} \qquad (4\text{–}28)$$

Models in which a cobamide (or a related cobalt complex) participates directly in the reaction involve the rearrangement of the organic ligand of an organocobalt complex. Most of the examples of this type of model have been provided by studies of the chemistry of bis(dimethylglyoximato)-cobalt complexes (cobaloximes), a family of compounds that have certain properties in common with cobamides. In particular, organocobaloximes possessing a carbon-cobalt σ bond stable to air and water but dissociated by light are easily prepared. One of these compounds, β-hydroxyethylpyridinatocobaloxime, has been reported to undergo carbon-cobalt bond cleavage with the production of acetaldehyde and pyridinatocob(I)aloxime on treatment with base (4-29) (136). A compre-

$$\underset{\underset{B}{|}}{(Co)}\text{—}CH_2\text{—}\underset{H}{\overset{H}{C}}\text{—}O\text{—}H \;\; {}^-OH \rightarrow \underset{\underset{B}{|}}{(\ddot{C}o)} + CH_3CHO \qquad (4\text{–}29)$$

hensive mechanism of action of adenosylcobalamin-dependent rearrangements has been proposed based on this reaction and the OH^--catalyzed

β elimination of the unsaturated nucleoside 4′,5′-anhydroadenosine from adenosylpyridinatocobaloxime (134). (This mechanism is discussed briefly on page 185).

Yet another mechanism is suggested by studies of the methanolysis of β-acetoxypyridinatocobaloxime (4-30) (136, 139), a solvolysis in which

$$\begin{array}{c} CH_2{-}CH_2OAc \\ | \\ (Co) \\ | \\ B \end{array} \xrightarrow[HOAc]{H^+} \begin{array}{c} CH_2{=}CH_2 \\ \downarrow \\ (Co) \\ | \\ B \end{array} \xrightarrow{MeOH} \begin{array}{c} CH_2{-}CH_2OMe \\ | \\ (Co) \\ | \\ B \end{array} \qquad (4\text{–}30)$$

intramolecular migration of cobalt from one carbon atom to the other has been unequivocally demonstrated by isotopic labeling (140). This reaction has been proposed to go by way of an intermediate π complex, as shown. The resemblance between this reaction and the adenosylcobalamin-dependent migration of X, particularly where X is –OH or $-NH_2$, is very apparent.

The currently accepted theory holds that the mechanism of migration involves an alkylcobalamin formed at the active site by the attack of a substrate radical on cob(II)alamin (reaction 4-31). At present this mech-

$$\dot{[Co]} \quad \begin{array}{c} X \\ | \\ \cdot C{-}C \end{array} \rightleftharpoons \begin{array}{c} X \\ | \\ C{-}C \\ | \\ [Co] \end{array} \rightleftharpoons \begin{array}{c} X \\ | \\ C{-}C \\ | \\ [Co] \end{array} \qquad (4\text{–}31)$$

anism is based on logic rather than experiment. In the absence of data obtained with enzymatic systems, it is difficult to know whether this mechanism is correct, whether indeed a single mechanism explains all the reactions, and which if any of the reactions presented above best models the migration of group X.

RIBONUCLEOTIDE REDUCTASE

The Enzyme and the Reaction

Ribonucleotide reductase is the enzyme that catalyzes the conversion of ribonucleotides to 2′-deoxyribonucleotides, the building blocks from which DNA is assembled (see 4-32). The enzyme is likely to be universally

$$-O{-}\overset{\overset{O}{\|}}{\underset{\underset{O}{|}}{P}}{-}O{-}CH_2 \text{ (ribose ring, O, Base; HO, OH)} \rightarrow -O{-}\overset{\overset{O}{\|}}{\underset{\underset{O}{|}}{P}}{-}O{-}CH_2 \text{ (deoxyribose ring, O, Base; HO)} \qquad (4\text{–}32)$$

distributed, and has been demonstrated in a variety of species of microorganisms, in unicellular eukaryotes, as well as in cells from several vertebrate species (1). In most species investigated, it is an iron-containing protein whose activity appears to be totally independent of any form of cobamide. In a number of microorganisms and in certain Euglenophyta, however, ribonucleotide reduction is carried out by an adenosylcobamide-dependent enzyme (141, 142).

The cobamide-dependent ribonucleotide reductase was first found in *Lactobacillus leichmannii* (99, 143–147), a discovery which represented the outcome of investigations seeking to explain the cobamide requirement of this microorganism, and it is with the enzyme from this organism that most of the work on cobamide-dependent ribonucleotide reductase has been carried out. This enzyme has been purified to homogeneity, and can now be obtained in gram quantities. It consists of a single polypeptide chain of molecular weight 76,000 (148). The enzyme shows little tendency to polymerize or form aggregates, either by itself or in the presence of nucleoside or deoxynucleoside triphosphates, a feature of considerable significance with respect to the mechanism by which the activity of this enzyme is regulated.

The earliest report of ribonucleotide reduction by a cell-free system described the reduction of the ribosyl moiety of CMP by a crude extract from *L. leichmannii* (147). For this particular reaction, mercaptoethanol, ATP, and an NADPH-generating system were required in addition to adenosylcobalamin. Subsequent investigations, however, showed that the true substrates of *L. leichmannii* ribonucleotide reductase are not the nucleoside monophosphates, but the nucleoside triphosphates (149–153). The requirement for ATP in the reduction of CMP is thus at least in part explained by the necessity to form CTP before the reaction could take place.

Although NADPH was the first reductant shown to supply electrons for the conversion of NTP to dNTP, it was soon discovered that the most effective reducing agents for this purpose were certain thiols. In fact, with highly purified ribonucleotide reductase, reduced pyridine nucleotides did not work at all, thiols being the only compounds capable of reducing ribonucleotide triphosphates in this system (150). Dihydrolipoic acid was the most effective of the sulfhydryl compounds tested (150, 151, 154, 155), and is the reducing agent which has been used most extensively in studies of ribonucleotide reductase. Other sulfhydryl compounds which have been found suitable include dithiothreitol, dithioerythritol, 1,3-dithiopropane, and 1,3-dithiopropan-2-ol; but BAL (2,3-dithiopropan-1-ol) has little activity (99, 151, 154,155). Data on monothiols are somewhat conflicting, reported activities in this system varying from none to some, but are consistent in showing that monothiols are much less effective than appropriate dithiols as reducing agents for the ribonucleotide reductase system (99, 143, 151, 154, 155). The specificity shown by these studies indicates that the best reducing agents are the 1,3- and 1,4-dithiols—that is, those in which oxidation of the sulfhydryl groups is accompanied by the formation of an intramolecular five- or six-membered dithiolane ring.

The *natural* reducing agent is also a dithiol, though it is not one of the compounds listed above. It appears that *in vivo,* ribonucleotide triphosphates are reduced by thioredoxin, a sulfhydryl protein which was first discovered by Reichard to be the natural ribonucleotide reductant in *E. coli,* an organism possessing an iron-requiring ribonucleotide reductase. This enzyme has been purified to homogeneity from *E. coli* (156), and partially purified from *L. leichmannii* (157). Each of these thioredoxins is a protein of molecular weight 12,000. Both are equally effective with the ribonucleotide reductase system from *L. leichmannii,* showing similar Michaelis constants (K_m 4 μM) and reducing ribonucleotide triphosphates with similar maximum velocities (100, 154, 157, 158). It is not surprising that the affinity of thioredoxin for *L. leichmannii* ribonucleotide reductase is much greater than that of dihydrolipoate (K_m 10,000 μM) (100).

In vitro, reduced *E. coli* thioredoxin is able to reduce a stoichiometric amount of ribonucleotide to deoxyribonucleotide, using the *L. leichmannii* system as catalyst (154). *In vivo,* however, stoichiometric amounts of thioredoxin are not required for this reduction, because thioredoxin itself is reduced by a pyridine nucleotide in a reaction catalyzed by another enzyme, thioredoxin reductase (159). Thioredoxin reductase, like thioredoxin, was first discovered in *E. coli,* but has also been detected in *L. leichmannii* (157). The enzyme, purified to homogeneity from *E. coli* (159, 160), is a flavoprotein of molecular weight 65,800 containing 2 moles FAD/mole of enzyme. In the presence of catalytic amounts of thioredoxin reductase, oxidized thioredoxin, a protein with no detectable –SH groups, can be reduced by NADPH to reduced thioredoxin, a protein containing two –SH groups per mole of enzyme (159). Thioredoxin reductase thus appears to catalyze the NADPH-mediated reduction of a disulfide bridge. This reaction is readily reversible, with an equilibrium constant $K_{eq} = \mathrm{TSH \cdot NADP/TSST \cdot NADPH}$ of 4.8 at pH 8.0. Spectrophotometric experiments indicate that the flavin participates in this process, since this group can be reduced by both NADPH and reduced thioredoxin.

Thioredoxin reductase is quite specific for both the pyridine nucleotide and the sulfhydryl compound (159). With respect to pyridine nucleotides, K_m values indicate a much higher affinity for NADPH than for NADH (5 μM versus 1800 μM for NADPH and NADH, respectively). When, in addition, the turnover numbers for the two pyridine nucleotides are taken into consideration (1500 min^{-1} for NADPH, and 500 min^{-1} for NADH), it becomes clear that as substrate, NADPH is by far the better of the two nucleotides. With respect to sulfhydryl compounds, the enzyme reduces thioredoxin very efficiently (K_m 4.3 μM), but shows no activity at all with glutathione, lipoic acid, or the sulfhydryl redox enzymes glutathione reductase and lipoate dehydrogenase. On the other hand, reduced thioredoxin is relatively nonspecific in its choice of electron acceptors. Consequently, though thioredoxin reductase cannot transfer electrons from NADPH to any disulfide except thioredoxin, a system containing catalytic amounts of both thioredoxin reductase and thioredoxin has been shown to be capable of reducing insulin, glutathione, lipoic acid, and oxytocin, all at the expense of NADPH (16).

The electron transfer pathway in ribonucleotide reduction is summarized in 4-33. Reducing equivalents from the original electron donor,

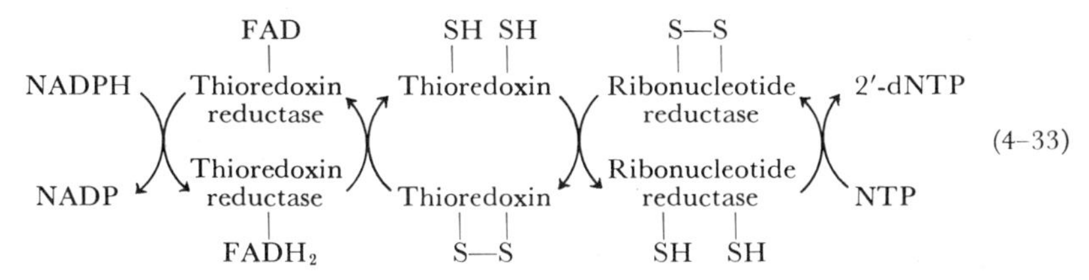

(4-33)

NADPH, are transferred to the final electron acceptor, ribonucleotide triphosphate, after successive passage through the enzymes thioredoxin reductase, thioredoxin, and ribonucleotide reductase. (The evidence that ribonucleotide reductase itself undergoes an oxidation-reduction cycle is discussed below.)

As mentioned previously, the best substrates for *L. leichmannii* ribonucleotide reductase are the ribonucleotide triphosphates (149, 150, 152, 153). Ribonucleotide diphosphates can also be reduced, but at much lower rates. The monophosphates and nucleosides are hardly touched. With regard to base specificity, any of the four common ribonucleotide triphosphates can be reduced by the enzyme (100, 150, 152), but the rates of reduction vary drastically, depending on the reaction conditions. The enzyme also reduces inosine triphosphate (153) and the riboside triphosphates of certain pyrrolopyrimidines (161) (4-34).

$$\text{R} = -\text{H},\ -\text{CN},\ -\overset{\text{O}}{\overset{\|}{\text{C}}}-\text{NH}_2 \qquad (4\text{-}34)$$

The fact that all major ribonucleotide triphosphates are reduced by the enzyme to metabolites whose tissue levels must be fairly closely regulated implies that ribonucleotide reductase activity should be subject to some sort of metabolic control. Such has proved to be the case. The first evidence for the existence of such regulation was the observation that ATP is required for the reduction of CTP by ribonucleotide reductase (149). The previously mentioned requirement of ATP for the reduction of CMP could be accounted for by the necessity to convert CMP to the true substrate for the reaction, CTP. Since substitution of CTP for CMP failed to abolish the ATP requirement, it is clear that that ATP must be performing a function apart from that of phosphorylating the nucleotide monophosphate. Investigation of the role of ATP and other nucleotides in the regulation of ribonucleotide reductase activity has revealed a pattern of control of great complexity. Only the outlines of this pattern are discussed here, since the details are as yet poorly understood. Basically, it has been found that a given regulator will alter the rate of reduction of every ribonucleoside triphosphate, affecting each substrate in a different way (100, 150, 152). As regulators, the deoxyribonucleotide triphosphates (i.e., the products of the reaction) are most effective, although the ribonucleotide triphosphates (i.e., the substrates) can also act to regulate rates of reduction by the enzyme. For each species of substrate, there is a corresponding regulator deoxyribonucleotide, termed the prime effector (152), in whose presence the rate of reduction of the substrate in question is maximized. The prime effector alters the rate of reduction of the other substrate species, in some cases accelerating the reaction and in other cases inhibiting it, but the major action of the prime effector is to accelerate the rate of reduction of its corresponding nucleotide. The effect of each prime effector on the rate of reduction of each of the four common ribonucleotide triphosphates under a single set of experimental conditions is shown in Table 4-9. Substrate-prime effector pairs are indicated by boldface numbers.

The kinetics which result from the regulatory pattern described above are exceedingly complex. Even in a reaction mixture containing only substrate and prime effector, three regulators are present: the substrate, the prime effector, and (after a little while) the product. Moreover, the concentration of two of these regulators is changing constantly with time. Added to this is the effect of Mg^{2+}, an ion frequently involved in reactions involving nucleotides. Although ribonucleotide reductase does not require a divalent cation for activity, the presence of such an ion greatly alters the regulatory interrelationships seen with the enzyme (150, 152). Typical kinetics obtained with the ribonucleotide reductase system are presented in Figure 4-15. These curves show reaction rates

Table 4-9 Influence of Various Effector Nucleotides on Reduction of Ribonucleotide Triphosphates

	Effector				
Substrate	None	dATP	dCTP	dGTP	dTTP
CTP	0.21[a]	**6.95**	0.51	0.80	0.76
UTP	0.25	0.40	**2.32**	0.48	0.54
ATP	2.48	1.15	0.80	**7.00**	0.73
GTP	2.91	3.30	3.35	1.42	**7.01**

[a] Nanomoles reduced per 20 min for every 0.64 μg of enzyme. From Beck (152).

with the four common substrates in the presence and absence of Mg^{2+} and prime effector. The complexities presented by this system are well illustrated.

Although no one as yet has attempted a complete quantitative analysis of regulation in the ribonucleotide reductase system, a qualitative interpretation of the effects of the regulators on the enzyme has been presented (152). It has been proposed that ribonucleotide reductase possesses two nucleotide binding sites, a catalytic site and a regulatory site. The

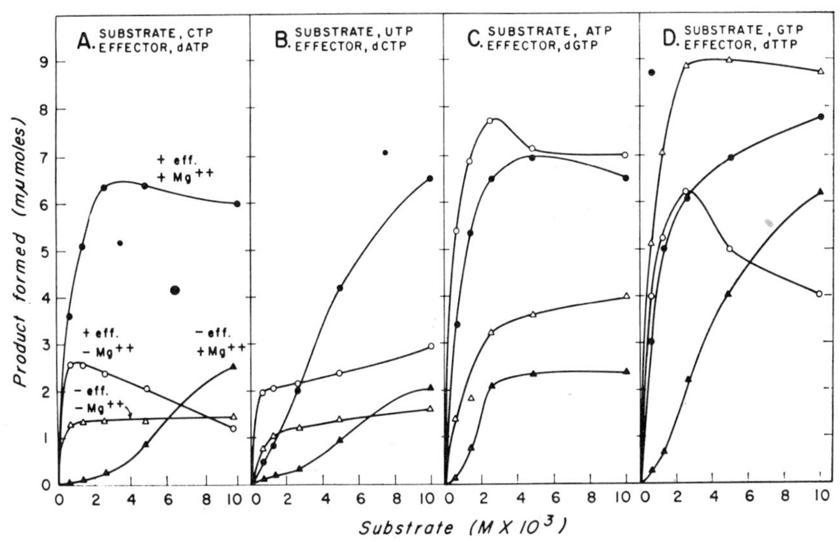

Fig. 4-15. Effects of substrate concentrations on the rates of reduction of ribonucleotide triphosphates in the presence and absence of Mg^{2+} (152).

binding of a nucleotide to the regulatory site induces a small conformational change in the enzyme which affects the rate of reduction of each of the ribonucleotide triphosphates by altering the properties of the catalytic site. The enzyme can thus exist in many forms with differing catalytic properties, each form corresponding to the complex between enzyme and one species of nucleotide. The extremely complex kinetics can be explained as a result of a situation in which various forms of the enzyme are all present in the reaction mixture at one time, the concentrations of these various forms depending on the concentrations of the various nucleotides present in the reaction mixture.

Ribonucleotide reduction is unique among adenosylcobamide-dependent reactions in that protons from the solvent are incorporated into the product during the course of the reaction. Experiments based on this property of the reaction have shown that the reduction of the ribonucleotide involves the direct replacement by hydrogen of the OH group on the 2′-carbon (162–165). In these experiments, it was demonstrated that labeled hydrogen transferred from solvent to deoxyribonucleotide is found exclusively on the 2′-carbon. Moreover, only one of the two stereochemically nonequivalent positions on this carbon atom becomes labeled. This is the C-2′α position, which in the unreduced ribonucleotide is occupied by the hydroxyl group. Displacement of hydroxyl by hydrogen therefore occurs with retention of configuration. Experiments with ^{18}O-labeled substrates (166) confirmed the direct displacement of the hydroxyl groups by hydrogen, since during the reduction of [2′-^{18}O]ATP and [3′-^{18}O]ATP, label was lost only from substrate labeled in the 2′ position. Failure of the 2′-^{18}O to exchange with water in the absence of reducing agent (in this case, dihydrolipoic acid) indicated that there was no labilization of the oxygen-carbon bond in the absence of ribonucleotide reduction, at least under the conditions of those experiments.

Binding of Cofactor to Ribonucleotide Reductase

Experiments with AdoCbl and related analogs have shown that the binding of corrinoids to ribonucleotide reductase resembles their binding to the enzymes which catalyze adenosylcobamide-dependent rearrangements (99–102). While a Coβ ligand closely related to the 5′-deoxyadenosyl group is required for cofactor activity, *Coβ*-adenosylcobamides with a variety of Coα ligands can serve as cofactor. Cobamides with various Coβ ligands, and with various alkyl substituents on the nitrogen of the *e*-amide group at the periphery of the ring, are effective inhibitors of the ribonucleotide reductase reaction. Hydrolysis of the *e*-amide group, however, gives rise to a compound which is without effect on the ribonucleotide reductase reaction. Ribonucleotide reductase therefore appears to possess one site to which the corrin ring is bound, at least in part by interactions involving the amide groups at the periphery of the ring, and another site to which a *Coβ*-adenosyl ligand can attach.

Very strong evidence for the existence of the *Coβ*-adenosyl binding site has been obtained by a study of the interaction between ribonucleotide reductase, cob(II)alamin, and various adenosyl analogs (167, 168). Cob(II)alamin is able to bind to the active site of ribonucleotide reductase. It has been shown both by direct binding studies and by esr spectroscopy that the affinity of the enzyme for cob(II)alamin is increased in the presence of adenosine, 5′-deoxyadenosine, or 4′,5′-anhydroadenosine (4-35). Similarly, the affinity of the enzyme for 5′-deoxyadenosine is in-

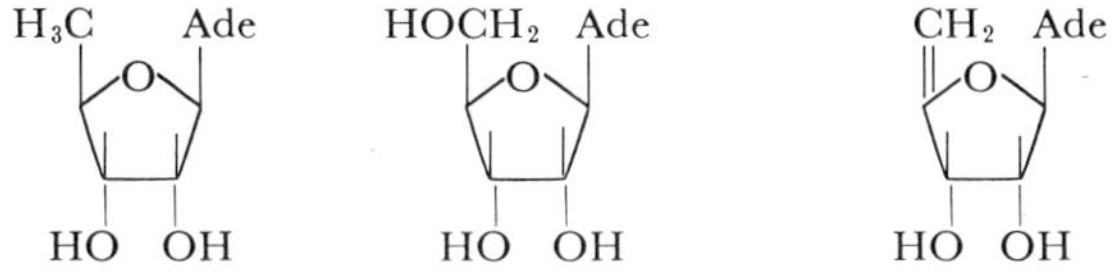

H_3C Ade / HO OH — 5′-Deoxyadenosine; $HOCH_2$ Ade / HO OH — Adenosine; CH_2 Ade / HO OH — 4′,5′-Anhydroadenosine (4-35)

creased in the presence of cob(II)alamin. For enzyme incubated at 37° with an effector nucleotide, dihydrolipoate, cob(II)alamin, and 5′-deoxyadenosine, direct binding studies have shown the dissociation constants for the latter two compounds to be 10.5 and 13 μM, respectively. When inhibition of ribonucleotide reductase by cob(II)alamin and 5′-deoxyadenosine was investigated (167), the K_i for cob(II)alamin was 37 μM in the absence of 5′-deoxyadenosine and 3.0 μM in its presence. The nucleoside failed to inhibit the enzyme in the absence of cob(II)alamin, but became quite an effective inhibitor (K_i 14 μM) when cob(II)alamin was present as well. Both inhibitors were competitive with respect to adenosylcobalamin. With this enzyme, then, the existence of both binding sites is well established.

The Mechanism of the Reaction

As discussed previously, the cofactor is an intermediate hydrogen carrier in adenosylcobalamin-dependent rearrangements. In these reactions, the mechanism appears to involve homolysis of the carbon-cobalt bond followed by the transfer of hydrogen from the substrate to the 5′-deoxyadenos-5′-yl radical to form 5′-deoxyadenosine. Evidence obtained with *L. leichmannii* ribonucleotide reductase suggests that the participation of adenosylcobalamin in ribonucleotide reduction may involve similar processes.

The initial evidence for the direct participation of adenosylcobalamin in hydrogen transfer was the demonstration of a ribonucleotide reductase-catalyzed exchange reaction involving the 5′-hydrogen atoms of the cofactor (169–172). The ribonucleotide reductase-dependent hydrogen exchange differed from that catalyzed by other adenosylcobalamin-dependent enzymes in that tritium on the 5′-carbon of the cofactor was transferred not to product, but to water (171). On the other hand, label from tritiated water could be incorporated into both cofactor and product. These results are consistent with the scheme shown in 4-36,

$$\begin{array}{ccccc} {}^{3}HOH & & AdoCbl & & [{}^{3}H]2'\text{-}dNTP \\ & \text{fast} & & \text{slow} & \\ H_2O & & [{}^{3}H]AdoCbl & & NTP \end{array} \qquad (4\text{–}36)$$

provided exchange between water and cofactor is rapid compared with the transfer of hydrogen from cofactor to product.

In addition to enzyme and cofactor, a thiol and an effector molecule are required for the exchange to take place (169–172). Suitable thiols are those which are able to reduce ribonucleotides in the enzymatic system. In the exchange reaction, however, the thiols act catalytically—that is, a single molecule of thiol is sufficient for the exchange of hydrogen into several molecules of cofactor (172). The requirement for effector probably indicates the need for the enzyme to bind a molecule of nucleotide in order to be converted to a catalytically active form. Whether the effector must be bound at the catalytic site, the regulator site, or both in order to promote the exchange reaction is not known. The effector does not itself appear to participate in the exchange reaction, since hydrogen is not incorporated into effector during the exchange of tritium from water into cofactor (172). (This is true if deoxyribonucleotide is used as effector. With ribonucleotides, which serve as both effectors and substrates, hydrogen will be transferred from water to the product, but only as a consequence of catalytic reduction.)

As mentioned before, the two hydrogen atoms attached to the 5′-carbon of the *Coβ*-adenosyl group can in principle be distinguished from each other by the enzyme. When the exchange of hydrogen from tritiated water into cofactor is permitted to reach equilibrium, however, the cofactor is found to have acquired 1.4 (i.e., > 1) g-atom of tritium per mole (172), suggesting that both the hydrogen atoms can be replaced in the exchange reaction. Using cofactor in which both C-5′ positions have been chemically tritiated, ribonucleotide reductase was found to catalyze the transfer of all the label to water (169). Complete exchange was obtained even at cofactor/enzyme ratios greatly exceeding unity, indicating that enzyme-bound cofactor readily dissociates and equilibrates with cofactor free in solution. These results indicate that the enzyme does not distinguish between the two C-5′ hydrogen atoms.

A similar observation made with enzymes catalyzing adenosylcobalamin-dependent rearrangements was explained by invoking the transient existence at the active site of 5′-deoxyadenosine, a species in which the C-5′ hydrogen atoms can become equivalent through rotation of the methyl group about the C-4′–C-5′ bond. This hypothesis was based in part on experiments showing that in the presence of substrate or an appropriate substrate analog, several of these enzymes are able to split the carbon-cobalt bond of adenosylcobalamin to give 5′-deoxyadenosine as one of the products. Ribonucleotide reductase too has been found to split the carbon-cobalt bond of the cofactor with the production of 5′-deoxyadenosine (167, 168). Though 5′-deoxyadenosine production is a slow reaction, requiring an hour or so to reach completion, its mechanistic significance is suggested by the fact that, like the hydrogen exchange reaction, the cleavage reaction takes place only in the presence of thiol (dihydrolipoate was used in these experiments) and effector. To the extent that irreversible cleavage of the carbon-cobalt bond with the production of 5′-deoxyadenosine constitutes support for the participation of that deoxynucleoside in catalysis, these experiments suggest that ribonucleotide reductase may generate 5′-deoxyadenosine during the reduction of a ribonucleotide to a deoxyribonucleotide.

In a further resemblance between the mechanism of action of ribonucleotide reductase and other adenosylcobalamin-dependent enzymes, cleavage of the carbon-cobalt bond appears to be a homolytic process (167, 168, 173). The first evidence for homolysis was the observation that in the slow reaction in which 5′-deoxyadenosine is produced, the cobamide residue was converted to cob(II)alamin, identified by its characteristic optical and esr spectra (Figure 4-16). This process took place without inactivation of the enzyme; in fact, when adenosylcobalamin was present in stoichiometric excess over enzyme, the cleavage of cofactor with the production of cob(II)alamin occurred catalytically (167). The possibility was considered that carbon-cobalt bond cleavage was heterolytic, generating cob(I)alamin as the immediate product, with cob(II)alamin arising from the oxidation of the former. The fact that the rate of cob(II)alamin production did not increase as the pH was lowered, even though the rate of oxidation of cob(I)alamin in water is known to increase with a fall in pH, and the failure of the cobalamin to be alkylated by iodoacetamide, methyl iodide, or acetylene present in the reaction mixture, were construed as suggestive evidence that cob(II) alamin was produced directly rather than through oxidation of cob(I)alamin (167).

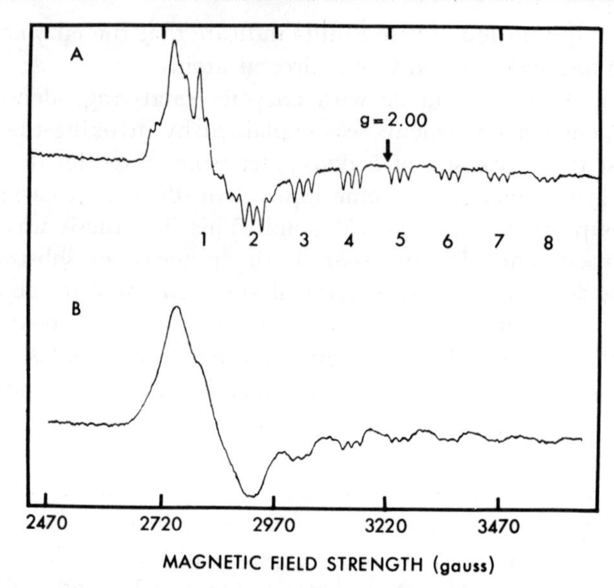

Fig. 4-16 Esr spectra of cobamides formed from AdoCbl by ribonucleotide reductase. [Reprinted with permission from Hamilton, J. A., Blakley, R. L., Looney, F. D., and Winfield, M. E. (1969). *Biochim. Biophys. Acta* **177,** 374–376.]

Shortly after cob(II)alamin production was reported, a second paramagnetic species was found to be produced by ribonucleotide reductase (174, 175). This species was characterized by a doublet signal in the free radical region of the esr spectrum, similar in appearance to the doublet signals observed with diol dehydrase, glycerol dehydrase, and ethanolamine ammonia-lyase (Figure 4-17). The kinetic properties of the doublet signal differed in several ways from those of the cob(II)alamin signal. First, the doublet signal was not generated in the absence of substrate. While cleavage of the carbon-cobalt bond and production of cob(II)alamin required only a thiol and an effector (i.e., deoxyribo-) nucleotide, the doublet signal only appeared if a reducible (i.e., ribo-) nucleotide was present as well. Second, the rate of appearance of the doublet signal was faster than that of the cob(II)alamin signal. Finally, the size of the doublet signal reached a maximum and then decreased with time, the disappearance of signal correlating with either the destruction of AdoCbl or the exhaustion of the substrate, depending on the reaction conditions, while the size of the cob(II)alamin signal increased monotonically with time.

Apart from the fact that the unpaired electron is not on the sulfur atom of the thiol, nothing is known about the identity of the species

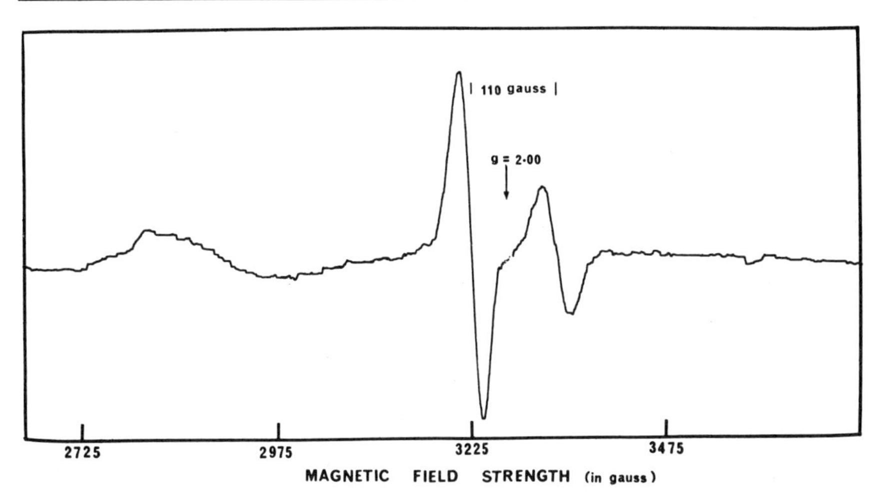

Fig. 4-17. Esr spectrum of the second paramagnetic species produced by ribonucleotide reductase. [Reprinted with permission from Hamilton, J. A., and Blakley, R. L. (1969). *Biochim. Biophys. Acta* **184**, 224–226.]

responsible for the doublet signal. The configuration of the signal varies in subtle ways depending on the identity of the thiol, the effector, and the substrate, but these variations in configuration have been attributed to slight differences in the conformation of the active site under the various reaction conditions. The major feature of the signal, the splitting between the components of the doublet, is not affected by the microwave frequency of the instrument used to measure the spectrum, indicating that the splitting represents an interaction between the electron which gives rise to the doublet signal and another magnetic species in its vicinity. The only evidence regarding the identity of the nearby magnetic species is the failure of the signal to be altered when the reaction is conducted in deuterated water, showing that the interaction which splits the signal is not with an exchangeable proton.

The irreversible cleavage reaction which produces 5′-deoxyadenosine and cob(II)alamin is too slow to participate as a step in catalysis. Similarly, the low rate of appearance of the doublet signal excludes the participation of the doublet species in the catalytic reaction. However, a third paramagnetic species was found which was produced at a catalytically competent rate. The first clue to the existence of this species was the discovery of an extremely rapid spectral change which occurred on adding cofactor to a reaction mixture containing enzyme, thiol, and an effector nucleotide (178). The new spectrum, resembling that of cob(II)-

alamin (Figure 4-18), appeared at a rate of about 40 sec^{-1}, well in excess of the turnover number for ribonucleotide reductase of 2 to 3 sec^{-1}. At 37°, the extent of the spectral change after the rapid reaction had reached steady state indicated cleavage of about half the enzyme-bound cofactor. Cooling the reaction mixture to 5° caused the spectrum to revert to that of adenosylcobalamin; the cob(II)alamin spectrum reappeared when the mixture was rewarmed to 37° (Figure 4-19). These findings indicated a situation in which cofactor at the active site existed in two forms: a form with an intact carbon-cobalt bond, and a form in which the carbon-cobalt bond was broken. Changing the temperature of the reaction mixture shifted the equilibrium between the two forms, low temperatures favoring the intact form and higher temperatures the form with the ruptured bond:

$$\underset{[\text{Co}]}{\overset{\text{CH}_2\text{R}}{|}} \underset{5^\circ}{\overset{37^\circ}{\rightleftharpoons}} \underset{[\text{Co}]}{\overset{\text{CH}_2\text{R}}{\vdots}} \qquad (4\text{–}37)$$

When the reaction mixture contained substrate in addition to effector nucleotide, or contained substrate alone, the amplitude of the spectral change varied according to a two-step process, reaching a maximum and

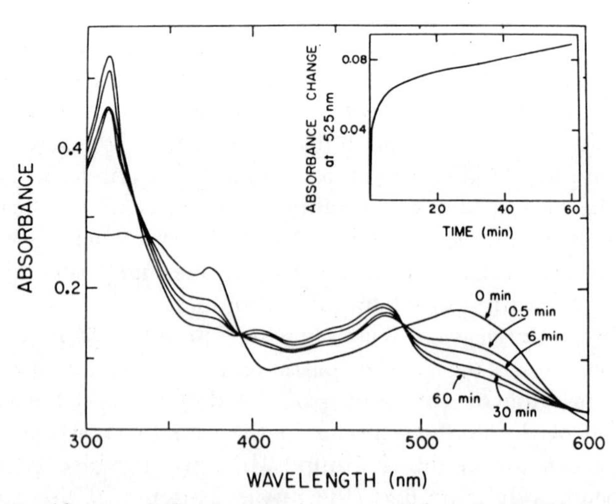

Fig. 4-18. Rapid spectral change occurring on addition of AdoCbl to a ribonucleotide reductase-containing reaction mixture [Reprinted with permission from Tamao, Y., and Blakley, R. L. (1973). *Biochemistry* **12**, 24–34. Copyright by the American Chemical Society.]

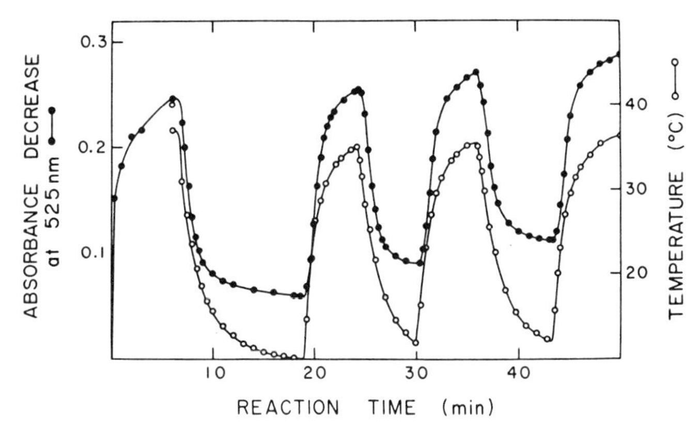

Fig. 4-19. Reversible absorbance changes due to shifts in the equilibrium of the ribonucleotide reductase-induced cleavage of AdoCbl with changes in temperature. [Reprinted with permission from Tamao Y., and Blakley, R. L. (1973). *Biochemistry* **12**, 24–34. Copyright by the American Chemical Society.]

then subsiding to an intermediate steady-state level (Figure 4-20). Similarly, addition of substrate to a reaction mixture containing enzyme, adenosylcobalamin, thiol, and effector (i.e., a reaction mixture in which the spectral change had reached a maximum) led to a diminution in the amplitude of the spectral change. The rate of the first step was relatively constant from nucleotide to nucleotide, although its amplitude varied, while both the rate and amplitude of the second step varied with the identity of the nucleotide. The second step was slower than the first step, but was still faster than the rate of turnover of substrate in the catalytic reaction. The involvement of a hydrogen abstraction in the second step was suggested by the isotope effect of 2.2 observed when [5′,5′-D_2]adenosylcobalamin was used as substrate in place of unlabeled adenosylcobalamin.

By rapid mixing quick freeze esr experiments, it was possible to confirm that one of the species generated by the rapid reaction was cob(II)-alamin, but in an environment leading to a very atypical esr spectrum (Figure 4-21) (177). This species was not seen before because the methods previously used to freeze the reaction mixture were so slow that the reconstitution of the carbon-cobalt bond that accompanied a decrease in the temperature of the reaction mixture had reached completion before freezing took place. The kinetic behavior of the esr signal was virtually identical to that of the rapid spectral change, so there is little doubt that the optical and esr spectra are produced by the same species. The rate of appearance of the esr spectrum excluded the possibility that

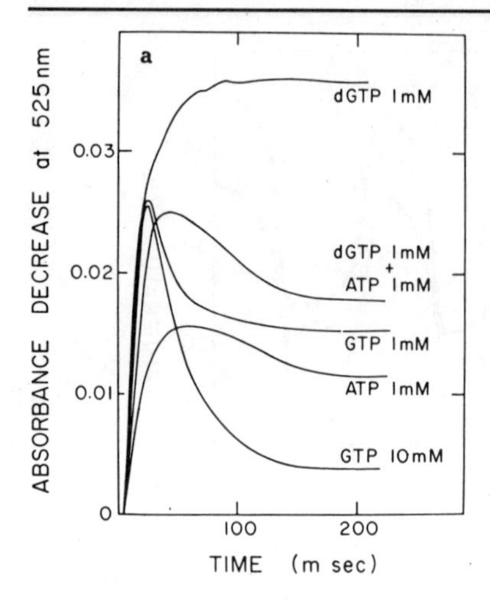

Fig. 4-20. Absorbance changes associated with the catalytic activity of ribonucleotide reductase. [Reprinted with permission from Tamao, Y., and Blakley, R. L. (1973). *Biochemistry* **12**, 24–34. Copyright by the American Chemical Society.]

cob(II)alamin was produced in a side reaction by the nonenzymatic oxidation of enzyme-generated cob(I)alamin. Cob(II)alamin seems to be the primary species produced by enzymatic cleavage of the carbon-cobalt bond, and it is produced at a rate and to an extent that strongly implies its involvement in catalysis of the ribonucleotide reductase reaction.

Sulfhydryl compounds appear to participate in a step which precedes the cleavage of the carbon-cobalt bond (178). Both dihydrolipoic acid and reduced thioredoxin are able to reduce ribonucleotide reductase, producing 2 moles of –SH for every mole of reductase. The reductase can then be reoxidized in the presence of adenosylcobalamin and substrate, with the production of 1 mole of deoxyribonucleotide for every mole of enzyme reoxidized. Production of both the doublet signal (175) and the rapid spectral change discussed above requires the reduced enzyme. The observations suggest that a disulfide bond on the enzyme undergoes cyclic reduction and reoxidation during catalysis of ribonucleotide reduction,[2] although the possibility cannot be excluded that the oxidized form is an inactive species that must be reduced for catalysis to take place.

The first three steps in the ribonucleotide reductase reaction can therefore be tentatively formulated as the reduction of a disulfide group

[2] Kinetic experiments and hydrogen exchange data have been interpreted to indicate that catalysis does not involve cyclic reduction and reoxidation of the enzyme. In a system as complex as the ribonucleotide reductase system, however, kinetic results are difficult to interpret, and the hydrogen exchange data may be affected by a tritium isotope effect as well as by the slow exchange rates sometimes seen with enzyme-bound protons.

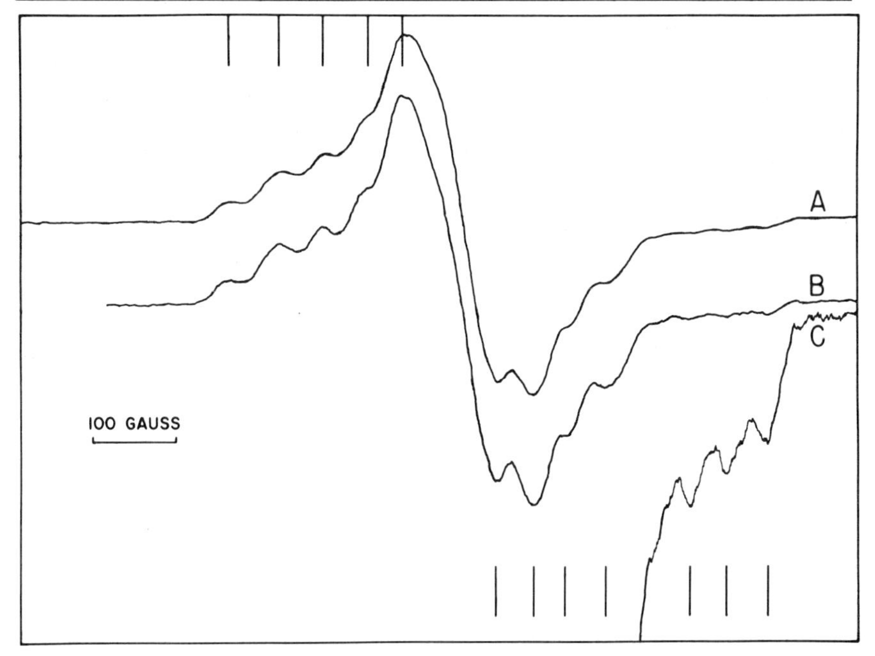

Fig. 4-21. Esr spectra of a ribonucleotide reductase-containing reaction mixture frozen quickly in isopentane at 130°K (177).

on the enzyme, the homolytic cleavage of the carbon-cobalt bond of the cofactor, and the abstraction of a hydrogen atom from one of the sulfhydryl groups to produce 5′-deoxyadenosine (4-38). Although this formu-

$$\left|\begin{array}{ll} CH_2R & S\text{—} \\ | & | \\ [Co] & S\text{—} \end{array}\right| \rightleftharpoons \left|\begin{array}{ll} CH_2R & HS\text{—} \\ | & \\ [Co] & HS\text{—} \end{array}\right| \rightleftharpoons \left|\begin{array}{ll} CH_2R & HS\text{—} \\ \cdot & \\ [\dot{C}o] & HS\text{—} \end{array}\right| \rightleftharpoons \left|\begin{array}{ll} CH_3R & \cdot S\text{—} \\ & \\ [\dot{C}o] & HS\text{—} \end{array}\right| \qquad (4\text{–}38)$$

lation accounts for all the experimental results to date, including the exchange of hydrogen between solvent and adenosylcobalamin, only the homolytic cleavage step is on really solid experimental footing. The location of the other unpaired electron produced when the hydrogen atom is abstracted by the 5′-deoxyadenos-5′-yl radical, if such a step is involved in catalysis of the reduction, and indeed the whole sequence of reactions that follow these first few steps, are unknown territory.

SUMMARY

Adenosylcobamide-dependent reactions fall into two classes: a group of rearrangements, and a reduction. In the rearrangements, a hydrogen atom moves from one carbon atom to an adjacent one in exchange for an alkyl, acyl, or electronegative group, which migrates in the opposite direction. The reduction involves the adenosylcobamide-dependent conversion of a ribonucleotide triphosphate to a deoxyribonucleotide triphosphate at the expense of NADPH in a reaction characterized by the displacement of the 2′-hydroxyl group of the nucleotide by a hydrogen atom. In all these reactions adenosylcobalamin is an intermediate hydrogen carrier, the migrating hydrogen atom residing for a time on the 5′-carbon of the *Coβ*-adenosyl group before it is transferred to the product. Transfer of the hydrogen atom to and from the cofactor involves species with unpaired electrons. Other aspects of the reaction mechanism are obscure.

REFERENCES

1. Reichard, P. (1968). *Eur. J. Biochem.* **3**, 259-266.
2. Barker, H. A., Weissbach, H., and Smyth, R. D. (1958). *Proc. Natl. Acad. Sci. U. S. A.* **44**, 1093–1097
3. Ohmori, H. Ishitani, H., Sato, K., Shimizu, S., and Fukui, S. (1971). *Biochem. Biophys. Res. Commun.* **43**, 156–162.
4. Barker, H. A., Rooze, V., Suzuki, F., and Iodice, A. A. (1964). *J. Biol. Chem.* **239**, 3260–3266.
5. Barker, H. A., Suzuki, F., Iodice, A., and Rooze, V. (1964). *Ann. N. Y. Acad. Sci.* **112**, 644–654.
6. Munch-Petersen, A., and Barker, H. A. (1958). *J. Biol. Chem.* **230**, 649–653.
7. Wachman, J. T. (1966). *J. Biol. Chem.* **223**, 19–27.
8. Barker, H. A., Smyth, R. D., Wawszkiewicz, E. J., Lee, M. N., and Wilson, R. M. (1958). *Arch. Biochem. Biophys.* **78**, 468-476.
9. Sprecher, M., Switzer, R. L., and Sprinson, D. B. (1966). *J. Biol. Chem.* **241**, 864–867.
10. Sprecher, M., and Sprinson, D. B. (1964). *Ann. N. Y. Acad. Sci.* **112**, 655–660.
11. Suzuki, F., and Barker, H. A. (1966). *J. Biol. Chem.* **241**, 878–888.
12. Switzer, R. L., and Barker, H. A. (1967). *J. Biol. Chem.* **242**, 2658-2674.
13. Swick, R. (1962). *Proc. Natl. Acad. Sci. U. S. A.* **48**, 288–293.
14. Eggerer, H., Stadtman, E. R., Overath, P., and Lynen, F. (1960). *Biochem. Z.* **333**, 1–9.
15. Kellermeyer, R. W., and Wood, H. G. (1962). *Biochemistry* **1**, 1124–1131.
16. Eggerer, H., Overath, P., Lynen, F., and Stadtman, E. R. (1960). *J. Am. Chem. Soc.* **82**, 2643–2644.
17. Phares, E. F., Long, M. V., and Carson, S. F. (1962). *Biochem. Biophys. Res. Commun.* **8**, 142–146.
18. Wood, H. G., Kellermeyer, R. W., Stjernholm, R., and Allen, S. H. G. (1964). *Ann. N. Y. Acad. Sci.* **112**, 661–679.

19. Phares, E. F., Long, M. V., and Carson, S. F. (1964). *Ann. N. Y. Acad. Sci.* **112**, 680–683.

20. Sprecher, M., Clark, M. J., and Sprinson, D. B. (1966). *J. Biol. Chem.* **241**, 872–877.

21. Sprecher, M., Clark, M. J., and Sprinson, D. B. (1964). *Biochem. Biophys. Res. Commun.* **15**, 581–587.

21a. Rétey, J., and Zagalak, B., (1973). *Angew. Chem. (Int. Ed.)* **12**, 671–672.

22. Cannata, J. J. B., Focesi, A., Jr., Mazumder, R., Warner, R. C., and Ochoa, S. (1965). *J. Biol. Chem.* **240**, 3249–3257.

23. Overath, P., Kellerman, G. M., Lynen, F., Fritz, H. P., and Keller, H. J. (1962). *Biochem. Z.* **335**, 500–518.

24. Kellermeyer, R. W., Allen, S. H. G., Stjernholm, R., and Wood, H. G. (1964). *J. Biol. Chem.* **239**, 2562–2569.

24a. Zagalak, B., Rétey, J., Sund, H. (1974). *Eur. J. Biochem.* **44**, 529–535.

25. Overath, P., Stadtman, E. R., Kellerman, G. M., and Lynen, E. (1962). *Biochem. Z.* **336**, 77–98.

26. Lengyel, P., Mazumder, R., and Ochoa, S. (1960). *Proc. Natl. Acad. Sci. U. S. A.* **46**, 1312–1318.

27. Beck, W. S., Flavin, M., and Ochoa, S. (1957). *J. Biol. Chem.* **229**, 997–1010.

28. Mazumder, R., Sasakawa, T., and Ochoa, S. (1963). *J. Biol. Chem.* **238**, 50–53.

29 Kung, H. F., Cedarbaum, S., Tsai, L., and Stadtman, T. C. (1970). *Proc. Natl. Acad. Sci. U. S. A.* **65**, 978–984.

30. Kung, H. F., and Tsai, L. (1971). *J. Biol. Chem.* **246**, 6436–6443.

31. Kung, H. F., Cederbaum, S., and Tsai, L. (1970). *Fed. Proc.* **29**, 343Abs.

32. Kung, H. F., and Stadtman, T. C. (1971). *J. Biol. Chem.* **246**, 3378–3388.

33. Abeles, R. H., and Lee, H. A., Jr. (1961). *J. Biol. Chem.* **236**, 2347–2350.

34. Lee, H. A., Jr., and Abeles, R. H. (1963). *J. Biol. Chem.* **238**, 2367–2373.

35. Zagalak, B., Frey, P. A., Karabatsos, G. L., and Abeles, R. H. (1966). *J. Biol. Chem.* **241**, 3028–3035.

36. Frey, P. A., Karabatsos, G. L., and Abeles, R. H. (1965). *Biochem. Biophys. Res. Commun.* **18**, 551–556.

37. Rétey, J., Umani-Ronchi, A., and Arigoni, D. (1966). *Experientia* **22**, 72–73.

38. Rétey, J., Umani-Ronchi, A., Seibl, J., and Arigoni, D. (1966). *Experientia* **22**, 502–503.

39. Manners, J. P., Morallee, K. G., and Williams, R. J. P. (1970). *J. Chem. Soc. Chem. Commun.* **1970**, 965–966.

40. Toraya, T., Sugimoto, Y., Tamao, Y., Shimizu, S., and Fukui, S. (1971). *Biochemistry* **10**, 3475–3484.

41. Toraya, T., Sugimoto, Y., Tamao, Y., Shimizu, S., and Fukui, S. (1970). *Biochem. Biophys. Res. Commun.* **41**, 1314–1320.

42. Toraya, T., and Fukui, S. (1972). *Biochim. Biophys. Acta* **284**, 536–548.

43. Toraya, T., Kondo, M., Isemura, Y., and Fukui, S. (1972). *Biochemistry* **11**, 2599–2606.

44. Essenberg, M. K., Frey, P. A., and Abeles, R. H. (1971). *J. Am. Chem. Soc.* **93**, 1242–1251.

45. Toraya, T., Uesaka, M., Kondo, M., and Fukui, S. (1973). *Biochem. Biophys. Res. Commun.* **52**, 350–355.
46. Bradbeer, C. (1965). *J. Biol. Chem.* **240**, 4675–4681.
47. Kaplan, B. H., and Stadtman, E. R. (1968). *J. Biol. Chem.* **243**, 1787–1793.
48. Kaplan, B. H., and Stadtman, E. R. (1968). *J. Biol. Chem.* **243**, 1794–1800.
49. Carty, T. J., Babior, B. M., and Abeles, R. H. (1974). *J. Biol. Chem.*, **249**, 1683-1688.
50. Babior, B. M., Carty, T. J., and Abeles, R. H. (1974). *J. Biol. Chem.*, **249**, 1689–1695.
51. Babior, B. M., (1969). *J. Biol. Chem.* **244**, 449–456.
52. Rétey, J., Suckling, C. J., Arigoni, D., and Babior, B. M. (1974). *J. Biol. Chem.*, **249**, 6359–6360.
53. Carty, T. J., Babior, B. M., and Abeles, R. H. (1971). *J. Biol. Chem.* **246**, 6313–6317.
54. Babior, B. M., and Li, T. K. (1969). *Biochemistry* **8**, 154–160.
55. Babior, B. M. (1969). *J. Biol. Chem.* **244**, 2917–2926.
56. Babior, B. M. (1969). *J. Biol. Chem.* **244**, 2927–2934.
57. Smiley, K. L., and Sobolov, M. (1964). *Ann. N. Y. Acad. Sci.* **112**, 706–712.
58. Smiley, K. L., and Sobolov, M. (1962). *Arch. Biochem. Biophys.* **97**, 538–543.
59. Pawełkiewicz, J., and Zagalak, B. (1965). *Acta Biochim. Polon.* **12**, 207–218.
60. Schneider, Z., Pech, K., and Pawełkiewicz, J. (1966). *Bull. Acad. Polon. Sci., Sér. Sci. Biol.* **14**, 7–12.
61. Schneider, Z., and Pawełkiewicz, J. (1966). *Acta Biochim. Polon.* **13**, 311–328.
62. Schneider, Z., Larsen, E. G., Jacobson, G., Johnson, B. C., and Pawełkiewicz, J. (1970). *J. Biol. Chem.* **245**, 3388–3396.
63. Zagalak, B. (1963). *Acta Biochim. Polon.* **10**, 387–398.
64. Stroiński, A., Schneider, Z., and Pawełkiewicz, J. (1967). *Bull. Acad. Polon. Sci. Sér. Sci. Biol.* **15**, 727–731.
65. Hong, S. L., and Barker, H. A. (1973). *J. Biol. Chem.* **248**, 41–49.
66. Morley, C. G. D., and Stadtman, T. C. (1972). *Biochemistry* **11**, 600–605.
67. Dekker, E. E., and Barker, H. A. (1968). *J. Biol. Chem.* **243** 3232–3237.
68. Tsai, L., and Stadtman, T. C. (1968). *Arch. Biochem. Biophys.* **125**, 210–225.
69. Morley, C. G. D., and Stadtman, T. C. (1971). *Biochemistry* **10**, 2325–2329.
70. Stadtman, T. C., and Tsai, L. (1967). *Biochem. Biophys. Res. Commun.* **28**, 920–926.
71. Dyer, J. K., and Costilow, R. N. (1970). *J. Bacteriol.* **101**, 77–83.
72. Tsuda, Y., and Friedmann, H. C. (1970). *J. Biol. Chem.* **245**, 5914–5926.
73. Somack, R. L., Bing, D. H., and Costilow, R. N. (1971). *Anal. Biochem.* **41**, 132–137.
74. Somack, R., and Costilow, R. N. (1973). *Biochemistry* **12**, 2597–2604.
75. Morley, C. G. D., and Stadtman, T. C. (1970). *Biochemistry* **9**, 4890–4900.
76. Stadtman, T. C. (1963). *J. Biol. Chem.* **238**, 2766–2773.
77. Dyer, J. K., and Costilow, R. N. (1968). *J. Bacteriol.* **96**, 1617–1622.
78. Baker, J. J., van der Drift, C., and Stadtman, T. C. (1973). *Biochemistry* **12**, 1054–1063.

79. Stadtman, T. C. (1964). *Ann. N. Y. Acad. Sci.* **112**, 728–734.
80. Costilow, R. N., Rochovausky, O. M., and Barker, H. A. (1966). *J. Biol. Chem.* **241**, 1573–1580.
81. Stadtman, T. C. (1962). *J. Biol. Chem.* **237**, PC2409–PC2411.
82. Stadtman, T. C., and Renz, P. (1968). *Arch. Biochem. Biophys.* **125**, 226–239.
83. Bray, R. C., and Stadtman, T. C. (1968). *J. Biol. Chem.* **243**, 381–385.
84. Zagalak, B., and Pawełkiewicz, J. (1965). *Acta Biochim. Polon.* **12**, 103–113.
85. Zagalak, B., and Pawełkiewicz, J. (1965). *Acta Biochim. Polon.* **12**, 219–228.
86. Pawełkiewicz, J., and Zagalak, B. (1964). *Ann. N. Y. Acad. Sci.* **112**, 703–705.
87. Zagalak, B., and Pawełkiewicz, J. (1964). *Acta Biochim. Polon.* **11**, 49–59.
88. Barker, H. A., Smyth, R. D., Weissbach, H., Toohey, J. I., Ladd, J. N., and Volcani, B. E. (1960). *J. Biol. Chem.* **235**, 480–488.
89. Hogenkamp, H. P. C., and Oikawa, T. G. (1964). *J. Biol. Chem.* **239**, 1911–1916.
90. Toohey, J. I., Perlman, D., and Barker, H. A. (1961). *J. Biol. Chem.* **236**, 2119–2127.
91. Stjernholm, R., and Wood, H. G. (1961). *Proc. Natl. Acad. Sci. U. S. A.* **47**, 303–313.
92. Yamane, T., Shimizu, S., and Fukui, S. (1965). *Biochim. Biophys. Acta* **110**, 616–618.
93. Uchida, Y., Hayashi, M., and Kamikubo, T. (1973). *Vitamins (Japan)* **47**, 27–32.
94. Tamao, Y., Morikawa, Y., Shimizu, S., and Fukui, S. (1967). *Biochem. Biophys. Res. Commun.* **28**, 692–698.
95. Kerwar, S. S., Smith, T. A., and Abeles, R. H. (1970). *J. Biol. Chem.* **245**, 1169–1174.
96. Tamao, Y., Morikawa, Y., Shimizu, S., and Fukui, S. (1968). *Biochim. Biophys. Acta* **151**, 260–266.
97. Friedrich, W., Heinrich, H. C., Konigk, E., and Schulze, P. (1964). *Ann. N. Y. Acad. Sci.* **112**, 601–614.
98. Jayme, M., and Richards, J. H. (1971). *Biochem. Biophys. Res. Commun.* **43**, 1329–1333.
99. Blakley, R. L. (1965). *J. Biol. Chem.* **240**, 2173-2180.
100. Vitols, E., Brownson, C., Gardiner, W., and Blakley, R. L. (1967). *J. Biol. Chem.* **242**, 3035–3041.
101. Morley, C. G. D., Blakley, R. L., and Hogenkamp, H. P. C. (1968). *Biochemistry* **7**, 1231–1239.
102. Blakley, R. L. (1966). *Fed. Proc.* **25**, 1633–1638.
103. Weisblat, D. A., and Babior, B. M. (1971). *J. Biol. Chem.* **246**, 6064–6071.
104. Babior, B. M., Kon, H., and Lecar, H. (1969). *Biochemistry* **8**, 2662–2669.
105. Erfle, J. D., Clark, J. M., Nystrom, R. F., and Johnson, B. C. (1964). *J. Biol. Chem.* **239**, 1920–1924.
106. Iodice, A. A., and Barker, H. A. (1963). *J. Biol. Chem.* **238**, 2094–2097.
107. Brownstein, A. M., and Abeles, R .H. (1961). *J. Biol. Chem.* **236**, 1199–1200.
108. Jencks, W. P. (1969). *Catalysis in Chemistry and Enzymology,* McGraw-Hill, New York, pp. 204–207.

109. Abeles, R. H., and Zagalak, B. (1966). *J. Biol. Chem.* **241**, 1245–1246.

110. Frey, P. A., Essenberg, M. K., and Abeles, R. H. (1967). *J. Biol. Chem.* **242**, 5369–5377.

111. Frey, P. A., and Abeles, R. H. (1966). *J. Biol. Chem.* **241**, 2732–2733.

112. Abeles, R. H., and Frey, P. A. (1966). *Fed. Proc.* **25**, 1639–1641.

113. Rétey, J., and Arigoni, D. (1966). *Experientia* **22**, 783-784.

114. Switzer, R. L., Baltimore, B. G., and Barker, H. A. (1969). *J. Biol. Chem.* **244**, 5263–5268.

115. Babior, B. (1968). *Biochim. Biophys. Acta* **167**, 456–458.

116. Cardinale, G. J., and Abeles, R. H. (1967). *Biochim. Biophys. Acta* **132**, 517–518.

117. Rétey, J., Kunz, F., Stadtman, T. C., and Arigoni, D. (1969). *Experientia* **25**, 801–802.

118. Miller, W. W., and Richards, J. H. (1969). *J. Am. Chem. Soc.* **91**, 1498–1507.

119. Eagar, R. G., Jr., Baltimore, B. G., Herbst, M. M., Barker, H. A., and Richards, J. H. (1972). *Biochemistry* **11**, 253–264.

120. Ingraham, L. L. (1964). *Ann. N. Y. Acad. Sci.* **112**, 713–720.

121. Wagner, O. W., Lee, H. A., Jr., Frey, P. A., and Abeles, R. H. (1966). *J. Biol. Chem.* **241**, 1751–1762.

122. Babior, B. M. (1970). *J. Biol. Chem.* **245**, 1755–1766.

123. Babior, B. M. (1970). *J. Biol. Chem.* **245**, 6125–6133.

124. Babior, B. M., Woodams, A. D., and Brodie, J. D. (1973). *J. Biol. Chem.* **248**, 1445–1450.

125. Finlay, T. H., Valinsky, J., Sato, K., and Abeles, R. H. (1972). *J. Biol. Chem.* **247**, 4197–4207.

126. Babior, B., and Gould, D. C. (1969). *Biochem. Biophys. Res. Commun.* **34**, 441–447.

127. Abeles, R. H., and Lee, H. A., Jr., (1964). *Ann. N. Y. Acad. Sci.* **112**, 695–702.

128. Babior, B. M. (1969). *Biochim. Biophys. Acta* **178**, 406–408.

129. Babior, B. M., Moss, T. H., and Gould, D. C. (1972). *J. Biol. Chem.* **247**, 4389–4392.

130. Law, P. Y., Brown, D. G., Lien, E. L., Babior, B. M., and Wood, J. M. (1971). *Biochemistry* **10**, 3428–3435.

131. Finlay, T. H., Valinsky, J., Mildvan, A. S., and Abeles, R. H. (1973). *J. Biol. Chem.* **248**, 1285–1290.

132. Cockle, S. A., Hill, H. A. O., Williams, R. J. P. Davies, S. P., and Foster, M. A. (1972). *J. Am. Chem. Soc.* **94**, 275–277.

133. Babior, B. M., Orme-Johnson, W. H., Beinert, H., and Moss, T. H. (1974). *J. Biol. Chem.* **249**, 4537–4544.

134. Schrauzer, G. N., and Sibert, J. W. (1970). *J. Am. Chem. Soc.* **92**, 1022–1030.

135. Schrauzer, G. N., and Windgassen, R. J. (1967). *J. Am. Chem. Soc.* **89**, 143–147.

136. Schrauzer, G. N., Holland, R. J., and Seck, J. A. (1971). *J. Am. Chem. Soc.* **93**, 1503–1505.

136a. Schrauzer, G. N., Seck, J. A., and Holland, R. J. 1973). *Z. Naturforsch.* **28C**, 1.

137. Frey, P. A., Essenberg, M. K., Abeles, R. H., and Kerwar, S. S. (1970). *J. Am. Chem. Soc.* **92**, 4488–4489.

138. Lowe, J. N., and Ingraham, L. L. (1971). *J. Am. Chem. Soc.* **93**, 3801–3802.

139. Golding, B. T., Holland, H. L., Horn, U., and Sakrikar, S. (1970). *Angew. Chem. (Int. Ed.)* **9**, 959–960.

140. Silverman, R. B., Dolphin, D., and Babior, B. M. (1972) *J. Am. Chem. Soc.* **94**, 4028–4030.

141. Gleason, F. K., and Hogenkamp, H. P. C. (1972). *Biochim. Biophys. Acta* **277**, 466–470.

142. Cowles, J. R., Evans, H. J., and Russell, S. A. (1969). *J. Bacteriol.* **97**, 1460–1465.

143. Beck, W. S., and Hardy, J. (1965). *Fed. Proc.* **24**, 421.

144. Dowling, M., Adams, A., and Helleuga, L. (1965). *Biochim. Biophys. Acta* **108**, 233–242.

145. Beck, W. S., and Hardy, J. (1965). *Proc. Natl. Acad. Sci. U. S. A.* **54**, 286–293.

146. Abrams, R., and Duraiswami, S. (1965). *Biochem Biophys. Res. Commun.* **18**, 409–414.

147. Blakley, R. L., and Barker, H. A. (1964). *Biochem. Biophys. Res. Commun.* **16**, 391–397.

148. Panagou. D., Orr, J. R., and Blakley, R. L. (1972). *Biochemistry* **11**, 2378–2388.

149. Abrams, R. (1965). *J. Biol. Chem.* **240**, PC3697.

150. Goulian, M., and Beck, W. S. (1966). *J. Biol. Chem.* **241**, 4233–4242.

151. Gleason, F. K., and Hogenkamp, H. P. C. (1970). *J. Biol. Chem.* **245**, 4894–4899.

152. Beck, W. S. (1967). *J. Biol. Chem.* **242**, 3148–3158.

153. Blakley, R. L., Ghambeer, R. K., Nixon, P. F., and Vitols, E. (1965). *Biochem. Biophys. Res. Commun.* **20**, 439–445.

154. Vitols, E., and Blakley, R. L. (1965). *Biochem, Biophys. Res. Commun.* **21**, 466–472.

155. Cowles, J. R., and Evans, H. J. (1968). *Arch. Biochem. Biophys.* **127**, 770–778.

156. Laurent, T. C., Moore, E. C., and Reichard, P. (1964). *J. Biol. Chem.* **239**, 3436–3444.

157. Orr, M. D., and Vitols, E. (1966). *Biochem. Biophys. Res. Commun.* **25**, 109–115.

158. Beck, W. S., Goulian, M., Larsson, A., and Reichard, P. (1966). *J. Biol. Chem.* **241**, 2177–2179.

159. Moore, E. C., Reichard, P., and Thelander, L. (1964). *J. Biol. Chem.* **239**, 3445–3452.

160. Thelander, L. (1967). *J. Biol. Chem.* **242**, 852–859.

161. Suhadolnik, R. J., Finkel, S. J., and Chassy, B. M. (1968). *J. Biol. Chem.* **243**, 3532–3537.

162. Blakley, R. L., Ghambeer, R. K., Batterham, T. J., and Brownson, C. (1966). *Biochem. Biophys. Res. Commun.* **24**, 418–426.

163. Gottesman, M. M., and Beck, W. S. (1966). *Biochem. Biophys. Res. Commun.* **24**, 353–359.

164. Batterham, T. J., Ghambeer, R. K., Blakley, R. L., and Brownson, C. (1967). *Biochemistry* **6**, 1203–1208.

165. Griffin, C. E., Hamilton, F. D., Hopper, S. P., and Abrams, R. (1968). *Arch. Biochem. Biophys.* **126**, 905–911.

166. Follman, H., and Hogenkamp, H. P. C. (1969). *Biochemistry* **8**, 4372–4375.

167. Yamada, R., Tamao, Y., and Blakley, R. L. (1971). *Biochemistry* **10**, 3959-3968.

168. Hamilton, J. A., Yamada, R., Blakley, R. L., Hogenkamp, H. P. C., Looney, F. D., and Winfield, M. E. (1971). *Biochemistry* **10**, 347–355.

169. Abeles, R. H., and Beck, W. S. (1967). *J. Biol. Chem.* **242**, 3589–3593.

170. Hogenkamp, H. P. C., Ghambeer, R. K., Brownson, C., and Blakley, R. L. (1967). *Biochem J.* **103**, 5C–7C.

171. Beck, W. S., Abeles, R. H., and Robinson, W. G. (1966). *Biochem. Biophys. Res. Commun.* **25**, 421–425.

172. Hogenkamp, H. P. C., Ghambeer, R. K., Brownson, C., Blakley, R. L., and Vitols, E. (1968). *J. Biol. Chem.* **243**, 799–808.

173. Hamilton, J. A., Blakley, R. L., Looney, F. D., and Winfield, M. E. (1969). *Biochim. Biophys. Acta* **177**, 374–376.

174. Hamilton, J. A., Tamao, Y., Blakley, R. L., and Coffman, R. E. (1972). *Biochemistry* **11**, 4696–4705.

175. Hamilton, J. A., and Blakley, R. L. (1969). *Biochim. Biophys. Acta* **184**, 224–226.

176. Tamao, Y., and Blakley, R. L. (1973). *Biochemistry* **12**, 24–34.

177. Orme-Johnson, W. H., Beinert, H., and Blakley, R. L. (1974). *J. Biol. Chem.* **249**, 2338–2343.

178. Vitols, E., Hogenkamp, H. P. C., Brownson, C., Blakley, R. L., and Connellan, J. (1967). *Biochem. J.* **104**, 58C–60C.

PATHOPHYSIOLOGY

CHAPTER FIVE

ABSORPTION AND TRANSPORT OF COBALAMIN

INTRINSIC FACTOR AND THE TRANSCOBALAMINS

LEON ELLENBOGEN

Lederle Laboratories
Pearl River, New York

CONTENTS

ABSORPTION

The absorption of physiological amounts of cobalamin has attracted more interest than the absorption of any other vitamin. It is probably the only vitamin that requires facilitation of its absorption by another substance normally secreted in the stomach. It is of interest that so special an absorption mechanism is needed for a vitamin for which the daily requirement is only about 1 μg. In discussing the relationship of intrinsic factor, cobalamin absorption, and pernicious anemia, Castle (1) said: "Thus this disease would not develop if the patient could effect daily the transfer of a millionth of a gram of vitamin B_{12} the distance of a small fraction of a millimeter across the intestine and into the blood stream. This he cannot do, principally as a result of failure of his stomach to secrete into its lumen some essential but still unknown substance." The existence of this essential substance called intrinsic factor was postulated by Castle and his associates (2, 3) more that 40 years ago.

Since the pioneer work of Castle, some progress has been made in understanding the mechanism by which ingested cobalamin is transported across the intestinal wall, a process mediated by intrinsic factor, and enters the blood bound to specific carrier proteins (transcobalamins) from which cobalamin is discharged to needed tissues. This chapter is concerned with this problem. The role of these carrier proteins in this absorption and transport can be much better appreciated, if, in addition, their properties, including their sites of secretion, assay, purification, and so on, are also discussed. There has been major progress recently in some of these areas.

It is widely accepted today that cyanocobalamin, also referred to as vitamin B_{12}, is not the naturally occurring form of the vitamin, but is rather an artifact which arises from the original isolation procedure. Cyanocobalamin, however, is the most widely used form of the cobamides in clinical practice because of its stability. Most of the studies on gastrointestinal absorption and transport studies utilized cyanocobalamin. However, as is discussed in the relevant portions of this chapter, there is no indication from the limited data available so far that the mechanism of absorption and transport is significantly different than that observed with the naturally occurring cobalamins. The relative rates of absorption of these cobalamins as well as the conversion of cyanocobalamin to the naturally occurring forms are also discussed herein.

SITE OF SECRETION OF INTRINSIC FACTOR

Human

Castle (2) was the first to show that the ingestion of normal human gastric juice mixed with beef muscle protein caused an erythropoietic response in pernicious anemia patients. This basic and pioneering work of Castle established the concept of the gastric origin of intrinsic factor.

Although saliva binds cobalamin (4, 5), Castle's early work (3) showed that saliva has no intrinsic factor activity. Castle et al. (3) also showed that normal human duodenal juice did not exhibit intrinsic factor activity and this was confirmed by Landboe-Christensen and Bohn (6). Further studies by Landboe-Christensen et al. (7) showed that desiccated preparations of normal human duodenum, but not jejunum, contained a small amount of intrinsic factor activity. This was undoubtedly due to contamination of the tissue by gastric juice, as patients who have undergone total gastric resection cannot absorb cobalamin, and intestinal juices of such patients are devoid of intrinsic factor activity (8–10).

Normal human stomach possesses intrinsic factor activity (11). More specifically, the pyloric region of the stomach is inactive whereas the fundic (corpus) and cardiac regions have been identified as the sites of intrinsic factor formation (12–14). These results are in agreement with histological findings that atrophy of the stomach, characteristically seen in pernicious anemia, occurs in the fundic area, but not in the pyloric region (15–17). Additional evidence for the belief that the fundic part of the human stomach is responsible for intrinsic factor production comes from the report of impaired cobalamin absorption in a patient who underwent a selective resection of this portion of the stomach (18).

The stomach seems to be the only source of intrinsic factor, as indicated by numerous reports on the development of pernicious anemia after several years of total gastrectomy (10, 19–23). After partial gastrectomy, in which the fundus is left intact, megaloblastic anemia or malabsorption of cobalamin does not develop (24).

With the use of autoradiographic and immunological procedures, Hoedemaeker et al (25, 26) showed that intrinsic factor is elaborated by the hydrochloric acid-secreting parietal cells. Autoradiographs were prepared from sections of human stomach incubated with CN[^{57}Co]Cbl. The uptake of cyanocobalamin by the stomach was abolished by prior incubation of the gastric section with antiintrinsic factor serum. Intrinsic factor is also produced by the parietal cells in the cat, rabbit, monkey, and ox (25, 26). Cellular fractionation studies of human gastric musocal extracts indicate that intrinsic factor activity is present in both the mitochondria and particle-free supernatant fluid (27).

Since the nutrition of the fetus takes place transcapillarily through the placenta, there appears no reason to believe that the fetus could make use of gastric intrinsic factor. By means of radioimmune assays, however, Schwartz and Weber (28) have recently shown that intrinsic factor is produced in the fetal stomach from about the eleventh to the thirteenth week. While intrinsic factor is found mostly in the fundus following delivery, a considerable amount of intrinsic factor is found in the pyloric part of the stomach of the fetus. It therefore appears that the production of intrinsic factor begins several months before the substance can be presumed to be of any value to the individual, and intrinsic factor may be one of the first important substances to appear in gastric juice and one of the last to disappear.

Hog

Extensive studies have been performed with hog intrinsic factor because it is more readily available than human intrinsic factor in large quantities and has been shown to be an effective source of intrinsic factor for the oral treatment of pernicious anemia (29). The effectiveness of hog intrinsic factor in humans was confirmed by various investigators who used fresh, whole hog stomach which was desiccated (30–32). In contrast to man, intrinsic factor is not found in the fundic area of the stomach, but mainly in the pyloric region, possibly with a trace in the cardiac portion (12, 13, 33, 34). This difference between the hog and human is rare in the study of comparative physiology.

Meulengracht (14, 33, 35) showed that the mucosal portion of the pyloric section of the hog is the richest source of intrinsic factor. The trace of activity found in the muscularis was probably due to contamination from the mucosa (35). Although the assay of intrinsic factor in the early years following Castle's classical pioneering studies was not as quantitative as in the last decade, nothing has appeared in the recent literature which would change these conclusions.

Intrinsic factor activity is also present in hog gastric juice and duodenal juice and in the duodenum itself (6, 34, 36–43). On the other hand, Heinrich (44) reported that crude lyophilized duodenum from the pylorectomized hog did not exhibit intrinsic factor activity in pernicious anemia patients. Dexter et al. (45) presented evidence that intrinsic factor activity which Uotila (46, 47) had reported to be present in the hog ileum, was due to adsorption on the ileal wall of intrinsic factor which had been secreted higher up in the gastrointestinal tract.

The most recent studies on the isolation of intrinsic factor have been done with hog pyloric mucosa and human gastric juice (48–53) (see the section on isolation and properties of intrinsic factor).

Rats

While the hog has been the major source of intrinsic factor for therapeutic purposes, a multitude of studies have been performed with rat preparations for the purpose of studying the mechanisms of cobalamin absorption.

Gastric juice, whole-stomach homogenates, and extracts of the glandular portion of the stomach of the rat possess intrinsic factor activity in the rat (54–57). Rat gastric juice appears to be as effective as human gastric juice in patients with pernicious anemia, even though normal human gastric juice and different preparations derived from hog stomach are ineffective in the gastrectomized rat (58). The anatomical distribution of intrinsic factor in the rat is similar to that observed in man; the fundic portion of the rat stomach contains most, if not all of the intrinsic factor activity (59). This was determined by measuring the effect of different areas of the rat stomach on cobalamin absorption in the gastrectomized rat.

These latter investigators as well as Hoedemaeker (25) showed by autoradiographic studies that the chief (pepsinogen) cells of the fundus appeared to secrete intrinsic factor, in contrast to findings that in humans intrinsic factor is produced by the parietal cells.

The studies of Boass and Wilson (60), who observed a simultaneous development of intrinsic factor and pepsinogen in growing rats, are consistent with the hypothesis that intrinsic factor and pepsinogen are produced by the same cell.

Recently, however, Håkanson et al. (61) presented evidence that the system of enterochromaffin-like cells (cells which also store histamine and various monoamines), which is restricted to the basal portion of the oxyntic gland area, may be the source of intrinsic factor. This evidence was based on the correlation between the distribution of this cell system and the appearance of gastric cobalamin binding proteins in neonatal rats. Håkanson et al. (61) feel it would be most unusual for such a physiologically important substance as intrinsic factor to be produced by different cell types in various species.

Further study by these investigators (62) showed that reserpine, which mobilizes histamine from the enterochromaffin-like cells in the oxyntic gland, also caused an almost total depletion of gastric cobalamin binding protein in the rat. However, the intestinal absorption of cobalamin was only slightly impaired. Thus most of the intrinsic factor activity persisted. Either intrinsic factor constitutes a minor portion of the cobalamin binding proteins in the stomach, or the formation and continuous release of intrinsic factor are sufficient, under conditions of impaired storage, to facilitate cobalamin absorption. The investigators believe the latter possibility to be the more likely one.

Other Species and Species Specificity

The studies of Schwartz et al. (63), Abels (64), and especially the work of Hippe and Schwartz (65) and Wilkinson (66–68) have provided evidence for the existence or absence of intrinsic factor in many different animal species. This evidence is based on the ability of different parts of the stomach of these animals to facilitate the absorption of cobalamin in pernicious anemia patients. The purpose of these studies was to gather further information and understanding of intestinal cobalamin absorption and possible species specificities. The activity of the different animal preparations is summarized in Table 5-1.

Table 5-1 Intrinsic Factor Activity in Various Species, Based on Enhancement of Cobalamin Absorption in Patients with Pernicious Anemia

Intrinsic Factor Detected		No Intrinsic Factor Detected		
Man	Rabbit	Dog	Hippopotamus	Lobster
Monkey	Hamster	Guinea pig	Deer	Alligator
Pig	Wild boar	Horse	Chicken	Lizard
Rat	Silver fox	Sheep	Toad	Crocodile
Cow	Lion	Elephant	Frog	Eel
Ferret	Tiger	Giraffe	Black slug	Sea trout
	Leopard	Rhinoceros	Edible snail	

Most of the studies have been carried out on a small portion of the gastrointestinal tract; therefore, the possibility exists that intrinsic factor could be produced elsewhere in the animals. Failure of a preparation to augment cobalamin absorption in the human does not necessarily mean that these animals lack an intrinsic factor mechanism; these negative results may be the results of species specificity. For example, Kaplan and Stein (69) suggest that the eel has an intrinsic factor mechanism based on the fact that a crude homogenate of eel gastric mucosa enhanced the cobalamin uptake of the distal portion of the eel intestine. In this regard, it is of interest that rat intrinsic factor is active in pernicious anema patients and in monkeys. Human gastric juice and monkey intrinsic factor are only weakly active in gastrectomized rats.

In addition to the study of the activity of these materials in the enhancement of cobalamin absorption, some studies have been initiated with regard to species specificity of cobalamin uptake *in vitro* by intestinal sacs and intestinal mucosal homogenates [see the review by Glass (70)]. These studies, although of interest in connection with the mechanism of action of intrinsic factor, may have no bearing in the role of species differences.

Studies have been initiated to determine cross-reactivity of these intrinsic factor preparations by immunological techniques. For example, in various preparations from animals there is almost complete correlation between the ability to function as intrinsic factor in man and the ability of human autoimmune anti-intrinsic factor antibody to inhibit the specific cobalamin binding to these preparations (65). One noteworthy exception is the reaction of guinea pig mucosa with human intrinsic factor antibody. This guinea pig preparation is inactive in the intestine of pernicious anemia patients. There may be immunological similarities among various preparations but there is evidence that there are two important "sites" on the intrinsic factor molecule, the cobalamin binding site and the intestinal receptor site (see the sections pertaining to these). This means that the guinea pig preparation has a similar cobalamin binding site to human intrinsic factor but that the intestinal receptor site is different.

Although the chemical structure of intrinsic factor remains to be elucidated, it appears as suggested by Hippe and Schwartz (65) that intrinsic factor of man, monkey, pig, and rat would prove to be different from dog, cat, and chicken.

Investigations with various animal intrinsic factor preparations have yielded important information with regard to human absorption. Further studies, however, should note the differences discussed above.

ISOLATION AND PROPERTIES OF INTRINSIC FACTOR

Although progress has been made toward understanding intrinsic factor-facilitated cobalamin absorption, the detailed mechanism and unequivocal understanding of cobalamin absorption have been limited by the problems of isolation and characterization of intrinsic factor.

Various laboratories have recently succeeded in isolating human and hog intrinsic factor in homogeneous form (48, 49, 53, 71, 72). The methods used were laborious and only small amounts of material were isolated. Many earlier purification attempts were also very laborious and usually did not yield homogeneous preparations (70).

The development of affinity chromatography proved to be an attractive method for the isolation of significant amounts of intrinsic factor and other cobalamin binding proteins; the isolation of cobalamin binding proteins by affinity chromatography has recently been described by Allen and Majerus (73–75), Allen and Mehlman (50, 51), and Christensen et al. (52). In the studies by Allen's group the affinity ligand was prepared by partial acid hydrolysis of the amide group of the unsubstituted propionamide side chains of the corrin ring of cyanocobalamin. The resultant mixture of mono-, di- and tricarboxylic cobalamin derivatives was separated by chromatography and the monocarboxylic derivatives of cyanocobalamin were coupled covalently to the free amino group of 3,3′-diaminodipropylamine-substituted Sepharose, thereby regenerating native cyanocobalamin stably coupled to Sepharose. The cobalamin Sepharose proved to be an effective absorbent and was used to isolate human granulocyte cobalamin binding protein (74), human transcobalamin II (75), human intrinsic factor (50), and hog intrinsic factor (51).

Holdsworth and Ellenbogen's groups were able to separate the cobalamin binding proteins of hog gastric mucosa into two fractions by ion exchange chromatography. Only one fraction contained intrinsic factor activity.* In Allen's laboratory the two cobalamin binding proteins were separated by a method of selective affinity chromatography with an affinity absorbent containing covalently bound derivatives of cobamide which lack the nucleotide portion of the native vitamin. The active fraction was not absorbed to the substituted Sepharose on the column.

Ellenbogen and Highley (48) and Highley et al. (76) reported a molecular weight of about 50,000 for both of their protein fractions in the absence of cobalamin, while formation of an oligomer of molecular weight about 100,000 was observed in the presence of CN-Cbl. Similar observations were made by Allen and Mehlman (51) for hog intrinsic factor, but no oligomer formation was observed with the nonintrinsic factor binder by these workers. CN-Cbl and intrinsic factor react in approximately equimolar amounts.

Both groups of workers have shown that hog intrinsic factor has a lower affinity for (Ade)CN-Cba relative to CN-Cbl than does the hog nonintrinsic factor preparation. The differences in the carbohydrate and amino acid composition between these two proteins were much more marked in the studies by Allen and Mehlman (51) than were observed in the studies by Ellenbogen and Highley (48).

* The fraction without intrinsic factor activity (i.e., unable to promote intestinal cobalamin absorption) is one of a number of related cobalamin-binding proteins collectively called "R binders". They are discussed further on p. 264 ff.

Hog intrinsic factor and hog nonintrinsic factor are probably distinct and separate proteins; such properties as facilitation of cobalamin absorption and of cobalamin binding by guinea pig ileal musosal homogenates, and blockade of binding by antiintrinsic factor antibody are not shared by the hog nonintrinsic factor fraction. The function of hog nonintrinsic factor is not known. Since it has been obtained in three different laboratories using different techniques, it would appear that it is not an artifact of isolation.

Isolation of human intrinsic factor from normal human gastric juice has been accomplished by Gräsbeck's laboratory (49, 53), Allen's laboratory (50), and by Christensen et al. (52). A molecular weight of 119,000 has been obtained for the intrinsic factor–CN-Cbl complex in Gräsbeck's early work. These investigators suggested that this complex is a dimer. A molecular weight of 60,000 to 73,000 (monomeric form) was obtained in their later work (53).

The elegant work on the isolation of human intrinsic factor by Allen and Mehlman (50) by affinity chromatography also suggests a monomer to oligomer formation, but these investigators obtained a molecular weight of 44,000 for their monomer. A nonintrinsic factor binder was not obtained by Allen and Mehlman whereas Gräsbeck's group isolated a nonintrinsic factor binder from human gastric juice.

As was observed with the homogeneous hog preparation, the homogeneous human preparation obtained by Allen's group (50) facilitated binding to homogenates of guinea pig and human distal ileum. Preparations of both pure hog and human intrinsic factor have single cobalamin binding sites and both bind about 30 μg of CN-Cbl per milligram of protein. This is approximately 1 mole of CN-Cbl per mole of intrinsic factor. The spectral maximum of 361 nm for unbound CN-Cbl shifts to 362 nm when the CN-Cbl is bound to either protein, suggesting that determination of CN-Cbl content using the extinction coefficient at 361 nm would result in falsely elevated values for CN-Cbl (50).

Preparations of homogeneous hog and human intrinsic factor displayed certain differences in properties. Antiintrinsic factor antibody from a pernicious anemia patient's serum has a lower affinity for hog intrinsic factor than for human intrinsic factor. The affinity of human intrinsic factor for (Ade)CN-Cba is less than that of hog intrinsic factor, although both proteins bind (Ade)CN-Cba to a much lesser extent than CN-Cbl. The amino acid and carbohydrate content of the two proteins are also different. Both proteins contain relatively large amounts carbohydrate as was suspected from the early fractionation studies. Human intrinsic factor has a molecular weight of 46,000 (50) and an approximate value of 55,000 was obtained for hog intrinsic factor (51, 76). Hog intrinsic factor was isolated from stomach mucosa, and human intrinsic factor was isolated from gastric juice. The isolation of human intrinsic factor from gastric mucosa would be of interest in order to determine if the differences in their properties are truly a species difference.

Christensen et al. also isolated intrinsic factor from human gastric juice by affinity chromatography (52). Hydroxocobalamin was insolubilized by covalent coupling to albumin which in turn was coupled to bromoacetyl-activated cellulose. This allowed specific adsorption of cobalamin-binding proteins from the human gastric juice. The cobalamin-binding proteins were eluted by increase of temperature and addition of cyanocobalamin. In contrast to the studies of Allen and Mehlman (50), the molecular weight of the isolated human intrinsic factor was about 60,000, and the amino acid composition was similar to hog intrinsic factor reported by Ellenbogen's group (48, 76). The properties of the intrinsic factor preparations reported by the various laboratories are shown in Table 5-2.

Table 5-2 Properties of Intrinsic Factor

	Human	Hog
Molecular weight	44,000 (50)[a] 59,900 (52) 63,000 (53)	55,000 (51, 76)
Carbohydrate (%)	15.0 (50) 8.3 (53)	17.5 (51) 35.0 (76)
Cyanocobalamin binding (μg/mg)	30.1 (50) 18.6 (53)	25.0 (48) 30.3 (51)
Stokes radius (nm)	3.3 (52)	
Association constant for cyanocobalamin (M^{-1})	1.5×10^{10} (50)	1.5×10^{10} (51)
Association constant for (Ade) CN-Cba	Weak (50)	Weak (51, 76)

[a] The numbers in parentheses are the references.

BINDING OF COBALAMIN TO INTRINSIC FACTOR

Introduction

The proteins involved in the transfer of cobalamin from ingested food to which it is bound to its ultimate site of action in the human body have specific binding affinities for various corrinoids. In man, five major proteins which bind endogenous cobalamin have attracted the most attention. These are intrinsic factor, the three different transcobalamins, and a protein associated with intrinsic factor which lacks intrinsic factor activity but which binds cobalamin. The ability to bind cobalamin is an essential property of intrinsic factor. In an effort to understand more fully the detailed function of the binding proteins, various investigators have attempted to identify the group on the vitamin in contact with the binding proteins as well as the residues in the protein in contact with the vitamin or its analogs.

Different groups of workers have used different methods for measuring cobalamin binding capacities, and the results have sometimes been confusing. The following methods have been used: (1) microbial growth inhibition; (2) microbial adsorption inhibition; (3) charcoal adsorption; (4) paper chromatography; (5) electrophoresis; (6) gel filtration; and (7) dialysis. Details on each of the methods have been given by Simons (77) and Chanarin (78).

Another difficulty encountered in binding assays is the dependence of the cobalamin binding capacity on the concentration of cobalamin and the material to be assayed. Cyanocobalamin in dilute solutions is transformed to aquocobalamin under the influence of light, and the aquocobalamin is bound more strongly to protein. Finally no exact data exist, since most of the studies have been done with impure preparations of the protein.

Structural Specificity

Cyanocobalamin, aquocobalamin, adenosylcobalamin, and methylcobalamin each have the same binding constant to intrinsic factor, which indicates that the substituent in the β coordination position is not important in the formation of the cobalamin-intrinsic factor complex (79–84).

Although substitution of the cyanide with sulfate, nitrate, or chloride ions reduced cobalamin absorption in man and in the rat (85, 86), human gastric juice showed no preferential binding of cyanocobalamin over sulfato- nitro- and chlorocobalamin (87). Since these binding studies were performed with "unfractionated" gastric juice, the binding of analog to the nonintrinsic factor binder contained in the gastric juice was also measured. The studies of Perlman and Toohey (82) clearly show that the binding of semipurified preparations of hog intrinsic factor by various cobalamin derivatives is different qualitatively and quantitatively from that observed by human gastric juice.

Support for the finding that the nature of the ligand attached to the cobalt atom is of no significant importance also comes from studies with cyanocobalamin labeled with ^{14}CN. Intrinsic factor does not displace the cyanide group during the binding (88).

Several studies have been performed with alterations in the corrin nucleus. The studies by Gottlieb et al. (89) showed that anilide-cyanocobalamin and anilide-aquocobalamin did not compete for the binding site on an intrinsic factor fraction of human gastric juice but appeared to bind to the nonintrinsic factor fraction of the gastric juice. Perlman and Toohey (82) also found that the anilide could compete with cyanocobalamin for binding to a more purified intrinsic factor from hog gastric juice, whereas it did not compete with cyanocobalamin for binding to a less purified hog intrinsic factor or to human gastric juice.

The lactone and lactam derivatives (see Chapter 1) could bind to the serum transcobalamins as efficiently as cyanocobalamin but did not bind human gastric juice as efficiently as cyanocobalamin (79, 87). Some descobaltocorrins effectively compete for the cobalamin binding site of hog intrinsic factor but are almost ineffective with human gastric juice and other nongastric juice binders (82). Methylation or ethylation of one of the propionamide side chains did not destroy the capacity of the resulting compound to compete with CN-Cbl binding to human gastric juice (83).

The *e*-amido group on the corrin ring appears to play a significant role in binding to intrinsic factor, since removal of this single amide group markedly reduces its affinity (84).

In general, the differences in the extent to which various analogs interfere with cobalamin binding to various other proteins such as whey, serum, milk, and so on, indicate that the cobalamin binding to intrinsic factor is the most specific. This is most readily seen with analogs in which there is variation in the nucleotide portion of the molecule. Neither (Ade)CN-Cba nor the 5,6-dimethylbenzimidazole moiety competes with cyanocobalamin for the binding to human gastric juice (87, 90, 91). Any (Ade)CN-Cba binding by human gastric juice and hog gastric mucosa concentrates is due to their nonintrinsic factor content (48, 89).

The binding by serum, on the other hand, does not manifest such a selectivity for CN-Cbl in the presence of excess (Ade)CN-Cba. (See the section on the binding of cobalamin to transcobalamins for detailed studies.) When equal amounts of CN-Cbl, (Ade)CN-Cba, (2-MeAde)CN-Cba, and $(CN)_2Cbi$ were added to purified intrinsic factor preparations from hog gastric mucosa, cyanocobalamin was preferentially bound (92).

Serum, in contrast to intrinsic factor, binds both (2-MeAde)CN-Cba and $(CN)_2Cbi$ as well as the ethylamide, methylamide, monobasic acid, and anilide analogs of cyanocobalamin (92, 93).

Cobinamide, which lacks the nucleotide, and cobyrinamide, which lacks both the nucleotide and the D-1-amino-2-propanol moiety of cobalamin, do not bind to intrinsic factor but do bind to the nonintrinsic factor portions of gastric juice (89). Analogs containing trimethylbenzimidazole or methoxybenzimidazole instead of 5,6-dimethylbenzimidazole appear to have little affinity for intrinsic factor (94, 95). Cyanocobalamin carbanilide, in which the carbanilide is joined to one of the ribose hydroxyl groups, competes only slightly with cyanocobalamin for intrinsic factor (83). All of these findings are consistent with the concept of Gräsbeck (96) that the cobalamin molecule fits into a "pit" of the intrinsic factor molecule with the nucleotide portion facing inward and the –CN group facing outward (Figure 5-1).

Modifications in the nature of the nucleotide portion of cobamides evidently have profound effects on the binding of cobamides to intrinsic factor and suggest that the mode of attachment to intrinsic factor is quite different from that of other vitamin binding proteins. Hippe et al. (79), however, argue that differences in binding of cobamide derivatives by intrinsic factor and by transcobalamins do not necessarily allow the assumption of differences in the region of contact between respective combining sites.

These investigators feel that the *B*-pyrrole ring with its side chain together with some overlap to the dimethylbenzimidazole is the most likely area of contact for the combining sites of the transcobalamins and intrinsic factor. These conclusions are based on the relatively poor binding of the *B*-pyrrole ring lactone of cyanocobalamin to intrinsic factor.

Lien et al. (97, 98) believe this conclusion is questionable. The *B*-pyrrole ring of cyanocobalamin may not bind to intrinsic factor because of the additional bulky lactone ring. This radical change in the geometry of the corrin ring may be enough to prevent the lactone from forming a "close fit" to intrinsic factor. Therefore, the lack of strong binding of the lactone to intrinsic factor does not prove the importance of the *B*-pyrrole ring in the binding of intrinsic factor. The position of the cobalamin binding site on intrinsic factor was determined by the use of cobalamin

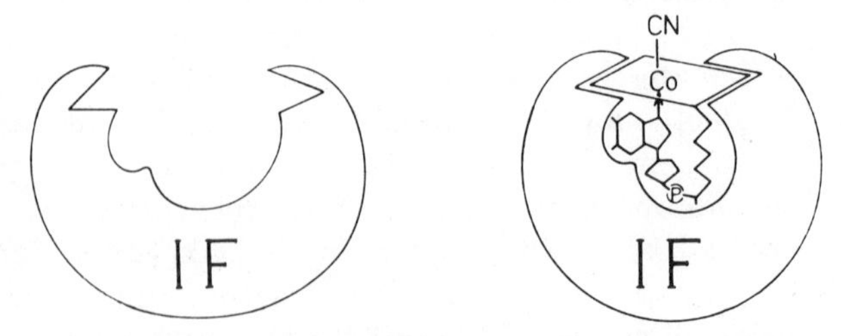

Fig. 5-1. The binding of cyanocobalamin to intrinsic factor according to Gräsbeck. [Reprinted with permission from Gräsbeck, R. (1967). *Scand. J. Clin. Lab. Invest. Supp.* **95**, 7. Copyright Universitetsforlaget, Oslo.]

affinity-labeled Sepharose (97, 98). Sepharose-aminoethylcobalamin binding to intrinsic factor was as good as cyanocobalamin binding to intrinsic factor. This result provided proof for the cobalamin binding site being near the surface of intrinsic factor, probably no more than 5 Å from the surface of the protein. The coordination sphere in the region of 5,6-dimethylbenzimidazole must be extremely hydrophobic because CN will not displace the base from the cobalt atom when cobalamin is bound to intrinsic factor.

Circular dichroism studies on the methylcobalamin-intrinsic factor complex, together with studies on the carbethoxylation of histidine residues in this complex, indicate that 5-6-dimethylbenzimidazole is probably displaced by a histidine residue of intrinsic factor which coordinates to the cobalt atom at the sixth coordination site. In addition, the stability of the Co–C bond for methylcobalamin bound to intrinsic factor is identical to that stability measured for the free methylcobinamide-histidine complex, but considerably different from the stability of this bond in free methylcobalamin (97, 98).

Physical and Chemical Nature of Cobalamin Binding

Cobalamin binding to human intrinsic factor is in part pH dependent. Some minor discrepancies between different workers exist probably because of different techniques and degree of purity of the intrinsic factor preparations. Maximal binding occurs at a pH between 7 and 9; binding at a pH below 7.0 is about 50% of that observed at alkaline pH according to Goldberg and Fudenberg (99). Rose and Chanarin (100) found no significant difference in binding between pH 3 and pH 7.5 while Wagstaff et al. (101) and Shum et al. (102) found that binding is only moderately affected at extreme pH. This view is not in conflict with the studies of Cooper and Castle (103) who showed that more cobalamin is removed from tissues and bound to intrinsic factor at lower pH than at higher pH. The effect was due to increased removal of cobalamin from food proteins at lower pH. The work of Cooper and Castle (103) indicates that the acid environment of the stomach facilitates removal of cobalamin from food, whereas the work of Goldberg and Fudenberg (99) suggests that the alkaline pH of the intestinal fluid enhances the binding of cobalamin to intrinsic factor.

The rate of binding of cyanocobalamin to intrinsic factor is rapid (76, 78). The reaction is complete in less than 1 min. Hippe and Olesen (80) determined certain thermodynamic parameters for binding of cyanocobalamin to a purified preparation of human intrinsic factor, transcobalamin I, and transcobalamin II. Equilibrium constants in the order of 10^9 M^{-1} were obtained and are similar to those reported by McGuigan

(104), Allen and Mehlman (50, 84), and Wagstaff et al. (101). The reaction between cyanocobalamin and intrinsic factor is exothermic, $\Delta H°$ being -22.7 kcal/mole (80).

The binding of cyanocobalamin to intrinsic factor causes certain physical changes in the intrinsic factor molecule (49, 72, 76, 96, 105, 106). Bromer and Davisson (72) reported an increase in sedimentation constant from 3.1 to 4.4 Syedberg units after addition of CN-Cbl to their purified hog intrinsic factor, and speculated that association had occurred or that the complex was "usually dense." Highley et al. (76, 105) demonstrated very clearly that following CN-Cbl binding, hog intrinsic factor undergoes a slow dimerization, although the rate of binding is instantaneous.

Hippe (107, 108) found a decrease in the Stokes radius of human intrinsic factor and transcobalamin II, indicating a conformational change following the binding by CN-Cbl. Following the binding to CN-Cbl, the behavior of the complex in gel filtration is slightly different from that of free intrinsic factor.

Since the complex is resistant to various enzymes (109–112), cobalamin, through some conformational change, protects the peptide bonds from enzymatic attack. In addition, intrinsic factor does not bind to cobalamin analogs in which the nucleotide is modified. From these observations, Gräsbeck (96) inferred that the nucleotide of the vitamin faces inward in the complex and the CN^- side of its planar structure outward. Chemical and immunological findings also suggest that cobalamins fit into a "pit" in the molecule (106, 113) and that intrinsic factor has an open structure which closes around the cobalamin after it is bound (Figure 5-1). (Also see the section on structural specificity.)

The bond between cyanocobalamin and purified hog intrinsic factor is quite strong (114–116), particularly at lower temperatures. Similar findings were reported by Bunge and Schilling (87) for human gastric juice. However, Donaldson and Katz (117) studied the temperature dependence of the exchange between free cobalamin and gastric juice-bound cobalamin, and they noted appreciable exchange at 37°; however, the exchange was much smaller at lower temperatures.

Comparatively little has been published on attempts to detach cobalamin from intrinsic factor or transcobalamin. At pH 12.3, no dissociation occurs, but at pH 12.6 there is almost complete dissociation (118). The complex can also be dissociated slowly by 5 *M* guanidinium chloride (76). The treatment is undoubtedly due to unfolding of the protein molecule. The rate constant for dissociation at pH 7.4 and 30°C is very low (101).

Cobalamin Binding and Intrinsic Factor Activity—General Comments

One of the most important properties of intrinsic factor is its binding by cobalamin, first observed by Ternberg and Eakin (119). The isolation of hog and human intrinsic factor (49–51, 76, 105) not only confirmed the hypothesis regarding the binding properties of intrinsic factor but also helped explain some of the confusion regarding the correlation between intrinsic factor activity and binding. These investigators were able to purify two fractions from hog pylorus and/or human gastric juice. Both fractions combined with cyanocobalamin, but only one possessed intrinsic factor activity. Many substances other than intrinsic factor bind cobalamin—for example, lysozyme (120, 121), saliva, (4, 122–124), and cerebrospinal fluid (125).

In addition to these findings, it is important to emphasize that cobalamin binding by intrinsic factor is only one of its properties. There may be several sites on the intrinsic factor molecule necessary for its activity (87, 111, 126–129). Intrinsic factor probably contains a site for cobalamin binding and at least one site concerned with "absorption" or attachment to ileal receptors in the intestine. Both groups would be necessary for activity. If the "absorption site" were absent or destroyed, the material would be inactive but would retain its ability to bind cobalamin.

Despite these facts, cyanocobalamin binding measurements have aided almost all investigators during the purification of intrinsic factor. Preparations that do not bind CN–Cbl do not enhance the absorption of cobalamin. Furthermore, intrinsic factor does not bind (Ade)CN-Cba to any significant extent, whereas (Ade)CN-Cba competes with CN-Cbl for nonintrinsic factor binders (51, 87, 89).

ASSAY OF INTRINSIC FACTOR

In vivo Techniques

The final criterion of biological activity of any intrinsic factor preparation is its ability to enhance the intestinal absorption of cobalamin. Therefore, the most unequivocal methods for measuring hog and human intrinsic factor are those identical techniques used to measure cobalamin absorption in man, namely, the urinary excretion, fecal excretion, hepatic uptake, blood plasma, and total body radioactivity. (See the section on methods of determining cobalamin absorption in man). These methods are used with patients who lack intrinsic factor (pernicious anemia or total gastric resection).

In 1960 in the United States, the Anti-Anemia Preparations Advisory Board of the National Formulary (N.F.) adopted a standardized procedure and a standard intrinsic factor sample for use in the urinary excretion test for assay of commercial intrinsic factor preparations (*National Formulary,* 11th ed.). An outline of the procedure is shown in Table 5-3. This procedure requires the use of human volunteers. It would be most desirable if unequivocal *in vitro* techniques could be used. In the remainder of this section are described techniques which, although not always unequivocal or quantitative, have directly and indirectly provided useful information about intrinsic factor.

Table 5-3 Outline of Urinary Excretion Method for Assay of Intrinsic Factor as Adopted by the National Formulary (N.F.)

Day	Labeled Oral Cyanocobalamin (μg)	Parenteral Cyanocobalamin (μg)	Intrinsic Factor	
1	2	1000	—	Baseline response
2	—	1000	—	
3	—	—	—	
4	2	1000	N.F. standard (1 N.F. unit)	Standard response
5	—	1000	—	
6	—	—	—	
7	2	1000	Unknown $sample_1$	Sample response
8	—	1000	—	
9	—	—	—	
10	2	1000	Unknown $sample_2$	Sample response
11	—	1000	—	
12	—	—	—	

Source. Leon Ellenbogen, in *Vitamins and Hormones,* Vol. 21 (1963), Table II, by permission of Academic Press, Inc.

Assay of Intrinsic Factor by Enhancement of Cobalamin Uptake by Tissue Preparations

Numerous investigators have explored the possibility of measuring intrinsic factor activity by using the *in vitro* enhancement of cobalamin uptake by intrinsic factor from various intestinal preparations (everted sacs, loops, rings, homogenates, and brush borders) and liver slices and homogenates (130–150). More unequivocal evidence is needed to show that these assays do measure quantitatively only intrinsic factor activity. Intrinsic factor preparations of varying potency from hog and human sources need to be tested by these methods. These techniques, however, have helped shed some light on the mechanism of cobalamin absorption and transport and are very useful in determining the presence of intrinsic factor in human gastric juice.

Assay of Intrinsic Factor by Immunological Techniques

The findings by Taylor (151) and Schwartz (152, 153) that the sera from pernicious anemia patients inhibited the gastrointestinal absorption of cobalamin and the demonstration that these sera contain antibodies to intrinsic factor (154, 155) provided the impetus for the immunological studies with intrinsic factor.

Two types of intrinsic factor antibodies are now recognized (156–164). These are the "blocking" antibody (type 1), which prevents the combination of cobalamin with intrinsic factor, and the "binding" antibody (type 2), which combines with cobalamin-intrinsic factor complex, thereby preventing the uptake of the complex by the intestinal mucosa (see Figure 5-2). These antibodies have made possible the development of the assay of intrinsic factor by immunological techniques. Assay of intrinsic factor by radioimmune procedure was first introduced by Ardeman and Chanarin (165) and Abels et al. (166). The amount of intrinsic factor is determined by its ability to bind radiolabeled cyanocobalamin. Binding of cyanocobalamin to intrinsic factor is prevented by the addition of intrinsic factor antibody from pernicious anemia patients. In general, the binding of the cyanocobalamin by nonintrinsic factor binders is unaffected by the antibody. The difference in the cyanocobalamin binding of intrinsic factor with and without antibody is a measure of the intrinsic factor content of the sample.

A variety of techniques employing the blocking antibody have been described; they are usually applicable to free intrinsic factor. In general, they differ mostly in the methods used to separate free from bound cobalamin. Albumin-treated, hemoglobin-treated (167, 168), as well as serum-treated charcoal (169) have been used to separate the free from the bound vitamin. In addition, electrophoresis (170, 171), dialysis (166), zirconyl phosphate gel (172), and sodium sulfate precipitation (173, 174) as means for separation of free and bound vitamin have also been described. Wolff and Nichols (175) utilized an antibody to a nonintrinsic factor binder (saliva) to prevent cobalamin uptake by saliva. Therefore, the antiserum was used to permit cobalamin binding to the intrinsic factor only.

Many of these techniques are very complicated and are not used widely. They have been used successfully as research tools or as diagnostic tools to determine the presence or absence of intrinsic factor, as well as type of antibody. The methods of Chanarin (78) and Gottlieb et al. (167) are probably the most widely used.

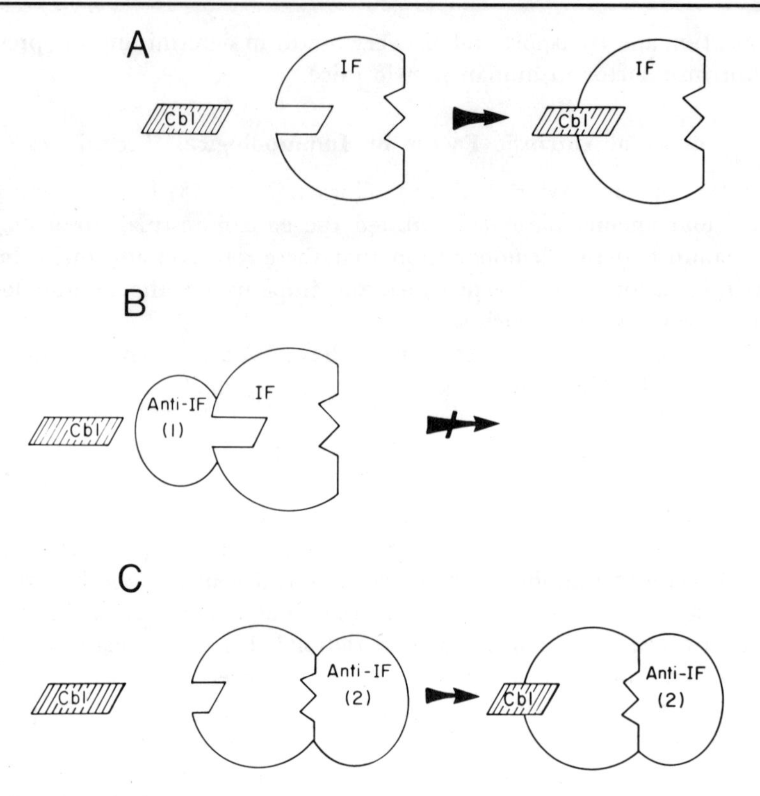

Fig. 5-2. Hypothetical illustration of intrinsic factor antibodies according to Schade et al. (162). (*A*) Normal absorption. (*B*) Blocking type antibodies (type 1). (*C*) Binding type antibodies (type 2).

Inhibition of Cobamide Coenzyme Activity

Intrinsic factor preparations have been found to inhibit several cobamide coenzyme-dependent reactions, presumably by combining with the added coenzyme to decrease the effective concentrations. As shown in Table 5-4, good correlation between the clinical activity of intrinsic factor preparations and their specific activity as inhibitors of the adenosylcobamide-dependent glutamate mutase reaction has been found (176).

As an assay of intrinsic factor, this method would have the same disadvantages as the cyanocobalamin binding techniques (see the section on cobalamin binding and intrinsic factor activity), since one is still measuring binding to a cobamide. However, inhibition of cobamide coenzyme activity is much more rapid than dialysis binding or other nonenzymatic binding techniques.

Table 5-4 Comparison of Inhibitor and Intrinsic Factor Activities of Intrinsic Factor Preparations*

	Activity as Inhibitor		Intrinsic factor activity		
IF Prep.	Protein Conc. at 50% Inhibition μg./ml.	Specific Activity as Inhibitor 100 μg./ml. (a)	Min. Effective Dose mg.	Specific Activity 1/mg. (b)	Activity ratio (a)/(b)
1	61	1.64	1.0	1.0	1.6
2	139	0.72	2.5	0.4	1.8
3	136	0.74	2.5	0.4	1.9
4	60	1.67	1.0	1.0	1.7
5	105	0.95	1.5	0.7	1.4
6	106	0.94	2.5	0.4	2.3
9	3.6	0.32	8.0	0.12	2.6
16	390	0.25	100.0	0.01	25.0

*Ellenbogen, et. al., (177)

The inhibition of cobamide coenzyme activity by intrinsic factor is the basis for its usefulness in the study of the role and mechanisms of action of the cobamide coenzymes in several metabolic reactions. Intrinsic factor has been helpful in providing additional evidence that the cobamide structure is involved in specific enzymatic reactions. In certain instances where separation of the coenzyme from the apoenzyme is difficult, specific inhibition of the coenzyme such as that obtained with intrinsic factor may be the only means of providing proof of a cobamide coenzyme requirement. Lengyel et al. (177) and Stadtman (178), for example, have utilized intrinsic factor to study the adenosylcobamide dependency of methylmalonyl CoA mutase and lysine mutase, respectively.

Assay of Intrinsic Factor in Gastrectomized Rats

Because the anatomical distribution of intrinsic factor in the stomach of the rat is similar to that of man, it was originally hoped that the use of the gastrectomized rat might serve as a means of measuring hog intrinsic factor activity. The gastrectomized rat fails to absorb orally administered cyanocobalamin (54–57), and rat gastric juice or rat stomach preparations restore cobalamin absorption to normal. This absorption defect cannot, however, be corrected with hog intrinsic factor. In fact, in many instances cobalamin absorption is depressed in the normal rat by hog intrinsic factor (179).

There is promise that the gastrectomized rat may be useful as an assay animal for human intrinsic factor. However, further work is needed to extend the findings by Taylor et al. (180) that low concentrations of purified human intrinsic factor promoted absorption of cyanocobalamin in the gastrectomized rat.

Miscellaneous Methods

Two additional methods based on electrophoresis of human gastric juice or hog intrinsic factor concentrates have been introduced (171, 181). Gullberg found good agreement between intrinsic factor activity and content of a major electrophoretic cobalamin binding component in hog intrinsic factor concentrates. However, there is some evidence that some of the major cobalamin binding components did not have intrinsic factor activity (139).

By paper electrophoresis combined with autoradiography, two cobalamin binding components were demonstrated in normal human gastric juice (171). Only one cobalamin binder was related to intrinsic factor. The use of this technique as a qualitative test for the presence of intrinsic factor was recommended. Although it might be useful for *in vitro* diagnosis of pernicious anemia, it would not be useful for quantitative assays of intrinsic factor preparations. An excellent review of all published *in vitro* assays can be found in the monograph by Glass. (70).

METHODS OF DETERMINING COBALAMIN ABSORPTION IN MAN

Before the use of labeled cyanocobalamin (182), estimates of absorption were based mostly on indirect evidence. Microbiological assays were used to measure deposition of the absorbed vitamin in tissues. Qualitatively, the degree of hematopoietic response in cobalamin-deficient patients (183) was used as an index of absorption. Due to the synthesis of cobamides by bacteria in the intestine, it has not been possible to measure microbiologically the unabsorbed vitamin by measuring fecal cobamide after an unlabeled oral dose.

Several different methods are available for determining the absorption of an oral dose of radioactive cobalamin in man. Various radioactive isotopes of cobalt have been used, including ^{56}Co, ^{57}Co, ^{58}Co, and ^{60}Co. Detailed reviews of these techniques have been presented (78, 184, 185). An excellent summary of the various methods for clinical purposes is also given by Mollin (186).

SITE OF COBALAMIN ABSORPTION

Human

The absorption of physiological doses of cobalamin in the human alimentary tract takes place almost exclusively in the ileum. This evidence is based on

1. Defective cobalamin absorption in diseases of the lower ileum or after resection of the ileum.
2. Measurement of intestinal radioactivity after oral administration of radioactive cobalamin.
3. Studies following instillation of radioactive cobalamin in various segments of intestine.

Clinical studies in patients who have undergone resection of various regions of the small intestine have provided firm support for the localization of the site of cobalamin absorption. Impaired absorption of the vitamin is found in patients who have undergone resection of the ileum but not following resection of the duodenum or jejunum (187, 192). The addition of intrinsic factor or antibiotics did not correct the impaired cobalamin absorption.

Various investigators have studied the site of cobalamin absorption by intestinal instillation of cobalamin and cobalamin-intrinsic factor complex. Direct instillation of cobalamin into the jejunum only (segment closed at its proximal and distal ends) did not result in any significant absorption in humans (193). Citrin et al. (194) showed that in humans, the vitamin was absorbed, whether it was instilled in the duodenum, jejunum, or ileum. Cobalamin instilled in the upper intestine probably moved down to the ileum before it was absorbed. In these studies, the ileum absorbed cobalamin in pernicious anemia patients only when intrinsic factor was added. Similar experiments were conducted by Best et al. (195). Johnson and Berger (196) implicated the ileum as the site of absorption following the use of delayed release capsules containing labeled cyanocobalamin. Experiments using polyethylene glycol as a marker, a test meal containing labeled vitamin, and sampling of intestinal juices from different parts of the intestine also led to the conclusion that most of the absorption occurred in the terminal portion of the small intestine (197). Various investigations have provided ample evidence that other parts of the gastrointestinal tract are of no significance in the absorption of physiological doses of cobalamin (70).

Further studies are needed to extend or confirm the recent report that an adaption of absorption of cobalamin occurs following ileal bypass

(198). The vitamin under these conditions may be absorbed from more proximal parts of the small intestine.

No absorption takes place following buccal administration, and the large intestine has been shown to be incapable of absorbing small doses of cobalamin (199, 200) although some absorption can take place following rectal administration of very large doses, such as 2000 μg (201). The mucosa of the large bowel does not secrete intrinsic factor or a substance that would mediate the absorption of the vitamin.

Other Species

The site of absorption of small amounts of cobalamin in the dog is the ileum (202–204). Little or no absorption occurs in the stomach, jejunum, or colon. Some absorption can take place in the upper part of the small intestine when large amounts of cobalamin are given (205).

In rats, the absorption of cobalamin takes place in the midportion and distal ends of the small intestine [lower part of the jejunum and upper part of the ileum (206–211)]. This conclusion was reached by both *in vivo* absorption studies and *in vitro* studies using everted sacs from various segments of the small bowel of rats.

Less extensive studies have been done with other species. The site of absorption appears to be the ileum in the guinea pig, rabbit, hamster, and monkey (60, 148, 212–214). Labeled cyanocobalamin placed directly into the caeca of rats and pigs and the colon of pigs was poorly absorbed (215).

SEQUENCE OF EVENTS DURING INTRINSIC FACTOR-MEDIATED COBALAMIN ABSORPTION

Introduction

Cobalamin can be absorbed by two different mechanisms (216–220). The active mechanism is intrinsic factor mediated. Absorption occurs in the ileum, and is of primary importance in the absorption of physiological doses of cobalamin (approximately 2 μg or less).

The events that take place in the intestine following ingestion of physiological amounts of cobalamin require about 8 to 10 hr for the completion of absorption in man. The sequence of events can be classified as follows:

1. Release of cobalamin from dietary sources.
2. Binding of the vitamin to intrinsic factor.
3. Intestinal transit through the small intestine to the ileum.

4. Attachment of cobalamin-intrinsic factor complex to specific receptors on the absorptive surface of the ileum.
5. Transfer of the vitamin across the small-intestinal epithelial absorptive cell to the portal plasma.
6. Release of the vitamin from the cobalamin-intrinsic factor complex.

Release of Cobalamin from Dietary Sources

Essentially all dietary cobalamin is attached to protein in coenzyme form. Unless cyanocobalamin in medicinal form is ingested, the main cobalamins in food are adenosylcobalamin and methylcobalamin (221–224). Significant amounts of the coenzymes are converted to hydroxocobalamin during the preparation of the food (225). Most studies concerned with cobalamin absorption have been performed with cyanocobalamin; therefore, the use of this form will be implied in discussing these studies except in those instances where different analogs were specifically used. Contrary to earlier reports, however, there is no evidence that liver-bound cobalamins are absorbed in a manner different than that of free cyanocobalamin (226, 227). Absorption of liver-bound cobalamins requires intrinsic factor and the absorption is not superior to that of free hydroxocobalamin or cyanocobalamin.

Cobalamin is released by heating, acid, and proteolytic enzymes (103, 228). In contrast to cobalamin-intrinsic factor complexes, proteolytic enzymes readily release cobalamin from food protein linkage. The release of cobalamin does not appear to be rate limiting. Because of the low dissociation constant (50, 84, 104) intrinsic factor readily binds the released vitamin. Beyond these facts nothing more definitive is known about this stage.

Binding of Cobalamin to Intrinsic Factor

A prerequisite for intestinal absorption of physiological amounts of cobalamin is binding to intrinsic factor. Upon release of the vitamin from food, it is bound to intrinsic factor secreted by the parietal cells of the stomach. Any cobalamin attached to nonintrinsic factor proteins is transferred to intrinsic factor in the intestine (115).

The intrinsic factor molecule probably has at least two functional sites, a cobalamin binding site, and a receptor site which attaches to a receptor mechanism on the surface of the intestinal mucosa (87, 111, 126–129). There is excellent evidence that the cobalamin binding sites are protected against enzymatic digestion and heating by the bound vitamin (110, 229). The presence of the vitamin also diminishes or prevents inactivation of human intrinsic factor by various means (72, 230). This

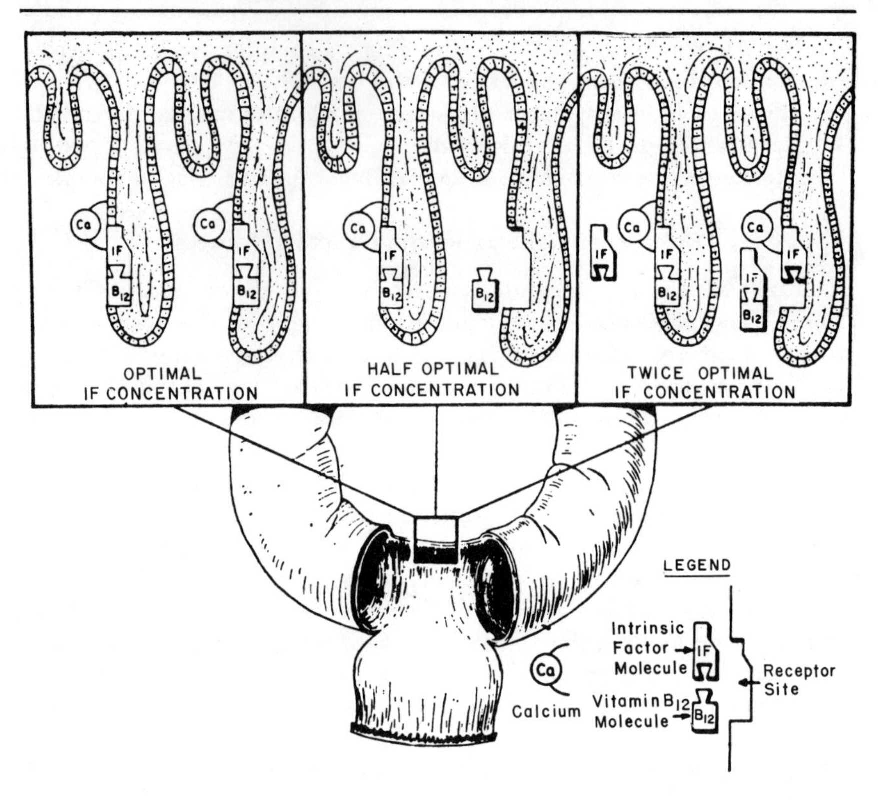

Fig. 5-3. Schematic illustration of the "key-in-lock" hypothesis of attachment of cobalamin-intrinsic factor complex to receptors on the brush border of the ileal mucosa. [Reprinted with permission from Herbert, V., Streiff, R. R., and Sullivan, L. W. (1964). *Medicine* 43, 679. © 1964 The Williams & Wilkins Co., Baltimore.]

stabilizing effect of the vitamin facilitates the isolation of intrinsic factor from both hog and human sources but more importantly, it has a useful physiological effect. It protects intrinsic factor against autodigestion during transport from the stomach to the ileum, where the intrinsic factor can then promote the uptake of the bound vitamin. Intrinsic factor also probably protects cobalamin from uptake by intestinal bacteria.

Passage along the Small Intestine

The cobalamin-intrinsic factor complex is carried to the ileum by peristalsis. The complex is resistant to hydrolysis by various proteolytic enzymes as mentioned above. *In vitro,* very few enzymes can liberate cobalamin from the complex or digest the complex (109, 110, 127), whereas the free intrinsic factor is more spontaneously labile as well as more susceptible to enzymatic breakdown. In view of the fact that the intrinsic factor molecule dimerizes (48–51) it would appear that polymerization causes some conformational change which prevents the enzymes from attacking susceptible bonds. In addition, the dimerization must also protect the vitamin from bacterial utilization, as suggested by the large number of studies mentioned above. Callender et al. (231) has shown the importance of binding of intrinsic factor in the upper part of the intestine. Patients with esophageal occlusion whose stomachs were bypassed by a Roux-Y type of anastomosis of the esophagus to the jejunum showed decreased absorption of cobalamin in the intestine. In these patients, the intrinsic factor reached the ingested vitamin in the jejunum. Citrin et al. (194) performed intestinal intubations on normal persons and patients with pernicious anemia. A test dose of labeled cyanocobalamin delivered to the duodenum resulted in greater absorption than when the vitamin was delivered to the jejunum or the ileum. In pernicious anemia patients, administration of cobalamin with intrinsic factor into the duodenum also resulted in greater absorption than when both substances were administered into the ileum.

The method of Jacob and O'Brien (232), which can measure intrinsic factor bound to cobalamin, was used to study the fate of intrinsic factor-cobalamin complex during intestinal transit in two normal subjects. Approximately 50 to 70% of orally administered radioactivity recovered from the ileum was bound to intrinsic factor, 15 to 25% was free cobalamin, and 10 to 30% was bound to nonintrinsic factor binders. This may explain why in normal subjects a test dose of radioactive cobalamin is never completely absorbed

Attachment of Cobalamin to Ileal Receptors

Several studies indicate that the cobalamin-intrinsic factor complex is transiently attached to an ileal receptor. This is probably the first characterized process that occurs in the ileum. *In vitro* studies utilizing everted sacs, (129, 142, 233), whole homogenates (144, 147, 234), including ileal homogenates (235), and brush borders (236) demonstrated that intrinsic factor enhanced the uptake of cobalamin The proximal small intestine does not have the ability to enhance absorption of the vitamin; only the ileum has this property. Free cobalamin is attached to the ileal receptors only weakly if at all (73, 210). This uptake is also specific for intrinsic factor. Cobalamin binding proteins of serum, saliva, colostrum, or tears do not enhance the uptake of the vitamin (77). In addition, there is a species specificity as shown by the fact that only intrinsic factor from hamsters, rats, and rabbits promoted uptake of the vitamin by hamster brush borders (236). Similar species specificities were observed with everted sacs (142).

The studies of Donaldson et al. (84, 236) showed that uptake of intrinsic factor-bound cyanocobalamin was inhibited when brush borders were preincubated with intrinsic factor.

Several other characteristics of this attachment to ileal receptor have been found.

1. Energy or a specific enzymatic step is probably not required (236). This attachment occurs in the absence of glucose or oxygen and despite poisoning with sodium azide and dinitrophenol (103).
2. Intrinsic factor enhancement of cobalamin uptake was not affected by changes in temperature from 7 to 37°C (236). These facts suggest that the attachment of intrinsic factor-bound vitamin, although specific, results from adsorption and not from a specific energy-requiring enzymatic process. The observation that increasing the time for incubation did not increase uptake of intrinsic factor-bound vitamin supports this view (236).
3. Attachment of the complex to the receptor occurs preferentially between pH 6.0 and 8.0 (144, 147, 236). Lowering the pH below 5.6 impairs uptake of cobalamin by everted sacs, (144) homogenates (147) and isolated brush borders (236). This pH effect is important only for the attachment of the intrinsic factor-cobalamin complex to the ileal receptor and is unimportant in the nonintrinsic factor-mediated cobalamin adsorption (144).
4. Divalent cations are required for attachment. Removal of calcium ions from an *in vitro* medium or addition of a chelating agent to the medium, reduces intrinsic factor-dependent uptake of cobalamin by liver slices (140), everted sacs of rat ileum (144, 237), homogenates of guinea pig intestine (147), and intestinal microvillous membranes (238). *In vivo* studies confirm the divalent cation dependence of cobalamin absorption. In humans (239) and rats (240), EDTA diminishes cobalamin absorption. This inhibition is reversed by calcium (239). Administration of calcium to some patients with steatorrhea enhances intestinal absorption of cobalamin (112), although Rosado-Rodriquez and Sheehy (241) did not find enhanced absorption in tropical sprue patients following calcium administration.

The precise mechanism by which calcium ions influence the attachment to the ileal receptors remains unknown. Calcium could enhance attachment of intrinsic factor-cobalamin to ileal receptor by directly linking anionic groups on the intrinsic factor molecule with those on ileal receptor or by forming salt bridges which alter the conformation of the intrinsic factor-cobalamin complex (238).

A schematic representation of the intrinsic factor complex on the ileal receptor as described by Herbert et al. (242) is shown in Figure 5-3. The complex is thought to fit into the ileal receptor by molecular complementarity in a key-in-lock fashion. This model can explain why large pharmacological doses of cobalamin are poorly absorbed and why large or excessive doses of intrinsic factor do not increase the absorption of cobalamin proportionately. In addition the species specificity of intrinsic factor can be explained by the failure of the specific receptor to bind a heterologous intrinsic factor.

To determine whether the molecular configuration of a corrinoid influences the attachment of intrinsic factor-corrin complex to ileal microvillous membrane receptor sites, Mathan et al. (84) have examined the kinetics of uptake of intrinsic factor-bound cyanocobalamin by brush borders and microvillous membranes isolated from guinea pig ileum and have compared this uptake with that of intrinsic factor alone and with that of intrinsic factor complexed with various analogs of cyanocobalamin. Attachment of intrinsic factor-bound cyanocobalamin to microvillous membranes showed saturation kinetics with a dissociation constant of 0.25 nM. Attachment was rapid and was 70% complete within 5 min; the second-order rate constant for attachment was 1.3×10^6 M^{-1} sec^{-1}. The half-time for dissociation of intrinsic factor-bound cyanocobalamin from the ileal receptor was approximately 35 min. Free intrinsic factor inhibited the attachment of intrinsic factor-bound cyanocobalamin.

When intrinsic factor was complexed with various analogs of cyanocobalamin, the affinities of these complexes for ileal microvillous membranes were similar to that of intrinsic factor-bound cyanocobalamin. These findings suggest that the molecular configuration of the corrin is not a major determinant in the interaction between intrinsic factor-corrin complex and the ileal receptor site.

The precise nature of the ileal receptor is unknown. The receptor is probably antigenic and is specific for the distal end of the intestine. Antisera obtained from rabbits injected with purified microvillous membrane preparations isolated from the distal but not proximal hamster small intestine inhibited attachment of the intrinsic factor-cobalamin complex to brush borders of distal small intestines (243). When antibodies to the microvillous membrane are conjugated with ferritin, the ferritin-conjugated antibodies can be shown to attach to the mucopolysaccharide surface coat of the microvilli. The antibody was probably attached to the receptor in some manner and blocked the attachment of the intrinsic factor-cobalamin complex.

It has been speculated that the receptor may be a glycoprotein or a mucopolysaccharide since these substances are present in the absorptive surface of the intestinal cell (236). The partial isolation of the receptor from a particulate fraction of the hamster ileum which sedimented at 54,000 but not at 28,500 $\times$ *g* has been claimed (244). This 54,000 $\times$ *g* particulate fraction was found to consist largely of ribosomes and some membranous fragments. Whether the ribosomes or the membranes are responsible for uptake and how they mediate intrinsic factor enhancement of cobalamin absorption remain to be determined.

The partial isolation of a soluble macromolecular factor from guinea pig ileum has been reported by Rothenberg (245). Although the exact nature of this substance has not been defined, it is believed to be a high-molecular-weight protein since it is nondialyzable, is excluded from G-200 Sephadex gel, is precipitated by 15% Na_2SO_4, and is heat and acid labile. Immunologically intact intrinsic factor with different solubility and electrophoretic properties was identified in fractions of ileal extract after the intact ileum was incubated with normal human gastric juice bound to cyanocobalamin. This suggested that the ileal factor is a binder of intrinsic factor and not the vitamin. The inhibition of the formation of a complex between the ileal extract and normal human gastric juice bound to cyanocobalamin by sodium EDTA and its reversal by calcium EDTA are consistent with certain known observations of cobalamin absorption.

Although direct extrapolation from results of animal experiments to explain human biologic phenomena may be questioned, it should be realized that intrinsic factor-enhanced absorption of cobalamin in guinea pig ileum can be mediated by many sources of intrinsic factor (142). More importantly, it has been shown that human intrinsic factor enhances the absorption of cobalamin by guinea pig ileum (237). More recently, very preliminary evidence (246) indicates that there is indeed a receptor for intrinsic factor-cyanocobalamin complex in human ileum. These investigators have partially isolated a factor from human intestinal ileal mucosa which does not sediment at 100,000 $\times$ *g*. Reaction of the intrinsic factor-cobalamin complex with the binding factor had characteristics previously reported for the receptor, namely, temperature independence, divalent cation requirement, and pH optimum near neutrality.

Transport of Cobalamin across the Intestinal Epithelial Cell (Enterocyte)

In contrast to the attachment of cobalamin to receptors on the brush borders, passage of the vitamin into the epithelial cell is a slow, energy-requiring process (103, 143, 247).

The attachment of cobalamin to the receptors occurs within about 2 to 3 hr after ingestion. An additional 3 to 4 hr are required before significant amounts of the vitamin appear in the blood in humans (219). Peak blood levels are reached 8 to 12 hr after ingestion. This indicates that a slow and probably complex process is involved in the transport of cobalamins across the cell membrane. Prolonged interaction between the cobalamin-intrinsic factor complex and the membrane receptor would explain the slow rate of absorption of the vitamin (248). Recent important studies (249–251) using the guinea pig have shown that cobalamin accumulates in the mitochondrial fraction of the mucosal epithelium during absorption. This delay within the mitochondria is longer than the time required for attachment to the brush borders. It therefore appears that the vitamin could be concentrated in mitochondria during the period of mucosal delay.

A number of possible explanations for this localization were postulated. Latner et al. (252) claimed, on the basis of fluoroacetate inhibition studies, that conversion of cyanocobalamin to adenosylcobalamin occurs during ileal transport. The mitochondrial delay might therefore be due to the interconversion of the two forms, and the adenosyl derivative should therefore be more rapidly absorbed than cyanocobalamin. Recent detailed studies have shown that a similar period of delay within the mitochondria occurs for both of these cobalamins (253, 254). Similar delay and mitochondria localization were also observed for methylcobalamin, although the uptake of the methyl- and adenosylcobalamin by the ileum was less than that of cyanocobalamin. These studies would therefore suggest that all three cobalamins are localized in the ileal mitochondria during their absorption of the vitamin.

If the mitochondria are the site of synthesis of adenosylcobalamin from cyanocobalamin, it is clearly not the sole function of the mitochondria in the transport of the vitamin. The conversion of cyanocobalamin to adenosylcobalamin is not obligatory (253–255). Moreover, significant amounts of cyanocobalamin are absorbed unchanged into portal plasma in the guinea pig (255, 256) and in man (257).

The principal form of the vitamin in guinea pig ileal mucosa is adenosylcobalamin. The localization of the adenosylcobalamin in the mitochondria is consistent with the known mitochondrial localization of methylmalonyl CoA (258, 259).

The finding of mitochondrial localization of cobalamin during its absorption does not support the hypothesis that pinocytosis (260) is involved in cobalamin absorption. Wilson (260) concluded that membrane vesiculation and micropinocytosis are probable means of entry of cobalamin into the cell while attached to intrinsic factor. This conclusion was based on the observation that the uptake of the vitamin diminished as rat yolk sacs matured. These indirect experiments have not been extended nor has further evidence for pinocytosis been presented. If pinocytosis does occur, it would suggest that intrinsic factor enters the cell. Studies by Hines et al. (247) appear to show, however, that intrinsic factor remains on the surface of the brush borders while the vitamin enters the epithelial cell. Hines et al. (247) however, used a heterologous system (hog intrinsic factor and guinea pig ileum). Peters and Hoffbrand (256), using a double-labeled complex (CN[^{57}Co]Cbl, [^{125}I]-human intrinsic factor), could find labeled vitamin but no labeled intrinsic factor in the enterocyte. Other investigators (261, 262) using similar techniques could not find any evidence of the absorption of intrinsic factor (see the following section).

Another hypothesis, proposed by Gräsbeck (106) and supported by Rothenberg et al. (251) suggests that the vitamin or a fragment thereof enters the cell by carrier-mediated transport after attachment of the complex to a surface receptor site. A significant fraction of cobalamin absorbed into the ileal cell is reported as localized in the mitochondrial fraction bound to a macromolecule with immunological identity to intrinsic factor (251).

Using electron microscope autoradiography (263), the intracellular localization of cobalamin was studied in dogs during gastrointestinal absorption. Localization of radioactive vitamin was demonstrated initially in the mucus of the goblet cells, then in brush border membranes, and then in the Golgi apparatus and endoplasmic reticulum in the later stages of absorption. The significance of these findings to absorption in man can be questioned since these same investigators (264) have reported that cobalamin absorption in the dog is not intrinsic factor mediated.

During the delay in the ileal enterocyte, cobalamin appears to be transferred to a relatively low-molecular-weight tissue binding protein (256). Despite other preliminary reports (249, 265), further studies on the form in which the vitamin enters the cell and the identity of protein binders in the cell are clearly needed.

Release of Cobalamin from Cobalamin-Intrinsic Factor Complex

Cobalamin is bound to intrinsic factor when it reaches the ileum. It is probably bound to intrinsic factor when it is attached to the ileal brush borders. However, when it enters the portal blood, it is no longer bound to intrinsic factor but to specific transport proteins, namely, transcobalamins. At some point in the absorption process, the vitamin must separate from intrinsic factor. Most of the recent evidence supports the concept that intrinsic factor is not absorbed (247, 256, 262, 266–268). Further support for the belief that the vitamin is detached from intrinsic factor comes from the data showing that very little vitamin is found in the lymphatics (269) which are known to carry large molecular complexes. The unknown factor or factors responsible for the release of the vitamin have been termed "releasing factor." This release probably is the last phase of cobalamin absorption; unequivocal evidence for its presence has not yet been presented.

Initial support for a releasing factor came from the work of Cooper and Castle (103) who found that an extract of a rat intestine made cobalamin bound to rat stomach intrinsic factor dialyzable. This extract was species specific since it did not cause release of cobalamin bound to human or hog intrinsic factor. This finding was confirmed and extended by various investigators (270–273). Since the major site of cobalamin absorption in the rat is the midileum, and only the proximal end of rat small intestine had "releasing factor" activity, its physiological role was questioned (272).

It was subsequently found (274) that the bulk of releasing factor could be due to endogenous, nonradioactive cobalamin present in the intestinal extracts. This endogenous cobalamin could exchange with intrinsic factor-bound vitamin as described by Donaldson and Katz (117). Mackenzie and Donaldson(275) proposed that the cobalamin-intrinsic factor complex is split at the intestinal cell surface, yielding a small "active fragment" which is then transported across the cell.

A recent preliminary study (276) suggested that a mechanism other than, or in addition to, exchange by endogenous vitamin in guinea pig ileal mucosa does indeed participate in the dissociation of cobalamin from human intrinsic factor. This conclusion was based on the observations that

1. Ileal cytosol dissociated cobalamin-intrinsic factor complex more rapidly at 37°C that at 4°C.
2. A macromolecular factor obtained from ileal cytosol but not jejunal cytosol dissociated the complex.

3. Dissociation of complex was slower when incubated with cytosol than with the free vitamin.
4. Addition of excess human intrinsic factor to bind any endogenous vitamin did not prevent dissociation of complex.
5. The cobalamin which dissociated from intrinsic factor remained free in the cytosol, indicating no exchange with endogenous macromolecule-associated cobalamin.

The unequivocal demonstration of the existence of a factor which releases cobalamin from intrinsic factor is still needed, despite these suggestive reports. The fact that the vitamin in blood is bound to a molecule which has neither the biological nor the immunological characteristics of intrinsic factor, and the fact that intrinsic factor is not absorbed, provide strong evidence that some process is occurring which separates the vitamin from intrinsic factor.

STUDIES ON THE POSSIBLE ABSORPTION OF INTRINSIC FACTOR

Until recently it was not certain whether cobalamin is accompanied by intrinsic factor during its passage into blood. Although there was never any unequivocal evidence that intrinsic factor enters the circulation from the gastrointestinal tract, several investigators suggested an extraintestinal function of intrinsic factor (130, 277, 278). Some investigators claimed an intestinal absorption of intrinsic factor, based on the finding of circulating material after administration of intrinsic factor which gave positive immunological reactions with intrinsic factor antibody (171, 279). However, when large doses of intrinsic factor were given orally no effect on antibody titer was observed (267). This is regarded as good indirect evidence against the absorption of intrinsic factor. In addition, Schilling and Schloesser (268) showed that 200 to 400 ml of human plasma failed to enhance the gastrointestinal absorption of cobalamin in pernicious anemia patients.

Approaching the problem more directly, Yamaguchi et al. (262) labelled human intrinsic factor with ^{51}Cr. This labeled intrinsic factor was active in man as judged by urinary, fecal, plasma, hepatic, and whole-body absorption techniques. After oral ingestion, only trace amounts of ^{51}Cr could be found in the plasma. No ^{51}Cr absorption could be detected by whole-body counting. Therefore, the intrinsic factor molecule did not enter the circulation in an intact state. Similar studies were performed by Rosenblum and Geoffroy (261) and Peters and Hoffbrand (256) (see the preceding two sections). The latter investigators used a CN[^{57}Co]Cbl, [^{125}I]-human intrinsic factor complex. This material was purified and fed to guinea pigs. Although both isotopes were found attached to brush border membrane, there was no significant localization of cobalamin-intrinsic factor complex in the mitochondria, suggesting that the complex is dissociated at the brush border.

In another study, the CN[^{57}Co]Cbl activity in blood obtained from the portal vein during CN[^{57}Co]Cbl absorption did not have the biological or the immunological feature of intrinsic factor from normal human gastric juice (266).

Recent studies (280) indicate that the properties of the binding material thought to be intrinsic factor, in the early studies which suggested the absorption of intrinsic factor, were due to nonintrinsic factor materials present in the preparations used. No definite proof exists today for the contention that intrinsic factor is absorbed. The evidence is to the contrary. The only well-proven function of intrinsic factor remains its facilitation of the gastrointestinal absorption of various corrinoids.

THE ROLE OF THE PANCREAS IN COBALAMIN ABSORPTION

The suggestion that the pancreas might be of importance in the regulation of cobalamin absorption was first made by McIntyre et al. (187), Perman et al. (281), and Veeger et al. (282). The latter investigators and LeBauer et al. (283) reported that malabsorption, due to pancreatic disease, could be corrected by sodium bicarbonate and/or pancreatin. It was originally believed that the impaired cobalamin absorption was due to the lowering of the pH of the intestinal contents resulting from the lack of bicarbonate secretion. However, it was also observed that the pancreatic extract alone, without sodium bicarbonate, could also correct the cobalamin malabsorption.

The recent work of Toskes, Deren, and their co-workers has been instrumental in clarifying the relationship between the pancreas and intestinal absorption of cobalamin. The pH hypothesis of Veeger et al. (282) was studied (284) and it was found that the pH of the ileal contents in patients with pancreatic insufficiency and malabsorption of cobalamin did not differ from the pH of patients with pancreatic insufficiency without cobalamin malabsorption. Known causes of cobalamin malabsorption, such as intrinsic factor deficiency, intrinsic factor antibodies, ileal disease, or bacterial overgrowth of the bowel, were also excluded. Intrinsic factor was present in the gastric juice of the patients studied by Toskes et al. (284) and hog intrinsic factor did not improve absorption. Antibodies could not be detected in any of the patients, and tetracycline did not improve cobalamin absorption. The administration of a single dose of pancreatic extract concurrently with labeled cyanocobalamin to patients with pancreatic insufficiency enhanced the absorption of the vitamin. It was not necessary to give pancreatic extract for several days to restore cobalamin absorption to normal. That this response was not due to intrinsic factor was shown by the failure of the extract to correct cobalamin malabsorption in patients with pernicious anemia (285).

Studies with rats subjected to 80 to 90% pancreatectomy (286) also indicate the importance of the pancreas in gastrointestinal absorption of cobalamin. The partially depancreatectomized rat appeared healthy, and had a growth rate comparable to control rats, but absorbed significantly less orally administered labeled cyanocobalamin compared to control and sham-operated animals. Furthermore, the defect in the absorption was corrected by the administration of exogenous pancreatic extract.

The active constituent in pancreatic extract is soluble following centrifugation at 50,000 × *g*, heat labile, and has a molecular weight of approximately 20,000 to 25,000 (287). These are physical-chemical properties in common with several pancreatic proteolytic enzymes. Crystalline trypsin proved to be as effective as the pancreatic extracts when administered to partially pancreatectomized rats. Whether the other proteolytic enzymes contained in the active subfractions (chymotrypsin and cathepsin) also possess the capacity to promote cobalamin absorption has not been evaluated. The precise mechanism by which pancreatic extract enhances cobalamin absorption remains to be determined. The pancreatic factor may function within the lumen of the gastrointestinal tract to maintain the gastric intrinsic factor-cobalamin complex in a form readily available for absorption, perhaps by secretion of a proteolytic enzyme which digests an endogenous binder which in turn binds the intrinsic factor-cobalamin complex. Such a binder, if undigested, might prevent the complex from attaching to the ileal receptor.

The pancreas may also play an important role in cobalamin absorption in the dog as well as the human. The dog appears to have an intrinsic factor-like cobalamin binder which is highly species specific and is produced by the pancreas (288).

INTESTINAL ABSORPTION OF COBALAMIN ANALOGS

Human

Except for hydroxocobalamin, and possibly adenosylcobalamin, most of the other corrinoids tested appear poorly absorbed compared to cyanocobalamin (81). It is of physiological significance that hydroxocobalamin, the main cobalamin in food, is absorbed as well as cyanocobalamin. The intestinal absorption and tissue deposition of hydroxocobalamin in rats were also found to be similar to those of cyanocobalamin (289). Erroneous results can be obtained by using the urinary excretion test to measure absorption, since hydroxocobalamin is more strongly bound and retained in tissues than cyanocobalamin (81, 290, 291). Direct measurement of absorption by fecal, hepatic, or whole-body counting techniques is necessary in this type of study.

Similar findings were obtained when adenosylcobalamin absorption was compared to that observed with cyanocobalamin. Measurement of absorption by the urinary excretion test gave low values with this form of cobalamin because of greater tissue retention (81, 292), but assessment of adenosylcobalamin absorption by hepatic uptake or fecal excretion technique shows that it did not differ from cyanocobalamin absorption (292–294). In contrast to these results, Heinrich and Gabbe (295) found that in man, adenosylcobalamin was not absorbed as well as cyanocobalamin, as measured by the fecal excretion test. These investigators presented evidence that, in contrast to rats, man has an absorption capacity for adenosylcobalamin about 50% of that for cyanocobalamin or hydroxocobalamin. In pernicious anemia patients, the same amount of intrinsic factor was required to facilitate the absorption of these corrinoids.

In man, cyanocobalamin in physiological doses (2 μg) is absorbed significantly better than chlorocobalamin, sulfatocobalamin, nitrocobalamin, and thiocyanatocobalamin (85, 86). Absorption was determined by the urinary excretion test, as well as fecal execretion and hepatic uptake methods. *In vitro* it is likely that the sulfato, chloro, and nitro analogs are converted to a common structural analog, since all the analogs are indistinguishable in their capacity to induce cobalamin excretion upon injection. All of these analogs have equal affinity for binding to gastric juice (87), suggesting a lack of correlation between strength of binding of a cobalamin to gastric juice and its subsequent absorption.

With regard to substitution on the benzimidazole group, only CN-Cbl and (5-OMeBza)CN-Cba were absorbed well, while benzimidazole, 5,6-dichlorobenzimidazole, 5-hydroxybenzimidazole, and trimethylbenzimidazole analogs were poorly absorbed (212, 295). In addition, 5,6-dimethylbenzimidazole itself does not interfere with cobalamin absorption.

(Ade)CN-Cba does not compete with CN-Cbl for the physiological absorption of cobalamins (90,296). This is not surprising, since (Ade)-CN-Cba does not bind strongly to intrinsic factor, nor does it have any hematopoietic activity when administered parenterally.

Rat

The effect of various analogs on the intestinal absorption of labeled cobalamin was also studied in rats (297). Neither (Ade)CN-Cba nor the cobalamin lactone was absorbed. The desdimethyl and a mixture of monoethylamide analogs were absorbed. This would seem to indicate that some binding to intrinsic factor by these latter analogs probably occurs *in vivo*.

Gallagher et al. (83) studied the specificity of various analogs of cyanocobalamin with regard to transport in the small intestine of the neonatal rat. The ethylamide and methylamide derivatives interefered with cobalamin transport, whereas the anilide, ethylamide monocarboxylic, and dicarboxylic acids had no effect.

Hydroxocobalamin, methylcobalamin, adenosylcobalamin, and sulfitocobalamin appeared to compete with cyanocobalamin for transport across the neonatal rat intestine. The latter compound has not been tested in man, but hydroxocobalamin, methylcobalamin, and adenosylcobalamin are naturally occurring cobalamins and are absorbed in man. Analogs which differ from cyanocobalamin in the nucleotide structure, such as (5-MeBza)CN-Cba, and a cyanocobalamin carbanilide in which the carbanilide is joined to one of the ribose hydroxyl groups, interfered with the transport of cyanocobalamin, whereas (Ade)CN-Cba was without effect.

PASSIVE ABSORPTION OF COBALAMIN

A passive mechanism (not mediated by intrinsic factor) is operative when the amount of administered vitamin is large, usually in excess of the amount available from a normal diet. Absorption probably occurs by diffusion in this passive process.

Only about 1% of an oral dose of 100 to 5000 μg is absorbed in pernicious anemia patients (298). In addition the site of cobalamin absorption with these large pharmacological doses is nonspecific (299); some absorption of the vitamin occurs through the nasal mucosa, rectal mucosa, and skin (299–301). It is not known exactly at what dose level one mechanism stops and the other starts. It is probable that at certain levels of intake some overlapping between the two mechanisms occurs.

The amount of cobalamin that can be absorbed from a single oral dose in normal and pernicious anemia subjects is shown in Tables 5-4 and 5-5, respectively. It appears from these data that the average maximum absorption is about 1.5 μg in normal subjects regardless of the size of the

Table 5-4 The Absorption of A Single Oral Dose of Cyanocobalamin in Normal Human Subjects

	Amount Absorbed	
Oral Dose (μg)	μg	%
0.1	0.08	80
0.25	0.19	76
0.5	0.35	70
1.0	0.56	56
2.0	0.92	46
5.0	1.4	28
10.0	1.6	16
20.0	1.2	6
50.0	1.5	3

Data from various authors and summarized by Chanarin (78).

dose. If doses in excess of 100 μg are given orally to pernicious anemia patients, significant amounts of cobalamin will be absorbed. Serum cobalamin levels in the pernicious anemia patients receiving these high doses are in the normal range (303). Since intrinsic factor is absent or deficient in pernicious anemia, large doses must be absorbed by a mechanism independent of intrinsic factor.

The patterns of plasma radioactivity following the administration of small and large oral doses of labeled cyanocobalamin have been studied (188, 218, 220, 304). The level of radioactivity in the blood starts to increase 3 to 6 hr after the dose. In addition, some of the vitamin appears in the lymph, in contrast to that observed with small physiological doses (269, 305).

In contrast to the intrinsic factor-mediated absorption, the nonintrinsic factor-mediated absorption of cobalamin does not appear to require calcium ions (103, 129, 237) and is independent of pH (103). Beyond these data, the mechanism of the nonintrinsic factor-mediated absorption is not well defined. It is not certain whether receptors for the large amounts of vitamin exist in the intestine. The mechanism of this absorption is assumed to be simple diffusion. Nevertheless, large oral doses of cyanocobalamin have been used for the treatment of pernicious anemia. Spies et al. (306) and Ungley (307) have noted good hematopoietic

Table 5-5 The Absorption of a Single Oral Dose of Cyanocobalamin in Pernicious Anemia Subjects

	Amount Absorbed	
Oral Dose (μg)	μg	%
1	0.057	5.7
3	0.108	3.6
10	0.260	2.6
100	1.1	1.1
200	2.0	1.0
400	4.0	1.0
800	8.8	1.1
5000	50.0	1.0

Data from Berlin et al. (302).

response in patients receiving in excess of 300 μg/day. Several investigators have confirmed this finding (303, 308–313).

THE INTESTINAL ABSORPTION OF COBALAMIN IN THE HUMAN INFANT AND NEWBORN RAT

There is only limited information on the intestinal absorption of cobalamin in the newborn human. Following a tracer oral dose of labeled cobalamin almost all the vitamin was absorbed, whereas only 27 to 70% of an oral tracer dose of the vitamin is absorbed in pregnant women (314).

There is some question whether intrinsic factor is secreted in the newborn infant in the first week of life (315), while other investigators have reported that normal infants secrete intrinsic factor (316, 317). Rarely, an infant will have intrinsic factor antibody transplacentally acquired from a mother with pernicious anemia and will not have intrinsic factor secretion during the first few weeks of life (316, 317). Intrinsic factor has been demonstrated (see the section on site of secretion in the human) in the human fetus at 11 to 13 weeks (28).

Studies with newborn rats provide a good model for studying the intrinsic factor-independent transport of physiological amounts of cobalamin across the gut (318). Negligible amounts of cobalamin binding protein are found in their stomachs (60). The transport of the vitamin across the intestinal epithelium can therefore be studied independently of the fate of intrinsic factor. The studies were performed in fasting animals in order to be certain that the transport of the vitamin was not due to the cobalamin binding protein which is present in milk, although the absorption of the vitamin independently of milk has been shown to occur (319). These studies gave results which suggest that free cobalamin is absorbed by pinocytosis. Pinocytotic absorption of cobalamin in the infant rat was first proposed by Boass and Wilson (60), who found that the uptake of large amounts of free vitamin by rings of distal small intestine decreased abruptly at a time when morphological evidence of pinocytosis disappeared.

Additional evidence in favor of pinocytotic absorption of the free vitamin in the young rat also comes from electron microscope and autoradiographic studies. Using tritiated cobalamin, the highest density of the labeled vitamin was found in the absorptive vacuoles and in the giant supranuclear vacuole of the ileal epithelium. A smaller amount was present in the cytoplasm and was associated mainly with mitochondria (38). It was found that small intestine took up a constant percentage of each of four oral doses of cobalamin (318). This is consistent with a mechanism of pinocytosis. The readiness with which large quantities of the vitamin entered the intestinal epithelium contrasts with the small amounts which left the intestine (318). The vitamin is still present in the distal small-intestinal wall 24 hr after ingestion and appears to return to the intestinal lumen. These results seem to indicate that while the neonatal rat has an unlimited capacity for cobalamin absorption, only small amounts of the vitamin reach the carcass, suggesting that young rats do contain a specific transport system for the vitamin.

THE TRANSPLACENTAL PASSAGE OF COBALAMIN

Early indirect evidence for an active transport of cobalamin across the placenta came from the data showing that there is a transplacental concentration gradient of the vitamin in favor of the fetus. This observation has been confirmed in a large number of laboratories (320–331).

More direct evidence has been obtained by tracer studies. Following parenteral administration of cyanocobalamin in rats, significant amounts of the injected dose was found in the fetuses (332, 333). Similar findings were obtained following oral and parenteral administration in pregnant guinea pigs (334).

In the human, labeled cyanocobalamin administered near the end of pregnancy accumulates in the placenta during the first 2 days and then goes into the fetus over the next 2 to 3 weeks. As much as 45% of the administered dose accumulates in the fetus (314, 335, 336). These investigators showed that only the newly administered vitamin was available for placental transfer. Any radioactive vitamin present in the maternal liver was not transported to the fetus.

It has recently been shown that methylcobalamin is the cobalamin most efficiently transported in the fetus and that cyanocobalamin is almost as well absorbed (337). The transplacental transport, like the transplacental gradient, does not appear to be unique. Other vitamins and minerals such as iron, vitamin C, pyridoxine, and riboflavin are transferred to the placenta against a concentration gradient (338, 339).

QUANTITATIVE ASPECTS OF COBALAMIN ABSORPTION

The total amount of cobalamin absorbed from the intestine of a normal human is limited to about 1×10^{-9} moles or about 1/100,000 the amount of iron absorbed (340–342). This small amount of vitamin is sufficient under normal conditions to meet the daily requirements of an adequately fed individual.

Various systematic studies of cobalamin in healthy persons have been performed with varying levels of the vitamin. The quantitative aspects of measuring absorption have been made possible by the use of radiolabeled vitamin and the various absorption techniques referred to in an earlier section. From these studies it has become clear that an inverse relationship exists between the percentage of the vitamin absorbed and the dose ingested (341–343). The absolute amount absorbed is very small. The amount of vitamin absorbed from an oral dose as high as 50 μg is only about 1.5 μg (Table 5-4). At physiological dose levels of 0.5 to 2.0 μg the efficiency of absorption is high; the percentage of cobalamin absorbed is about 45 to 80%. Thus a 100-fold increase in ingested dose from 0.5 to 50 μg results in an increase in absorption of about 1.2 μg. This regression of efficiency of absorption has been noted in experimental animals as well as in humans (142, 213, 344).

The regression of absorption of cobalamin has also been noted using labeled food (345). The whole-body counter was utilized for these measurements. Muscle meat was administered to healthy young men as hamburgers. Absorption was very efficient from the meat containing 1 and 3 μg of cobalamin. The additional absorption with an increase to 5 μg of vitamin was negligible.

Since the regression of cobalamin absorption has been found in normal subjects as well as pernicious anemia patients, it is obvious that this limitation of absorption is not due to an inadequate amount of intrinsic factor. It is now believed that the limitation of the efficiency of intestinal absorption of physiological amounts of cobalamin is due to saturation of intestinal receptors (see the section on the sequence of events during intrinsic factor-mediated cobalamin absorption).

The restricted capacity for cobalamin absorption in the intact hamster may be explained by the small number of receptor sites and the slow rate of release of cobalamin from the receptor (248). These membrane receptors maximally bind approximately the same quantity of cobalamin-intrinsic factor complex that the hamster is capable of absorbing *in vivo*.

GENERAL COMMENTS, CONCLUSIONS, AND CRITIQUE ON COBALAMIN ABSORPTION

Some of the factors complicating the interpretation of the results of studies concerned with the mechanism of cobalamin absorption have been clearly cited here and by others (346). Many of the problems still exist and any conclusions from experimental results must take certain factors into consideration. Some of these factors are (1) species specificity —for example, hog intrinsic factor does not facilitate the absorption of cobalamin in the rat (57, 344). In contrast to earlier observations, more recent studies have shown an awareness of this problem. Homologous systems are used where species specificity has been observed. (2) Cyanocobalamin is not a naturally occurring vitamin. Most of the information concerned with the absorption of cobalamins has accumulated from work done with cyanocobalamin. A limited number of investigations with some of the coenzyme forms of cobalamin suggest that the mechanisms involved in the absorption and transport of the coenzymes are similar to that in the vitamin. However, cyanocobalamin is converted into coenzymatically active cobalamins and the latter are present in the foods of a normal diet. (3) The importance of effects of endogenous cobalamins in the tissues: This factor has been taken into account in studies concerned with the "release" of cobalamin prior to its transport into the blood. (4) Use of crude intrinsic factor preparations: More recently, investigators have become aware of the facts that crude preparations can contain proteins which bind cobamides but which lack the ability to promote cobalamin absorption. The isolation of pure hog and human intrinsic factor has recently been accomplished (48–53, 71, 72) (see the section pertaining to this). The validation of crucial experiments will take some time. A limited number of studies with homogeneous intrinsic factor preparations have confirmed findings obtained with cruder preparations. On the other hand, some differences in binding properties between pure and semipurified preparations have been noted (see the section on the binding of cobalamin to intrinsic factor).

With all of these complications, nevertheless excellent progress has been made in understanding many characteristics of cobalamin absorption. In many respects, considerably more is known about the absorption of cobalamin, whose history is relatively short, than with substances whose history is much older and whose concentrations in man are considerably higher.

TRANSPORT

INTRODUCTION

Cobalamin is rarely found in free state, but is usually attached to specific proteins. There are at least two types of cobalamin binding proteins in plasma. They were intially distinguished by their electrophoretic mobility (347–350). The binding proteins were found in the α-globulin and β-globulin fractions. See Beal et al. (351) for an excellent review on this early work.

These two binders were studied more extensively by Hall and Finkler using DEAE-cellulose and CM-cellulose chromatography and were named transcobalamin I and transcobalamin II, respectively (44, 352, 353). These terms transcobalamin I (TC I) and transcobalamin II (TC II), originally introduced for the cobalamin binding components of plasma separated by ion exchange chromatography and electrophoresis, have also been used for the serum components separated by gel filtration mainly according to difference in molecular size (354–356). Gullberg (357–360) has proposed a separate nomenclature for the two major cobalamin binding protein components in human blood plasma. Transcobalamin "large" (TC L) and transcobalamin "small" (TC S), referring to molecular size, were proposed for the cobalamin binding proteins separated by ion exchange chromatography. However, the recent studies on the isolation and physical characterization of the transcobalamins (74, 75) indicate that these two proteins have very similar molecular weights. Gel filtration studies can give falsely high values for molecular weights with proteins containing a high percentage of carbohydrate (361).

Evidence for a third binder (transcobalamin III) in normal human serum has been presented by several investigators (354, 355, 362–372).

TC III seems to resemble TC II in its behavior on DEAE-cellulose and in electrophoretic mobility, and TC I in antigenic characteristics and in its inability to stimulate uptake of cobalamin by reticulocytes.

TC III also appears to bind cobalamin to a greater extent than TC I. In addition, cobalamin-labeled TC II and TC III prepared from normal human and pernicious anemia patients exhibited a higher uptake by perfused rat liver than labeled TC I (373).

The evidence for this third protein, however, is not unequivocal and it has not been established whether this third protein is a distinct protein with chemical and structural differences from that seen in TC I and II, or merely an artifact of isolation. It has been suggested that such a third cobalamin binder is a polymer of TC II (374–376) or a part of TC I complex showing immunological identity with other binders in tissue fluids (377).

Whether the TC III is similar to the fetal binder described by Kumento and co-workers (371, 378) or to that found in polycythemia vera (370) also remains to be determined.

SYNTHESIS AND SOURCE OF TRANSCOBALAMINS

In the mouse and rat, TC II is probably synthesized in the liver (379, 380). This evidence is based on liver perfusion studies and on the fact that the synthesis of transcobalamin II is decreased following carbon tetrachloride poisoning. In contrast, the concentration and rate of synthesis of transcobalamin II in dogs are not influenced by removal of the liver, implying that this organ is not the only tissue responsible for synthesis of protein (381). This interpretation assumes that cobalamin does not dissociate from TC II, and that TC II does not recirculate (380).

It has been speculated that TC II could possibly come from the intestine since cobalamin absorbed from the intestine is found attached to TC II (106). The possibility that TC II represents partially degraded intrinsic factor has also been presented (171). This fragment of intrinsic factor could be absorbed from the intestinal lumen, and man is normally immunologically tolerant to this degraded intrinsic factor because it has lost its immunological determinants. Since TC II is believed to be free of carbohydrate (75), and to be different from intrinsic factor, based on its amino acid composition and subunit structure, this hypothesis seems unlikely (50, 51).

It has been suggested that granulocytes synthesize transcobalamin I and other "R"-type proteins* (382, 383). This evidence is based on the fact that (1) antibodies to cobalamin binding protein of granulocytes cross-react with transcobalamin I (384); (2) the content of transcobalamin I is proportional to the total content of granulocytes; and (3) two patients who had a low level of the granulocyte cobalamin binding protein were

* See p. 264 ff.

deficient in transcobalamin I (385). Recent studies support the concept that leukocytes are a major source of the TC III (386). This protein is elevated in leukocytosis. If leukocytes are the source of both TC I and TC III, some transformation of the protein must occur inside or outside of the cell (387). Alternatively, only certain cells, perhaps the immature ones, may contain TC I, the rest containing TC III.

In tissue culture, synthesis of R-type salivary binders occurs in serous glandular cells from both submandibular and parotid glands (388).

PURIFICATION AND PROPERTIES OF TRANSCOBALAMINS

The magnitude of the problem with regard to the isolation of transcobalamin I and II is apparent when it is realized that 1000 liters of human plasma contain approximately 80 and 20 mg, respectively, of each of these proteins. The isolation of each of these proteins in homogeneous form requires, therefore, that transcobalamin I be purified almost one million-fold and transcobalamin II in excess of two million-fold (96).

There are about 60 μg of transcobalamin I per liter of normal plasma and this can carry about 700 to 800 pg of cyanocobalamin per milliliter of plasma, but about half of the binding capacity of transcobalamin is normally saturated with endogenous cobalamin. The plasma content of transcobalamin II is estimated to be about 15 to 20 μg/l and transcobalamin II can carry an average of 1000 pg of cyanocobalamin per milliliter when fully saturated.

Transcobalamin II has been extensively purified (approximately one million-fold) as its cobalamin complex from human plasma Cohn Fraction III by ion exchange cellulose and gel filtration (389). A limited amount of chemical data was obtained because of the small amount of final product; there appeared to be some uncertainty regarding its purity. This preparation was reported to contain at least 13% carbohydrate but no sialic acid.

Transcobalamin II has been isolated in homogeneous form from Cohn Fraction III derived from pooled human plasma by a combination of conventional purification techniques and affinity chromatography (75). The overall purification relative to human plasma was about two million-fold. This apparently pure transcobalamin II binds 28.6 μg of cyanocobalamin per milligram of protein and contains one cobalamin binding site per 59,550 g of protein as determined by amino acid analysis. The molecular weight determined by sedimentation equilibrium ultracentrifugation was 53,900 and by gel filtration on Sephadex G-150 was 60,000. Transcobalamin II is a dimer consisting of one subunit of approximately 25,000 molecular weight and one of molecular weight 39,000. When cyanocobalamin binds to transcobalamin II there is a shift in the peak of cobalamin absorption from 361 to 364 nm. No carbohydrate residues were detected following amino sugar analysis and gas-liquid chromatography, suggesting that in contrast to human and hog intrinsic factor and transcobalamin I ("R" protein), transcobalamin II is not a glycoprotein.

The method of affinity chromatography described by Allen and Majerus (73) was also used to isolate the cobalamin binding proteins from granulocytes obtained from patients with chronic granulocytic leukemia (74). In fact, affinity chromatography was the sole purification technique employed in isolating this protein in homogeneous form in 90% yield. (The relationship and/or identity between transcobalamin I and the granulocyte cobalamin binding protein is discussed in the following section —see the preceding section also.)

The granulocyte cobalamin binding protein binds 34.9 μg of cyanocobalamin per milligram of protein and has a single cobalamin binding site. Since removal of the vitamin from the protein results in some denaturation of the protein, it was difficult to determine the binding constant between cyanocobalamin and the granulocyte binding protein. The molecular weight of this protein determined by sedimentation equilibrium ultracentrifugation was 56,000, by amino acid and carbohydrate analysis was 58,200, and by gel filtration and sodium dodecyl sulfate polyacrylamide gel electophoresis 121,000 to 138,000. Proteins that contain large amounts of carbohydrate often give falsely elevated values for molecular weight when determined by gel filtration (361).

Transcobalamin II has the following properties in common with the granulocyte cobalamin binding protein:

a. Both proteins have single cobalamin binding sites.
b. Both proteins have molecular weights of 50,000 to 60,000.
c. Neither protein has free sulfhydryl groups.
d. When cyanocobalamin binds to the proteins there is a shift and an enchancement of the absorption spectrum above 300 nm.

The differences between the two proteins are

a. Transcobalamin has two subunits (a 39,000 and 25,000 molecular weight subunit) whereas the granulocyte binding protein has a single polypeptide chain.
b. Transcobalamin II is not a glycoprotein.
c. The amino acid compositions differ.
d. When cyanocobalamin is bound to transcobalamin II the 361-nm spectral peak for unbound vitamin shifts to 364 nm. No shift occurs

when cyanocobalamin is bound to granulocyte cobalamin binding protein.

e. The difference spectra between the individual protein-cyanocobalamin complexes and unbound vitamin are different, suggesting that the cobalamin binding sites for the two proteins are not the same.

f. Pure transcobalamin II appears to facilitate the uptake of cobalamin by human diploid fibroblasts, whereas a homogeneous preparation of granulocyte cobalamin binding protein does not (75).

Other workers have produced artifacts by some phase of their separation technique and claims for the isolation of new cobalamin binders must be examined carefully. For example, repeated freezing and thawing can change the electrophoretic mobility of plasma binders. Perchloric acid can change certain characteristics of transcobalamin II to those of transcobalamin I. Transcobalamin II forms large-sized complexes when exposed to buffers of low ionic strength (363). Heparin will complex with transcobalamin II (390) and affect elution properties using DEAE-cellulose chromatography (377). EDTA affects the properties of the transcobalamins especially when separated by gel filtration (390).

RELATIONSHIP OF TRANSCOBALAMIN I TO OTHER TISSUE COBALAMIN BINDERS

Transcobalamin I has been classified as belonging to a group of proteins called "R proteins" (96). The term R protein originally signified rapid binder (77) or fast binder (391) in human gastric juice, indicating its electrophoretic mobility in relation to gastric juice. The term is now used for cobalamin binding proteins found in various locations such as saliva, erythrocytes, leukocytes, granulocytes, bile, and so on, which are immunologically identical even though they may differ in molecular weight and electrophoretic mability (77, 96, 356, 367). The R proteins are separate and distinct both functionally and immunologically from transcobalamin II and intrinsic factor (96, 356,).

These immunological studies, which used crude or partially purified preparations, suggest that cobalamin binding proteins found in human body fluids can be divided into three distinct groups: intrinsic factor, transcobalamin II, and the R group of proteins to which the human granulocyte cobalamin binding protein, transcobalamin I, and gastric nonintrinsic factor belong.

Antitranscobalamin II serum was found to react only with transcobalamin II regardless of the source. It did not react with any of the other binders. Antitranscobalamin I and antisaliva cobalamin binding protein reacted identically with all the R binders. No cross-reaction between the three groups could be demonstrated (see Table 5-6). Some properties of the various groups of cobalamin binders are shown in Table 5-7. In addition to these data, recent experiments (377) suggest that TC I and TC III give reactions of identity with antiserum to salivary R binder, suggesting that these two proteins are very closely related to each other.

Table 5-6 Classification of Cobalamin Binders by Immunological Properties

	Antiserum			
Antigen	Saliva	TC I	TC II	Intrinsic Factor
Saliva	+	+	−	−
Normal TC I	+	+	−	−
Normal TC II	−	−	+	−
Leukemic TC II	−	−	+	−
CSF TC II	−	−	+	−
Seminal fluid TC II	−	−	+	−
CSF TC I	+	+	−	−
Seminal fluid TC I	+	+	−	−
Leukemic TC I	+	+	−	−
Leukocyte binder	+	+	−	−
Nonintrinsic factor	+	+	−	−
Intrinsic factor	−	−	−	+

Much of the data taken from Hall and Finkler (384).

The variations in sialic acid content of these R proteins are believed to be the cause of the heterogeneity seen in these R proteins when subjected to electrofocusing (387, 388, 391, 392). During the purification of TC I and the cobalamin binding protein in leukocytes, a reduction in their electrophoretic mobilities occurs, and the same change is produced by sialidase treatment (387). It has been found that TC I from patients with chronic granulocytic leukemia differs in its properties from normal TC I, although it is possible that the differences are only in their carbohydrate content.

The difference between the estimated molecular weight of 138,000 obtained for the granulocyte cobalamin binding protein by gel filtration and the value of approximately 56,000 obtained by sedimentation equilibrium centrifugation and chemical composition is also related to the

Table 5-7 Comparative Properties of Three Groups of Cobalamin Binders

	Intrinsic Factor	Granulocyte Binder	TC II
Molecular weight	45,000 (human) 55,000 (hog)	58,000 (human)	60,000 (human)
CN-Cbl (μg/mg)	30.1 (human) 30.3 (hog)	34.9	28.6
Glycoprotein	Yes	Yes	No
Uptake ileal tissue	Yes	No	Yes
Uptake HeLa cells	No	No	Yes
Reaction with Anti-TC II	No	No	Yes
Reaction with anti-intrinsic factor	Yes	No	No
Reaction with anti-TC I	No	Yes	No

carbohydrate content of these proteins (74) (the section on purification and properties of transcobalamins). This suggests that the different molecular weights reported for the individual R-type cobalamin binding proteins by gel filtration may be the results of differences in carbohydrate content rather than in amino acid composition or sequence. This recent observation also lends support to the finding of similar immunological properties of R-type proteins.

Before the detection and study of cobalamin transport proteins in plasma, there were many attempts to show that intrinsic factor or a related substance functioned in the circulation as well as in the intestine. Subsequently it was shown that the component of hog intrinsic factor active in these systems was an R binder and not intrinsic factor (139). Use was made of this fact in the fractionation of hog intrinsic factor. "Nonintrinsic factor" was removed from intrinsic factor preparations by treating the latter with an antiserum prepared against hog leukocyte binder, an R-type binder (393).

Despite the common immunochemical features with TC I, it has been claimed that the salivary cobalamin binder is different from R binders (394), and is an exception to the similarity of all R-type binders. However, the studies by Carmel (395) using gel filtration and sucrose density gradient ultracentrifugation support the identity of all the R binders including the salivary binder.

The isolation of plasma transcobalamin II has been accomplished (see the preceding section). Several other R-type proteins, including nonintrinsic factor cobalamin binding proteins from hog pylorus and human granulocyte cobalamin binding protein, have been isolated (see the preceding section and also the one on the isolation and properties of intrinsic factor). None of these homogeneous R proteins has been subjected to immunological analysis with other homogeneous R proteins; however, their molecular characteristics, including molecular weight, are very similar. The isolation of transcobalamin I and various other R-type cobalamin binding proteins and their physical and chemical characterization is needed in order to elucidate fully the relationships among these proteins.

There is genetic evidence that the R proteins are closely related (385). These proteins are apparently coded by one gene, and it has been reported that two brothers lacked R proteins (385). Variability among the R proteins could be due to differences in gene expression in different cells. The protein could possibly receive its sialic acid at a late stage of synthesis, perhaps just before it leaves the cell (96).

It is evident that the R binders have enough properties in common to form a third group which is quite distinct both functionally and immunologically from transcobalamin II and intrinsic factor.

BINDING OF COBALAMIN TO TRANSCOBALAMINS

The difficulties encountered in measuring and interpreting the affinity of cobalamin to intrinsic factor (see the earlier section pertaining to this) are also present with the transcobalamins.

Somewhat more methylcobalamin than cyanocobalamin is bound to TC I and TC II when excess amounts of these cobalamins are added *in vitro* to these proteins (396). The ratio of the binding capacity for methylcobalamin to cyanocobalamin was 1.32 to 1.36 for TC I and 1.34 for TC II. The lactone and lactam derivatives can bind to the transcobalamins as efficiently as cyanocobalamin (79). Other than these studies, little has been done on the binding of various cobalamins to these different serum cobalamin binding proteins. Most of the studies have been performed with whole serum.

Unlike the case with intrinsic factor, the nature of the nucleotide in the corrin is relatively unimportant in serum binding. (Ade)CN-Cba and (2-MeAde)CN-Cba were bound by serum as well as cyanocobalamin (87, 92). $(CN)_2$ Cbi bound well to chicken serum, though poorly to human (89, 92). The ethylamide, methylamide, monobasic acid, and anilide analogs also bind to serum to about the same degree as cyanocobalamin (93).

As mentioned earlier, the binding to intrinsic factor is more specific than to transcobalamins. That intrinsic factor has greater specificity than transcobalamin for the various cobalamins is pertinent because specificity at the stage of intestinal absorption efficiently prevents the absorption of undesired compounds. High selectivity is no longer needed during further stages of transport.

Cobalamin can be dissociated from TC I and TC II by raising the pH to 12.9 (118) but TC II proved to be quite labile—80% of its cobalamin binding power was destroyed.

The association constant for binding of cyanocobalamin to the salivary binder was found to be $15 \times 10^9 M^{-1}$ at 26° and pH 7.4, about one-tenth the value for binding to transcobalamin I and II (108). Affinity constants with nonpure proteins, however, can be difficult to interpret. In view of the possible identity between transcobalamin I and salivary binder, the tenfold difference is surprising. Equilibrium studies of cobalamin binding to pure transcobalamins are also difficult to interpret since some denaturation of the protein occurs in an attempt to remove all endogenous vitamin from the protein (74).

SEQUENCE OF EVENTS DURING PLASMA TRANSPORT

Initial State of Plasma Transport

Relative to the intrinsic factor-mediated gastrointestinal transport of cobalamin, the state of knowledge regarding the molecular events during plasma transport is fairly poor. Most of the data deal with kinetics rather than molecular and/or physical changes.

Following the absorption of physiological amounts of cobalamin, the vitamin enters the blood at a slow rate. The vitamin first appears in the blood within 3 to 4 hr after ingestion and reaches a peak level at about 8 to 12 hr (44).

The cobalamin entering the blood after oral ingestion as the vitamin or in food appears bound to TC II, according to the work of Hall and Finkler (384). More than 8 hr after administration, an increasing proportion of the labeled vitamin becomes bound to TC I. The amount of vitamin bound to TC I increases as the amount of vitamin in TC II decreases. However, the uptake of the vitamin by transcobalamin II is probably still going on at 24 hr. Since a normal person usually ingests food several times a day, a small but constant amount of cobalamin-TC II complex is circulating. The amount of vitamin adsorbed to TC II probably never exceeds 20 pg/ml of plasma (397).

The transfer of the absorbed cobalamin from one plasma protein to another has been compared to the fate of copper after oral ingestion. The absorbed copper is bound initially to albumin, and then transported to the liver. In the liver copper is gradually incorporated into another protein, ceruloplasmin, and copper appears bound to plasma ceruloplasmin (398). Alternate explanations, however, have been offered to account for the fact that endogenous cobalamin is bound primarily to an α-globulin (TC I), but cobalamin fed or injected attaches first to a β-globulin (TC II) (399). Ingested or injected cobalamin could attach to both α and β-globulin cobalamin binding proteins; that which is attached to β-globulin is delivered to tissues rapidly whereas that attached to α-globulin is retained in serum.

Very recently, it has been shown that all three serum cobalamin binders (TC I, TC II, and TC III) took up vitamin absorbed from the intestine at about the rate (368, 377). Thus, there appears to be an increasing amount of evidence that the transport of the vitamin from the intestine to the blood is not the exclusive function of a single blood protein.

The plasma turnover of radioactive cyanocobalamin bound *in vitro* to transcobalamin II has been studied after intravenous injection. The half-life of the injected dose has been reported to be as low as 5 min (400). Simultaneously with the rapid clearance of the cobalamin to transcobalamin II, the vitamin appears rapidly in the liver. The uptake by other tissues must be very rapid also.

Other investigators have clearly demonstrated that cobalamin-transcobalamin II is cleared much more rapidly than cobalamin-transcobalamin I, with a half-life on the order of about 1.5 hr (374, 401–403). The work of England et al. (377) indicates that the half-life of the TC II complex is 8 to 28 hr. These latter workers labeled TC II by way of absorbed cyanocobalamin and followed the rate of disappearance of radioactivity. It is possible that the rapid disappearance observed by the former investigators could be due to denaturation of the isolated and reinjected TC II.

When physiological quantities of methylcobalamin and cyanocobalamin are injected intravenously in man, both methylcobalamin and cyanocobalamin are cleared at approximately the same rate and distributed in the same way between TC I and TC II (396). It would not have been surprising if methylcobalamin, which is the more physiological form of plasma cobalamin, had been cleared in a different way. The overall behavior of methylcobalamin appears to be identical to that of cyanocobalamin, and previously obtained results with regard to transport of radioactive cyanocobalamin appear to be relevant.

Later Stages of Plasma Transport

Recent studies strongly suggest that TC II may play a major role in cobalamin transport in late as well as early distribution of the vitamin (371, 404–406). In addition to delivering the vitamin to tissues from the gut, it appears that TC II may also transport the vitamin from storage sites to other tissues. In the congenital absence of TC II, there is impaired delivery of cobalamin from "storage" sites to bone marrow (406) (see the following section).

Using a bioassay procedure, it was shown that some endogenous cobalamin is indeed bound to TC II. This was demonstrated also in patients who were on a cobalamin-free diet for extended periods (405).

Approximately 30% of the serum cobalamin is bound to TC II, so the bulk of the vitamin is attached to additional proteins. Nevertheless, it is obvious that the function of TC II must extend beyond the postabsorption phase.

FUNCTION OF TRANSCOBALAMIN

The function of transcobalamin I is not unequivocally known. It has been suggested that this protein serves the role of a circulating cobalamin reservoir in equilibrium with tissue stores of cobalamin (407). Cobalamin bound to TC I in blood is removed slowly by tissues (401). The protein does not promote cell uptake *in vitro* (400, 408).

As is pointed out in Chapter 6, most of the endogenous cobalamin bound to TC I has been identified as methylcobalamin (409). This finding has led to the speculation that transcobalamin I as well as other R proteins are apoenzymes (96, 410) for methylcobalamin. This cobalamin is known to act as a coenzyme in several enzymatic reactions.

A deficiency of the transcobalamin I produces no apparent abnormality in cobalamin metabolism. Persistent deficiency of serum transcobalamin I was demonstrated in two brothers manifesting primarily low serum cobalamin levels (385). Despite injection of cyanocobalamin, one subject maintained low serum cobalamin levels. The α-globulin binder normally present was virtually absent from saliva and peripheral leukocyte extracts of both subjects, further suggesting the close relationship between R proteins and transcobalamin I.

There must be some way for the body to use TC I-bound cobalamin. Possibly TC I-bound cobalamin is taken up slowly by some tissues not yet studied. There may be a release of cobalamin initially bound to TC I and a rebinding to TC II, and then uptake by the cells. Some evidence that TC I carries cobalamin out of cells comes from the study of the uptake of the vitamin by HeLa cells (400). The labeled vitamin bound to TC II was incorporated into the HeLa cells. The labeled vitamin that comes out of the cells was shown to be bound to a protein resembling TC I. The cells could have converted TC II to TC I or could have synthesized TC I *de novo*.

In contrast to transcobalamin I, transcobalamin II is necessary for normal cellular maturation of the hematopoietic system (406, 411). The function of TC II is to carry cobalamin to recipient tissues. This is supported by its facilitation of cobalamin uptake by HeLa cells, liver, reticulocyte, erythrocyte and so on. (See Chapter 8.) Hereditary transcobalamin II deficiency is a rare inborn error of protein metabolism which is inherited as an autosomal recessive condition (406). The transcobalamin II-mediated uptake of cobalamin by mammalian tissues is a biphasic process involving initial attachment of the cobalamin-TC II complex to the cell followed by transport of the vitamin across the cell membrane. This process thus resembles intrinsic factor-mediated transport of cobalamin across the intestinal epithelium. Uptake of cobalamins by certain microorganisms also appears to a biphasic process, consisting of an initial rapid phase which is independent of the energy metabolism of the cell, followed by a slower, secondary, energy-dependent phase. With these organisms, however, no protein binder is required to assist transport (412–415).

TRANSPORT SYSTEM IN ANIMALS

If a model resembling the human cobalamin transport system could be found, the exploration of various aspects of transport would be facilitated. Unfortunately it is only recently that model systems in other animals have been studied. The plasma transport of cobalamin in the dog was studied using the same approaches as applied to human cobalamin transport (416). A cobalamin binding protein was found in canine plasma which eluted from DEAE-cellulose with the β-globulins, had β-mobility on paper electrophoresis, and carried cobalamin as it was absorbed. It also reacted immunologically with antihuman transcobalamin II. The counterpart of human transcobalamin I appears to be present but in lesser amounts than in man. The time, amount, and binding of cobalamin entering the plasma after absorption in the dog were like that in man, which might give support to an intrinsic factor system in the dog.

The cobalamin binders in normal rhesus monkey serum appear to differ from human TC II is physical and chemical behavior (417). Sera from cats, rabbit, hog, cow, pig, sheep, rat and guinea pig are resolvable into two major components (418). The two components present in different animal sera are similar but vary with respect to concentration and relative elution pattern. Chickens have a single unsaturated serum cobalamin binder and an immunologically similar proventriculus binder (419, 420). High-molecular-weight binders in sera of amphibians have also been found (421), suggesting that the transport of cobalamin also requires specific factors in these species.

GENERAL COMMENTS, CONCLUSIONS, AND CRITIQUE ON COBALAMIN TRANSPORT

Major progress in our understanding of cobalamin transport may be facilitated by the recent isolation of various transcobalamins (72, 73). The differences in the chemical and physical properties of transcobalamin II and the granulocyte binding protein are very marked—much more so than the differences between intrinsic factor and other cobalamin binding proteins of gastric origin. Transcobalamin II is unique. It is the first cobalamin binding protein which has been shown to be devoid of carbohydrate. The significance of this remains to be determined. *In vivo,* however, it appears to be the most metabolically active of the cobalamin transport proteins.

As observed with the cobalamin binding proteins of gastric origin, the transcobalamins are quite labile. As mentioned earlier, the properties of the various transcobalamins can be readily changed by slight changes in isolation techniques. Artifacts can thus be readily produced. For example, heparin causes transcobalamin II and transcobalamin III to elute with transcobalamin I (377), and this has been observed in blood samples taken from patients given heparin.

Independent of the isolation studies has been the accumulation of recent data suggesting that transcobalamin II plays a major role in late as well as early stages of cobalamin transport. The original assumption that the function of transcobalamin II did not extend beyond the immediate postabsorption phase appears to be incorrect. Therefore, in addition to delivering the vitamin from the intestine to tissues it is becoming apparent that transcobalamin II can also transport the vitamin from storage tissues to other tissues.

Recently reported investigations support the critical importance of transcobalamin II for normal cellular maturation of the hematopoietic system (406, 411). In the absence of transcobalamin II and with normal cobalamin blood levels, megaloblastic anemia can develop. In contrast congenital deficiency of transcobalamin I does not lead to any sign of cobalamin deficiency (385). In addition, it appears from *in vitro* studies that transcobalamin I is not involved in transport of cobalamin from plasma to tissues.

These major developments in the study of transport of cobalamin have been reported only very recently. It is thus apparent that the acquisition of information about the transport of cobalamin has been more sudden and dramatic than the slow and steady growth of information about the gastrointestinal absorption of the vitamin.

ACKNOWLEDGMENTS

The author is grateful to Dr. Leo M. Meyer, Chief of Hematology, Veterans Administration Hospital, Brooklyn, New York for his helpful criticism of the manuscript and to Dr. William Pearl, Lederle Laboratories, for his valuable assistance in the many chores incident to final preparation of the manuscript.

REFERENCES

1. Castle, W. B. (1953). *N. Engl. J. Med.* **24**, 603.
2. Castle, W. B., and Townsend, W. C. (1929). *Am. J. Med. Sci.* **178**, 764.
3. Castle, W. B., Townsend, W. C., and Heath, R. W. (1930). *Am. J. Med. Sci.* **180**, 305.
4. Beerstecher, E., Jr., and Altgelt, S. (1951). *J. Biol. Chem.* **189**, 31.
5. Bertcher, R. W., Meyer, M. M., and Miller, I. F. (1958). *Proc. Soc. Expt. Biol. Med.* **99**, 513.
6. Landboe-Christensen, E., and Bohn, C. L. S. (1947). *Acta Med. Scand.* **127**, 116.
7. Landboe-Christensen, E., Berk, L., and Castle, W. B. (1952). *Am. J. Med. Sci.* **224**, 1.
8. MacDonald, R. M., Ingelfinger, F. J., and Belding, H. W. (1947). *N. Engl. J. Med.* **237**, 887.
9. Paulson, M., Conley, C. L., and Gladsden, E. S. (1950). *Am. J. Med. Sci.* **220**, 310.
10. Swendseid, M. E., Halsted, J. A., and Libby, R. L. (1953). *Proc. Soc. Exp. Biol. Med.* **83**, 226.
11. Wilkinson, J. F., Klein, L., and Ashfond, C. A. (1938). *Q. J. Med.* **7**, 555.
12. Fox, H. J., and Castle, W. B. (1942). *Am. J. Med. Sci.* **203**, 18.
13. Landboe-Christensen, E., and Plum, C. M. (1948). *Am. J. Med. Sci.* **215**, 17.
14. Meulengracht, E. (1952). *Acta Med. Scand.* **143**, 207.
15. Magnus, H. A., and Ungley, C. C. (1938). *Lancet* **I**, 420.
16. Meulengracht, E. (1939). *Am. J. Med. Sci.* **197**, 201.
17. Motteram, R. (1951). *J. Pathol. Bacteriol.* **63**, 389.
18. Wruble, L. D., Cole, G. W., Lessner, H. E., Haidri, S. Z. H., and Kalser, M. H. (1964). *Ann. Intern. Med.* **60**, 877.

19. Hurst, H. F. (1923). *Lancet* **1**, 111.
20. Paulson, M., and Harvey, J. C. (1954). *JAMA* **156**, 1556.
21. Schilling, R. F., Clatanoff, D. V., and Korst, D. R. (1955). *J. Lab. Clin. Med.* **45**, 926.
22. MacLean, L. D., and Sundberg, R. D. (1956). *N. Engl. J. Med.* **254**, 885.
23. MacDonald, R. M., Ingelfinger, F. J., and Belding, H. W. (1957). *N. Engl. J. Med.* **237**, 887.
24. MacLean, L. D. (1957). *N. Engl. J. Med.* **257**, 262.
25. Hoedmaeker, P. J., Abels, J., Wachters, J. J., Arends, A., and Nieweg, H. O. (1964). *Lab. Invest.* **13**, 1394.
26. Hoedemaeker, P. J., Abels, J., Wachters, J. J., Arends, A., and Nieweg, H. O. (1966). *Lab. Invest.* **15**, 1163.
27. Taylor, W. H., Mallett, B. J., and Taylor, K. B. (1961). *Biochem. J.* **80**, 342.
28. Schwartz, M., and Weber, J. (1971). *Scand. J. Gastroenterol. Suppl.* **9**, 57.
29. Sturgis, C. C., and Isaacs, R. (1929). *JAMA* **93**, 747.
30. Sharp, E. A. (1929). *JAMA* **93**, 749.
31. Wilkinson, J. F. (1930). *Brit. Med. J.* **I**, 236.
32. Snapper, I., and du Preez, J. D. G. (1931). *Nederl. Tijdschr. Geneesk.* **75**, 29.
33. Muelengracht, E. (1934). *Acta Med. Scand.* **82**, 352.
34. Meulengracht, E. (1935). *Acta Med. Scand.* **85**, 79.
35. Meulengracht, E. (1953). *Acta Med. Scand.* **144**, 290.
36. Sharp, E. A., McKean, R. M., and von der Heide, E. C. (1931). *Ann. Intern. Med.* **4**, 1282.
37. Gutzeit, K., and Herrman, J. (1931). *Münch. Med. Wochenschr.* **78**, 266.
38. Kühnau, J. (1933). *Münch. Med. Wochenschr.* **80**, 1772.
39. Braun, B. (1934). *Folia Haematol.* **53**, 27.
40. Thompson, J. C. (1937). *Ann. Intern. Med.* **11**, 39.
41. Bethell, F. H., Swendseid, M. E., Meyers, M. C., Neligh, R. B., and Richards, H. G. (1949). *Univ. Mich. Med. Bull.* **15**, 49.
42. Hall, B. E., Bethell, F. H., Morgan, E. H., Campbell, D. C., Swendseid, M. E., Miller, S., and Cintron-Rivera, A. A. (1950). *Proc. Staff Meetings Mayo Clinic* **25**, 105.
43. Heatley, N. G., Jennings, M. A., Florey, H., Watson, G. M., Turnbull A., Wakisaka, G., and Witts, L. J. (1954). *Lancet* **2**, 578.
44. Heinrich, H. C. (1957). *Vitamin B_{12} und Intrinsic Factor, 1. Europäisches Symposion, Hamburg, 1956,* Heinrich, H. C., ed., Enke, Stuttgart, p. 213.
45. Dexter, S. O., Heinle, R. W., Fox, H. J., and Castle, W. B. (1939). *J. Clin. Invest.* **18**, 473.
46. Uotila, U. (1936). *Acta Med. Scand.* **89**, 50.
47. Uotila, U. (1938). *Acta Med. Scand.* **95**, 415.
48. Ellenbogen, L., and Highley, D. R. (1967). *J. Biol. Chem.* **242**, 1004.
49. Gräsbeck, R., Simons, K., and Sinkkonen, I. (1966). *Biochim. Biophys. Acta* **127**, 47.
50. Allen, R. H., and Mehlman, C. S. (1973). *J. Biol. Chem.* **248**, 3660.

51. Allen, R. H., and Mehlman, C. S. (1973). *J. Biol. Chem.* **248**, 3670.
52. Christensen, J. M., Hippe, E., Olesen, H., Rye, M., Haber, E., Lee, L., and Thomsen, J. (1973). *Biochim. Biophys. Acta* **303**, 319.
53. Visuri, K., and Gräsbeck, R. (1973). *Biochim. Biophys. Acta* **310**, 508.
54. Watson, G. M., and Florey, H. W. (1955). *Brit. J. Exp. Pathol.* **36**, 479.
55. Nieweg, H. O., Arends, A., Mandema, P., and Castle, W. B. (1956). *Proc. Soc. Exp. Biol. Med.* **91**, 328.
56. Clayton, C. G., Latner, A. L., and Schofield, B. (1955). *J. Physiol. (London)* **129**, 56P.
57. Chow, B. F., Quattlebaum, J. K., Jr., and Rosenblum, C. (1955). *Proc. Soc. Exp. Biol. Med.* **90**, 279.
58. Abels, J., Woldring, M. G., Vegter, J. J. M., and Nieweg, H. O. (1957). *Science* **126**, 558.
59. Keuning, F. J., Arends, A., Mandema, E., and Nieweg, H. O. (1959). *J. Lab. Clin. Med.* **53**, 127.
60. Boass, A., and Wilson, T. H. (1963). *Am. J. Physiol.* **204**, 101.
61. Håkanson, R., Lindstrand, K., Nordgren, L., and Owman, Cr. (1969). *Eur. J. Pharmacol.* **8**, 315.
62. Håkanson, R., Lindstrand, K., Nordgren, L., and Owman, Ch. (1971). *Biochem. Pharmacol.* **20**, 1259.
63. Schwartz, M., Lous, P., and Meulengracht, E. (1958). *Lancet* **II**, 1200.
64. Abels, J. (1959). Intrinsic Factor Van Castle en Resorptie Van Vitamin B_{12} (Thesis), University of Groningen.
65. Hippe, E., and Schwartz, M. (1971). *Scand. J. Haematol.* **8**, 276.
66. Wilkinson, J. F. (1949). *Lancet* **I**, 249.
67. Wilkinson, J. F. (1949). *Lancet* **I**, 291.
68. Wilkinson, J. F. (1949). *Lancet* **I**, 336.
69. Kaplan, M. E., and Stein, A. (1969). *Proc. Soc. Exp. Biol. Med.* **131**, 790.
70. Glass, G. B. J. (1963). *Physiol. Rev.* **43**, 529.
71. Holdsworth, E. S. (1961). *Biochim. Biophys. Acta.* **51**, 295.
72. Bromer, W. W., and Davisson, E. O. (1961). *Biochem. Biophys. Res. Commun.* **4**, 61.
73. Allen, R. H., and Majerus, P. W. (1972). *J. Biol. Chem.* **247**, 7695.
74. Allen, R. H., and Majerus, P. W. (1972). *J. Biol. Chem.* **246**, 7702.
75. Allen, R. H., and Majerus, P. W. (1972). *J. Biol. Chem.* **247**, 7709.
76. Highley, D. R., Davies, M. C., and Ellenbogen, L. (1967). *J. Biol. Chem.* **242**, 1010.
77. Simons, K. (1964). *Soc. Sci. Fenn. Commenta. Biol.* (Commentat biol.) **27**, 1.
78. Chanarin, I. (1969). *The Megaloblastic Anemias,* Blackwell, Oxford.
79. Hippe, E., Haber, E., and Olesen, H. (1971). *Biochim. Biophys. Acta* **243**, 75.
80. Hippe, E., and Olesen, H. (1971). *Biochim. Biophys. Acta* **243**, 83.
81. Chosy, J. J., Killander, A., and Schilling, R. F. (1962). In *Vitamin B_{12} und Intrinsic Factor, 2. Europäisches Symposion, Hamburg, 1961,* Heinrich, H. C., ed., Enke, Stuttgart, p. 668.

82. Perlman, D., and Toohey, J. I. (1968). *Arch. Biochem. Biophys.* **124**, 462.
83. Gallagher, N. D., Foley, K., and Brown, J. (1972). *Gastroenterology* **63**, 83.
84. Mathan, V. I., Babior, B. M., and Donaldson, R. M., Jr., (1974). *J. Clin. Invest.* **54**, 598.
85. Rosenblum, C., Woodbury, D. T., Gilbert, J. P., Okuda, D., and Chow, B. F. (1955). *Proc. Soc. Exp. Biol. Med.* **89**, 63.
86. Rosenblum, C., Yamamato, R. S., Wood, R., Woodbury, D. T., Okuda, K., and Chow, B. F. (1956). *Proc. Soc. Exp. Biol. Med.* **91**, 364.
87. Bunge, M. B., and Schilling, R. F. (1957). *Proc. Soc. Exp. Biol. Med.* **96**, 587.
88. Gregory, M. E., and Holdsworth, E. S. (1953). *Biochem. J.* **59**, 335.
89. Gottlieb, C. W., Retief, F. P., and Herbert, V. (1967). *Biochim. Biophys. Acta* **141**, 560.
90. Bunge, M. B., Schloesser, L. L., and Schilling, R. F., (1956). *J. Lab. Clin. Med.* **48**, 735.
91. Toporek, M. (1960). *Am. J. Clin. Nutr.* **8**, 297.
92. Gregory, M. E., and Holdsworth, E. S. (1960). *Biochim. Biophys. Acta* **42**, 462.
93. Meyer, L. M., Reizenstein, P. G., Cronkite, E. P., Miller, I. F., and Mulzac, C. W. (1963). *Brit. J. Haematol.* **9**, 158.
94. Heinrich, H. C. (1958). *Z. Vitam. Horm. ü. Fermentforsch.* **9**, 385.
95. Heinrich, H. C. (1958). *Naturwissenschaften* **45**, 269.
96. Gräsbeck, R. (1969). in *Progress in Hematology,* Vol. VI, Brown, E. B., and Moore, C. V., Eds., Grune & Stratton, New York, pp. 223–260.
97. Lien, E. L., Ellenbogen, L., Law, P. Y., and Wood, J. M. (1973). *Biochem. Biophys. Res. Commun.* **55**, 730.
98. Lien, E. L., Ellenbogen, L., Law, P. Y., and Wood, J. M. (1974). *J. Biol. Chem.* **249**, 890.
99. Goldberg, L. S., and Fudenberg, H. H. (1969). *J. Lab. Clin. Med.* **73**, 469.
100. Rose, M. S., and Chanarin, I. (1969). *Brit. Med. J.* **1**, 468.
101. Wagstaff, M., Broughton, A., and Jones, F. R. (1973). *Biochem. Biophys. Acta* **320**, 406.
102. Shum, H. J., O'Neill, B. J., and Streeter, A. M. (1971). *J. Clin. Pathol.* **24**, 239.
103. Cooper, B. A., and Castle, W. B. (1960). *J. Clin. Invest.* **39**, 199.
104. McGuigan, J. E. (1967). *J. Lab. Clin. Med.* **70**, 666.
105. Highley, D. R., Davies, M. C., and Ellenbogen, L. (1966). *Fed. Proc.* **25**, 277.
106. Gräsbeck, R. (1967). *Scand. J. Clin. Lab. Invest. Suppl.* **95**, 7.
107. Hippe, E. (1970). *Biochim. Biophys. Acta* **208**, 337.
108. Hippe, E. (1972). *Scand. J. Clin. Lab. Invest.* **29**, 59.
109. Okuda, K., and Fujii, T. (1966). *Arch. Biochem. Biophys.* **115**, 302.
110. Ellenbogen, L., and Highley, D. R. (1970). *Fed. Proc.* **29**, 633.
111. Gräsbeck, R. (1958). *Acta Chem. Scand.* **12**, 142.
112. Gräsbeck, R., Kantero, I., and Siurala, M. (1959). *Lancet* **1**, 234.
113. Schade, S. G., Abels, J., and Schilling, R. F. (1967). *J. Clin. Invest.* **46**, 615.
114. Gregory, M. E., and Holdsworth, E. S. (1957). *Biochem. J.* **66**, 456.

115. Highley, D. R., and Ellenbogen, L. (1962). *Arch. Biochem,* **99**, 126.

116. McGuigan, J. E., and Peterson, M. L. (1966). *Clin. Res.* **14**, 302.

117. Donaldson, R. M., and Katz, J. H. (1963). *J. Clin. Invest.* **42**, 534.

118. Gräsbeck, R., Stenman, U. H., Puutula, L., and Visuri, K. (1968). *Biochim. Biophys. Acta* **158**, 292.

119. Ternberg, J. L., and Eakin, R. E. (1949). *J. Am. Chem. Soc.* **71**, 38.

120. Bird, O. D., and Hoevet, B. (1951). *J. Biol. Chem.* **190**, 181.

121. Meyer, C. E., Eppstein, S. H., Bethell, F. H., and Hall, B. E. (1950). *Fed. Proc.* **9**, 205.

122. Beerstecher, E., Jr., and Edmonds, E. J. (1951). *Science* **114**, 412.

123. Beerstecher, E., Jr., and Edmonds, E. J. (1951). *Proc. Soc. Exp. Biol. Med.* **77**, 563.

124. Gregory, M. E., and Holdsworth, E. S. (1955). *Biochem. J.* **59**, 329.

125. Meyer, L. M., Bertcher, R. W., and Mulzac, C. (1959). *Proc. Soc. Exp. Biol. Med.* **100**, 607.

126. Glass, G. B. J., Stephason, L., Rich, M., and Laughton, R. W. (1957). *Brit. J. Haematol.* **3**, 401.

127. Gräsbeck, R. (1959). *Acta Physiol. Scand.* **45**, 88.

128. Gräsbeck, R. (1959). *Acta Physiol. Scand.* **45**, 116.

129. Herbert, V. (1959). *J. Clin. Invest.* **38**, 102.

130. Miller, O. N., and Hunter, I. M. (1957). *Proc. Soc. Exp. Biol. Med.* **96**, 39.

131. Herbert, V., and London, I. M. (1957). *Clin. Res. Proc.* **5**, 289.

132. Latner, A. L., and Raine, L. C. D. P. (1957). *Biochem. J.* **66**, 53p.

133. Latner, A. L., and Raine, L. C. D. P. (1959). *Biochem. J.* **71**, 344.

134. Johnson, P. C., and Driscoll, T. B. (1958). *Proc. Soc. Exp. Biol. Med.* **98**, 73.

135. Minard, F. N., and Wagner, C. L. (1958). *Proc. Soc. Exp. Biol. Med.* **98**, 684.

136. Herbert, V., Castro, Z., and Wasserman, L. R. (1960). *Proc. Soc. Exp. Biol. Med.* **104**, 160.

137. Herbert, V. (1958). *Proc. Soc. Exp. Biol. Med.* **97**, 668.

138. Simons, R., Kvist, G., and Sauren-Lindfors, M. (1966). *Ann. Med. Exp. Biol. Fenn.* **44**, 361.

139. Gullberg, R. (1969). *Scand. J. Clin. Lab. Invest.* **24**, 391.

140. Herbert, V. (1958). *J. Clin. Invest.* **37**, 901.

141. Wilson, T. H., and Wiseman, G. (1954). *J. Physiol. London* **123**, 116.

142. Wilson, T. H., and Strauss, E. W. (1959). *Am. J. Physiol.* **197**, 926.

143. Strauss, E. W., and Wilson, T. H. (1960). *Am. J. Physiol.* **198**, 103.

144. Herbert, V., and Castle, W. B. (1961). *J. Clin. Invest.* **40**, 1978.

145. Wolff, R., and Nabet, P. (1961). *Bull. Acad. Natl. Med.* **145**, 150.

146. Wolff, R., Nicholas, J. P., and Jeanmaire, N. (1972). *Ann. Biol. Clin.* **30**, 271.

147. Sullivan, L. W., Herbert, V., and Castle, W. B. (1963). *J. Clin. Invest.* **42**, 1443.

148. Castro-Curel, Z., and Glass, G. B. J. (1963). *Proc. Soc. Exp. Biol. Med.* **112**, 715.

149. Castro-Curel, Z., and Glass, G. B. J. (1964). *Clin. Chim. Acta* **9**, 317.

150. Schjönsby, H., and Peters, T. J. (1971). *Scand. J. Gastroenterol.* **65**, 441.

151. Taylor, K. B. (1959). *Lancet* **2**, 106.

152. Schwartz, M. (1958). *Lancet* **2**, 61.

153. Schwartz, M. (1960). *Lancet* **2**, 1263.

154. Jeffries, G. H., Hoskins, D. W., and Sleisenger, M. H. (1962). *J. Clin. Invest.* **41**, 1106.

155. Abels, J., Bouma, W., Jansz, A., Martinus, G. W., Bakker, A., and Nieweg, H. O. (1965). *J. Lab. Clin. Med.* **61**, 893.

156. Roitt, I. M., Doniach, D., and Shapland, C. (1965). *Ann. N. Y. Acad. Sci.* **124**, 664.

157. Imrie, M. H., and Schilling, R. F. (1965). *J. Lab. Clin. Med.* **66**, 680.

158. Garrido-Pinson, G. C., Turner, M. D., Crookston, J. H., Samloff, I. M., Miller, L. L., and Segal, H. L. (1966). *J. Immunol.* **97**, 897.

159. Carmel, R., and Herbert, V. (1966). *Clin. Res.* **14**, 482.

160. Hansen, H. J., Miller, O. N., and Tan, C. H. (1966). *Am. J. Clin. Nutr.* **19**, 10.

161. Cooper, B. A., and Yates, T. (1967). *Brit. J. Haematol.* **13**, 687.

162. Schade, S. G., Feick, P. L., Imrie, M. H., and Schilling, R. F. (1967). *Clin. Exp. Immunol.* **2**, 399.

163. Samloff, I. M., Kleinman, M. S., Turner, M. D., Sobel, M. V., and Jeffries, G. H. (1967). *Gastroenterology* **52**, 1141.

164. Bardhan, K. D., Hall, J. R., Spray, G. H., and Callender, S. T. (1968). *Lancet* **I**, 62.

165. Arderman, S., and Chanarin, I. (1963). *Lancet* **2**, 1350.

166. Abels, J., Bouma, W., and Nieweg, H. O. (1963). *Biochim. Biophys. Acta* **71**, 227.

167. Gottlieb, C., Lau, K. S., Wasserman, L. R., and Herbert, V. (1965). *Blood* **25**, 875.

168. Herbert, V., Gottlieb, C. W., and Lau, K. S. (1966). *Blood* **28**, 130.

169. Rodbro, P., Christiansen, P. M., and Schwartz, M. (1965). *Lancet* **2**, 1200.

170. Jeffries, G. H., and Sleisenger, M. H. (1963). *J. Clin. Invest.* **42**, 442.

171. Gullberg, R. (1966). *Acta Med. Scand. Suppl.* **463**, 11.

172. Hansen, H. J., Miller, O. N., Gallo-Torres, H., and Goldsmith, G. A. (1966). *Anal. Biochem.* **16**, 287.

173. Rothenberg, S. P. (1966). *J. Lab. Clin. Med.* **67**, 879.

174. Rothenberg, S. P., Kajani Kontha, K. R., and Ficarra, A. (1971). *J. Lab. Clin. Med.* **77**, 476.

175. Wolff, R., and Nichols, J. P. (1967). *Lancet* **1**, 1008.

176. Ellenbogen, L., Highley, D. R., Barker, H. A., and Smyth, R. D. (1960). *Biochem. Biophys. Res. Commun.* **3**, 178.

177. Lengyel, P., Mazumder, R., and Ochoa, S. (1960). *Proc. Natl. Acad. Sci. U. S. A.* **46**, 1312.

178. Stadtman, T. C. (1962). *J. Biol. Chem.* **237**, 2409.

179. Holdsworth, E. S., and Coates, M. E. (1956). *Nature* **177**, 701.

180. Taylor, K. B., Mallett, B. J., Witts, L. J., and Taylor, W. H. (1958). *Brit. J. Haematol.* **4**, 63.

181. Barlow, G. H., and Frederick, K. J. (1959). *Proc. Soc. Exp. Biol. Med.* **101**, 400.

182. Chaiet, L., Rosenblum, C., and Woodbury, D. T. (1950). *Science* **3**, 601.

183. Minot, G. R., and Castle, W. B. (1935). *Lancet* **2**, 319.

184. Ellenbogen, L., and Highley, D. R. (1963). *Vitam. Horm.* **21**, 1.

185. Ellenbogen, L. (1963). *Newer Methods of Nutritional Biochemistry,* Albanese, A., Ed., Academic Press, New York, p. 235.

186. Mollin, D. L. (1959). *Brit. Med. Bull.* **15**, 8.

187. McIntyre, P. A., Sachs, M. V., Krevans, J. R., and Conley, C. L. (1956). *Arch. Intern. Med.* **95**, 541.

188. Booth, C. C., and Mollin, D. L. (1957). *Lancet* **2**, 1007.

189. Cooke, W. T., Cox, E. V., Meynell, M. J., and Gaddie, R. (1959). *Lancet* **2**, 123.

190. Clark, A. C. L., and Booth, C. C. (1960). *Arch. Dis. Child.* **35**, 595.

191. Cornell, G. N., Gilder, H., Moody, F., Frey, C., and Beal, J. M. (1961). *Bull. N. Y. Acad. Med.* **37**, 675.

192. Allcock, E. (1961). *Gastroenterology* **40**, 81.

193. Ungley, C. C. (1950). *Brit. Med. J.* **2**, 915.

194. Citrin, Y., DeRosa, C., and Halsted, J. A. (1957). *J. Lab. Clin. Med.* **50**, 667.

195. Best, W. R., Frenster, J. H., and Zolot, M. M. (1957). *J. Lab. Clin. Med.* **50**, 793.

196. Johnson, P. C., and Berger, E. S. (1958). *Blood* **13**, 457.

197. Rønnov-Jessen, V., and Hansen, J. (1965). *Blood* **25**, 224.

198. Nygaard, K., Helsingen, N., and Rootwelt, K. (1970). *Scand. J. Gastroenterol.* **5**, 349.

199. Ross, G. I. M., Mollin, D. L., Cox, E. V., and Ungley, C. C. (1954). *Blood* **9**, 473.

200. Okuda, K., and Sasayama, K. (1965). *Proc. Soc. Exp. Biol. Med.* **120**, 17.

201. Okuda, K., and Takedatsu, H. (1966). *Proc. Soc. Exp. Biol. Med.* **123**, 504.

202. Baker, S. J., Mackinnon, N. L., and Vasudevia, P. (1958). *Indian J. Med. Res.* **46**, 812.

203. Fleming, W. H., and King, E. R. (1962). *Gastroenterology* **42**, 164.

204. Drapanas, T., Williams, J. S., McDonald, J. C., Heyden, W., Bow, T., and Spencer, R. P. (1963). *JAMA* **184**, 337.

205. Rosenthal, H. L., and Hampton, J. K. (1955). *J. Nutr.* **56**, 67.

206. Booth, C. C., Chanarin, I., Anderson, B. B., and Mollin, D. L. (1957). *Brit. J. Haematol.* **3**, 253.

207. Latner, A. L., and Raine, L. (1957). in *Vitamin B_{12} und Intrinsic Factor, 1. Europäisches Symposium, Hamburg, 1956,* Heinrich, H. C., Ed., Enke, Stuttgart, p. 243.

208. Moertel, C. G., Scudamore, C. A., Owen, C. A., Jr., and Bollman, J. L. (1960). *Am. J. Physiol.* **199**, 289.

209. Wolff, R. (1958). *C. R. Soc. Biol.* **246**, 3101.

210. Strauss, E. W., and Wilson, T. H. (1958). *Proc. Soc. Exp. Biol. Med.* **99**, 224.

211. Okuda, K. (1960). *Am. J. Physiol.* **199**, 84.

212. Heinrich, H. C. (1961). *In Kunstliche, Radioactive Isotope in Physiologie und Therapie,* Schwiegk, H., and Turba, F., Eds., Springer-Verlag, Berlin, and New York, p. 660.

213. Rosenthal, H. L. (1959). *Am. J. Physiol.* **197**, 1048.

214. Wilson, T. H., Strauss, E. W., and Hotchkiss, A. (1959). *Am. J. Physiol.* **197**, 926.
215. Holdsworth, E. S., and Coates, M. E. (1961). *Clin. Chim. Acta* **6**, 44.
216. Ungley, C. C. (1955). *Vitam. Horm.* **13**, 137.
217. Doscherholmen, A., and Hagen, P. S. (1956). *J. Clin. Invest.* **35**, 699.
218. Doscherholmen, A., Hagen, P. S., and Liu, M. (1957). *Blood* **12**, 336.
219. Doscherholmen, A., Hagen, P. S., and Olin, L. (1959). *J. Lab. Clin. Med.* **54**, 434.
220. Doscherholmen, A., and Hagen, P. S. (1957). *J. Clin. Invest.* **36**, 1551.
221. Weissbach, H., Toorey, J. I., and Barker, H. A. (1959). *Proc. Natl. Acad. Sci. U. S. A.* **45**, 521.
222. Toohey, J. I., and Barker, H. A. (1961). *J. Biol. Chem.* **231**, 560.
223. Lindstrand, K. (1964). *Nature* **204**, 188.
224. Ståhlherg, K. G. (1967). *Scand. J. Haematol. Suppl.* **1**.
225. Lindstrand, K. (1971). In *The Cobalamins,* Arnstein, H. R. V., and Wrightson, R. J., Eds., Churchill-Livingstone, Edinburgh, p. 41.
226. Sullivan, L. W., Herbert, V., and Reizenstein, P. (1962). *Am. J. Clin. Nutr.* **11**, 568.
227. Okuda, K., Takara, I., and Fujii, T. (1958). *Blood* **32**, 313.
228. Schade, S. G., and Schilling, R. F. (1967). *Am. J. Clin. Nutr.* **20**, 636.
229. Okuda, K. (1962). *Clin. Chim. Acta* **7**, 780.
230. Abels, J., and Schilling, R. F. (1964). *J. Lab. Clin. Med.* **60**, 375.
231. Callender, S. T., Witts, L. J., Allison, P. R., and Gunnig, A. (1961). *Gut* 2, 150.
232. Jacob, E., and O'Brien, H. A. W. (1972). *J. Clin. Pathol.* **25**, 320.
233. Cooper, B. A. (1964). *Medicine* **43**, 689.
234. England, J. M., and Taylor, K. B. (1966). *Clin. Res.* **14**, 135.
235. Carmel, R., Rosenberg, A. H., Law, K.-S., Streiff, R. F., Herbert, V. (1969). *Gastroenterology* **51**, 548.
236. Donaldson, R. M., Jr., MacKenzie, I. L., and Trier, J. S. (1967). *J. Clin. Invest.* **46**, 1215.
237. Cooper, B. A., Paranchych, W., and Lowenstein, L. (1962). *J. Clin. Invest.* **41**, 370.
238. Mackenzie, I. L., and Donaldson, R. M., Jr. (1972). *J. Clin. Invest.* **51**, 2465.
239. Gräsbeck, R., and Nyberg, W. (1958). *Scand. J. Clin. Lab. Invest.* **10**, 448.
240. Okuda, K., and Sasayama, P. (1965). *Am. J. Physiol.* **208**, 14.
241. Rosado-Rodriguez, A. L., and Sheehy, T. W. (1961). *Am. J. Med. Sci.* **242**, 548.
242. Herbert, V., Streiff, R. R., and Sullivan, L. W. (1964). *Medicine* **43**, 679.
243. Mackenzie, I. L., Kopp, W. L., Donaldson, R. M., Jr., and Trier, J. S. (1967). *Clin. Res.* **15**, 239.
244. Donaldson, R. M., Jr., Mackenzie, I. L., and Trier, J. S. (1966). *J. Clin. Invest.* **45**, 1001.
245. Rothenberg, S. P. (1968). *J. Clin. Invest.* **47**, 913.
246. Katz, M., and Cooper, B. A. (1974). *J. Clin. Invest.* **54**, 733.

247. Hines, J. D., Rosenberg, A., and Harris, J. W. (1968). *Proc. Soc. Exp. Biol. Med.* **129**, 653.

248. Donaldson, R. M., Small, D. M., Robins, S., and Mathan, V. I. (1973). *Biochem. Biophys. Acta* **311**, 477.

249. Rosenthal, H. L., Cutler, L., and Sobieszczanska, W. (1970). *Am. J. Physiol.* **218**, 358.

250. Peters, T. J., and Hoffbrand, A. V. (1970). *J. Haematol.* **19**, 369.

251. Rothenberg, S. P., Weisberg, H., and Ficarra, A. (1972). *J. Lab. Clin. Med.* **79**, 587.

252. Latner, A. L., Hodson, A. W., and Smith, P. A. (1962). *Lancet* **II**, 230.

253. Peters, T. J., Quinlan, A., and Hoffbrand, A. V. (1971). *Brit. J. Haematol.* **20**, 123.

254. Peters, T. J., Linnell, J. C., Matthews, D. M., and Hoffbrand, A. V. (1971). *Brit. J. Haematol.* **20**, 299.

255. Hoffbrand, A. V., Linnell, J. C., Matthews, D. M., and Peters, T. J. (1970). *J. Physiol.* **208**, 66P.

256. Peters, T. J., and Hoffbrand, A. V. (1971). In *The Cobalamins,* Arnstein, H. R. V., and Wrightman, R. J., Eds., Churchill-Livingstone, Edinburgh, pp. 61–77.

257. Linnell, J. C., Hoffbrand, A. V., Peters, T. J., and Matthews, D. M. (1969). *J. Clin. Pathol.* **22**, 742.

258. Gurani, S., Mistry, S. P., and Johnson, B. C. (1960). *Biochim. Biophys. Acta* **38**, 187.

259. Stadtman, E. R., Overath, P., Eggerer, H., and Lynen, F. (1960). *Biochem. Biophys. Res. Commun.* **2**, 1.

260. Wilson, T. H. (1963). *Physiologist* **6**, 11.

261. Rosenblum, C., and Geoffroy, R. F. (1969). In *Proceedings, International Congress on Radioisotopes in Pharmacology,* Wason, P. G., and Glasson, B., Eds., Wiley, London, p. 349.

262. Yamaguchi, N., Rosenthal, W. S., and Glass, G. B. J. (1970). *Am. J. Clin. Nutr.* **23**, 156.

263. Weisberg, H., Rhodin, J., and Glass, G. B. J. (1968). *Lab. Invest.* **19**, 516.

264. Yamaguchi, N., Weisberg, H., and Glass, G. B. J. (1969). *Gastroenterology* **56**, 914.

265. Frentz, G. D., and Miller, O. N. (1966). *Clin. Res.* **14**, 432.

266. Cooper, B. A., and White, J. J. (1960). *Brit. J. Haematol.* **14**, 73.

267. Ardeman, S., Chanarin, I., and Berry, V. (1965). *Brit. J. Haematol.* **11**, 11.

268. Schilling, R. F., and Schloesser, L. L. (1957). In *Vitamin B_{12} and Intrinsic Factor, 1. Europäisches Symposium, Hamburg, 1956,* Heinrich, H. C., Ed., Enke, Stuttgart, pp. 194–201.

269. Reizenstein, P. G., Cronkite, E. P., Meyer, L. M., and Usenik, E. A. (1960). *Proc. Soc. Exp. Biol. Med.* **105**, 233.

270. Nyberg, W., Saarni, M., and Gräsbeck, R. (1962). In *Vitamin B_{12} und Intrinsic Factor, 2. Europäisches Symposion, Hamburg, 1961,* Heinrich, H. C., Ed., Enke, Stuttgart, pp. 469–472.

271. Doscherholmen, A. (1962). In *Vitamin B_{12} und Intrinsic Factor, 2. Europäisches Symposion, Hamburg, 1961,* Heinrich, H. C., Ed., Enke, Stuttgart, pp. 472–485.

272. Herbert, V., Cooper, B. A., and Castle, W. B. (1962). *Proc. Soc. Exp. Biol. Med.* **110**, 315.

273. Ellenbogen, L., and Highley, D. R. (1963). *Proc. Soc. Exp. Biol. Med.* **113**, 229.

274. Highley, D. R., Streiff, R. R., Ellenbogen, L., Herbert, V., and Castle, W. B. (1965). *Proc. Soc. Exp. Biol. Med.* **119**, 565.

275. Mackenzie, I. L., and Donaldson, R. M. (1969). *Fed. Proc.* **28**, 41.

276. Rothenberg, S. P., and Ficarra, A. (1973). *Fed. Proc.* **32**, 892.

277. Callender, S. T., and Lajtha, L. G. (1951). *Blood* **6**, 1234.

278. Toporek, M., and Philipp, L. J. (1961). *Am. J. Physiol.* **200**, 557.

279. Ukyo, S., Inada, M., Uchino, H., and Wakisaka, G. (1967). *Proc. 3rd World Congr. Gastroenterol.* **2**, 416.

280. Toporek, M. (1966). *Fed. Proc.* **25**, 429.

281. Perman, G., Gullberg, R., Reizenstein, P. G., Snellman, B., and Allgen, L. G. (1960). *Acta Med. Scand.* **168**, 117.

282. Veeger, W., Abels, J., Hellemans, N., and Nieweg, H. O. (1962). *N. Engl. J. Med.* **267**, 1341.

283. LeBauer, E., Smith, K., and Greenberger, N. J. (1965). *Arch. Intern. Med.* **122**, 423.

284. Toskes, P. P., Hansell, J., Cerda, J., and Deren, J. J. (1971). *N. Engl. J. Med.* **284**, 627.

285. Toskes, P. P., Deren, J. J., Fruiterman, J., and Conrad, M. E. (1973). *Gastroenterology* **65**, 199.

286. Toskes, P. P., and Deren, J. J. (1972). *J. Clin. Invest.* **51**, 216.

287. Toskes, P. P., Deren, J. J., and Conrad, M. E. (1973). *J. Clin. Invest.* **52**, 1660.

288. Abels, J., and Mucherheide, M. M. (1970). *Clin. Res.* **17**, 530.

289. Rosenblum, C. (1962). In *Vitamin B_{12} und Intrinsic Factor, 2. Europäisches Symposion, Hamburg, 1961,* Heinrich, H. C., Ed., Enke, Stuttgart, p. 306.

290. Killander, A., and Schilling, R. F. (1961). *J. Lab. Clin. Med.* **57**, 553.

291. Glass, G. B. J., Skeggs, H. R., Lee, D. H., Jones, E. L., and Hardy, W. W. (1962). In *Vitamin B_{12} und Intrinsic Factor 2. Europäisches Symposion, Hamburg, 1961,* Heinrich, H. C., Ed., Enke, Stuttgart, p. 673.

292. Herbert, V., and Sullivan, L. W. (1964). *Ann. N. Y. Acad. Sci.* **112**, 855.

293. Okuda, K., and Tantengco, V. (1962). *Proc. Soc. Exp. Biol. Med.* **110**, 396.

294. Lee, D. H., and Glass, G. B. J. (1961). *Proc. Soc. Exp. Biol. Med.* **107**, 293.

295. Heinrich, H. C., and Gabbe, E. E. (1964). *Ann. N. Y. Acad. Sci.* **112**, 871.

296. Toporek, M. (1960). *Am. J. Clin. Nutr.* **8**, 297.

297. Latner, A. L., and Raine, L. (1957). *Nature* **180**, 1197.

298. Shinton, W. K., and Singh, A. K. (1967). *Brit. J. Haematol.* **13**, 75.

299. Monto, R. W., and Rebuck, J. W. (1954). *Arch. Intern. Med.* **93**, 219.

300. Monto, R. W., and Rebuck, J. W. (1955). *Blood* **10**, 1151.

301. Israels, M. C. G., and Shubert, S. (1954). *Lancet* **1**, 341.

302. Berlin, H., Berlin, R., Bronte, G., and Sjöberg, S.-G. (1962). In *Vitamin B_{12} und Intrinsic Factor 2 Europäisches Symposion, Hamburg, 1961,* Heinrich, H. C., Ed., Enke, Stuttgart, p. 485.

303. Chalmers, J. N. M., and Shinton, N. K. (1958). *Lancet* **II**, 1069.

304. Booth, C. C., and Mollin, D. L. (1956). *Brit. J. Haematol.* **2**, 223.

305. Taylor, K. B., and French, J. E. (1960). *Q. J. Exp. Physiol.* **45**, 72.

306. Spies, T. D., Lopez, G. G., Milanes, F., Toca, R. L., and Aramburu, T. (1949). *South. Med. J.* **42**, 528.

307. Ungley, C. C. (1950). *Brit. Med. J.* **2**, 905.

308. Conley, C. L., Krevans, J. R., Chow, B. F., Barrows, C., and Lang, C. A. (1951). *J. Lab. Clin. Med.* **38**, 84.

309. Chalmers, J. N. M., and Hall, Z. M. (1954). *Brit. Med. J.* **1**, 1179.

310. Unglaub, W. G., and Goldsmith, G. A. (1955). *South. Med. J.* **48**, 261.

311. Reisner, E. H., Weiner, L., Schittone, M. T., and Henck, E. A. (1955). *N. Engl. J. Med.* **253**, 502.

312. Brody, E. A., Estren, S., and Wasserman, L. R. (1959). *N. Engl. J. Med.* **260**, 361.

313. McIntyre, P. A., Hahn, R., Masters, J. M., and Krevans, J. R. (1960). *Arch. Intern. Med.* **106**, 280.

314. Luhby, A. L., Cooperman, J., Donnenfeld, A., Herrero, J., Teller, J., and Wenig, J. (1958). *J. Dis. Child.* **96**, 532.

315. Agunod, M. B., Yomaguchi, N., Lopez, R., Cooperman, J. M., and Lubby, A. L. (1967). *Fed. Proc.* **26**, 338.

316. Goldberg, L. S., Barnett, E. V., and Desai, R. (1967). *Pediatrics* **40**, 851.

317. Bar-Shany, S., and Herbert, V. (1967). *Blood* **30**, 777.

318. Gallagher, N. D., and Foley, K. (1971). *Gastroenterology* **61**, 332.

319. Williams, D. L., and Spray, G. H. (1968). *Brit. J. Nutr.* **22**, 297.

320. Killander, A., and Vahlquist, B. (1954). *Nord. Med.* **51**, 777.

321. Karlin, R., and Dumont, M. (1955). *C. R. Soc. Biol. (Paris)* **149**, 1986.

322. Okuda, K., Hellegers, A. E., and Chow, B. F. (1956). *Am. J. Clin. Nutr.* **4**, 440.

323. Boger, W. P., Wright, L. D., and Bayne, G. M. (1957). In *Vitamin B_{12} und Intrinsic Factor I. Europäisches Symposion, Hamburg, 1956,* Heinrich, H. C., Ed., Enke, Stuttgart, p. 443.

324. Boger, W. P., Bayne, G. M., Wright, L. D., and Beck, C. D. (1957). *N. Engl. J. Med.* **256**, 1085.

325. Dixit, C. H., Mody, B. N., Jhala, H. I., Parekh, J. G., and Romasarma, G. B. (1957). *Indian J. Med. Sci.* **11**, 145.

326. Baker, H., Ziffer, H., Pasher, I., and Sobotka, H. (1958). *Brit. Med. J.* **I**, 978.

327. Prystowski, H., Hellegers, A. E., Ranke, E., Ranke, B., and Chow, B. F. (1959. *Am. J. Obstet. Gynecol.* **77**, 1.

328. Sadovsky, A., Bercovici, B., Rachmilewitz, M., Grossowicz, N., and Aronovitch, J. (1959). *Obstet. Gynecol.* **13**, 346.

329. Lowenstein, L., LaLonde, M., Deschenes, E., and Shapiro, L. (1960). *Am. J. Clin. Nutr.* **8**, 265.

330. Luhby, A. L., Feldman, R., Marley, J., Odang, D., and Cooperman, J. (1961). *Fed. Proc.* **20**, 451.

331. Zachau-Christiansen, B., Hoff-Jørgensen, E., and Østergard-Kristensen, H. (1962). *Danish Med. Bull.* **9**, 157.

332. Chow, B. F., Barrows, L., and Ling, C. T. (1951). *Arch. Biochem.* **34**, 151.

333. Hellegers, A., Okuda, K., Nesbitt, R. E. L., Jr., Smith, D. W., and Chow, B. F. (1957). *Am. J. Clin. Nutr.* **5**, 327.

334. Karlin, R. (1956). *C. R. Soc. Biol.* **150**, 1211.

335. Luhby, A. L., Cooperman, J. M., Herrero, J. M., and Donnenfeld, A. M. (1959). *J. Dis. Child.* **98**, 619.

336. Luhby, A. L., Cooperman, J. M., Stone, M. L., and Slobody, L. B. (1961). *Am. J. Dis. Child.* **102**, 753.

337. Flodh, H. (1968). *Acta. Radiol. Suppl.* **284**, 4.

338. Chow, B. F., and Okuda, K. (1960). *JAMA* **172**, 422.

339. Lust, J. E., Hagerman, D. D., and Villee, C. A. (1954). *J. Clin. Invest.* **33**, 38.

340. Callender, S .T., and Evans, J. R. (1955). *Clin. Sci.* **14**, 295.

341. Okuda, K., Gräsbeck, R., and Chow, B. F. (1958). *J. Lab. Clin. Med.* **51**, 17.

342. Swendseid, M. E., Gasster, M., and Halsted, J. A. (1954). *Proc. Soc. Exp. Biol. Med.* **86**, 834.

343. Glass, B. J., Boyd, L. J., and Stephanson, L. (1954). *Fed. Proc.* **13**, 54.

344. Clayton, C. G., Latner, A. L., and Shofield, B. (1957). *Brit. J. Nutr.* **2**, 339.

345. Bozian, R. C., Heyssel, R. M., and Darby, W. J. (1964). *Am. J. Clin. Nutr.* **14**, 239.

346. Donaldson, R. M., Jr. (1964). *Gastroenterology* **46**, 609.

347. Pitney, W. R., Beard, M. F., and Van Loon, E. J. (1954). *J. Biol. Chem.* **207**, 143.

348. Miller, A. (1958). *J. Clin. Invest.* **37**, 556.

349. Heinrich, H. C., and Erdmann-Oehlecker, S. (1956). *Clin. Chim. Acta* **1**, 326.

350. Hall, C. A., and Finkler, A. E. (1962). *J. Lab. Clin. Med.* **60**, 765.

351. Beal, R. W., Gizis, E., and Meyers, L. M. (1969). *Scand. J. Haematol. Suppl.* **10**, 3.

352. Hall, C. A., and Finkler, A. E. L. (1963). *Biochim. Biophys. Acta* **78**, 233.

353. Hall, C. A., and Finkler, A. E. L. (1966). *Proc. Soc. Exp. Biol. Med.* **123**, 55.

354. Hom, B. L. (1967). *Clin. Chim. Acta* **18**, 315.

355. Hom, B. L., and Ahluwalia, B. K. (1968). *Scand. J. Haematol.* **5**, 64.

356. Finkler, A. E., Green, P. M., and Hall, C. A. (1970). *Biochim. Biophys. Acta* **200**, 151.

357. Gullberg, R. (1970). *Clin. Chim. Acta.* **29**, 97.

358. Gullberg, R. (1971). *Clin. Chim. Acta* **33**, 173.

359. Gullberg, R. (1972). *Scand. J. Haematol.* **9**, 639.

360. Gullberg, R. (1972). *Scand. J. Rheum.* **1**, 129.

361. Andrews, P. (1965). *Biochem. J.* **96**, 595.

362. Hom, B. L., Olesen, H., and Lous, P. (1966). *J. Lab. Clin. Med.* **68**, 958.

363. Hom, B. L., and Olesen, H. (1967). *Scand. J. Clin. Invest.* **19**, 269.

364. Lawrence, C. (1969). *Blood* **33**, 899.

365. Gizis, E. J., Dietrich, M. F., Ohoi, G., and Meyer, L. M. (1970). *J. Lab. Clin. Med.* **75**, 673.

366. Bloomfield, F. J., and Scott, J. M. (1972). *Brit. J. Haematol.* **22**, 33.

367. Carmel, R. (1972). *Brit. J. Haematol.* **22**, 53.

368. Chanarin, I., England, J. M., Rowe, K. L., and Stacey, J. A. (1972). *Brit. Med. J.* **I**, 441.

369. Hall, C. A., and Finkler, A. E. (1967). *Biochim. Biophys. Acta* **147**, 186.

370. Hall, C. A., and Finkler, A. E. (1969). *J. Lab. Clin. Med.* **73**, 60.

371. Kumento, A. (1969). *Acta Paediatr. Scand. Suppl.* **194**, 9.

372. Hayem-Levy, A., Carlier-Vanneuville, A., Vandewalle, P., and Havez, R. (1971). *Clin. Chim. Acta* 35, 151.

373. Toporek, M., Gizis, E. J., and Meyer, L. M. (1971). *Proc. Soc. Exp. Biol. Med.* **136**, 1119.

374. Hom, B. L. (1967). *Scand. J. Haematol.* **4**, 321.

375. Cooper, B. A. (1969). *Blood* **34**, 856.

376. Gizis, E. J., and Meyer, L. M. (1971). *Proc. Soc. Exp. Biol. Med.* **137**, 1213.

377. England, J. M., Clarke, H. G. M., Down, M. C., and Chanarin, I. (1973). *Brit. J. Haematol.* **25**, 737.

378. Kumento, A., Lopez, R., Luhby, A. L., and Hall, C. A. (1967). *Clin. Res.* **15**, 283.

379. Tan, C. H., and Hansen, H. J. (1968). *Proc. Soc. Exp. Biol. Med.* **127**, 740.

380. England, J. M., Tavill, A. S., and Chanarin, I. (1973). *Clin. Sci. Mol. Med.* **45**, 479.

381. Sonneborn, D. W., Abouna, G., Mendez-Picon, G. (1972). *Biochim. Biophys. Acta* **273**, 283.

382. Simons, K., and Weber, T. (1966). *Biochim. Biophys. Acta* **117**, 201.

383. Corcino, J., Krauss, S., Waxman, S., and Herbert, V. (1970). *J. Clin. Invest.* **49**, 2250.

384. Hall, C. A., and Finkler, A. E. (1971). *Methods Enzymol.* **18**, 108.

385. Carmel, R., and Herbert, V. (1969). *Blood* **33**, 1.

386. Carmel, R., and Herbert, V. (1972). *Blood* **40**, 542.

387. Stenman, V. H., Simons, K., and Gräsbeck, R. (1968). *Scand. J. Clin. Lab. Invest.* **21**, 202.

388. Hurlimann, J., and Zuber, C. (1969). *Clin. Exp. Immunol.* **4**, 141.

389. Puutula, L., and Gräsbeck, R. (1972). *Biochim. Biophys. Acta* **263**, 734.

390. Cooper, B. A. (1970). *Blood* **35**, 829.

391. Hurlimann, J., and Zuber, C. (1969). *Clin. Exp. Immunol.* **4**, 123.

392. Gräsbeck, R., Visuri, K., and Stenman, V.-H. (1972). *Biochim. Biophys. Acta* **263**, 721.

393. Gräsbeck, R., and Aro, H. (1971). *Biochim. Biophys. Acta* **252**, 217.

394. Hippe, E. (1972). *Scand. J. Clin. Lab. Invest.* **29**, 59.

395. Carmel, R. (1972). *Biochim. Biophys. Acta* **263**, 747.

396. Gräsbeck, R., and Puutula, L. 1971). In *The Cobalamins,* Arnstein, H. R. V., and Wrighton, R. J., Eds., Churchill-Livingstone, Edinburgh, p. 143.

397. Hall, C. A. (1969). *Brit. J. Haematol.* **16**, 429.

398. Bearn, A. G., and Kunkel, H. G. (1954). *Proc. Soc. Exp. Biol. Med.* **85**, 64.
399. Herbert, V. (1968). *Blood* **32**, 305.
400. Finkler, A. E., and Hall, C. A. (1967). *Arch. Biochem. Biophys.* **120**, 79.
401. Hall, C. A., and Finkler, A. E. (1965). *J. Lab. Clin. Med.* **65**, 459.
402. Meyer, L. M., Osofsky, M., and Miller, I. F. (1967). *Scand. J. Haematol.* **4**, 301.
403. Gizis, E. J., Arkun, S. N., Miller, I. F., Choi, G., Dietrich, M. F., and Meyer, L. M. (1969). *J. Lab. Clin. Med.* **74**, 574.
404. Rachmilewitz, M., Rachmilewitz, B., and Moshkowitz, B. (1971). *J. Lab. Clin. Med.* **78**, 275.
405. Benson, R. E., Rappazzo, M. E., and Hall, C. A. (1972). *J. Lab. Clin. Med.* **80**, 488.
406. Hakami, N., Neiman, P. E., Canellos, G. P., and Lazerson, J. (1971). *N. Engl. J. Med.* **285**, 1163.
407. Lawrence, C. (1966). *Brit. J. Haematol.* **12**, 569.
408. Retief, F. P., Gottlieb, C. W., and Herbert, V. (1967). *Blood* **29**, 837.
409. Ståhlberg, K. G. (1964). *Scand. J. Haematol.* **1**, 220.
410. Simons, K. (1968). *Prog. Gastroenterol.* **1**, 195.
411. Scott, C. R., Hakami, N., Teng, C. C., and Sagerson, R. N. (1972). *J. Pediatr.* **81**, 1106.
412. Di Girolamo, P. M., and Bradbeer, C. (1971). *J. Bacteriol.* **106**, 745.
413. Di Girolamo, P. M., Kadner, R. J., and Bradbeer, C. (1971). *J. Bacteriol.* **106**, 751.
414. White, J. C., Di Girolamo, P. M., Fu, M. L., Preston, Y. A., and Bradbeer, C. (1973). *J. Biol. Chem.* **248**, 3978.
415. Giannella, R. A., Broitman, S. A., and Zamcheck, N. (1971). *J. Clin. Invest.* **50**, 1100.
416. Rappazo, M. E., and Hall, C. A. (1972). *Am. J. Physiol.* **222**, 202.
417. Meyer, L. M., Sood, S. K., and Gizis, E. J. (1972). *Proc. Soc. Exp. Biol. Med.* **140**, 99.
418. Rosenthal, H. L., Haessler, I. E., and Hill, R. C. (1965). *Biochim. Biophys. Acta* **104**, 46.
419. Sonneborn, D. W., and Hansen, H. J. (1970). *Science* **168**, 591.
420. Kidroni, G., and Grossowicz, N. (1969). *Biochim. Biophys. Acta* **188**, 113.
421. Sonneborn, D. W., and Hansen, H. J. (1971). *Proc. Soc. Exp. Biol. Med.* **136**, 903.

CHAPTER SIX

THE FATE OF COBALAMINS *in vivo*

JOHN C. LINNELL, Ph.D.

Dept. of Experimental Chemical Pathology
Vincent Square Laboratories
Westminster Hospital
London, England

CONTENTS

INTRODUCTION

The bulk of cobalamin in the body occurs in the form of the two coenzymatically active *Co*-alkyl derivatives AdoCbl and MeCbl, but in most tissues the extremely low concentrations of each have until recently precluded their separate estimation. In man, AdoCbl and MeCbl remained undiscovered for more than a decade after the vitamin was first isolated as CN-Cbl (1, 2), and in retrospect two factors seem largely to account for this. Not only are tissue concentrations of the cobalamins very much lower than those of any other known vitamin, but in addition the coenzyme forms are sensitive to light and are rapidly converted to OH-Cbl. Furthermore, the early finding that inorganic cyanides facilitated the recovery of the vitamin from fermentation liquors and liver homogenates set a pattern for subsequent work which for a number of years ensured that CN-Cbl was the form most likely to be isolated and hence for a time that considered to be the "active principle." More recently these same factors of low tissue concentration and light sensitivity have hampered attempts to estimate the cobalamin coenzymes in blood and organs, and many workers have been deterred from setting up suitable assay methods.

The distribution of cobalamins in the body has long attracted attention; even before the coenzyme forms were known, total cobalamins were being estimated in tissues by microbiological assay (3–7). The introduction of radioisotopically labeled cobalamins (8–12) provided a new and important means of studying cobalamin metabolism; since then a considerable body of information on the fate of administered cobalamins has been amassed both for man and other species. Neither of these techniques can, however, provide direct information about the metabolic interconversion of endogenous or administered cobalamins. At least four cobalamins are now known to coexist in the body and ideally, perhaps, specific assays should be available for each.

Enzymic methods have been described for AdoCbl, of which the assay based on the diol dehydrase reaction (13) is probably the most sensitive, and in aqueous solution as little as 16 pg/ml AdoCbl can be detected (14). This method has been used to estimate AdoCbl in liver, kidney, heart, and brain (15, 16) but is barely sensitive enough to estimate the very low concentrations of AdoCbl in plasma and erythrocytes. No specific assay methods for the other cobalamins have yet been described.

Lindstrand and Ståhlberg approached the problem in a different manner which proved successful in more ways than one. Using paper chromatography and bioautography they separated several cobalamins from human plasma (17), including AdoCbl, the coenzyme which had then only recently been isolated from human liver (18), and OH-Cbl. In addition they detected a new cobalamin which was later identified both in plasma and liver as MeCbl (19, 20). Chromatography has a long historical association with cobalamin and previously facilitated the isolation of a number of biologically important corrins, but this was the first time a chromatographic technique had successfully been applied to plasma. Some idea of the magnitude of the problem is perhaps indicated by the samples initially required. To obtain the few milligrams of purified MeCbl required for spectroscopic identification, it was necessary to extract at least 100 kg of calf liver. Identification of the cobalamins in plasma was less demanding but still required at least 50 ml of blood from each patient or volunteer.

The method which has been developed from this work is considerably more modest in its requirements, yet it can separate and semiquantitate the cobalamins detectable in any tissue in the body. One to two milliliters of plasma or 10 to 100 mg of most other tissues is sufficient for the estimation (21–24). All samples should be taken by dim light, and particular care should be used to avoid exposing CSF and plasma samples to white light at any time. Blood is withdrawn in a syringe covered with aluminum foil or paper and ejected into a foil-covered heparinized tube. Plasma should be separated from the cells within 1 hr by the light of a red photographic lamp and stored frozen in a foil-covered tube. Under these conditions the cobalamins are stable for many months.

Details of the technique have previously been described (22–24) but in principle the method consists of a multiple solvent extraction procedure which yields a final extract of cobalamins, concentrated approximately 100-fold. The cobalamins are separated by one- or two-dimensional thin-layer chromatography and located bioautographically by overlayering the chromatogram with agar containing a cobalamin-sensitive *E. coli* mutant and a tetrazolium growth indicator. The crimson growth zones which appear on the bioautogram after incubation are quantitated by photometric scanning and comparison with standards. The plasma or tissue concentration of each cobalamin is then calculated on a proportionate basis using the total cobalamin estimated by radioisotopic assay (25).

A method which allows the estimation of each cobalamin at physiological concentrations is likely to be of value in any center concerned with studies of cobalamin metabolism. The chromatobioautographic technique provides the only means at present available for estimating individual cobalamins in plasma and tissues, and the results obtained so far suggest that this is a technique of considerable potential in the diagnosis and investigation of possible disturbances of cobalamin metabolism.

DISTRIBUTION OF COBALAMINS IN HEALTH

The ultimate fate of cobalamin in the body is unknown since no catabolic pathway for the degradation of cobalamins has yet been found, though small amounts are lost daily in the urine and feces. The total cobalamin content of the body has been estimated by a number of workers using either microbiological assay of tissues at autopsy or dilution of radioisotopically labeled cobalamins administered during life. Mean values for the body content calculated for human adults range from 2 to 5 mg (26, 27), of which approximately 1.5 mg is contained in the liver. The total cobalamin content of other organs is discussed in a subsequent section.

Plasma

In healthy adults and hospital control subjects, MeCbl is the major cobalamin and usually accounts for between 60 and 80% of the plasma total cobalamin (Figure 6-1). Smaller amounts of AdoCbl and OH-Cbl are present, of which the former predominates (23, 24, 28). In about one-third of all control subjects there is a small amount of CN-Cbl (Figure

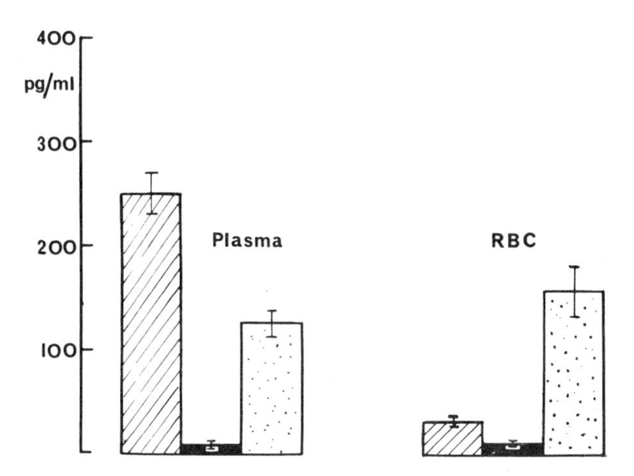

Fig. 6-1 Plasma and erythrocyte cobalamins in healthy adults. ▨ = MeCbl, ■ = CN-Cbl, ⬚ = AdoCbl + OHCbl (mean values ± SEM). Plasma total cobalamin 385 ± 31 pg/ml ($n = 22$); erythrocyte total cobalamin 202 ± 25 pg/ml ($n = 12$). [Reprinted from Linnell et al. (1974). *Clin. Sci. Molec. Med.* **46,** 163.]

6-2) which on the average represents 2% of the total cobalamin, although in many plasma samples CN-Cbl is undetectable. The appearance of this cobalamin was once thought to be the result of the increased cyanide intake caused by smoking (29), but it is not confined to smokers. In children the plasma contains a higher proportion of MeCbl than in adults and up to 90% of the total cobalamin may be in this form (130). Similarly, a study of the cobalamins in maternal and fetal blood has revealed (Figure 6-3) that the plasma MeCbl is higher in the neonate than in the mother, and this largely accounts for the higher total plasma cobalamin in infants (31). During pregnancy the plasma total cobalamin declines and more steeply so toward term. Estimation of the mother's plasma cobalamins shows that this fall coincides with a significant reduction in plasma MeCbl, a feature which is rapidly reversed during the postpartum period (31).

Erythrocytes

In normal subjects the total cobalamin concentration of erythrocytes is about half that of plasma or serum. Estimation of the red cell total cobal-

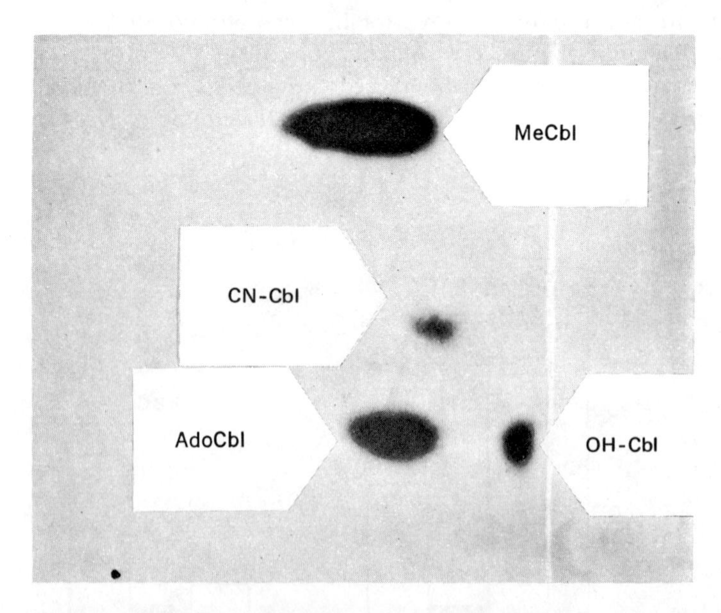

Fig. 6-2 Separation of plasma cobalamins from a normal subject using two-dimensional chromatography and bioautography. A trace of CN-Cbl is present in this sample. The origin is at the lower left-hand corner and marked by a dot (Reprinted from Linnell et al. (1974). *Clin. Sci. Molec. Med.* **46**, 163 by Courtesy of the editor of Clinical Science and Molecular Medicine).

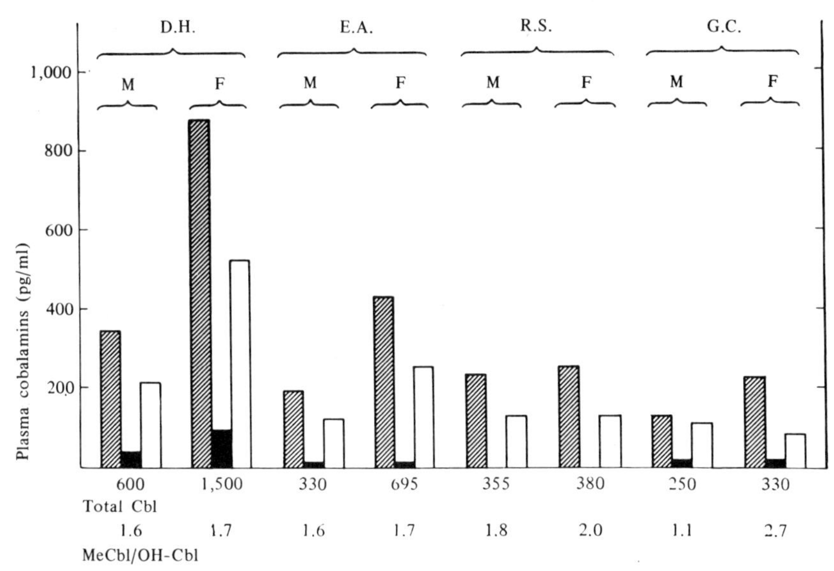

Fig. 6-3 Plasma cobalamins in maternal and fetal blood. ▨ = MeCbl, ■ = CN-Cbl, □ = AdoCbl + OH-Cbl. In each case the plasma MeCbl concentration is higher in fetal than maternal blood. [Reprinted from Craft, I. L., Matthews, D. M., and Linnell, J. C. (1971) *J. Clin. Path.* **24**, 449, by courtesy of the Editor of the *Journal of Clinical Pathology*.]

amin by *Lactobacillus leichmannii* in a series of 50 subjects gave a mean value of 217 pg/ml (range 110–500 pg/ml) (32). Similar results were obtained in a more recent study (33) in which erythrocytes from 50 normal subjects were assayed by Euglena (mean value 205 pg/ml) and by a radioisotopic method (mean 209 pg/ml).

The distribution of cobalamins in erythrocytes differs markedly from that in plasma (30, 34). AdoCbl is the predominant form and accounts for more than half of the total cobalamin, while OH-Cbl represents another quarter. Unlike plasma only 10 to 15% is in the form of MeCbl and this largely accounts for the difference in total cobalamin between these two tissues (Figure 6-1). In healthy persons the CN-Cbl concentration of red cells is similar to that of plasma (34) but as a percentage the mean value is twice as great in erythrocytes as plasma and may represent up to 15% of the total cobalamin. The reason for this is not known although an enzyme system has been demonstrated in human erythrocytes which oxidizes thiocyanate to cyanide *in vitro* (35). If the intracellular cyanide concentration is increased by this or other means, it is possible that this may be sufficient to convert a small proportion of the other erythrocyte cobalamins to CN-Cbl.

Leukocytes and Bone Marrow

In leukocytes from normal subjects the total cobalamin concentration is about 10 times the concentration in plasma, and it has been estimated that the cellular content of leukocytes is approximately 38 pg/10^7 cells (36, 37). The total cobalamin concentration of normal bone marrow is approximately 12,000 pg/ml (38), about four times the concentration in normal leukocytes. A recent study of the cellular distribution of cobalamins revealed that in both leukocytes and bone marrow AdoCbl accounted for almost half the total cobalamin, OH-Cbl for between one-third and one-quarter, and MeCbl for one-fifth or less (Figure 6-4). CN-Cbl accounted for 5% or less in either tissue (24, 34).

Bile, CSF, and Aqueous Humor

Estimates of the total cobalamin concentration in human bile are very variable. Values reported by Okuda et al. (39) ranged from 4.5 to 8.8 ng/ml. Ardeman and co-workers (40) reported values of 1.1 to 11.5 ng/ml in five specimens and 0.09 ng/ml in a sixth. Gräsbeck (41) found that bile binds approximately 1 ng/ml of added cobalamin and that no more than a trace of the endogenous vitamin remains unbound. AdoCbl is the major cobalamin in bile; the proportion of MeCbl is small. There is some evidence that the distribution of cobalamins resembles that in liver though their concentration is lower (24; Linnell, unpublished observations).

In CSF the total cobalamin concentration is less than a tenth of that in plasma or serum, and mean values of between 17 and 32 pg/ml have been reported (42–45). In two subjects investigated more recently (24) approximately three-quarters of the total cobalamin was found to be AdoCbl and no other cobalamin accounted for more than 11% (Table 6-1). The cobalamin content of aqueous humor appears to be very similar to that of CSF. In 16 subjects undergoing cataract removal, total cobalamin concentration ranged from 12 to 60 pg/ml (mean 29.9 pg/ml) (46). Chromatobioautographic analysis of a single sample of aqueous humor showed that 75% of the total cobalamin was present as AdoCbl + OH-Cbl and 25% as CN-Cbl. No MeCbl was detected (Linnell 1969, unpublished observation).

Breast Milk

Human milk contains between 100 and 1500 pg of total cobalamin per milliliter (47). The concentration is inversely related to the volume of

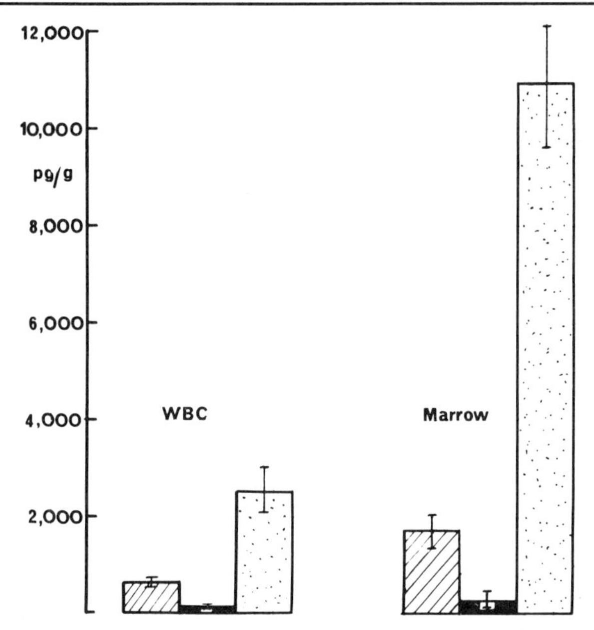

Fig. 6-4 Leukocyte and bone marrow cobalamins in healthy subjects. ▨ = MeCbl, ■ = CN-Cbl, ⬚ = AdoCbl + OH-Cbl (mean values ± SEM). Leukocyte total cobalamin 3830 ± 437 pg/g; bone marrow total cobalamin 12960 ± 1440 pg/g. [Reprinted from Linnell et al (1974) *Clin. Sci Molec. Med.* **46,** 163.]

milk secreted. In a study involving 12 mothers, Karlin (48) found that the total cobalamin in the milk fell from a mean of 660 pg/ml shortly after delivery to 260 pg/ml 8 months later, although the average milk produced by each mother increased from 190 g to 625 g/day during the same period, which more than compensated for the decline in total cobalamin concentration. About 1 week postpartum the predominant cobalamin in breast milk is MeCbl (Figure 6-5), a finding similar to that reported in only one other tissue or body fluid-that is, plasma. In addition to MeCbl the milk also contains a substantial proportion of AdoCbl and the concentration of Ado- + OH-Cbl is on the average more than twice as great in milk as in plasma (31). Little is known about the membrane transport of cobalamins although there is evidence that when orally administered, both CN-Cbl and MeCbl appear in the blood unchanged (28), though some CN-Cbl is converted to AdoCbl during absorption by the ileal mucosal cells (49). Parenterally administered CN-Cbl appears in the milk in high concentration (31). If the cobalamin coenzymes are absorbed substantially unchanged, it is possible that breast milk may provide the infant with a form of the vitamin immediately usable by the rapidly growing tissues.

Table 6-1 Cobalamins in CSF (24)

Subject	Plasma Total Cbl (pg/ml)	CSF Total Cbl (pg/ml)	MeCbl		CN-Cbl		AdoCbl		OH-Cbl	
			pg/ml	%	pg/ml	%	pg/ml	%	pg/ml	%
Control	280	18	1.4	8	1.8	10	13.4	74	1.4	8
Early pernicious anemia	150	20	0.8	4	0	0	17.0	85	2.2	11

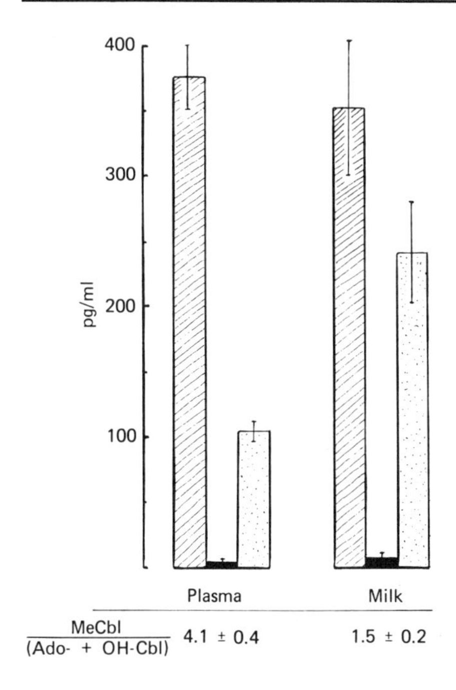

Fig. 6-5 Plasma and milk cobalamins in 10 healthy mothers 4 to 8 days postpartum (mean values ± SEM). Plasma total Cbl 490 ± 59 pg/ml; milk total Cbl 605 ± 185 pg/ml (31). ▨ = MeCbl ■ = CNCbl ⛆ AdoCbl + OHCbl.

Internal Organs

The total cobalamin content of the organs varies widely. In adult man, concentrations are highest in liver, kidney, and pituitary and lowest in tissues which include lung and bone marrow. The liver contains more total cobalamin than any other organ and the majority of this occurs in the mitochondria (50–52). For many years the organs of high cobalamin content have been considered principally as storage depots. More recently it has been found that high levels of the cobalamin-dependent enzymes occur in liver and kidney (53–56), which suggests that these organs at least have more than a storage function for the vitamin.

In adults, AdoCbl is probably the major cobalamin in all cellular tissues and certainly predominates in liver, kidney, spleen, brain, and pituitary (Table 6-2). AdoCbl has been estimated in rat liver by the diol dehydrase method and the results indicate that this form accounts for at least 43 to 72% of the total cobalamin (18). More recently, cobalamins have been estimated by the chromatobioautographic method in liver samples from 10 rats, and the proportion of AdoCbl was found to be 71.6 ± 6.5% (Linnell 1974, unpublished observation). Quadros and coworkers (135) have since reported a mean value of 68 ± 2.3 percent. In man, the total cobalamin concentration of liver is higher than that in the rat, but the proportions of AdoCbl appear to be very similar (34). It may be calculated that the liver in a healthy man contains about 1000 μg of AdoCbl, which accounts for between one-fifth and one-half of the total cobalamin content of the body. The proportions of AdoCbl in the other organs so far analyzed are similar to those in the liver, although the actual concentration varies considerably between organs. OH-Cbl follows a similar pattern and accounts for some 20 to 30% of the total cobalamins in these tissues. The significance of this relatively high proportion of OH-Cbl is not known although it is unlikely to be the result of photolysis except perhaps in the case of small biopsy samples which may not have been adequately protected from white light. One explanation is that OH-Cbl may represent a degradation product of the less stable metabolic intermediates including cob(II)alamin and cob(I)alamin which are involved in the interconversion of cobalamins and synthesis of the cobalamin coenzymes in the tissues.

The organ distribution of MeCbl stands in marked contrast to that of AdoCbl or OH-Cbl and in adults the proportion of MeCbl is highest in the spleen (30–40% of the total cobalamins) and lowest in the liver (1–5%). Although low as a percentage, the actual concentration of MeCbl in liver is far from low and is for example about 100 times the concentration in plasma. The highest MeCbl levels have been found in kidney and pituitary. It may be calculated that in an adult, the liver, kidneys, spleen, and brain together contain about 50 μg of MeCbl or some 2% of the total cobalamins in the body. About 25 to 35 μg of MeCbl is in the liver. One kidney weighing perhaps 150 g would contain more than six times as much MeCbl as there is in the whole volume of plasma in the body. In man the total cobalamin is significantly higher in the cortical region of the kidney than the medulla (57). The same is true in the rat, and preliminary studies on the kidney in this species show that the distribution of cobalamin differs markedly between regions, MeCbl in the medulla being proportionately higher than in the cortex (135).

The distribution of cobalamins in fetal tissues and tissues from young children differs considerably from that in adults. In liver and in certain other organs total cobalamin increases as the child develops, but at birth the concentration of total cobalamin may be no more than a tenth of that in adults. In the liver, the most rapid increase in total cobalamin concentration occurs shortly after birth (Table 6-3). A similar change occurs in kidney, heart, and brain, the fetal levels being significantly lower than those in adults (57). In spleen, lung, and colon, this pattern is reversed and fetal levels are higher than those in adults. In other

Table 6-2 Cobalamins[a] in Normal Liver, Kidney, Spleen, Brain, and Pituitary Gland (24)

Tissue		Total Cobalamin	MeCbl		AdoCbl		OH-Cbl	
		ng/g	ng/g	%	ng/g	%	ng/g	%
Liver biopsies (n = 7)	Mean	1048	12.7	1.2	656	61.1	380	37.7
	(SEM)	(161)	(2.6)	(0.07)	(117)	(1.6)	(42.2)	(1.6)
Autopsy samples								
Liver		670	20	3.0	565	84.3	95	12.7
		440	23	5.2	243	55.2	174	39.6
	Mean	555	21.5	4.1	404	69.8	130	26.2
Kidney		115	31	30.0	55	47.8	29	25.2
		152	36	23.7	86	56.6	30	19.7
	Mean	134	34	26.9	71	52.2	29	22.5
Spleen		70	21	30.0	29	41.4	20	28.6
		56	25	44.6	25	44.6	6	10.7
	Mean	63	23	37.3	27	43.0	13	19.7
Brain		35	9	25.7	20	57.1	6	17.2
		138	6	4.3	86	62.4	46	33.3
		69	3	4.3	44	63.8	22	31.9
	Mean	81	6	11.4	50	61.1	25	27.5
Pituitary		230	47	21	111	48	72	31

[a] No cyanocobalamin was detected in any sample.

Table 6-3 Liver Total Cobalamin Concentration at Different Ages (58)

Age	Number of Cases	Total Cobalamin (ng/g wet wt.)
Stillborn	5	56 ± 28[a]
4–48 hr	3	108 ± 16
0.5–2 yr	2	383 ± 120
20–40 yr	5	310 ± 111
40–60 yr	6	598 ± 168
60–85 yr	6	775 ± 511

[a] Mean ± SD.

tissues there appears to be no consistent age-related change and total cobalamin concentrations in fetus and adult are very similar (57).

In the newborn, activities of many enzymes are low, but these increase, and often dramatically, during the first few days of life. In the liver the important urea cycle enzymes (for example, arginine synthetase) have an activity at birth which is only a few percent of that in adults, yet there is a fivefold increase in the activity during the first 10 days of independent life (59). Similarly, there is a significant increase in methionine-activating enzyme in the brain during this same period (60), which coincides with maximal DNA synthesis and glial cell multiplication (61). Not all enzymes increase in activity after birth as recent studies on N^5-methyltetrahydrofolate methyltransferase show. This enzyme, which is one of at least five utilizing homocysteine as substrate, has a considerably higher activity in fetal brain and liver during the second trimester of gestation than in adult brain or liver (Figure 6-6). N^5-Methyltetrahydrofolate methyltransferase is one of the only two enzymes known to have a coenzyme requirement for cobalamin in man. The evidence presently available suggests that in the fetus, the further metabolism of homocysteine to cysteine by way of the transsulfuration pathway is turned off in favor of the N^5-methyltetrahydrofolate-Cbl remethylation pathway (Figure 6-7). It has been suggested that the importance of this may be to form $N^{5,10}$-methylenetetrahydrofolate, a one-carbon precursor for the *de novo* synthesis of thymidylate, uniquely required for DNA but not for RNA. This suggests that the β-carbon of serine is being shunted into DNA synthesis during periods of rapid multiplication rather than having the entire carbon skeleton accept the sulfur from homocysteine to form

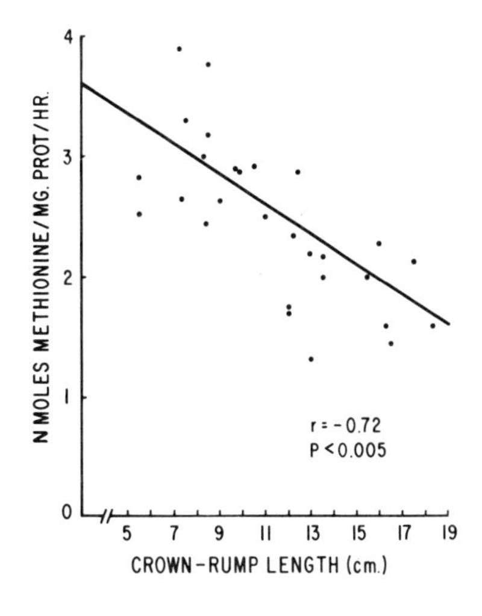

Fig. 6-6 N^5-Methyltetrahydrofolate homocysteine methyltransferase activity in human fetal brain. [From Gaull, G. E. (1973) in *Inborn Errors of Metabolism,* F. A. Hommes and C. J. van den Berg, eds., Academic Press, London.]

cysteine. Cysteine thus becomes an essential amino acid in organs undergoing rapid cellular multiplication (56). This hypothesis and the results on which it is based are particularly interesting in the light of recent cobalamin analyses. Tissues from infants and children have been found to contain higher proportions of MeCbl than adult tissues, but the highest proportions of all have been found in fetal tissues during the second trimester of gestation (Figure 6-8). In fetal liver, kidney, spleen, and brain, MeCbl is the major cobalamin and accounts for a higher proportion of the total tissue cobalamin than at any other stage in development yet investigated.

These findings provide direct evidence that changes in tissue MeCbl parallel the changes in tissue N^5-methyltetrahydrofolate methyltransferase activity. It is likely that comparative studies of tissue AdoCbl and the activity of methylmalonyl CoA mutase may be equally rewarding.

DISTRIBUTION OF COBALAMINS IN DISEASE

The extent to which the normal distribution of cobalamins is altered in human disease has been studied in only relatively few conditions.

Pernicious Anemia

In pernicious anemia, failure of the normal intrinsic factor mechanism greatly reduces the amounts of cobalamins absorbed from food. In

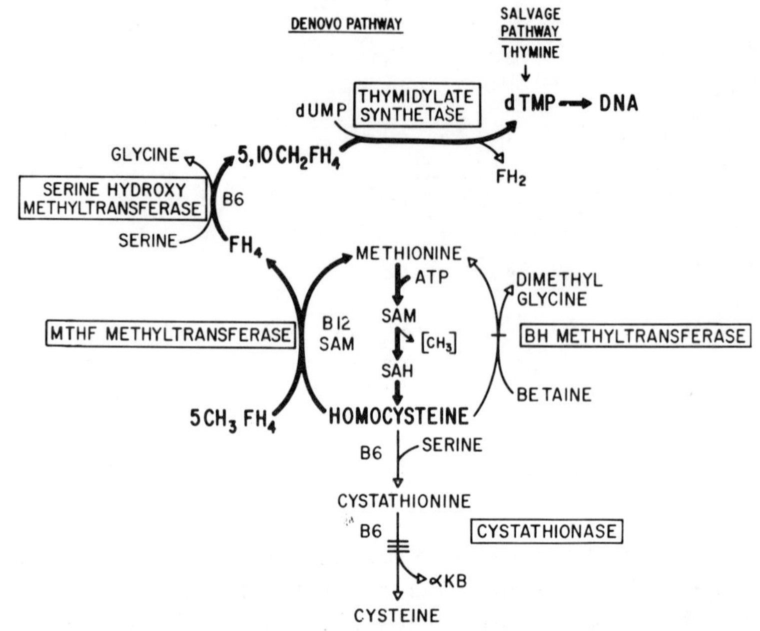

Fig. 6-7 Interrelationships of methionine metabolism, folate metabolism, and DNA synthesis in the human fetus. Abbreviations: ATP, adenosine triphosphate; DNA, deoxyribonucleic acid; SAM, *S*-adenosylmethionine; SAH, *S*-adenosylhomocysteine; dUMP, deoxyuridinemonophosphate; dTMP, deoxythymidinemonophosphate; αKB, ketobutyrate; 5 CH_3FH_4, N^5-methyltetrahydrofolate; FH_4, tetrahydrofolate; FH_2, dihydrofolate; 5,10 CH_2FH_4, $N^{5,10}$-methylenetetrahydrofolate. [From G. E. Gaull (1973), in *Inborn Errors of Metabolism,* F. A. Hommes and C. J. van den Berg, eds., Academic Press, London.]

addition, there is an increased loss of cobalamins from the bile which cannot be reabsorbed owing to the lack of intrinsic factor. It has been estimated that between 0.5 and 5 μg of total cobalamin may be excreted in the bile of healthy persons (63), of which between two-thirds and three-quarters is normally reabsorbed (64). In addition there is a continuing loss of cobalamins from desquamated epithelial cells of the intestinal mucosa which are normally reabsorbed in the lower ileum. In pernicious anemia cobalamins from either of these sources largely fail to be reabsorbed and are excreted. This failure to reabsorb cobalamins contributes to the relatively early onset of frank cobalamin deficiency in patients with anemia, in contrast to vegans and others on a low cobalamin diet who generally remain in good health for very many years (63).

The plasma cobalamin pattern in pernicious anemia differs markedly from that in healthy subjects (Figure 6-9) since the concentration of each

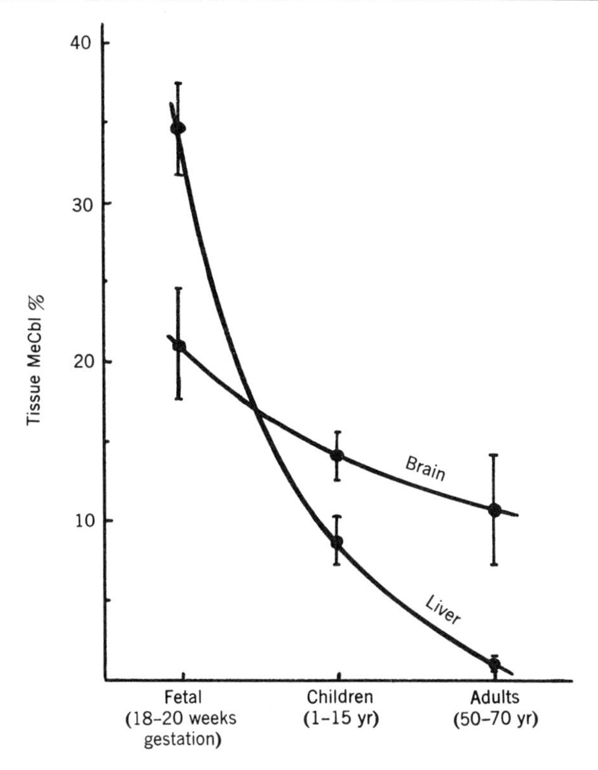

Fig. 6-8 Age-related changes in the proportion of MeCbl in liver and brain (mean values ± SEM).

cobalamin is not reduced to the same extent. MeCbl, normally the major form is disproportionately lowered to a value which is on the average less than 10% of that in normal subjects, but may in some cases be as little as 1% (Figure 6-10). The ratio of MeCbl to Ado- + OH-Cbl falls to less than unity and AdoCbl becomes the major plasma form of the vitamin in this disease (21, 22, 28). Another feature of the pattern in pernicious anemia concerns CN-Cbl. Whereas in healthy subjects this is a very minor component and in general accounts for only 2% of the total plasma cobalamin, in pernicious anemia the plasma CN-Cbl may be increased several-fold and in some cases accounts for as much as 40% of the plasma total cobalamin. The reason for this is not known although it is possible that CN-Cbl may represent a relatively inactive form of the vitamin or one less readily released from its plasma binders.

In erythrocytes, leukocytes, and bone marrow cells from patients with pernicious anemia there is, in contrast to the situation in plasma, no

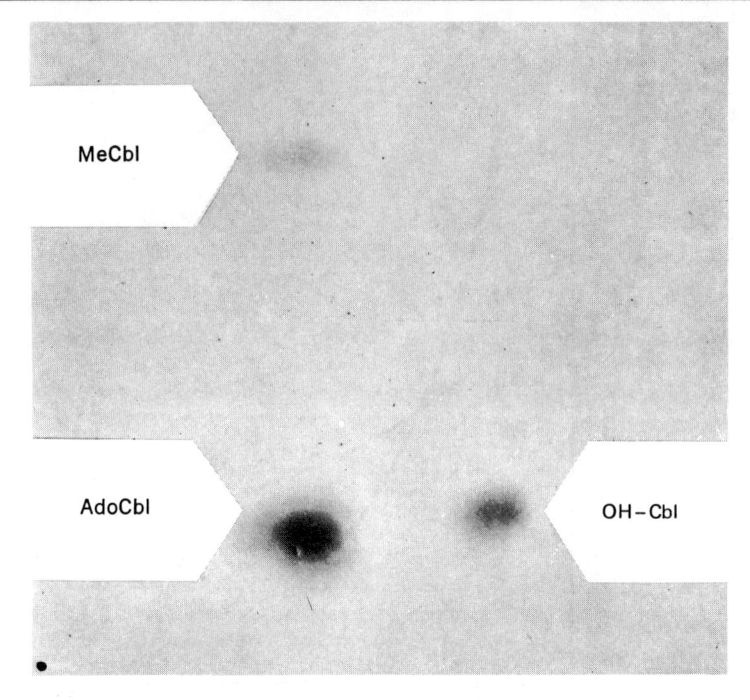

Fig. 6-9 Plasma cobalamins from a patient with untreated pernicious anemia (plasma total cobalamin 80 pg/ml). MeCbl is disproportionately reduced. No CN-Cbl is detectable in this sample. The origin is at the lower left-hand corner, marked by a dot. Reprinted from Linnell, J. C., Hoffbrand, A. V., Hussein, H. A-A., Wise, I. J. and Matthews, D. M. (1974) *Clin. Sci. & Molec. Med.* **46**, 163 by courtesy of the editor of *Clinical Science & Molecular Medicine.*

gross change in the cobalamin pattern (Figure 6-11). MeCbl, AdoCbl, and OH-Cbl are reduced in proportion, and the MeCbl to Ado- + OH-Cbl ratio is little changed. CN-Cbl is the exception, however, and tissue concentrations of this cobalamin are higher than those in corresponding normal tissues. Since concentrations of the other cobalamins are lowered, this means that the proportion of CN-Cbl in each tissue is increased and values are approximately twice those in normal erythrocytes, leukocytes, or bone marrow (30). In view of the dramatic effects of cobalamin deficiency on the normal functioning of bone marrow, it is perhaps surprising that in pernicious anemia the marrow cobalamin concentrations are not more severely depressed. Cobalamins are taken up preferentially by proliferating primitive cells rather than by mature cells (65, 66). The bone marrow in untreated pernicious anemia contains proportionately many more primitive cells than a normal marrow and it

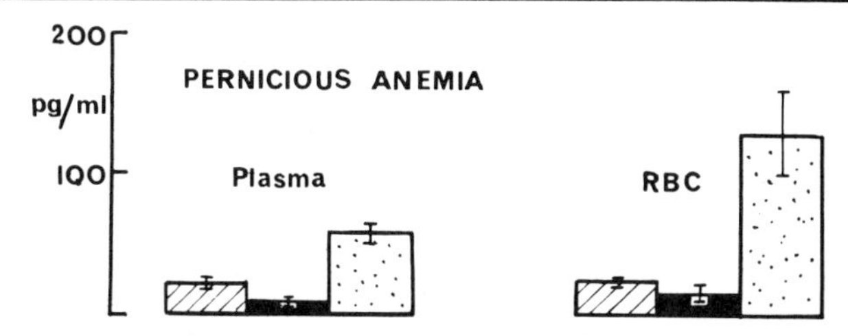

Fig. 6-10 Plasma and erythrocyte cobalamins in 12 patients with untreated pernicious anemia. ▨ = MeCbl, ■ = CN-Cbl, ⊡ AdoCbl + OH-Cbl (mean values ± SEM). Plasma total cobalamin 88 ± 11 pg/ml. Erythrocyte total cobalamin 202 ± 25 pg/ml. [Reprinted from Linnell et al. (1974) *Clin. Sci. Molec. Med.* **46**, 163.]

may be that in cobalamin deficiency the marrow cells are in fact more depleted of cobalamins for their stage of development than is apparent by direct comparison with normal marrow. The importance of the availability of an adequate concentration of MeCbl for the normal development of bone marrow is shown by the effect MeCbl has on cobalamin-deficient marrow cells. It is found that DNA synthesis is stimulated more by MeCbl than by any other cobalamin added to the culture medium, suggesting that the delivery rate of MeCbl and its direct activation of N^5-methyltetrahydrofolate homocysteine methyltransferase may be the critical determinant of megaloblastic inactivation (67).

In other cases of cobalamin deficiency due, for example, to atrophic gastritis, gastrectomy, intestinal malabsorption, or dietary deficiency, the plasma cobalamin pattern is usually similar to that in pernicious anemia. In some patients with intestinal disease or after gastrectomy the ratio of MeCbl to Ado- + OH-Cbl is subnormal in the presence of an unequivo-

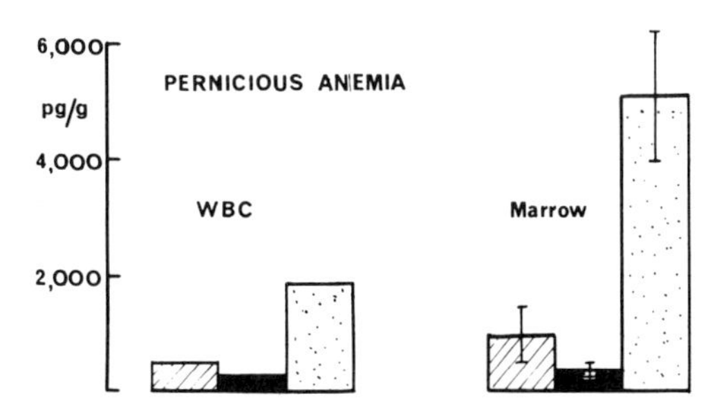

Fig. 6-11 Leukocyte and bone marrow cobalamins in patients with untreated pernicious anemia. ▨ = MeCbl, ■ = CN-Cbl, ⊡ = AdoCbl + OH-Cbl (mean values ± SEM). Leukocyte total cobalamin 2575 + 1467 pg/g; bone marrow total cobalamin 6450 ± 1404 pg/g. [Reprinted from Linnell et al. (1974) *Clin. Sci. Molec. Med.* **46**, 163.]

cally normal plasma total cobalamin (Figure 6-12). The reason for the reduction in plasma MeCbl is still not clear. It may be that this compound is the main transport form of the vitamin as has been suggested by Hall (69). Preliminary studies on the identity of cobalamins attached to the plasma protein binders show that the bulk of the plasma MeCbl is attached to TC I, the predominant binder in plasma (Linnell and England 1973, unpublished observations) although evidence accumulated from the study of cobalamins in conditions in which TC I is selectively elevated indicates that MeCbl may not be the only cobalamin attached to this plasma binder (see the section on leukemia and liver disease). It should be noted that the cobalamin findings in pernicious anemia suggest that plasma levels do not provide a reliable guide to cobalamin levels in the tissues.

Folate Deficiency

In folate deficiency it would not be surprising to find changes in the plasma MeCbl level owing to lack of 5-methyltetrahydrofolate as methyl donor in the methylation of homocysteine, in which MeCbl is required as a coenzyme (70–71). There is evidence that the synthesis of MeCbl is impaired in folate-deficient chicks (72). In fact, in patients with folate deficiency there appears to be no consistent change in the plasma cobalamins. Some patients have a normal pattern while in others the ratio of MeCbl to Ado- + OH-Cbl is subnormal. In a minority of samples a trace of CN-Cbl is present, but in most, CN-Cbl is undetectable (28).

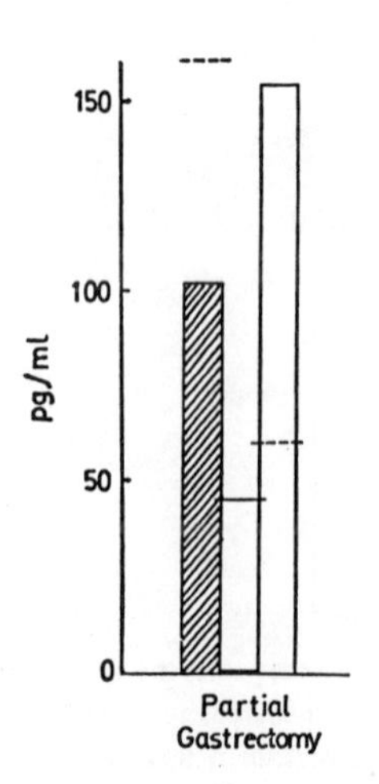

Fig. 6-12 Plasma cobalamins in a case of partial gastrectomy. Although the plasma total cobalamin (255 pg/ml) is normal, both MeCbl and the ratio MeCbl : AdoCbl + OH-Cbl are subnormal. ▨ = MeCbl, ■ = CN-Cbl, □ = AdoCbl + OH-Cbl, ––– = lower limit of normal. [From Arnstein, H. R. V. and Wrighton, R. J. (1971) *The Cobalamins (A Glaxo Symposium)*. Edinburgh: Churchill-Livingstone.]

Leukemia and Liver Disease

In chronic myeloid and other leukemias the plasma total cobalamin is abnormally raised. This is mainly associated with an increase in plasma cobalamin binding capacity attributable to an increase in circulating TC I (74).

Separation of the plasma cobalamins in chronic myeloid leukemia has shown that the pattern is not grossly altered (20) but values for MeCbl and Ado- + OH-Cbl may each be more than four times the values in normal subjects (28) (Figure 6-13). A disproportionate increase in serum OH-Cbl has been observed when the serum total cobalamin is extremely high (73). In erythroleukemia a low or borderline plasma total cobalamin is associated both with a decrease in the ratio of MeCbl to Ado- + OH-Cbl and an increase in plasma CN-Cbl. In both acute and chronic cases of hepatitis there is a disproportionate increase in Ado- + OH-Cbl, and although the plasma MeCbl is also raised, its level is only about half that in chronic myeloid leukemia (Figure 6-13). In a study of six cases of cirrhosis, the plasma MeCbl was elevated in three, and in only one case was the value for Ado- + OH-Cbl abnormally raised (28). In a recently studied case of hepatoma in which the plasma TC I was exceptionally high (75) the plasma AdoCbl was raised, but concentrations of the other cobalamins were within normal limits. In two cases of adolescent hepatoma, very high plasma total Cbl concentrations were

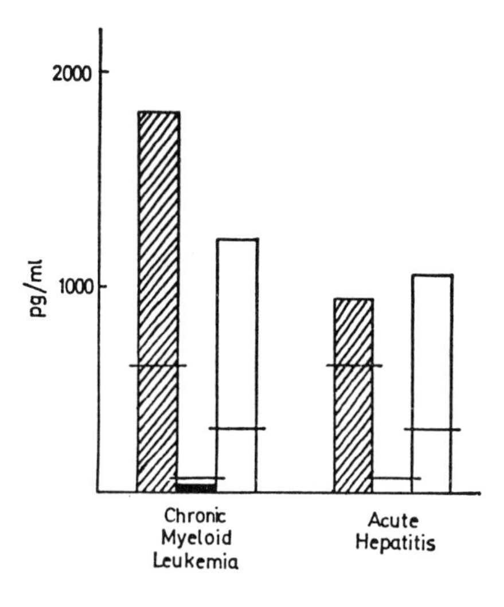

Fig. 6-13 Plasma cobalamins in a case of chronic myeloid leukemia and a case of acute hepatitis. ▨ = MeCbl, ■ = CN-Cbl, □ = AdoCbl + OH-Cbl, — = upper limit of normal (68). [From Arnstein, H. R. V. and Wrighton, R. J. (1971). *The Cobalamins (A Glaxo Symposium)*. Edinburgh: Churchill-Livingstone.]

found in association with high levels of an abnormal α-binding protein similar but apparently not identical to TC I (76).

The evidence at present available suggests that in both leukemia and liver disease, increases in the plasma transcobalamin concentrations, in particular of TC I, lead to concomitant increases of the plasma cobalamins. An additional factor in some cases of liver disease may be the release of AdoCbl from damaged hepatic cells into the blood leading to the disproportionate increase of plasma AdoCbl.

Neurological Disorders

It has long been known that in cobalamin deficiency, clinical signs of abnormal neurological function occur and may be the presenting feature. In the older literature, references to neurological sequelae were common, particularly in cases of advanced pernicious anemia. Osler and Gardner (77) mentioned numbness of the patient's extremeties and Cabot (78) observed that of the 1200 cases of pernicious anemia included in his survey, the majority progressed to subacute combined degeneration of the cord.

Unlike the sex distribution seen in uncomplicated pernicious anemia, retrobulbar neuritis developing as a complication of cobalamin deficiency displays a marked male preponderance; Hamilton (79) recorded 29 such cases, of whom 26 were men. This is very similar to the incidence in tobacco amblyopia, and Traquair (80) found only 7 women so afflicted in a total of 1525 cases. The similarity of the sex distribution in these two conditions suggests either a common etiology or at least a possible relationship between retrobulbar neuritis, tobacco amblyopia, and the metabolism of cobalamins and cyanide.

In tobacco amblyopia it is generally accepted that a nutritional deficiency is involved, and some hold that this is the primary cause of the disease, since other amblyopias of acknowledged toxic origin have different manifestations and lead rapidly to complete blindness (81). This view is at variance with the observations of other workers (82, 83), but although the exact role of cyanide in the pathogenesis of visual failure is not yet known it has been found that the administration of large doses of OH-Cbl is a most effective form of treatment.

The plasma distribution of cobalamins in subacute combined degeneration of the cord is very similar to that in uncomplicated pernicious anemia (21, 22, 28). Plasma MeCbl is disproportionately reduced so that the ratio of MeCbl to Ado- + OH-Cbl falls to less than unity. The proportion of the plasma total cobalamin present as CN-Cbl is increased and is significantly higher than that in normal subjects, though not higher than in uncomplicated pernicious anemia.

In tobacco amblyopia there is a slight reduction in plasma MeCbl but the plasma CN-Cbl concentration is significantly increased by comparison with normal subjects (84). Similar increases in plasma CN-Cbl have been recorded in Leber's optic atrophy, dominantly inherited optic atrophy, and optic atrophies of other origins (Figure 6-14). In a recent study of plasma and erythrocyte cobalamins in patients with visual failure, the most conspicuous alteration was in the erythrocyte MeCbl level, the mean value in the patients being more than three times that in the controls (Linnell 1973, unpublished observations). These findings suggest that in visual failure of various origins there is a redistribution of the plasma and erythrocyte cobalamins resulting in increased levels of MeCbl in the erythrocyte. The significance of this is at present unknown.

In West Africa, the West Indies, and in certain other countries where the incidence of various neuropathies is high, the staple diet for much of the population is "cassava," or "gari," prepared from the tuberous root of *Manihot utilissima,* a plant which flourishes in the poorest soil. Both sweet and bitter strains are grown, the latter giving the larger crop. An appreciable quantity of the cyanogenetic glucoside linamarin occurs

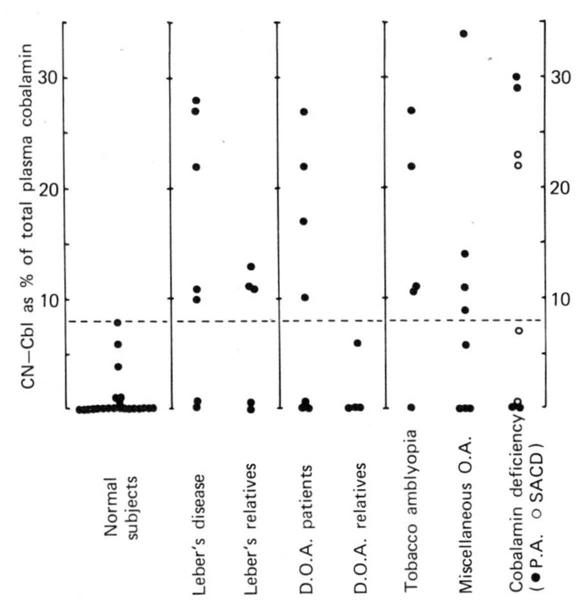

Fig. 6-14 Plasma CN-Cbl in neuroophthalmological diseases. D.O.A. = dominantly inherited optic atrophy; S.A.C.D. = subacute combined degeneration (84).

in the tuber, and the concentration is highest in the outer cortical layer of the bitter variety although there is no clear differentiation between strains. Linamarin is readily hydrolyzed to free cyanide and the dried cortex from bitter cassava root will yield up to 245 mg HCN/100 g (85). A causal relationship between cassava eating and amblyopia was first proposed by Moore (86), and Clark (87) suggested that cyanide might be the toxic constituent responsible. According to Montgomery (85), the poisonous properties of cassava were first recorded by Clusius in 1605, and certainly the comments of other travelers, including Charles Darwin (1832), indicate that this knowledge was widespread. Recent findings suggest that by no means all the cyanide is removed during preparation of the food. Patients subsisting on a diet of cassava have been found who suffer from tropical ataxic neuropathy (TAN), a condition in which optic atrophy, bilateral perceptive deafness, and sensory changes associated with loss of conduction in the posterior spinal columns are demonstrable. These patients have elevated plasma thiocyanate levels and greatly reduced plasma concentrations of the sulfur-containing amino acids cysteine and methionine; plasma cobalamin levels remain normal or are above normal (88, 89).

Estimation of plasma and liver cobalamins in patients with TAN showed that their plasma contained approximately twice as much AdoCbl and more than six times as much CN-Cbl as plasma from control subjects. A small proportion of CN-Cbl was detectable in the majority of the liver biopsies from TAN patients, but none in the controls (90).

These findings provide further evidence that patients with TAN may be suffering from a chronic form of cyanide intoxication but the changes in cobalamin distribution are likely to be of secondary importance.

Metabolic Errors

The metabolic reactions in man and other animals which are known to require cobalamin coenzymes are fully discussed in Chapters 3 and 4, and are mentioned here only in connection with findings related to the distribution of cobalamins in patients with inborn errors of cobalamin metabolism. The reactions believed to precede the synthesis of MeCbl and AdoCbl, normally carried out by all cells of the body, are indicated in Figure 6-15. A block at step 1, 2, or 3 prevents the synthesis of both MeCbl and AdoCbl whereas a failure at step 4 or 5 will interrupt the synthesis of only one of these coenzymes. Although in man these reactions are imperfectly understood, advances have recently been made through

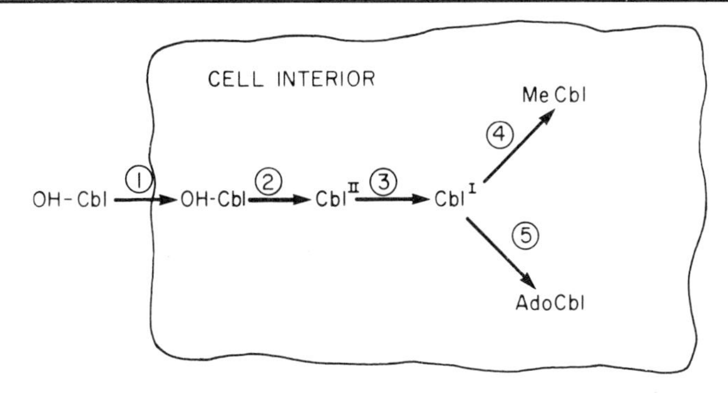

Fig. 6-15 Scheme of probable steps leading to synthesis of MeCbl and AdoCbl. (1) Transport into cell. Though uptake of OH-Cbl is shown, the form or forms in which cobalamin enters cells are uncertain. (2) and (3) Reduction steps. (4) Synthesis of MeCbl. (5) Synthesis of AdoCbl. Metabolic errors might occur in any of these steps (91) (Dillon et al. 1974, reprinted by courtesy of the editor of *Clinical Science and Molecular Medicine).*

studies of children suffering from one or other of the rare genetically determined errors in cobalamin metabolism. These studies are more fully discussed in Chapter 8.

During the last few years a number of patients have been described who excrete in their urine large amounts of methylmalonic acid or homocystine or both. The excretion of either of these compounds alone may be due to any one of several factors, including defective synthesis of apoenzyme proteins or a failure to synthesize MeCbl or AdoCbl (step 4 or 5, respectively, Figure 6-15). In patients who excrete both methylmalonic acid and homocystine the available evidence indicates that this is chiefly due not to a lack of apoenzymes but to a defect in the cellular uptake of synthesis of a common precursor of both cobalamin coenzymes (steps 1, 2, or 3, Figure 6-15). One such case which has recently been investigated was of a child with retarded physical and mental development who died at the age of 7 years after chronic ill health since birth, with recurrent megaloblastic anemia, methylmalonic aciduria, and evidence of abnormal homocysteine metabolism (91). Both serum total cobalamin and serum folate were normal or high. Estimation of her plasma cobalamins gave the first clear indication that there was a serious disturbance in either the cellular uptake or interconversion of cobalamins. The plasma MeCbl was abnormally low while levels of AdoCbl and OH-Cbl were raised, so that the ratio of MeCbl to Ado- + OH-Cbl assumed a very low value similar to that in cobalamin deficiency, although actual concentrations of the cobalamins were much higher (Figure 6-16). Total cobalamin in the erythrocytes was only a tenth of

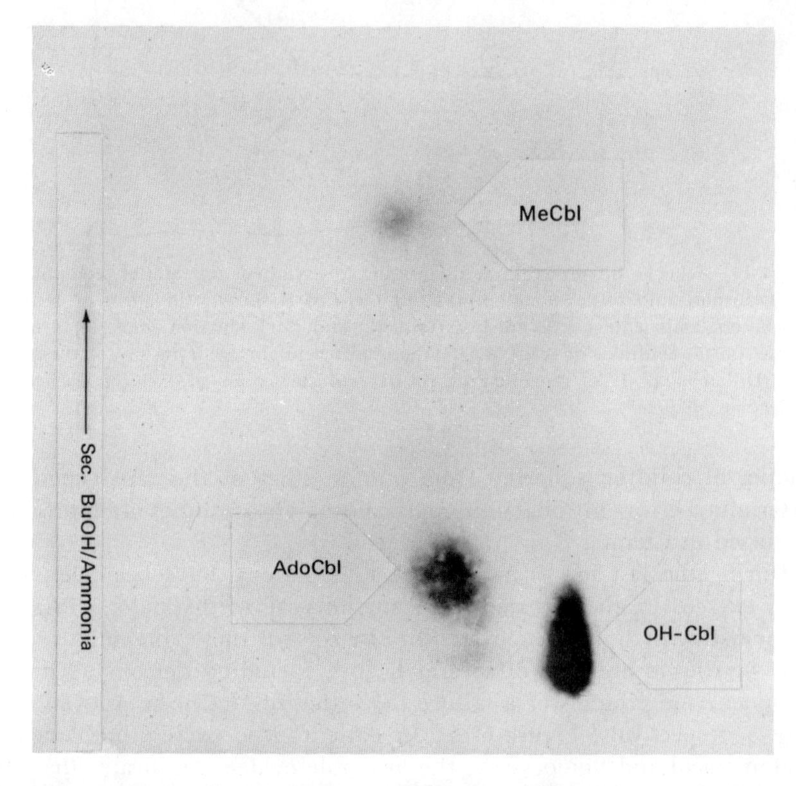

Fig. 6-16 First separation of plasma cobalamins in a child at the age of $5\frac{1}{2}$ years found to have an inborn error of cobalamin metabolism. Total cobalamin was 1090 pg/ml. Note the abnormally low proportion of MeCbl.

that in the plasma, and the erythrocyte MeCbl was at the low end of the normal range. Transcobalamins were normal. It is interesting that in previous otherwise similar cases, abnormal erythropoiesis was not detected.

In the first described case of methylmalonic aciduria with homocystinuria (71, 92) the metabolic disturbance was clearly very severe and the child died at $7\frac{1}{2}$ weeks. Had the patient survived longer it is possible that megaloblastosis might have developed. The two brothers investigated by Goodman (93) were, on the other hand, less severely affected and were in addition cobalamin responsive. Perhaps for this reason both children maintained normal marrow function. In the case described by Dillon et al. (91) the estimation of individual cobalamins in postmortem tissues showed that levels of both coenzyme and noncoenzyme forms of the vitamins were grossly reduced (Figure 6-17) and total cobalamin in the liver was as low as that in severe untreated pernicious anemia. The results were particularly striking since 5600 μg of AdoCbl had been administered to the child parenterally over a period of a few days only 3 weeks before she died. Evidently very little of this AdoCbl was retained by the tissues.

In the case described by Mudd, values for AdoCbl by enzymic assay were markedly subnormal in both liver and kidney (71). Later use of the chromatobioautographic assay confirmed this for AdoCbl and showed that MeCbl and OH-Cbl also were subnormal although less so than in

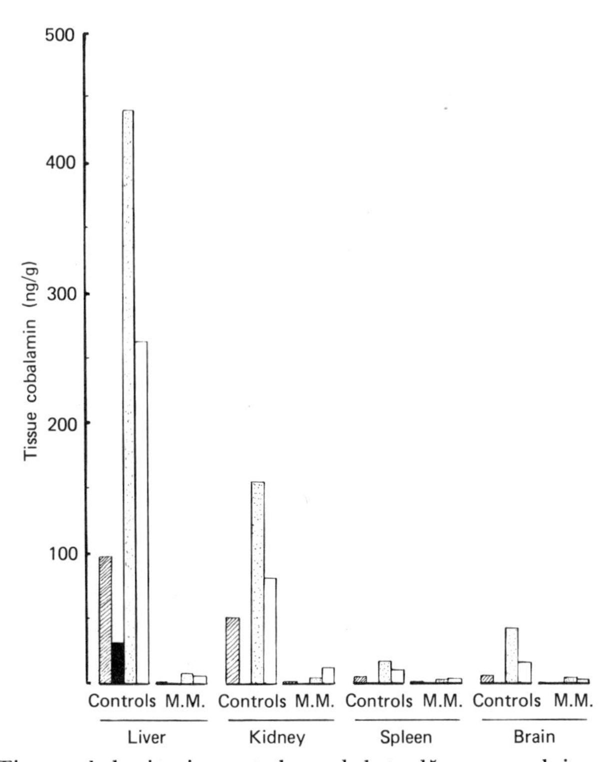

Fig. 6-17 Tissue cobalamins in controls aged 1 to 15 years and in a child with mental retardation, megaloblastic anemia, methylmalonic aciduria, and abnormal homocysteine metabolism due to an error in the body handling of cobalamins, who died at the age of 7 years (91). ▨ = MeCbl, ■ = CN-Cbl, ⬚ = AdoCbl, □ = OH-Cbl.

the case described by Dillon et al. (91). The low tissue levels of MeCbl and AdoCbl in these patients effectively produce the same end result as pernicious anemia, but megaloblastic erythropoiesis apparently results only when the disturbance or deficiency is extreme.

Estimations which so far appear to be of most value in the investigation of these cases include

1. Estimation of N^5-methyltetrahydrofolate homocysteine methyltransferase and methylmalonyl CoA mutase in any available tissue (e.g., leukocytes, liver obtained by needle biopsy, or fibroblasts cultured from a skin biopsy). Results of assays with and without added coenzyme allow apoenzyme and holoenzyme status to be assessed.
2. Estimation of cobalamins in plasma and erythrocytes before and after small parenteral doses (25–100 μg) of cobalamins, particularly OH-Cbl. This tests *in vivo* ability to convert OH-Cbl to MeCbl and AdoCbl.
3. Estimation of cobalamins in fibroblasts cultured with and without OH-Cbl added to the growth medium. Corparison of the results in cells from affected children with those in control cell lines provides good evidence of any failure to accumulate AdoCbl or MeCbl. Fibroblasts cultured from the children described by Mudd et al. (94) and by Dillon et al. (91) were found to accumulate much less AdoCbl and MeCbl than control cells. Cells from individuals known to have a defective apoenzyme, on the other hand, accumulated both cobalamin coenzymes at least as well as control cell lines (Linnell and Mudd 1973, unpublished observations).

It should be remembered that the exact location of the metabolic block has not with certainty been established in any of these patients, although the possibilities include a defect in cellular uptake, a defect in the formation of a common precursor or MeCbl and AdoCbl, the failure to retain cobalamins within the cell, or the possible enzymic destruction of the cobalamin coenzymes.

TISSUE UPTAKE AND INTERCONVERSION OF COBALAMINS

Cobalamins are absorbed from the intestine by an intrinsic factor (IF)-dependent and an IF-independent mechanism, of which the former is normally the more important, though IF-independent uptake accounts for absorption of approximately 1.2% of the dose over a wide range (95). Both of these mechanisms are fully described in Chapter 5 and are not further discussed here. In the guinea pig, orally administered CN-Cbl is partly converted to AdoCbl during absorption by the ileal mucosal cell, but much of an oral dose appears in the portal blood as unchanged CN-Cbl (49).

The interconversion of cobalamins is an aspect of cobalamin metabolism which has until recently been extremely difficult to study, either in animals or man, owing to the lack of suitable methods. A number of workers have compared the uptake and distribution of labeled cobalamins by measuring radioactivity in the plasma or tissues directly, or by whole-body monitoring (96–99), while in other studies use has been made of autoradiography (100, 101). There seems little doubt that for isotopes of cobalt the label detected in the body represents intact cobalamin (102, 103), although the chemical form of the cobalamin cannot be determined by a counting technique alone.

Tissue Studies

In 1961, Coates and his co-workers (104) showed that 18 hr after administration of AdoCbl or CN-Cbl the uptake from intact loops of rat intestine *in vivo* was very similar. More recently, although the intestinal uptake of radioactivity from MeCbl or CN-Cbl in rats was likewise found to be very similar (as was the uptake by the majority of the animals' tissues), the increase in activity in the liver was significantly greater from MeCbl than from CN-Cbl (105). Using double-labeled MeCbl it has been shown that in the rat, MeCbl is at least partly demethylated in the upper gastrointestinal tract, though the effectiveness of IF in protecting the molecule was not investigated (106).

In man the whole-body retention of radioactivity after 3 days was greater from parenterally administered AdoCbl than from OH-Cbl or CN-Cbl, but 25 days later more activity had been retained from the administration of OH-Cbl than from that of either AdoCbl or CN-Cbl (99). More recently it has been found that in man, MeCbl and CN-Cbl are cleared from the plasma at very similar rates (107). Administered cobalamins have been shown to accumulate preferentially in tissues of high metabolic activity or rapid cell division. Whole-body radiography after intravenous administration of CN[^{57}Co]Cbl to male adult mice showed that uptake of radioactivity was greatest in the kidney, reproductive and endocrine organs, gastric mucosa, and liver. In nonpregnant female mice the liver and kidneys accumulated more activity than any other tissue while in pregnant animals much activity was found in the placenta shortly after the dose, and within 4 days passed to the fetus (101). These and other studies show that uptake varies considerably between organs. Further evidence of the avidity many rapidly dividing tissues have for administered cobalamins is given by the finding that a soft fibroblastic sarcoma accumulated very high concentrations of administered CN-Cbl (108). This is not universal for all neoplastic tissues, however, since in a recently studied case of neuroblastoma, the tumor itself was found to take up less CN[^{57}Co]Cbl than apparently normal areas of brain adjacent to the tumor (109).

Interconversion of CN-Cbl to OH-Cbl has been studied in the guinea pig both by *in vivo* and *in vitro* techniques (110) and the results show that liver will convert 0.1 to 0.4 ng of CN-Cbl per gram of tissue each day. Further studies in the rat have revealed that an enzyme responsible for the decyanation of CN-Cbl in both liver and kidney is localized to the soluble fraction (111). Under anaerobic conditions the pH optimum is 7.2 and the reaction *in vitro* requires reduced pyridine and flavin nucleotides. Yagiri (112) studied the conversion both of CN-Cbl and OH-Cbl to AdoCbl in rat liver and kidney *in vivo,* and in human liver *in vivo* following parenteral administration of labeled cobalamins. In the rat, conversion of OH-Cbl to AdoCbl in both liver and kidney was about 50% faster than the conversion of CN-Cbl to AdoCbl. After chemically induced liver damage, the conversion rate was reduced only for CN-Cbl, that for OH-Cbl remaining unaffected. In cobalamin-deficient rats, conversion of either cobalamin proceeded at a faster rate than in control animals, the increase being more marked for OH-Cbl than for CN-Cbl. Human liver converted CN-Cbl and OH-Cbl to AdoCbl at about twice the rate in rat liver, and of a 2 to 4 μg dose administered parenterally, approximately 50% was converted to AdoCbl within 24 hr (112).

In the guinea pig, orally administered CN-Cbl is initially taken up unchanged by the ileal mucosal cells, and peak levels are detected within 1 to 2 hr. Thereafter, the specific activity of CN-Cbl in the mucosa declines as that of AdoCbl increases (Figure 6-18) until after 4 hr the specific activities of each are approximately equal. The time course of these changes suggests a precursor-product relationship (113). Uptake by the mitochondrion of the ileal mucosal cell appears to be an early step in the absorption of CN-Cbl, perhaps preceded by decyanation in the soluble fraction as occurs in certain tissues in the rat (111). The evidence suggests that synthesis of AdoCbl may be localized to the mitochondrion (Figure 6-19) although conversion to this form is clearly not a necessary prerequisite for absorption since a large proportion of the dose appears unchanged as CN-Cbl in the portal plasma. After 7 hr, CN-Cbl still accounts for more than half of the radioactivity in the plasma (Figure 6-20).

In rats, individual cobalamins have been estimated in the liver after intramuscular administration of small single doses of CN[^{57}Co]Cbl.

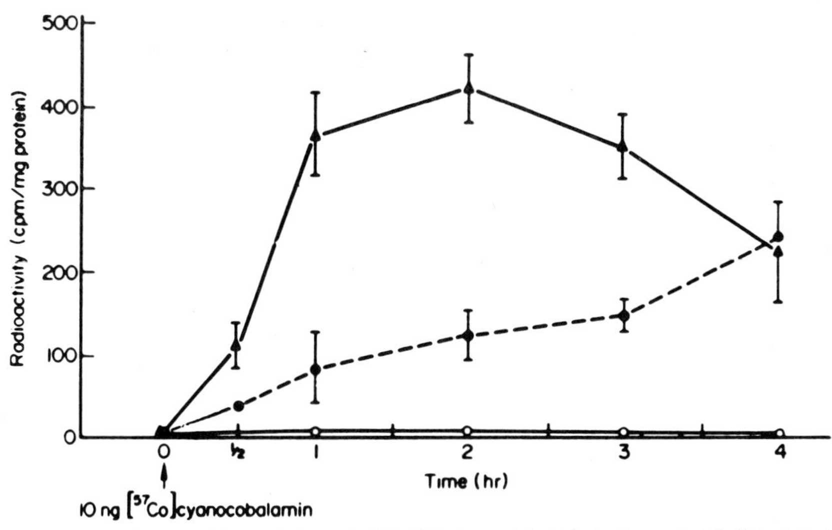

Fig. 6-18 The specific activity of CN-Cbl (▲), AdoCbl (●), and MeCbl (○) in guinea pig ileal mucosa at timed intervals after feeding radioactive CN-Cbl (49). Peters, Linnell, Matthews.

Animals were killed at timed intervals and the liver removed for cobalamin analysis by chromatography and bioautography followed by γ-scintillation spectrometry. During the first 20 hr, the rate of disappearance of CN-Cbl was faster than the rate of appearance of AdoCbl. Twenty hours after dosing, approximately 25% of the total activity in the liver corresponded to newly synthesized AdoCbl, 5 to 10% to MeCbl and 45 to 60% to CN-Cbl. Thereafter, while the proportions of MeCbl and OH-Cbl remained constant, the proportion of newly synthesized AdoCbl continued to increase and that of CN-Cbl to decrease. Three weeks after dosage the liver contained approximately 80% AdoCbl, 5% CN-Cbl and 5 to 10% each of MeCbl and OH-Cbl (Figure 6-21).[1]

Plasma and Milk

In man, there is a marked increase in the plasma CN-Cbl shortly after a small parenteral dose of this form of the vitamin. In one subject, 8 hr after 100 μg of CN-Cbl i.m., CN-Cbl accounted for 67% of the plasma total cobalamin, while the concentrations of MeCbl and Ado- + OH-Cbl remained essentially unchanged. In another subject, 12% of the total plasma cobalamin was present as CN-Cbl 24 hr after a dose of 100

[1] Linnell, Hussein, Tavill, and Cooksley, manuscript in preparation.

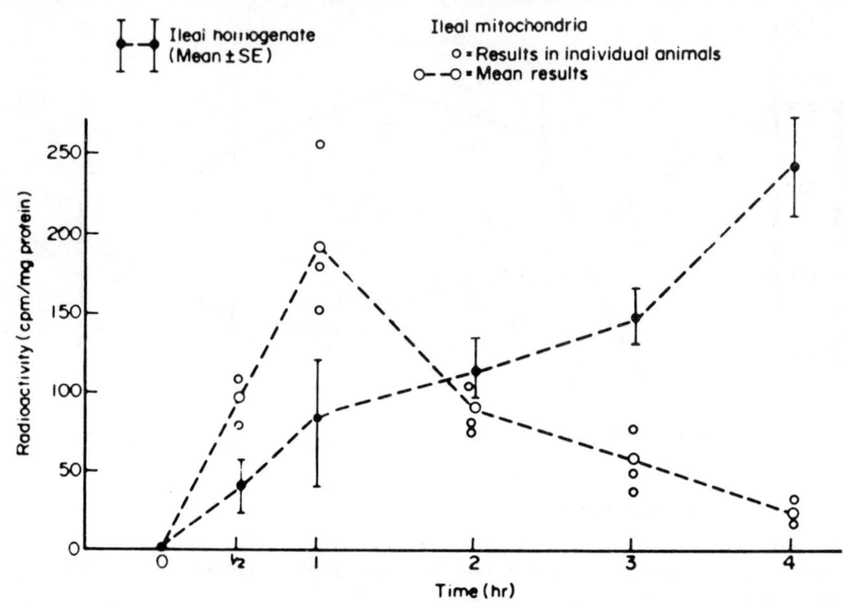

Fig. 6-19 The specific activity of AdoCbl in guinea pig ileal mitochondria at timed intervals after feeding radioactive CN-Cbl (49).

μg of CN-Cbl i.m., but by this time much of the dose had been converted to MeCbl, the level of which in the plasma had increased to over 150% of the predose value (28). Similarly, after oral dosage with CN-Cbl (10 μg), CN-Cbl was detected in the plasma, and in one healthy subject accounted for approximately 10% of the plasma total cobalamin 5 and 24 hr after the dose (Figure 6-22). Much smaller oral doses may still produce detectable plasma levels. Thus in a vegan, 5 and 8 days after 1 μg of CN-Cbl daily by mouth, small amounts of CN-Cbl appeared in the plasma. Much of the dose was evidently converted to MeCbl, however, since the plasma MeCbl increased to 10 times the predosage level (Figure 6-23). Small (1 μg) daily doses of MeCbl i.m. rapidly reverse the abnormal plasma cobalamin pattern in cobalamin deficiency, without apparently changing the levels of cobalamins other than MeCbl (28). Yagiri (112) has shown that OH-Cbl is more rapidly converted to AdoCbl by human liver than is CN-Cbl. The results in Table 6-4 show that unlike CN-Cbl, OH-Cbl does not remain unchanged in the plasma for long but is converted to both MeCbl and AdoCbl. Hence, 3 days after this large dose (1500 μg i.m.) only 14% was detected in the plasma as OH-Cbl.

To investigate the possible transport of cobalamins into breast milk, CN-Cbl (100 μg and 1 mg) was administered to two healthy lactating

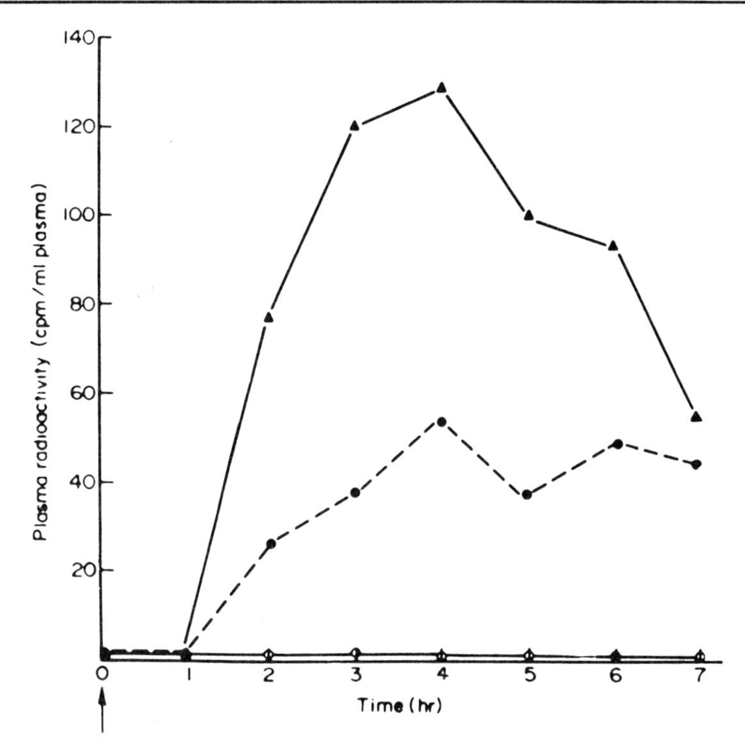

Fig. 6-20 The concentration of labeled CN-Cbl (▲), AdoCbl (●), and MeCbl (○) in portal blood from a healthy guinea pig at timed intervals after feeding radioactive CN-Cbl (49).

mothers about 1 week after delivery. Blood and milk samples were collected before the dose and 8 hr later. High levels of CN-Cbl were found in both milk and plasma postdosage (Figure 6-24). While it is uncertain to what extent the plasma and milk cobalamins are correlated, the experiment showed that parenterally administered CN-Cbl is transferred into the milk on a substantial scale (31).

Although the uptake and interconversion of cobalamins in plasma and tissues have as yet been very incompletely studied, there is evidence that at least a part of the dose administered, whether orally or parenterally, appears in the plasma unchanged, where, particularly for CN-Cbl, it may remain for a considerable time before conversion to the coenzymes is complete. It is also clear that cobalamins may be taken up by a number of tissues in the form administered. This may initially distort the tissue cobalamin pattern, and the time taken for this to revert to normal will depend on several factors, including enzymic activities in the tissue and the dose administered.

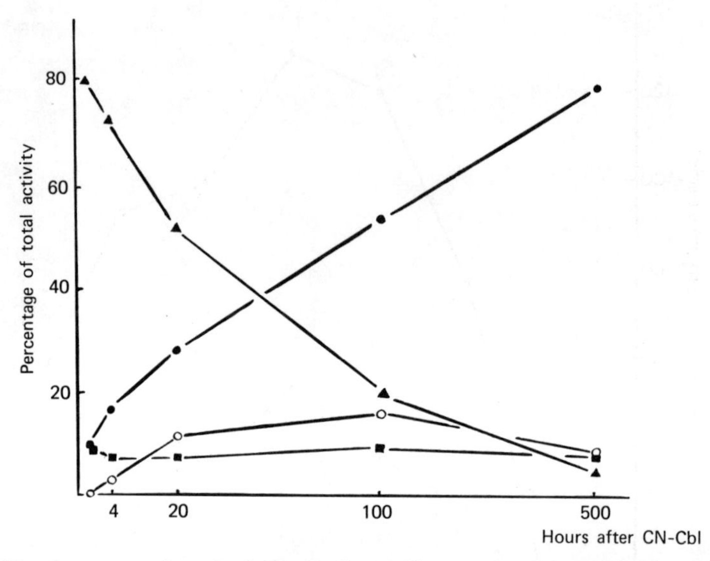

Fig. 6-21 Interconversion of cobalamins in rat liver at timed intervals after administration i.m. of labeled CN-Cbl. ▲ CN-Cbl; ● AdoCbl; ○ Me-Cbl; ■ OHCbl. Linnell, J. C., Hussein, H. A-A., Tavill, A. S., and Cooksley, G. Manuscript in preparation.)

Cells in Tissue Culture

It is at least highly probable that in man, all cells of the body possess the complete enzyme systems necessary to synthesize both cobalamin coenzymes. Initial uptake of cobalamin by the cell is facilitated by TC II, and binding to this or another protein may be a prerequisite for cellular uptake. The uptake of CN-Cbl by *E. coli* cells consists of an initial rapid phase during which the cobalamin is bound by specific receptors on the outer cell membrane, followed by a slower secondary phase when the

Table 6-4 Plasma Cobalamins in a Healthy Boy of 15 before and after OH-Cbl

	Total Cobalamin (pg/ml)	MeCbl pg/ml	MeCbl %	CN-Cbl %	AdoCbl pg/ml	AdoCbl %	OH-Cbl pg/ml	OH-Cbl %
Predose	500	416	83	0	66	13	18	4
3 days after 1500 μg OH-Cbl i.m.	2765	1781	64	0	592	21	392	14

Goodman, Mudd, and Linnell 1973, unpublished data.

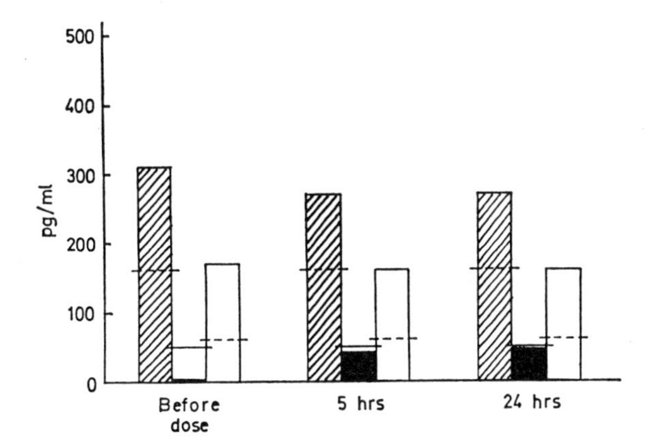

Fig. 6-22 Plasma cobalamins before and after an oral dose of 10 μg CN-Cbl in a healthy adult. Plasma CN-Cbl rises from less than 5 pg/ml to nearly 50 pg/ml. ▨ = MeCbl, ■ = CN-Cbl, □ = AdoCbl + OH-Cbl, ---- = lower limit of normal, —— = upper normal limit. [From Arnstein, H. R. V. and Wrighton, R. J. (1971). *The Cobalamins (A Glaxo Symposium)*. Edinburgh: Churchill-Livingstone.]

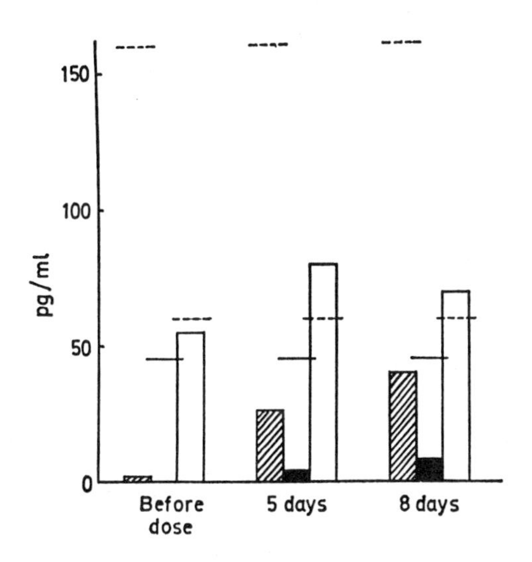

Fig. 6-23 Plasma cobalamins in a cobalamin-deficient vegan given 1 μg CN-Cbl daily. Before dosage the pattern is typical of cobalamin deficiency. At 5 and 8 days some plasma CN-Cbl was detected together with a large increase in MeCbl. ▨ = MeCbl, ■ = CN-Cbl, □ = AdoCbl + OH-Cbl, ---- = lower limit of normal, —— = upper limit of normal [From Arnstein, H. R. V. and Wrighton, R. J. (1971) *The Cobalamins (A Glaxo Symposium)*, Edinburgh: Churchill-Livingstone.]

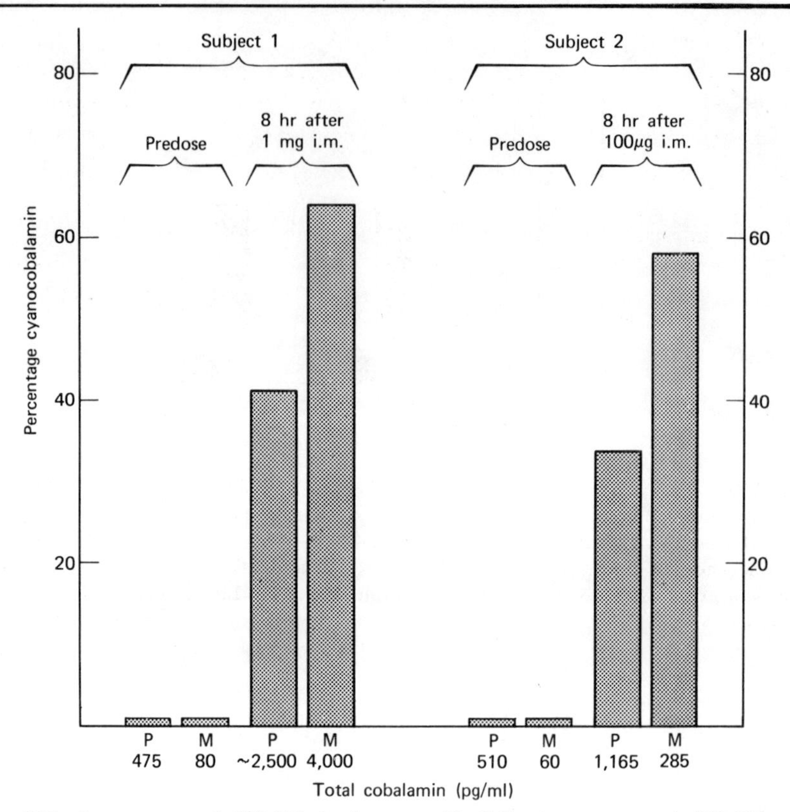

Fig. 6-24 Appearance of CN-Cbl in breast milk following parenteral CN-Cbl administration. P = plasma; M = milk. [Reprinted from Craft, I. L., Matthews, D. M., and Linnell, J. C. (1971). *J. Clin. Path.* **24**, 449, by courtesy of the editor of the *Journal of Clinical Pathology*.]

cobalamin is transferred to the cell interior. Kinetics of the initial phase of uptake are apparently similar for CN-Cbl, OH-Cbl, AdoCbl, and MeCbl (114).

The cellular uptake of cobalamins may conveniently be studied using either "cold" or radioactively labeled cobalamins added to the culture medium. Cobalamin uptake has been studied with a number of types of cells including *E. coli*, HeLa cells, and various cells from the hemapoietic system. HeLa cells readily convert OH-Cbl to AdoCbl, and in the presence of ATP and a reducing system, cell extracts alone have been found to carry out this conversion (115). Uptake of CN-Cbl by phytohemagglutinin (PHA)-transformed lymphocytes was reduced in cobalamin-deficient cells by comparison with normal cells and was related to the cellular rate of RNA synthesis (66). In contrast, cobalamin-deficient marrow cells have been found to take up more CN-Cbl than normoblastic cells (116). More recent studies suggest that under suitable conditions, cobalamin-deficient PHA-transformed lmyphocytes also take up CN-Cbl more avidly than do normal lymphocytes (Quadros and Linnell, 1974, unpublished observations).

Interconversion of cobalamins has been studied in fibroblasts cultured from human skin cells. Mahoney and Rosenberg (117) found that the addition of labeled OH-Cbl to the culture medium (140–200 pg/ml) resulted in cellular synthesis of 0.26 ± 0.09 pg AdoCbl and 0.38 ± 0.12 pg MeCbl/mg wet cells. Of the radioactivity recovered from the cells, 20 to 25% represented AdoCbl and 30 to 35% MeCbl. More recently it has been found that when higher concentrations of OH-Cbl are added to the culture medium there is a proportional increase in MeCbl and AdoCbl synthesis, normal fibroblasts apparently having a capacity for maintaining fairly constant proportions of the two coenzymes over a wide range of OH-Cbl concentrations without restricting uptake from the culture medium. Over the range 200 pg/ml to 100 ng/ml, 25 to 35% of the cellular total cobalamin corresponded to AdoCbl, 20 to 30% to MeCbl, and the remainder to OH-Cbl (Linnell, Matthews, Mudd, and Uhlendorf 1973, unpublished). It has been shown that fibroblasts from certain children with errors of cobalamin metabolism are deficient in N^5-methyltetrahydrofolate homocysteine methyltransferase activity (94, 118). Estimation of cobalamins in some of these cases has shown that their fibroblasts contain abnormally low levels of MeCbl which may be almost undetectable (72) (Figure 6-25). Almost normal activity of the holoenzyme is restored by addition of MeCbl to the assay medium though not by addition of OH-Cbl. Abnormally low levels of both cobalamin coenzymes have been detected in fibroblasts from a further child found to be suffering from an inborn error of cobalamin metabolism (91). This direct evidence of an intracellular deficiency of both coenzymes was discussed in an earlier section and is more extensively considered in Chapter 8.

Despite the many and varied studies on tissue uptake and interconversion of cobalamins which have already been completed, it is clear that much still remains to be done before the involvement of cobalamin coenzymes in the regulation of cellular metabolism is fully understood. Of the means at present available for investigating cobalamin metabolism in man, perhaps the most attractive model is that of human cells in tissue culture. Bone marrow cells *in vitro* continue to synthesize DNA under suitable condtions for about 24 hr and PHA-transformed lymphocytes survive longer, remaining viable for at least 3 days. Fibroblasts in

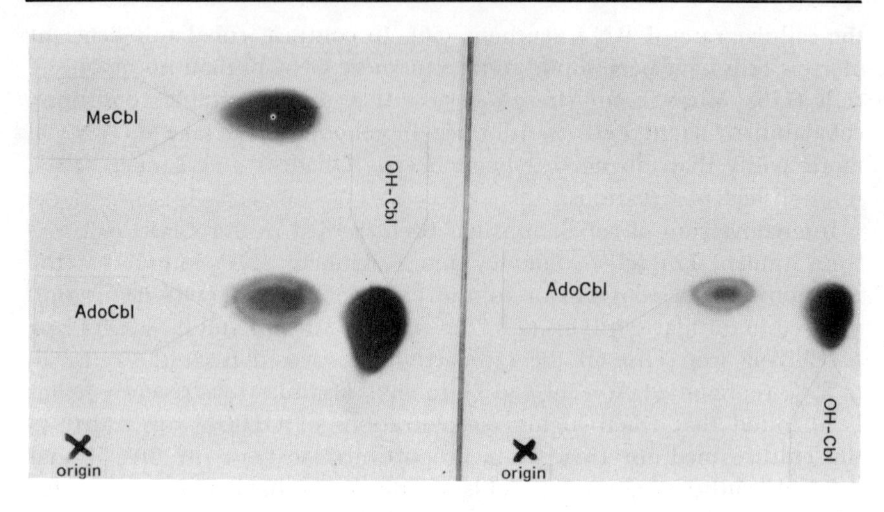

Fig. 6-25 Separation of cobalamins by two-dimensional chromatography and bio-autography in fibroblasts cultured from a normal child and from one with deranged cobalamin metabolism. Left: control; right: M.R.

many ways provide an even more amenable system since mitosis continues, and cultures can be maintained *in vitro* almost indefinitely, allowing long-term studies to be carried out which may provide answers to those problems inaccessible to other methods.

EXCRETION

Cobalamins are apparently ubiquitous in the body and it may be assumed that small amounts are continually eliminated in sweat, tears, and desquamated skin cells. Urinary, biliary, and fecal routes are probably the most important and possess special means of controlling losses of cobalamins from the body.

Urinary Excretion

Cobalamins circulate in the plasma attached to at least one binding protein (see Chapter 5), but there is in addition some residual unbound cobalamin binding protein; a recent estimate of this by an improved technique suggests that as much as 40% of the total plasma cobalamin binding capacity is normally unsaturated (119). Free cobalamin is virtually undetectable in the plasma and it is unlikely that more than very small quantities of the cobalamins are filtered by the kidney or appear in the glomerular filtrate. Although the origin of the urinary cobalamin is uncertain, some, it may be expected, is derived from desquamated tubular epithelial cells and some perhaps may be contributed by the lymph.

Estimates of the urinary total cobalamin excretion in healthy subjects are variable. Mean values in medical students on various diets ranged from 31 to 43 ng/24 hr, with higher values (mean 62 ng/24 hr) in six fasting students (120). Mollin and Ross (121) reported values ranging from 110 to 240 ng/24 hr in six subjects and Heinrich (122) found that the normal urinary cobalamin excretion ranged from 50 to 250 ng/day (mean 150 ng/day). Healthy smokers have been found to excrete more total cobalamin in their urine (mean 81 ± 8.7 ng/24 hr) than do nonsmokers (mean 60 ± 7.9 ng/24 hr), which correlated well with their thiocyanate excretion (123). In addition there was a negative correlation, particularly marked for smokers, between serum total cobalamin and urine thiocyanate (Figure 6-26), perhaps reflecting a redistribution of cobalamins in serum and tissues, as a result of the increased cyanide intake from smoking.

In pernicious anemia and other conditions associated with cobalamin deficiency, urinary excretion of the vitamin is reduced. Heinrich (122) reported a value of 8 ng Cbl/24 hr as the average excretion in cobalamin

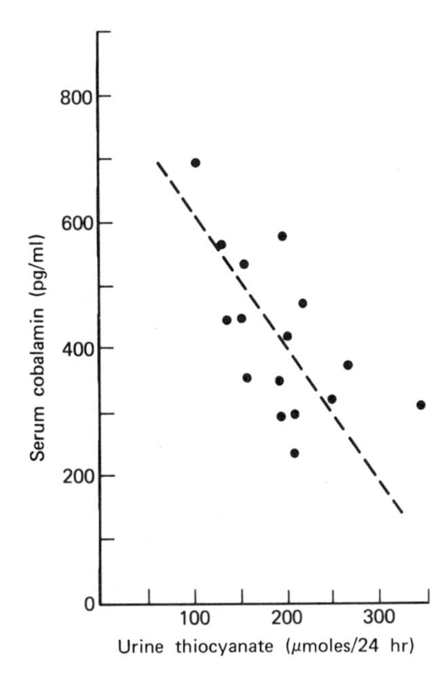

Fig. 6-26 Correlation between serum total cobalamin and urine thiocyanate excretion in healthy smokers. The regression line shown bisects the angle between the regression of y on x and that of x on y. [From Linnell, J. C., Smith, A. D. M., Smith, C. L., Wilson, J., and Matthews, D. M. (1968) *Brit. Med. J.* 2, 215, by courtesy of the editor of the *British Medical Journal.*]

deficiency and Mollin and Ross (121) found that in six of their seven cases of pernicious anemia, urinary excretion ranged from less than 20 to 30 ng Cbl/24 hr. In the seventh case, which also had the highest cobalamin at 100 pg/ml, cobalamin excretion was normal (150 ng/24 hr).

A number of investigators have found that urinary excretion of cobalamin correlates poorly with the doses of the radioactive vitamin. Sokoloff et al. (124) reported that when CN-Cbl (84–211 μg) was administered parenterally to pernicious anemia patients, 53 to 68% appeared in the urine within 18 hr but little of a smaller dose (42 μg) was excreted unless a flushing dose of CN-Cbl was administered. The results in normal subjects and patients were very similar. Likewise in 30 previously untreated pernicious anemia patients, parenteral administration of 54 μg to 30 mg CN-Cbl produced no consistent excretion pattern, though in most, excretion of the label was essentially complete within 24 hr (125). Patients with renal disease excrete labeled CN-Cbl administered intravenously more slowly than normal subjects, particularly during the first 24 hr. If urine is collected for periods up to 72 hr, differences between patients and controls are less apparent, suggesting that it may be only the rate of excretion, not the total amount, which is reduced (126). When the loss of radioactivity in renal disease was estimated by whole-body counting over periods of up to 99 days, total excretion by all routes was higher than normal in three of five patients. Serum total cobalamin in these three patients was higher than in the other two (127). In an earlier study of renal failure an abnormally high serum cobalamin was found in 14 of 32 patients (128).

The identity of the cobalamins excreted in the urine remains uncertain, although there is evidence that the radioactivity detected in urine after administration of labeled cobalamins corresponds to intact cobalamin (98). Kennedy and Adams (129) found that after parenterally administered OH[^{57}Co]Cbl, a substantial amount of CN-Cbl and anionic complexes ("red acids") appeared in the urine, but since a similar admixture resulted from adding OH-Cbl to collected urine they concluded that conversion may not have occurred *in vivo*. Analysis by chromatography and bioautography of the cobalamins in a single sample of urine from a healthy subject collected in a foil-covered vessel showed only Ado- + OH-Cbl; no MeCbl or CN-Cbl was detected (Linnell 1972, unpublished).

Biliary and Fecal Excretion

In both man and rat the total cobalamin content of bile is high, although values reported are variable. Thus Okuda and co-workers (130) found samples to contain between 4.5 and 8.8 ng/ml and Ardeman et al. (40)

reported values of 1.1 to 11.5 ng/ml in five specimens and 0.09 ng/ml in a sixth. After injection of labeled cobalamins most of the vitamin appearing in the feces originates in the bile, which is a major route of cobalamin excretion (40, 64, 105, 131, 132). Gräsbeck (41) found that bile has an unsaturated cobalamin binding capacity of approximately 1 ng/ml and that no more than a trace of the endogenous vitamin remains unbound. Separation of individual cobalamins in bile shows that AdoCbl is by far the largest component and that the proportions of MeCbl and OH-Cbl are small (Figure 6-27).

In man between 0.5 and 5 μg total cobalamin is secreted into the bile daily of which at least 65 to 75% is reabsorbed (64). Reabsorption depends on intrinsic factor (133). In the rat it has been shown that ligating the bile duct reduces fecal excretion of cobalamin by only about 25% (39), indicating that the remainder is extrabiliary in origin. Since the lower ileum is the site of cobalamin absorption this means that the gut also has an opportunity of reabsorbing cobalamins present in digestive secretions and desquamated cells (134). It has been suggested that the relatively rapid development of cobalamin deficiency (3–6 years) in subjects who have a reduced ability to absorb cobalamins by way of the

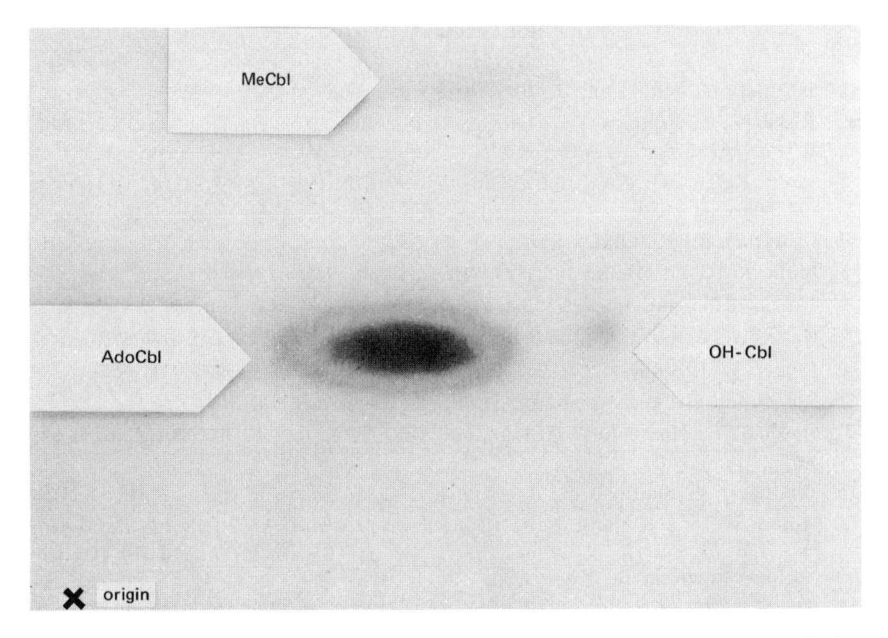

Fig. 6-27 Separation of cobalamins in bile from a gall bladder obtained during cholecystectomy. MeCbl and OH-Cbl were present only in traces. No CN-Cbl was detected.

intestine owing to failure to secrete intrinsic factor, is at least partly due to their inability to reabsorb biliary cobalamin (63). Subjects on a low cobalamin diet take many more years to become deficient owing to the normal functioning of their enterohepatic circulation so vital to the conservation of biliary cobalamins.

ACKNOWLEDGMENTS

It is a pleasure to acknowledge all those with whom I have collaborated, both in this department and elsewhere. In particular, I must thank Professor David Matthews for his continued interest and encouragement, and for constructively criticizing this manuscript. I am indebted to Dr. G. E. Gaull and Academic Press for permission to reproduce Figures 6-6 and 6-7, to Dr. J. M. Hsu and Nature (London) for the date reproduced in Table 6-3, and I thank the various editors and publishers cited elsewhere for permission to reproduce many of the other figures. Financial support was provided by the Wellcome Trust and is gratefully acknowledged.

REFERENCES

1. Smith, E. L., and Parker, L. F. J. (1948). *Biochem. J.* **43**, 8.
2. Rickes, E. L., Brink, N. G., Koniuszy, F. R., Wood, T. R., and Folkers, K. (1948). *Science* **107**, 396.
3. Wolff, R., Drouet, P.-L., and Karlin-Weissman, R. (1951). *C. R. Acad. Sci. (Paris)* **232**, 568.
4. Girdwood, R. H. (1952). *Biochem. J.* **52**, 58.
5. Blum, K.-U., and Heinrich, H. G. (1957). *Vitam. Horm.* **7**, 486.
6. Nelson, R. S., and Doctor, V .M. (1958). *Ann. Intern. Med.* **49**, 1361.
7. Halsted, J. A., Carroll, J., and Rubert, S. (1959). *N. Engl. J. Med.* **260**, 575.
8. Chaiet, L., Rosenblum, C., and Woodbury, D. T. (1950). *Science* **111**, 601.
9. Rosenblum, C., and Woodbury, D. T. (1951). *Science* **113**, 215.
10. Smith, E. L., Hockenhull, D. J. D., and Quilter, A. R. J. (1952). *Biochem. J.* **52**, 387.
11. Bradley, J. E., Smith, E. L., Baker, S. J., and Mollin, D. L. (1954). *Lancet* **2**, 476.
12. Mollin, D. L., and Smith, E. L. (1956). *Proceedings of the International Conference on the Peaceful Uses of Atomic Energy, Geneva 1955,* Vol. 10, United Nations Organization, Geneva, p. 475.
13. Abeles, R. H., Myers, C., and Smith, T. A. (1966). *Anal. Biochem.* **15**, 192.
14. Turner, M. K., and Mervyn, L. (1971). In *The Cobalamins. A Glaxo Symposium,* Arnstein, H. R. V., and Wrighton, R. J., Eds., Churchill-Livingstone, London, p. 35.

15. Yagiri, Y. (1967). *J. Vitaminol.* **13**, 197.
16. Cardinale, G. J., Dreyfus, P. M., Auld, P., and Abeles, R. H. (1969). *Arch. Biochem. Biophys.* **131**, 92.
17. Lindstrand, K., and Ståhlberg, K.-G. (1963). *Acta Med. Scand.* **174**, 665.
18. Toohey, J. I., and Barker, H. A. (1961). *J. Biol. Chem.* **236**, 560.
19. Lindstrand, K. (1964). *Nature* **204**, 188.
20. Ståhlberg, K.-G. (1964). *Scand. J. Haematol.* **1**, 220.
21. Linnell, J. C., Mackenzie, H. M., and Matthews, D. M. (1969). *J. Clin. Pathol.* **22**, 506.
22. Linnell, J. C., Mackenzie, H. M., Wilson, J., and Matthews, D. M. (1969). *J. Clin. Pathol.* **22**, 545.
23. Linnell, J. C., Hussein, H. A.-A., and Matthews, D. M. (1970). *J. Clin. Pathol.* **23**, 820.
24. Linnell, J. C., Hoffbrand, A. V., Hussein, H. A.-A., Wise, I. J., and Matthews, D. M. (1974). *Clin. Sci. Mol. Med.* **46**, 163.
25. Matthews, D. M., Gunasegaram, R., and Linnell, J. C. (1967). *J. Clin. Pathol.* **20**, 683.
26. Chanarin, I. (1969). *The Megaloblastic Anaemias,* Blackwell, Oxford, p. 56.
27. Adams, J. F., Tankel, H. I., and MacEwan, F. (1970). *Clin. Sci.* **39**, 107.
28. Linnell, J. C., Hoffbrand, A. V., Peters, T. J., and Matthews, D. M. (1971). *Clin. Sci.* **40**, 1.
29. Lindstrand, K., Wilson, J., and Matthews, D. M. (1966). *Brit. Med. J.* **2**, 988.
30. Linnell, J. C. (1972). *Scand. J. Clin. Lab. Invest.* **29**, *Suppl.* **126**, 6.6.
31. Craft, I. L., Matthews, D. M., and Linnell, J. C. (1971). *J. Clin. Pathol.* **24**, 449.
32. Biggs, J. C., Mason, S. L. A., and Spray, G. H. (1964). *Brit. J. Haematol.* **10**, 36.
33. Kelly, A., and Herbert, V. (1967). *Blood* **29**, 139.
34. Linnell, J. C., Hoffbrand, A. V., Hussein, H. A.-A., Matthews, D. M., and Wise, I. J. (1973). *Clin. Sci. Mol. Med.* **45**, 15P.
35. Goldstein, F., and Reiders, F. (1953). *Am. J. Physiol.* **173**, 287.
36. Thomas, J. W., and Anderson, B. B. (1956). *Brit. J. Haematol.* **2**, 41.
37. Kidd, H. M., and Thomas, J. W. (1962). *Brit. J. Haematol.* **8**, 64.
38. Cooke, W. T., Cox, E. V., Gaddie, R., Matthews, D. M., and Meynell, M. J. (1959). *J. Physiol.* **149**, 36P.
39. Okuda, K., Gräsbeck, R., and Chow, B. F. (1958). *J. Lab. Clin. Med.* **51**, 17.
40. Ardeman, S., Chanarin, I., and Berry, V. (1965). *Brit. J. Haematol.* **11**, 11.
41. Gräsbeck, R. (1960). *Adv. Clin. Chem.* **3**, 299.
42. Herbert, V., and Zalusky, R. (1961). *Fed. Proc.* **20**, 453.
43. Worm-Petersen, J. (1962). *Acta Neurol. Scand.* **38**, 241.
44. Kidd, H. M., Gould, C. E. G., and Thomas, J. W. (1963). *Can. Med. Assoc. J.* **88**, 876.
45. Basil, W., Brown, J .K., and Matthews, D. M. (1965). *J. Clin. Pathol.* **18**, 317.
46. Phillips, C. I., Ainley, R. G., van Peborgh, P., Watson-Williams, E. J., and Bottomley, A. C. (1968). *Nature* **217**, 67.

47. Collins, R. A., Harper, A. E., Schreiber, M., and Elvehjem, C. A. (1951). *J. Nutr.* **43**, 313.
48. Karlin, R. (1954). *C. R. Soc. Biol.* **148**, 371.
49. Peters, T. J., Linnell, J. C., Matthews, D. M., and Hoffbrand, A. V. (1971). *Brit. J. Haematol.* **20**, 299.
50. Strength, D. R., Alexander, W. F., and Wack, J. P. (1959). *Proc. Soc. Exp. Biol. Med.* **102**, 15.
51. Newman, G. E., O'Brien, J. R. P., Spray, G. H., and Witts, J. L. (1962). In *Vitamin B_{12} und Intrinsic Factor. 2. Europäisches Symposion, Hamburg,* Heinrich, H. C., Ed., Enke, Stuttgart, p. 424.
52. Wilson, K. A., Elliot, J. M., and Mathias, M. M. (1967). *J. Dairy Sci.* **50**, 1280.
53. Beck, W. S., Flavin, M., and Ochoa, S. (1957). *J. Biol. Chem.* **229**, 997.
54. Mudd, S. H., Finkelstein, J. D., Irreverre, F., and Laster, L. (1965). *J. Biol. Chem.* **240**, 4382.
55. Morrow, G., Barness, L. A., Cardinale, G. J., Abeles, R. H., and Flaks, J. G. (1969). *Proc. Natl. Acad. Sci. U. S. A.* **63**, 191.
56. Gaull, G. E., von Berg, W., Räihä, N. C. R., and Sturman, J. A. (1973). *Pediatr. Res.* **7**, 527.
57. Rappazzo, M. E., Salmi, H. A., and Hall, C. A. (1970). *Brit. J. Haematol.* **18**, 425.
58. Hsu, J. M., Kawin, B., Minor, P., and Mitchell, J. A. (1966). *Nature* **210**, 1264.
59. Räihä, N. C. R., and Schwartz, A. L. (1973). In *Inborn Errors of Metabolism,* Hommes, F. A., and van den Berg, C. J.. Eds., Academic Press, New York, p. 221.
60. Gaull, G. E., Räihä, N. C. R., and Sturman, J. A. (1972). *Pediatr. Res.* **6**, 538.
61. Dobbing, J. (1971). In *Handbook of Neurochemistry,* Vol. 6, Lajtha, Ed., Plenum Press, New York, Chapter 9.
62. Gaull, G. E. (1973). In *Inborn Errors of Metabolism,* Hommes, F. A., and van den Berg, C. J., Eds., Academic Press, New York, p. 133.
63. Herbert, V. (1968). *Am. J. Clin. Nutr.* **21**, 743.
64. Gräsbeck, R., Nyberg, W., and Reizenstein, P. (1958). *Proc. Soc. Exp. Biol. Med.* **97**, 780.
65. Schilling, R. F., and Meyer, O. O. (1964). *Trans. Assoc. Am. Physicians* **77**, 79.
66. Hoffbrand, A. V., Tripp, E., and Das, K. C. (1973). *Brit. J. Haematol.* **24**, 147.
67. Van der Weyden, M., Cooper ,M., and Firkin, B. G. (1973). *Blood* **41**,299.
68. Matthews, D. M., and Linnell, J. C. (1971). In *The Cobalamins. A Glaxo Symposium,* Arnstein, H. R. V., and Wrighton, R. J., Eds. Churchill-Livingstone, London, p. 23.
69. Hall, C. A. (1969). *Brit. J. Haematol.* **16**, 429.
70. Weissbach, H., and Taylor, R. T. (1968). *Vitam. Horm.* **26**, 395.
71. Mahoney, M. J., Rosenberg, L. E., Mudd, S. H., and Uhlendorf, B. W. (1971). *Biochem. Biophys. Res. Commun.* **44**, 375.
72. Lindstrand, K., Anderson, B. B., Cowan, J. D., Coates, M. E., and Hoffbrand, A. V. (1967). *Scand. J. Haematol.* **4**, 181.
73. Myasishcheva, N. V., Areshkina, L. Ya., Kutzeva, L. S. and Skorobogatova, E. P. (1968). *Voprosy Meditsinskoi Khimi* **14**, 273.

74. Hall, C. A., and Finkler, A. E. (1966). *Blood* **27**, 611.

75. Olesen, H., and Nexo, E. (1973). Personal communication.

76. Waxman, S., and Gilbert, H. S. (1973). *N. Engl. J. Med.* **289**, 1053.

77. Osler, W., and Gardner, W. (1877). *Can. Med. Surg. J.* **5**, 385.

78. Cabot, R. C. (1908). In *A System of Medicine,* Vol. 4, Osler, W., and McCrae, T., Eds., Hodder and Stoughton, London, Chapter 10.

79. Hamilton, H. E., Ellis, P. P., and Sheets, R. F. (1959). *Blood* **14**, 378.

80. Traquair, H. M. (1930). *Trans. Ophthalmol. Soc. U. K.* **50**, 351.

81. Victor, M. (1963). *Arch. Ophthalmol.* **70**, 73.

82. Foulds, W. S., Chisholm, I. A., Bronte-Stewart, J., and Wilson, T. (1969). In *Third Congress of the European Society of Ophthalmology, Amsterdam, 1968,* François, E. J., Ed., Karger, Basel.

83. Wilson, J., and Matthews, D. M. (1970). *Trans. Ophthalmol. Soc. U. K.* **90**, 733.

84. Wilson, J., Linnell, J. C., and Matthews, D. M. (1971). *Lancet* **1**, 259.

85. Montgomery, R. D. (1969). In *Toxic Constituents of Plant Foodstuffs,* Academic Press. New York, p. 144.

86. Moore, D. G. F. (1934). *Ann. Trop. Med. Parasitol.* **28**, 295.

87. Clark, A. (1936). *J. Trop. Med. Hyg.* **39**, 269.

88. Monekosso, G. L., and Wilson, J. (1966). *Lancet* **1**, 1062.

89. Osuntokun, B. O., Durowoju, J. E., McFarlane, H., and Wilson, J. (1968) *Brit. Med. J.* **2**, 647.

90. Osuntokun, B. O., Matthews, D. M., Hussein, H. A.-A., Wise, I. J., and Linnell, J. C. (1974). *Clin. Sci. Mol. Med.* **46**, 563.

91. Dillon, M. J., England, J. M., Gompertz, D., Goodey, P. A., Grant, D. B., Hussein, H. A.-A., Linnell, J. C., Matthews, D. M., Mudd, S. H., Newns, G. H., Seakins, J. W. T., Uhlendorf, B. W., and Wise, I. J. (1974). *Clin. Sci. Mol. Med.* **47**, 43.

92. Levy, H. L., Mudd, S. H., Schulman, J. D., Dreyfus, P. M., and Abeles, R. H. (1970). *Am. J. Med.* **48**, 390.

93. Goodman, S. I., Moe, P. G., Hammond, K. B., Mudd, S. H., and Uhlendorf, B. W. (1970). *Biochem. Med.* **4**, 500.

94. Mudd, S. H., Uhlendorf, B. W., Hinds, K. R., and Levy, H. L. (1970). *Biochem. Med.* **4**, 215.

95. Berlin, H., Berlin, R., and Brante, G. (1968). *Acta Med. Scand.* **184**, 247.

96. Cooperman, J. M., Luhby, A. L., Teller, D. N., and Marley, J. F. (1960). *J. Biol. Chem.* **235**, 191.

97. Adams, J. F. (1961). *Brit. Med. J.* **1**, 1735.

98. Adams, J. F. (1961). *J. Clin. Pathol.* **14**, 351.

99. Boddy, K., King, P., Mervyn, L., Macleod, A., and Adams, J. F. (1968). *Lancet* **2**, 710.

100. Ullberg, S., Kristofferson, H., Flodh, H., and Hanngren, A. (1967). *Arch. Int. Pharmacodynam. Ther.* **167**, 431.

101. Flodh, H. (1968). *Acta Radiol. Suppl.* **284**, 1.

102. Glass, G. B. J., and Laughton, R. W. (1958). *J. Lab. Clin. Med.* **52**, 875.

103. Reizenstein, P., Robertson, J. S., and Cronkite, E. P. (1961). In *Vitamin B_{12} und Intrinsic Factor. 2. Europäisches Symposion, Hamburg,* Heinrich, H. C., Ed., Enke, Stuttgart, p. 404.

104. Coates, M. E., Doran, B. M., and Harrison, G. F. (1961). In *Vitamin B_{12} und Intrinsic Factor. 2. Europäisches Symposion, Hamburg,* Heinrich, H. C., Ed., Enke, Stuttgart, p. 147.

105. Okuda, K., Yashima, K., Takara, I., Kitazaki, T., Kurashiga, M., and Takamatsu, M. (1969). *Vitamins (Japan)* **40**, 224.

106. Okuda, K., Yashima, K., Kitazaki, T., and Takara, I. (1973). *J. Lab. Clin. Med.* **81**, 557.

107. Gräsbeck, R., and Puutula, L. (1971). In *Cobalamins. A Glaxo Symposium,* Arnstein, H. R. V., and Wrighton, R. J., Eds., Churchill-Livingstone, London, p. 143.

108. Blomquist, L., Flodh, H., and Ullberg, S. (1969). *Experientia* **25**, 294.

109. Cooperman, J. M. (1972). *Cancer Res.* **32**, 167.

110. Reizenstein, P. (1967). *Blood* **29**, 494.

111. Cima, L., Levorato, C., and Mantovan, R. (1967). *J. Phar. Pharmacol.* **19**, 32.

112. Yagiri, Y. (1967). *J. Vitaminol.* **13**, 228.

113. Mahler, H. R., and Cordes, E. H. (1966). *Biological Chemistry,* Harper & Row, New York, p. 224.

114. White, J. C., Girolamo, M. D., Fu, M. L., Preston, Y. A., and Bradbeer, C. (1973). *J. Biol. Chem.* **248**, 3978.

115. Kerwar, S. S., Spears, C., McAuslan, B., and Weissbach, H. (1971). *Arch. Biochem. Biophys.* **142**, 231.

116. Wickramasinghe, S. N., and Carmel, R. (1972). *Brit. J. Haematol.* **23**, 307.

117. Mahoney, M. J., and Rosenberg, L. E. (1971). *J. Lab. Clin. Med.* **78**, 302.

118. Mudd, S. H., Uhlendorf, B. W., Freeman, J. M., Finkelstein, J. D., and Shih, V. E. (1972). *Biochem. Biophys. Res. Commun.* **46**, 905.

119. Scott, J. M., Bloomfield, F. J., Stebbins, R., and Herbert, V. (1974). *J. Clin. Invest.* **53**, 228.

120. Register, U. D., and Sarett, H. P. (1951). *Proc. Soc. Exp. Biol. Med.* **77**, 837.

121. Mollin, D. L., and Ross, G. I. M. (1952). *J. Clin. Pathol.* **5**, 129.

122. Heinrich, H. C. (1964). *Sem. Haematol.* **1**, 199.

123. Linnell, J. C., Smith, A. D. M., Smith, C. L., Wilson, J., and Matthews, D. M. (1968). *Brit. Med. J.* **2**, 215.

124. Sokoloff, M. F., Sanneman, E. H., and Beard, M. F. (1952). *Blood* **7**, 243.

125. Adams, J. F. (1964). *J. Clin. Pathol.* **17**, 31.

126. Edwards, T. L., Clason, W. P. C., and Reinfrank, R. F. (1962). *Am. J. Med. Sci.* **244**, 587.

127. Adams, J. F., and Boddy, K. (1968). *Lancet* **2**, 328.

128. Matthews, D. M., and Beckett, A. G. (1962). *J. Clin. Pathol.* **15**, 456.

129. Kennedy, E. H., and Adams, J. F. (1965). *Clin. Sci.* **29**, 417.

130. Okuda, K., Gräsbeck, R., and Chow, B. F. (1962). In *Vitamin B_{12} und Intrinsic Factor. 2. Europäisches Symposion, Hamburg,* Heinrich, H. C., Ed., Enke, Stuttgart, p. 17.

131. Halstead, J., Lewis, P. M., Hvolboll, E. E., Gasster, M., and Swendseid, M. E. (1956). *J. Lab. Clin. Med.* **48**, 92.

132. Wider, S., Wider, J. A., and Reinecke, E. P. (1958). *Proc. Soc. Exp. Biol. Med.* **98**, 180.

133. Booth, M. A., and Spray, G. H. (1960). *Brit. J. Haematol.* **6**, 288.

134. Loehry, G. A., and Creamer, B. (1969). *Gut* **10**, 662.

135. Quadros, E. V., Wise, I. J., Matthews, D. M., and Linnell, J. C. (1975). *Clin. Sci. Molec. Med.* **48**, 4P-5P.

CHAPTER SEVEN

MECHANISMS OF MALABSORPTION OF COBALAMIN

ROBERT M. DONALDSON, JR., M.D.

Professor of Medicine
Yale University School of Medicine
New Haven, Connecticut

CONTENTS

INTRODUCTION

As described in detail in Chapter 5, the normal process of cobalamin absorption can readily be divided into individual events occurring in sequence (1). After its "release" from dietary protein within the lumen of the stomach, the free vitamin binds rapidly and tightly to gastric intrinsic factor (IF) to form a stable macromolecular intrinsic factor-cobalamin complex (IF•Cbl). This complex is remarkably resistant to acid and digestive enzymes and apparently remains intact as it passes along the small-bowel wall until it reaches the distal ileum. Here the IF•Cbl complex attaches to specific receptor sites located on the outer surface of the plasma membrane lining the microvilli of ileal absorptive cells. Once this attachment has occurred, the subsequent fate of the IF•Cbl complex and the means whereby the vitamin traverses the ileal cell are not at all clear. There is general agreement, however, that transport of the vitamin across the ileal mucosa is an unusually slow process, that this process requires cellular energy, and that the vitamin is no longer associated with IF when it emerges from the ileal cell. Newly absorbed cobalamin is removed from the intestine by way of the portal blood where it is bound to a specific serum transport protein, transcobalamin II. This chapter is concerned with the various ways in which this normal sequence of events may be interrupted with consequent impaired absorption and ultimate deficiency of cobalamin. Indeed, it is important to recognize that present understanding of normal cobalamin absorption has evolved largely from investigations of patients with malabsorption of the vitamin.

COBALAMIN MALABSORPTION RELATED TO INTRAGASTRIC EVENTS

Impaired Gastric Secretion of Acid-Pepsin

Absorption of cobalamin is usually tested with crystalline cyanocobalamin whereas ordinarily the vitamin is ingested bound to animal protein. The precise nature of the interaction between the vitamin and dietary protein is unknown, although it has been shown that release of the vitamin from food is facilitated by acid (2), pepsin (2, 3), and other proteolytic enzymes (3). One might expect, therefore, that patients unable to secrete adequate quantities of acid and pepsin would absorb dietary cobalamin poorly even if they were able to secrete sufficient IF to absorb crystalline cyanocobalamin normally. Low serum levels of the vitamin together with normal absorption of crystalline cyanocobalamin have in fact been reported in patients with partial gastric resection (4) or simple achlorhydria without pernicious anemia (5). Furthermore, such patients absorb radioactively labeled vitamin which has been incorporated into egg protein poorly when compared to their ability to absorb crystalline cyanocobalamin (6).

Thus, malabsorption of dietary cobalamin can occur in patients with impaired acid-pepsin secretion because of inadequate release of the vitamin from egg protein. However, several observations tend to cast doubt on the importance of this mechanism. First, those patients with partial gastric resection who develop overt manifestations of cobalamin deficiency almost always lack IF and absorb crystalline cyanocobalamin poorly (7, 8). Moreover, the high affinity of cobalamin for IF allows transfer of the vitamin from meat protein to IF in the absence of acid or pepsin (1). In addition, patients with pernicious anemia who are fed exogenous IF absorb cobalamin from liver as efficiently as they absorb crystalline cyanocobalamin even though these patients lack acid-pepsin (9). Finally, there is no firm evidence that acid-pepsin is required for the absorption of cobalamin incorporated into meat protein (10), and apparently acid-pepsin does not even enhance absorption of cobalamin incorporated into egg protein when the eggs are simply homogenized (11) rather than homogenized and cooked (6) before they are fed.

Intrinsic Factor Deficiency

Intrinsic factor deficiency is the most common cause for cobalamin malabsorption and may result from atrophy of the glandular mucosa of the stomach, from a selective defect in gastric secretion of IF, from the secretion of an abnormal IF which fails to promote cobalamin absorption, or from surgical resection of the stomach. Thus, there are several ways in which a patient may develop IF deficiency, but in this chapter, the term "pernicious anemia" will be applied only to those patients who fail to secrete both acid-pepsin and intrinsic factor because of severe atrophy of the glandular mucosa of the fundus and body of the stomach. Deficiency of IF is clearly the cause and not the result of cobalamin malabsorption and subsequent deficiency of the vitamin. Lack of IF precedes the hematologic and neurologic consequences of cobalamin deficiency by several years (12, 13), and prolonged continuous therapy with cobalamin corrects neither gastric mucosal atrophy nor defective intrinsic factor secretion (13).

Pernicious Anemia

The fascinating history of this once mysterious and formidable disorder is detailed in the introduction to this volume, and has been previously recorded elsewhere (14, 15), so that only a brief summary is needed here.

It was only 15 years after Addison's initial account of what he called "idiopathic anaemia" in 1855 (16) that Fenwick (17) first described gastric mucosal atrophy encountered at the autopsy of a patient with this disease. Subsequent studies soon showed that atrophy of the glandular mucosa accompanied by a failure of the stomach to secrete acid-pepsin was a regular feature of pernicious anemia. A pathogenetic link between gastric secretory failure and the development of anemia was not recognized, however, until Castle (18) investigated the problem in 1929. The earlier demonstration by Minot and Murphy (19) that liver but not meat induced a brisk reticulocytosis and subsequent remission in patients with pernicious anemia led Castle to show that meat also induced a remission when the meat was fed together with normal human gastric juice. From his careful observations, Castle was able to conclude that an "extrinsic factor" present in meat combined with an "intrinsic factor" in gastric juice to form the "hemopoietic factor" necessary to induce a remission in pernicious anemia. Patients with the disease lacked the "intrinsic factor." When it was subsequently shown that cobalamin was in fact both the "extrinsic factor" as well as "hemopoietic factor," it became apparent that Castle's gastric intrinsic factor was necessary for efficient intestinal absorption of this vitamin, a fact which was clearly documented (20) once radioactively labeled cyanocobalamin became available.

The gastric mucosal atrophy characteristic of pernicious anemia is a diffuse lesion that involves the entire body and fundus of the stomach; in contrast, the antral mucosa appears remarkably normal (15). As shown in Figure 7-1, the atrophic mucosa is extremely thin because of extensive loss of gastric glands containing parietal and chief cells. On the other hand, the mucus cells lining the gastric pits and the luminal surface appear quite normal, although the surface epithelium may undergo intestinal metaplasia (21, 22) to a variable extent. Large numbers of lymphocytes, plasma cells, and eosinophils are regularly present in the lamina propria. It is difficult to separate in any arbitrary way the complete, or nearly complete, gastric atrophy that is regularly seen in patients with cobalamin malabsorption from the much more common atrophic gastritis associated with advancing age and a variety of chronic illnesses (15, 22, 23). In atrophic gastritis the lesion tends to be patchy rather than diffuse and frequently involves the gastric antrum (23, 24). Gastric secretion of acid and pepsin is diminished or absent, but sufficient IF is secreted to maintain cobalamin absorption.

Since pernicious anemia is a disorder which occurs with increasing frequency with advancing age, one is tempted to believe that the gastric atrophy of pernicious anemia merely reflects the end stage of a spectrum of pathology which begins as mild inflammatory changes in the glandular

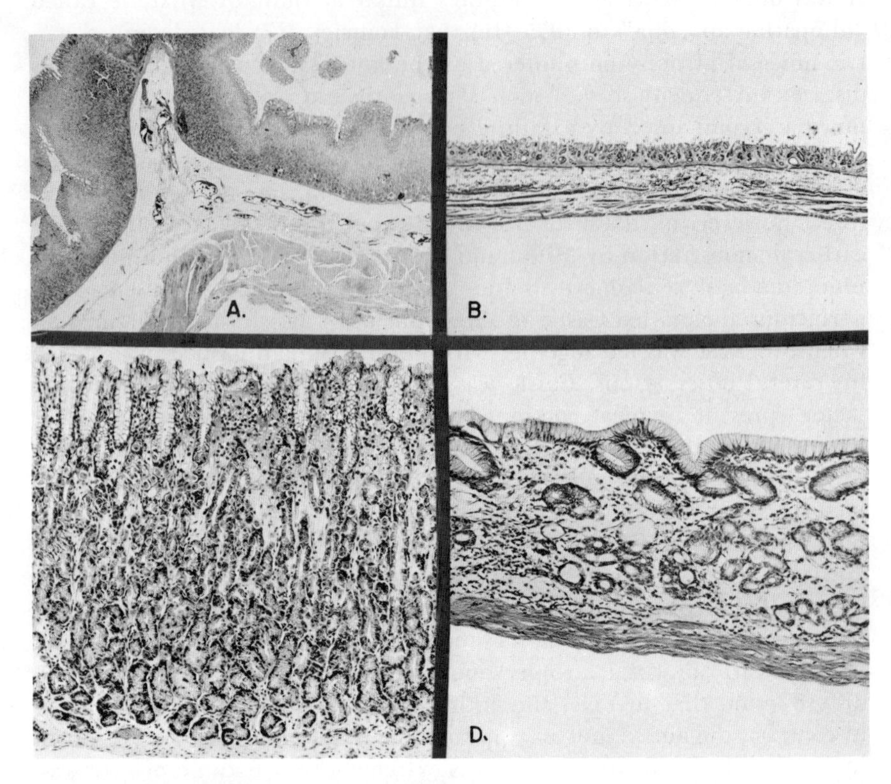

Fig. 7-1 The gastric atrophy of adult pernicious anemia. (*A*) *Sections* through the body of a normal stomach (X9). (*B*) Section through the body of the stomach of a patient with Addisonian pernicious anemia (X9). (*C*) Normal gastric mucosa (X 102). (*D*) Gastric mucosa with pernicious anemia (X 138). [This photograph was kindly provided by the widow of the late Prof. H. A. Magnus of Kings College Hospital Medical School.]

mucosa. This notion is consistent with, but not directly supported by, the observation (25) that patients generally fall into the following categories with respect to secretion of acid, pepsin, and IF: all three secreted, only pepsin and IF secreted, only IF secreted, or none of the three secreted. Since the parietal cells normally secrete much more IF than is needed for the absorption of physiological quantities of cobalamin (26, 27), total or nearly total loss of parietal cells is required before cobalamin absorption is significantly impaired (22, 23). Even then, 3 to 5 years are needed for depletion of hepatic stores of cobalamin before deficiency of the vitamin manifests itself (12). Nevertheless, the concept that pernicious anemia results from the slow, relentless progression of relatively mild atrophic gastritis still lacks definitive proof. Clearly needed are long-term continuous observations which convincingly document such a progression in a high-risk group of patients (15). It should be noted that many patients with simple atrophic gastritis have been carefully observed for as long as 20 to 30 years without evidence of subsequent development of total gastric atrophy and IF deficiency (15, 24, 28). Indeed, as discussed below, clinical and laboratory findings have led some investigators to conclude that pernicious anemia and atrophic gastritis are two separate and distinct entities.

Although much more common in middle-aged and elderly individuals, pernicious anemia has also been described in a few children (29). These patients are usually afflicted relatively late in childhood and have all of the characteristic features of adult pernicious anemia including gastric mucosal atrophy, total achlorhydria, and severe IF deficiency. The term "juvenile pernicious anemia" should be reserved for these patients rather than loosely applied to all children who absorb cobalamin poorly for whatever reason (13). It is of interest that the gastric antibodies discussed below have regularly been detected in the serum of these children with pernicious anemia and that abnormalities of thyroid, parathyroid, or adrenal function are commonly present (30).

The fundamental cause of total gastric atrophy and concomitant IF deficiency remains unkown. Castle (13) has accurately and succinctly summarized current understanding. No exogenous dietary or infectious agent or agents have been convincingly implicated. On the other hand, many observations emphasize an important role for constitutional, perhaps hereditary factors. Although pernicious anemia occurs throughout the world, the disorder appears to occur most frequently in temperate climates and to afflict chiefly Caucasians (15). The differences in prevalence rates are too great to be explained by differences in diagnostic accuracy. Pernicious anemia also develops in patients with early graying of the hair with sufficient frequency to suggest the influence of constitutional factors. In addition, the prevalence of gastric atrophy is distinctly increased among relatives of patients with pernicious anemia (15, 31–33) and is particularly common among relatives of patients with juvenile pernicious anemia (15, 29). These observations all support the concept of a constitutional, perhaps hereditary, predisposition to this disease. On the other hand, the finding that pernicious anemia can develop in one identical twin but not in the other (34) strongly suggests that a simple genetic mechanism cannot be the only constitutional factor involved.

Jeffries (30) has recently summarized the evidence which suggests that immunological phenomena may in some way be responsible for an individual's susceptibility to the development of pernicious anemia. It is now generally recognized that, as first described by Schwartz (35) and by Taylor (36), sera obtained from the majority of patients with pernicious anemia inhibit the biological activity of IF. Subsequent work has shown that inhibition results from the action of at least two distinct antibodies (37). Antibody I or "blocking antibody" prevents the formation of IF•Cbl complex by blocking the attachment of cobalamin to IF (38, 39). Antibody II or "binding antibody" attaches to the IF•Cbl complex (37, 40) and prevents its uptake by the ileal mucosa (41, 42). Thus blocking antibody interacts with those antigenic sites on IF concerned with the binding of cobalamin to IF, whereas binding antibody acts at distinctly different sites which are concerned with the attachment of IF•Cbl complex to the absorptive surface of the ileum. Antibody I is present in the sera of 60 to 75% of patients with pernicious anemia (30) and the frequency with which this antibody is found has increased with increasing sensitivity of the methods used to detect it (43). In contrast, type II antibody is found in only about 30% of patients with pernicious anemia (44), usually occurs in those patients who also have high titers of antibody I, but has also been described in serum apparently devoid of antibody I (37, 45). With very few exceptions (33, 46–48) these anti-IF antibodies have been described only in patients with IF deficiency due to gastric mucosal atrophy. Both the blocking and the binding antibodies belong to the class of IgG immunoglobulins when found in the serum (37). Antibodies with binding and blocking activity against IF have also been demonstrated in gastric juice (49, 50) and saliva (51) of patients with pernicious anemia. These secreted antibodies may be either IgG (49) or IgA (51) immunoglobulins. A complete secretory IgA antibody containing secretory piece and directed against IF has been identified in the gastric juice of one patient with pernicious anemia (50).

In addition to these anti-IF antibodies, the sera of nearly all patients with pernicious anemia contain organ-specific IgG or IgA immunoglobulins which bind to the cytoplasm of gastric parietal cells when tested by indirect immunofluorescence (52) and which react with gastric mucosal microsomes when examined by complement fixation (53). Unlike anti-IF antibodies, which are highly specific for patients with IF deficiency secondary to gastric atrophy, antiparietal cells antibodies are found not only in pernicious anemia but also in approximately 10% of healthy elderly subjects and in a significantly higher percentage patients with iron deficiency anemia, thyroid disease, Addison's disease, or diabetes mellitus (see Table 7-1). Furthermore, all of these disorders occur more frequently in patients with pernicious anemia than in the general population, and pernicious anemia occurs with increased frequency in each of these disorders (15, 30). The relation between pernicious anemia and thyroid disease is particularly interesting (30, 54). The two diseases occur in the same patient more frequently than should be expected, and the serum of patients with pernicious anemia frequently contains antibodies directed against thyroid tissue while patients with thyroid disease often have antibodies against gastric antigens.

Table 7-1 Percentage of Patients with Demonstrable Circulating Antibodies

	Parietal Cell Antibody	Intrinsic Factor Antibody	Thyroid Antibodies
Healthy controls	5–11	—	3–13
Adult pernicious anemia	76–91	33–70	27–47
Iron deficiency	17–18	—	—
Thyroid disease	24–33	3–4	34–99
Addison's disease	38	—	28

Adapted from data summarized by Jefferies (30).

Not only are circulating antibodies against gastric antigens frequently present in pernicous anemia, but there are also indications that cell-mediated immune mechanisms may be operative in this disease. Lymphocytic infiltration, a cardinal manifestation of delayed hypersensitivity, is a prominent feature of the gastric mucosal lesion. In patients with pernicious anemia associated with hypogammaglobulinemia, circulating antigastric antibodies are absent, but the capacity to develop delayed hypersensitivity appears to be intact (55). Furthermore, various gastric antigens stimulate lymphocytic mitosis (56) and inhibit leukocyte migration (57, 58) in most, but by no means all, patients with pernicious anemia. It remains to be determined, however, whether pernicious anemia lymphocytes when stimulated by gastric antigens can actually damage gastric mucosa or whether the phenomena observed to date are merely a result of the disease.

Further impetus for considering pernicious anemia to be an immunological disorder comes from studies of the effects of adrenocortical steroids. Prolonged treatment with these agents induces regeneration of gastric parietal cells and increased secretion of IF (59). The mechanism whereby steroids cause these effects has not been defined, however, and the observed improvement in gastric mucosal function does not necessarily result from suppression of an immunological process.

Most investigations designed to assess the role of immune phenomena in pernicious anemia have been primarily concerned with the circulating antibodies against gastric antigens which are so frequently observed in patients with this disease. Two questions need to be considered in some detail: Do these antibodies in fact cause gastric atrophy characteristic of pernicious anemia, and do the anti-IF antibodies themselves directly impair cobalamin absorption? Although the role of anti-IF and anti-parietal cell antibodies in the pathogenesis of gastric mucosal atrophy is far from clear, most of the available evidence suggests that these antibodies are a consequence rather than a cause of pernicious anemia. In many patients with well-documented pernicious anemia, anti-IF antibodies cannot be detected with currently available techniques (60, 61) and these antibodies have occasionally been observed in patients who do not have pernicious anemia (58, 62). Furthermore, antibodies against IF can cross the placental barrier without causing permanent damage to fetal gastric mucosa (63). Particularly obvious is the fact that parietal cell antibodies frequently occur in the absence of gastric atrophy and IF deficiency. Twomey et al. (55) have described a group of adults with hypogammaglobulinemia who have all of the clinical and laboratory features of pernicious anemia, yet lack both antiparietal-cell and anti-IF antibodies. Certainly it would be difficult to attribute gastric atrophy to a disturbance of humoral immunity in these patients.

In spite of evidence which argues against a primary role for circulating antibodies, one must also recognize that most patients with idiopathic atrophic gastritis who lack antiparietal cell and anti-IF antibodies do not develop clinically significant IF deficiency associated with total gastric atrophy even when followed for many years (22, 24, 64). Moreover, patients with documented gastric mucosal injury resulting from aspirin, alcohol, surgery, radiation, or freezing do not develop circulating parietal cell antibodies (30, 65). In addition, it is now clear that antibodies directed against IF or parietal cells are frequently observed among apparently healthy members of families of patients with pernicious anemia (66). Long-term studies are obviously needed to determine the extent to which frank IF deficiency ultimately develops in these potentially susceptible individuals. In any event, it seems quite likely that there are two separate and distinct kinds of gastric mucosal disease, that which tends to progress to total atrophy of glandular mucosa and that which does not. The role of circulating antibodies in this progression remains to be defined. Although antigastric antibodies certainly cannot be implicated in the pathogenesis of all cases of pernicious anemia, it remains possible that there may be more than one etiology for the total gastric atrophy characteristic of this disorder. Moreover, while there appears to be no evidence that circulating antibodies against IF or parietal cells directly damage gastric mucosal function, it is equally true that this possibility has not yet been convincingly tested.

Another question which has received considerable attention is whether antibodies directed against IF contribute in any meaningful way to cobalamin malabsorption in patients with pernicious anemia. When placed directly in the stomach of these patients, serum containing high titers of anti-IF antibodies is certainly capable of preventing the usual increase in cobalamin absorption induced by exogenous human or hog IF (35, 45). Whether the mere presence of these antibodies in the serum impedes the action of exogenous IF is another matter, however. When pernicious anemia patients with and without detectable anti-IF were compared, no differences in the absorption response to exogenous IF could be detected (67). Even when antibodies were purposely induced by subcutaneous immunization of pernicious anemia patients with hog IF, there was no decrease in the cobalamin absorption mediated by exogenous hog IF (68).

As indicated above, however, antibodies against IF are not only present in the serum but are also secreted into saliva (51) and gastric juice (49, 50). The possibility must therefore be considered that patients with pernicious anemia may secrete sufficient IF to promote cobalamin absorption, but that this IF is inactivated by secreted antibodies against IF. It is clear that secreted antibody is capable of blocking IF action, and in at least one patient with pernicious anemia the presence of binding (type II) antibody in gastric secretions prevented any IF-mediated increase in cobalamin absorption (49). In another group of patients (69), it was possible to show that the presence of antibody in gastric juice was clearly associated with a diminished response to hog IF, although the action of hog IF was not completely blocked by secreted antibodies in any of these patients. Minute amounts of IF can in fact be detected in the gastric juice of some patients with pernicious anemia, and when gastric juice from patients with this disorder is treated with acid to dissociate antigen-antibody complexes, the amount of immunologically detectable IF substantially increases (70, 71). The amount of IF presumably released from antibody in this way is of course very small when compared to the quantity of IF required to sustain normal cobalamin absorption (27). Nevertheless, these studies are consistent with the notion that secretion of anti-IF antibodies into the intestinal lumen impairs, at least to some extent, IF function and consequent cobalamin absorption. Further, but much less direct, support for this concept comes from the observation (72) that human IF promotes cobalamin absorption more effectively than hog IF in patients with total gastrectomy, while the reverse is true in patients with pernicious anemia. One interpretation of these findings would implicate antibodies secreted into the gut by pernicious anemia patients. Such antibodies would presumably have a higher affinity for human IF than for hog IF and would thus block the action of human IF more effectively. In patients with total gastrectomy, however, anti-IF antibody would not be present and the exogenous IF derived from humans would be more effective than IF derived from a heterologous species. Thus it appears that anti-IF antibodies secreted into the gastrointestinal lumen may directly interfere with IF function and further impair cobalamin absorption in patients with pernicious anemia. Except in unusual patients (49), however, this effect seems to be quantitatively small, and a deficiency of secreted IF still appears to be a much more important cause of cobalamin malabsorption in pernicious anemia than does inactivation of IF by secreted antibodies.

Under ordinary circumstances, then, exogenous IF substantially enhances cobalamin absorption in patients with pernicious anemia who lack this essential substance. Occasionally, however, IF may fail to increase absorption in these patients. First, as discussed above, the patient may secrete into his gastrointestinal tract sufficient quantities of anti-IF antibody to inactivate the administered IF. A pernicious anemia patient may also fail to respond to exogenous IF because of concomitant small-bowel bacterial overgrowth (73). The presence of large numbers of bacteria in the lumen of the small intestine is favored by achlorhydria (74), and occasionally IF will not promote cobalamin absorption in patients with pernicious anemia until enteric microorganisms are suppressed by broad spectrum antibiotics (75). Finally, if the patient with pernicious anemia has moderate or severe cobalamin deficiency at the time he is tested, IF may fail to increase absorption of the vitamin until cobalamin stores are repleted (76, 77). Presumably the intestinal mucosal lesion found in some patients who are depleted of cobalamin (78) prevents normal absorption of the vitamin even in the presence of exogenous IF.

Congenital Intrinsic Factor Deficiency

Deficiency of cobalamin may develop in infants or young children who specifically lack the capacity to secrete IF even though other aspects of gastric function appear to be normal (29, 79). These children resemble those with juvenile pernicious anemia in that normal human gastric juice or hog IF increases their absorption of cobalamin whereas their own gastric juice fails to promote cobalamin absorption in patients with pernicious anemia or total gastrectomy. Nevertheless, there are distinct differences. Unlike patients with juvenile pernicious anemia, patients with congenital IF deficiency secrete adequate quantities of acid-pepsin, and gastric mucosal histology appears to be normal (80). Although present in large numbers, gastric parietal cells in these patients appear totally incapable of producing normal IF. Cobalamin deficiency develops much earlier, usually before the age of 2 to 3 years, in children with congenital IF deficiency than in those with juvenile pernicious anemia, who usually do not become ill until they are 7 to 12 years old (29). Presumably the time required for the latter to develop gastric mucosal atrophy is responsible for this difference.

Patients with congenital IF deficiency have been followed for as long as 29 years without developing the gastric atrophy or total achlorhydria that is characteristic of acquired pernicious anemia (81). Furthermore, the serum antibodies against IF and gastric parietal cells regularly observed in patients with juvenile pernicious anemia are not found in patients with congenital lack of IF, nor do the latter seem to have the predisposition for endocrine disorders commonly found in patients with juvenile pernicious anemia (29, 81). Thus, juvenile pernicious anemia seems to be a more general disorder in which the essential defect is an atrophic gastric mucosa incapable of secreting adequate gastric juice, whereas in congenital IF deficiency an isolated, perhaps enzymatic, defect results in a specific failure to secrete IF. Although apparently distinct, these two disorders may nevertheless be interrelated in some way. Both early graying of hair and overt pernicious anemia have been reported in relatives of patients with congenital IF deficiency (29, 80, 81).

Secretion of Abnormal Intrinsic Factor

In 1972, Katz et al. (82) described a child who developed frank cobalamin deficiency at age 13. This child's gastric juice contained acid and normal quantities of immunoreactive IF, and his gastric mucosal biopsy was normal. In spite of the presence of adequate quantities of a cobalamin binding substance which reacted normally with anti-IF antibodies, the patient failed to absorb radioactive cyanocobalamin. Exogenous IF corrected the patient's malabsorption of cobalamin, and the patient's gastric juice failed to correct malabsorption in a subject with total gastrectomy. Thus, this child secreted an IF that bound cobalamin normally and reacted with anti-IF antibodies. Nevertheless, this IF was biologically inert in that it failed to promote cobalamin absorption either in the patient himself or in another subject who lacked IF. This abnormal IF also failed to enhance cobalamin uptake of homogenates of guinea pig ileal mucosa.

Although the patient's parents were first cousins, there was no family history of acquired pernicious anemia. Furthermore, neither the patient nor the members of his immediate family had circulating anti-IF antibodies. Subsequent studies (83), however, showed that both parents and the patient's only sibling all secreted a mixture of approximately equal amounts of normal and biologically inert IF. Although abnormal IF could be identified in the gastric juice of these relatives, they also secreted sufficient normal IF to sustain adequate cobalamin absorption. Highly purified preparations of both normal and abnormal IF have now been isolated from the gastric juice of the patient and his relatives by affinity chromatography (83). Detailed analyses of these preparations showed that biologically inert IF is indistinguishable from normal IF in almost all respects including affinity for cobalamin, reactions with binding and blocking antibodies present in pernicious anemia serum, molecular size, weight, and charge as well as the content of amino acids and amino sugars. Nevertheless, purified biologically inert IF failed to promote cobalamin absorption in a totally gastrectomized subject, failed to enhance uptake of the vitamin by guinea pig mucosa homogenates, and contained antigenic sites distinctly different from those of normal IF.

Thus far, an abnormal biologically inert IF has been reported in only one family. Nevertheless, it is possible that some patients previously labeled as congenital IF deficiency may in fact have been secretors of biologically inert IF. This cannot be true of all such patients, however, since at least some patients with congenital IF deficiency clearly lack demonstrable immunoreactive IF in their gastric juice (29, 81). In any event, the discovery by Katz and his colleagues of a biologically inert IF provides the strongest evidence yet available that the cobalamin binding and the absorption-promoting properties of IF are distinct and are mediated by separate sites on the IF molecule.

Gastric Resection

Total gastrectomy in humans (12, 84) or in rats (85) regularly leads to malabsorption of cobalamin which is correctable by exogenous IF. This fact warrants little discussion except to point out that total gastrectomy provides unequivocal identification of the stomach as the source of IF and allows the time required for the development of cobalamin deficiency to be established with certainty. The situation is somewhat more complex in patients subjected to partial gastric resection. As discussed above, such patients frequently develop low serum levels of cobalamin even though they appear to absorb crystalline cyanocobalamin normally. Nevertheless, only 15 to 30% of patients who develop anemia after partial gastric resection have overt cobalamin deficiency, and malabsorption of crystalline cyanocobalamin due to lack of IF is regularly present in such patients (7, 8). Thus, lack of acid-pepsin may play some role in depletion of cobalamin stores in patients with partial gastric resection, but clinically important cobalamin deficiency develops only in those patients who lack quantities of IF sufficient to maintain adequate absorption of the vitamin.

The minimal amount of IF required to bring about physiological absorption of the vitamin in patients with pernicious anemia or total gastrectomy has been established, and it is apparent that healthy individuals ordinarily secrete much more IF than is actually needed (26). Patients with a markedly reduced capacity to secrete IF may at some times produce the required amount of IF, while at other times they may be IF deficient. This situation leads to a marked variability in the results of Schilling tests performed at different times in the same gastrectomized patients (86). In addition, the quantity of IF secreted by patients with partial gastrectomy may vary depending on several factors. Obviously, reduction in IF secretion will be proportional to the amount of glandular tissue resected, and patients with resections of the proximal half of the stomach are particularly likely to develop clinically significant cobalamin malabsorption (87). Gastritis with consequent atrophic changes frequently develops in the residual gastric pouch following partial gastrectomy (88), and the extent to which this occurs will also determine the quantity of IF secreted. Patients operated on for gastric ulcer appear to become IF deficient more frequently than do those operated on for duodenal ulcer (7, 8, 15), possibly as a consequence of the greater tendency for patients with gastric ulcer to develop atrophic gastritis (89, 90). Finally, there are two mechanisms whereby cobalamin malabsorption may develop after partial gastric resection even though the patient continues to secrete adequate IF. First, small-bowel stasis resulting from a poorly functioning anastomosis may lead to bacterial overgrowth in the small-bowel lumen which in turn causes a cobalamin malabsorption that does not respond to IF administration. Second, an occasional patient may develop cobalamin malabsorption corrected only by the administration of normal intestinal juice (91). It seems likely that this phenomenon is related to pancreatic insufficiency (see below). In view of the multiple factors operative in patients subjected to partial gastric resection, it is not at all surprising that the proportion of such individuals who become frankly deficient in cobalamin varies considerably from one reported series to another (7, 8, 15, 92) and that the time required for the development of cobalamin deficiency varies greatly from one individual to another (15).

COBALAMIN MALABSORPTION RELATED TO EVENTS IN THE SMALL-BOWEL LUMEN

Small-Bowel Bacterial Overgrowth

Cobalamin malabsorption regularly accompanies marked proliferation of bacteria within the small-bowel lumen. In fact, the most consistent feature of clinically significant small-bowel bacterial overgrowth in patients and experimental animals is abnormal cobalamin absorption which is not corrected by IF (93, 94). In contrast, treatment with antibiotics or appropriate surgery for the cause of bacterial overgrowth regularly corrects the absorptive defect. Since the most effective mechanism for limiting the numbers of microorganisms present in the small-bowel lumen is the mechanical cleansing action of normal peristalsis (93), any small-bowel abnormality conducive to local stasis or recirculation of small-bowel contents is likely to be accompanied by marked intraluminal proliferation of bacteria. Such abnormalities occur most frequently after abdominal surgery and include "self-filling" or "blind" pouches of intestine, enteroenterostomies, afferent loop dysfunction following gastrojejunostomy, and partial intestinal obstruction due to adhesions. Stasis may also be seen in patients with small-bowel strictures due to regional enteritis or impaired intestinal motility resulting from scleroderma. Local stasis can also be produced surgically in experimental animals.

Whatever abnormality may be the cause of small-bowel stasis the result is the same: intraluminal proliferation of a large number of different bacterial species. Although bacteroides and anaerobic lactobacilli predominate, coliforms, enterococci, clostridia, and other species may be present. Characteristically the flora under these circumstances tends to be extremely complex and to resemble colonic flora, and investigations accomplished with adequate microbiologic techniques consistently report 8 to 20 different bacterial species in the small-bowel contents of patients and experimental animals with this "blind loop syndrome" (94–96).

The mechanism whereby bacteria proliferating within the small-bowel lumen impair the absorption of cobalamin has been the subject of many investigations. IF deficiency does not play a role since absorption is not improved by administration of IF and since the gastric juice of patients and experimental animals contains adequate quantities of IF (97, 98). Although a patchy, mild lesion of the small-bowel mucosa may be present in some patients with small-bowel bacterial overgrowth (99), it seems unlikely that the mucosal abnormality is sufficiently severe to account for the observed absorption abnormality. Furthermore, *in vivo* uptake of IF-bound cobalamin by the ileum of rats bearing blind pouches is not impaired if the vitamin does not come into direct contact with luminal bacteria (97), and *in vitro* uptake of the vitamin by ileal brush borders prepared from blind-loop rats is appropriately stimulated by IF (100). Thus there is no convincing evidence that the presence of bacteria in the small-bowel lumen directly impairs the processes normally involved in cobalamin absorption.

On the other hand, considerable evidence has now accumulated to suggest that cobalamin is poorly absorbed because bacteria proliferating within the small-bowel lumen take up the vitamin and thereby prevent its absorption by the intestine. Several different intestinal bacterial species, including those isolated from patients and animals with the blind-loop syndrome, are capable of removing cobalamin from the incubation medium (101–103). Bacterial uptake of the vitamin (102) appears to occur in two stages. First, there occurs a rapid attachment of the vitamin to the surface of the bacteria which is not temperature dependent and which does not require metabolic energy. The affinity of cobalamin for bacterial binding sites appears to be similar to the affinity of the vitamin for IF (102). The second stage occurs more slowly, requires metabolic energy, and probably represents transfer of the vitamin from the bacterial surface to the interior. Although bacterial uptake of cobalamin is considerably diminished when the vitamin is bound to IF (101, 102), substantial uptake of the IF•Cbl complex does occur particularly when large numbers of microorganisms are present and are in the log phase of growth and when sufficient time is allowed for bacterial uptake to proceed (101). Finally, it should be pointed out that both in experimental animals (97) and in patients (94, 103) with small-bowel bacterial growth, uptake of IF-bound cobalamin by intraluminal microorganisms has been documented. Such observations provide direct support for the notion that bacteria proliferating in the small-bowel lumen successfully compete with the ileal absorptive cell for ingested cobalamin.

Enteric microorganisms also synthesize cobalamin and thus are a rich source of this nutrient. Since the vitamin is not released by viable bacteria, however, it is totally unavailable to the host. When cobalamin is bound to dead microorganisms and fed to rats, it is absorbed normally (104). When bound to viable bacteria, however, the vitamin is poorly absorbed. Dead bacteria are digested and the vitamin becomes available to the host. It has been amply demonstrated, however, that within the small-bowel lumen intestinal bacteria are not killed but pass intact into the colon (105, 106), and any cobalamin taken up by viable intestinal bacteria is no longer available for absorption. Thus the patient with small-bowel bacterial overgrowth is faced with the paradoxical situation of cobalamin deficiency despite the presence of large quantities of the vitamin in the small-bowel lumen.

Fish Tapeworm Infestation

A relatively common cause of cobalamin deficiency, particularly in Scandinavian countries, is invasion of the intestine by *Diphyllobothrium latum*. Von Bonsdorff (107) has thoroughly reviewed the earlier literature. These early studies have clearly shown that cobalamin malabsorption is more likely to occur: (1) when the parasite is located in the more proximal portions of the small bowel, (2) when a large number of rapidly growing worms are present, and (3) when the quantity of IF produced by the host is limited. The fish tapeworm readily takes up cobalamin both *in vitro* (108) and *in vivo* (109), and the parasite contains large quantities of cobalamin, as much as 1 to 3 μg/g of dried worm (107). Thus most observations are perfectly consistent with the notion that the fish tapeworm successfully competes with the ileal absorption surface for the vitamin. Although uptake of the vitamins by the parasite is impaired when IF is present, the tapeworm, like enteric bacteria, is capable of removing limited amounts of IF-bound vitamin (108). In addition, the parasite is able to release cobalamin from its complex with human IF (110). It does seem likely, however, that the capacity to secrete abundant quantities of IF protects against cobalamin malabsorption in patients with fish tapeworm infestations, since the majority of such patients who become cobalamin deficient have impaired secretion of IF (111), and there is no evidence that eradication of the worms or repletion with cobalamin increases IF secretion in these patients.

Pancreatic Insufficiency

Within a few years after a practical method for measuring cobalamin absorption became generally available, investigators described impaired absorption of the vitamin in sporadic cases of pancreatic exocrine insufficiency (112). Veeger et al. (113) were the first to examine the problem in any detail. More recently Deren and his colleagues have conducted extensive investigations both in patients and experimental animals, and Toskes and Deren (114) have provided a detailed review of the relevant literature. It is now apparent that approximately one-third of patients with impaired pancreatic function, when tested at one time or another, have diminished absorption of cobalamin (113, 115). There appears to be no obvious relation between cobalamin malabsorption and the severity of pancreatic insufficiently as manifested by steatorrhea. The absorptive defect is not corrected by administration of IF. Although sodium bicarbonate increases the absorption of cobalamin in some patients (113), it is apparent that pH changes alone cannot explain the defect in pancreatic insufficiency. The ileal pH in such patients is not low enough to prevent ileal uptake of the vitamin (115). Furthermore, pancreatic extract either alone or in combination with sodium bicarbonate is more regularly effective than bicarbonate alone. Diminished absorption of the vitamin, correctable by pancreatic extract, has also been demonstrated in partially pancreatectomized rats (116). The factor in pancreatic extract which enhances cobalamin absorption has all the physical chemical properties of several pancreatic proteolytic enzymes (117), and it is now clear that highly purified trypsin is fully capable of correcting cobalamin malabsorption in partially pancreatectomized rats and in patients with pancreatic insufficiency.

Although malabsorption of cyanocobalamin can be demonstrated in a substantial proportion of patients with pancreatic insufficiency including those with cystic fibrosis, overt cobalamin deficiency hardly ever accompanies chronic pancreatic disease and has been reported in only a very few cases (114). Of interest in this regard is the fact that patients whose absorption of the vitamin is distinctly impaired on one occasion may show normal absorption when tested at a later date (115). Furthermore, only small amounts of pancreatic extract are required to correct cobalamin malabsorption in patients with pancreatic insufficiency. In addition, those patients with pancreatic insufficiency who absorb the vitamin poorly when tested in the fasting state have normal or near normal absorption when labeled cyanocobalamin is given with a cobalamin-free meal (118). Thus, the absence of cobalamin deficiency in these patients with abnormal Schilling tests is not surprising since only minimal pancreatic function seems to be required to provide the proteolytic enzyme needed for normal IF-mediated absorption of cobalamin. When stimulated by a meal, even the diseased pancreas is capable of delivering this minimal amount of enzyme activity. Other explanations for the rarity of cobalamin deficiency in patients with pancreatic insufficiency include the long time required to develop cobalamin deficiency in relation to the short survival of many patients with pancreatic disease and the frequency with which patients with pancreatic insufficiency are treated with pancreatic extract.

Although the abnormal cobalamin absorption observed in pancreatic insufficiency appears to be of minimal clinical importance, clarification of this phenomenon is needed for complete understanding of normal IF-mediated absorption of the vitamin. The mechanism whereby a pancreatic proteolytic enzyme, presumably trypsin, is required for this absorptive process remains obscure. It is clear that the pancreatic enzyme performs its function rapidly since prolonged pretreatment with pancreatic extract is not necessary to enhance absorption (115). Also apparent is the fact that pancreatic proteolysis is not directly involved in IF function. IF readily binds cobalamin and the IF•Cbl complex attaches to the ileal absorptive surface in the absence of pancreatic enzymes. Furthermore, small bowel mucosa prepared from pancreatectomized rats is fully capable of taking up IF-bound cobalamin (140). Toskes and Deren (114) have recently postulated the existence of a macromolecular inhibitory factor in intestinal contents. In the absence of pancreatic proteolytic enzymes, this factor would bind to the IF•Cbl complex and thereby impair its uptake by ileal absorptive cells (Figure 7-2). Digestion of the factor by one or more proteolytic enzymes would presumably prevent it from interacting with the IF•Cbl complex. Although supported to some extent by previous suggestions of a possible inhibitory factor present in bile (119), this interesting speculation at present lacks any direct experimental verification, and the precise role of the pancreas in cobalamin absorption remains unidentified.

FAILURE OF INTRINSIC FACTOR-BOUND COBALAMIN TO ATTACH TO ILEAL RECEPTORS

Ileal Resection

Since the initial observations of Booth and Mollin (120), many investigators have shown that resection of the terminal ileum in man is regularly

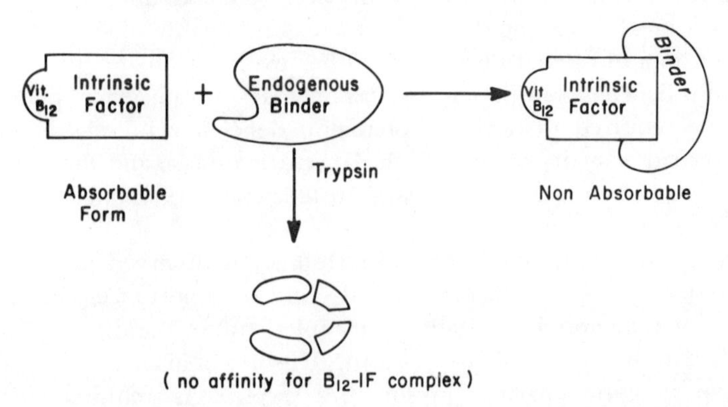

Fig. 7-2 Kindly provided by Dr. J. J. Deren and previously published in *Gastroenterology* (114).

accompanied by cobalamin malabsorption which cannot be corrected by IF (121–123). In our own series of cases (123), which has now expanded to 35 patients, severely impaired absorption of the vitamin was observed whenever more than 50 cm of terminal ileum had been surgically removed. On the other hand, ileal resections which did not include the terminal 15 to 20 cm of ileum did not result in cobalamin malabsorption. Furthermore, absorption of the vitamin remains normal after all of the small bowel except the distal ileum had been resected (124) or bypassed (125). Such observations serve to identify the terminal ileum as the site of cobalamin absorption in man. Similar findings in experimental animals (126) further support the concept that IF-mediated absorption of the vitamin occurs at the more distal portions of the small bowel. Since the surface receptors specific for IF•Cbl complex reside on absorptive cells of the ileum but not the jejunum of experimental animals (127) and man (128, 129), it is the loss of specific distal receptors which results in cobalamin malabsorption when the distal small bowel is resected. Although cobalamin absorption is regularly diminished following ileal resection, frank cobalamin deficiency is relatively uncommon in such patients (130). Although one might speculate that adaptive changes in the residual small bowel may gradually increase cobalamin absorption. there is no direct evidence for such adaptation in man, and we have observed the persistence of severe cobalamin malabsorption in a patient for 15 years after ileal resection.

Ileal Disease

Cobalamin malabsorption also ensues when ileal receptors specific for the IF-bound vitamin are lost or damaged by disease involving the ileal mucosa. Most patients subjected to ileal resection undergo this operation for *regional enteritis,* but cobalamin malabsorption has also been described in patients with regional enteritis involving the terminal ileum who have not undergone surgery (131). In many cases, however, assimilation of the vitamin appears to be normal even though the distal ileum is severely diseased. Nevertheless, once the diseased segment is resected, absorption immediately becomes markedly impaired. The fact that the abnormal ileal segment can often assimilate cobalamin efficiently may be related to the nature of the bowel lesion in regional enteritis. This disease affects all layers of the bowel wall, but predominantly involves the submucosa. The mucosa, and particularly the epithelial absorptive cells, is often spared and thus able to take up the vitamin even when the bowel appears grossly abnormal by X-ray examination or at surgery.

On the other hand, *celiac sprue* is a mucosal disease which primarily involves the epithelial absorptive cells, and extensive damage to the microvillous membrane lining the absorptive surface is an early and characteristic feature of this disease (132). In the ileum, the receptor for IF-bound cobalamin has been localized to this membrane (127, 133). Although most often a disease of the proximal small bowel, celiac sprue frequently extends to involve the ileum (132). Thus, in perhaps 30 to 40% of patients with celiac sprue, steatorrhea tends to be more severe, and cobalamin is poorly absorbed (134). Treatment with a gluten-free diet restores mucosal architecture and corrects the absorptive abnormalities, including cobalamin malabsorption.

Impaired absorption of cobalamin is a regular manifestation of *tropical sprue* (135). Although the etiology and pathogenesis of this disease remain unknown, it is clear that like celiac sprue, tropical sprue is a small-bowel mucosal disease. Although the mucosal lesion may be patchy, it appears to involve the ileum frequently, and this involvement appears sufficient to account for the impaired cobalamin absorption so frequently present in the disease (136). When patients with tropical sprue are treated with a broad spectrum antibiotic such as tetracycline, improvement in mucosal structure is accompanied by increased absorption of cobalamin (137). This improvement with antibiotics does not indicate that tropical sprue results from microorganisms proliferating in the small-bowel lumen. There are distinct differences between patients with this disease and those with cobalamin malabsorption due to small-bowel bacterial overgrowth. In tropical sprue the numbers of microorganisms present in small-bowel contents are not usually impressive (138). Furthermore, the intestinal mucosal lesion is more diffuse and severe than that seen in patients with bacterial overgrowth. Finally, correction of cobalamin malabsorption usually requires prolonged antibiotic therapy in tropical sprue (137), whereas this can ordinarily be accomplished in only a few days in patients with well-documented small-bowel bacterial overgrowth (93). Thus, although there is considerable evidence signifying an infectious etiology for tropical sprue (135), cobalamin malabsorption in this disease appears to be due to inflammatory changes in the structure and function of ileal mucosa rather than to increased numbers of microorganisms in the small-bowel lumen.

Increased cobalamin absorption has also been reported when patients with tropical sprue are treated with folic acid (139). Although such treatment is associated with improvement of the sprue lesion, the mechanisms involved are not understood and folate-induced remissions are less complete and more transient than those brought about by antibiotics (137). It should be pointed out that folic acid has also corrected cobalamin malabsorption in patients with folate deficiency completely unrelated to tropical sprue (140). A jejunal mucosal lesion which can be reversed by folic acid therapy has been delineated in alcoholic patients who are folate deficient (141). Deficiency of cobalamin, whether due to inadequate diet (142) or pernicious anemia (77), can also cause cobalamin malabsorption not correctable by IF. Patients who are severely cobalamin deficient, like those with marked folate deficiency, have a distinctly abnormal jejunal mucosa (78). Thus, it has generally been assumed that abnormal ileal mucosal function is responsible for cobalamin malabsorption in those patients who suffer from marked deficiency of either folic acid or cobalamin and who have impaired absorption of the vitamin even in the presence of exogenous IF. Although it is likely that the ileal mucosa, like the jejunal mucosa, is abnormal in these deficiency states, an ileal lesion has not yet in fact been documented in patients with severe cobalamin or folate deficiency.

Chelation of Divalent Cations

Although of no obvious clinical importance it is of considerable theorectical interest that the feeding of ethylenediaminetetraacetic acid (EDTA) impairs cobalamin absorption in experimental animals (143) and human subjects (144). This malabsorption cannot be overcome by IF but is corrected by the administration of calcium. The attachment of IF-bound cobalamin to its specific receptor on microvillous membranes requires the presence of divalent cations, particularly calcium (145). Presumably, the cobalamin malabsorption induced by EDTA results from calcium depletion and consequent failure of IF•Cbl complex to attach to its receptor.

MISCELLANEOUS CAUSES OF COBALAMIN MALABSORPTION

There remain several causes of cobalamin malabsorption which are difficult to classify in relation to the normal process of cobalamin transport. It should be stated, incidentally, that for the obvious or presumed ileal mucosal diseases described above it has been assumed that the major absorption defect is failure of attachment of IF-bound cobalamin to the ileal absorptive surface because of loss or alteration of membrane receptors. Although microvillous membranes are certainly lost or altered in mucosal disease, this is undoubtedly an oversimplification since any step in the subsequent transcellular transport of the vitamin could also be impaired. Until these subsequent steps are clarified, however, it is impossible to identify the specific defects which occur in various diseases of the ileal mucosa. Similarly, better understanding of the mechanisms whereby cobalamin is transported across the ileal cell will almost certainly clarify many of the causes of cobalamin malabsorption that are currently unclassifiable.

Familial Selective Cobalamin Malabsorption

In 1960 Imerslund (146) and Gräsbeck (147) independently described a familial disorder characterized by the onset of cobalamin deficiency in childhood. The frequency with which this disorder occurs in families, particularly among siblings, suggests that the condition is inherited, but the mode of inheritance has not been defined. Children afflicted with the disorder absorb the vitamin poorly from birth, but malabsorption is not related to IF deficiency. Exogenous IF fails to correct the absorptive defect, and gastric juice from these patients increases cobalamin absorption in patients who lack IF. These children have no clinical, radiological, or morphological evidence of small-bowel disease, and intestinal absorption of both fat and xylose is normal. Mild but persistent proteinuria is almost always present. More than 35 cases of familial selective cobalamin malabsorption have now been described (148), but the mechanism responsible for impaired absorption of the vitamin remains unknown. Since IF production is normal in these children and since cobalamin malabsorption cannot be corrected by the administration of antibiotics or pancreatic extract (129, 148), it has been assumed that a specific defect is present in the intestinal mucosa, possibly a congenital lack of the ileal membrane receptors for IF•Cbl complex (13). A recent study of one family has shown, however, that the ileal mucosa in this condition is normal by light and electron microscopy (129). Moreover, attachment of IF-bound cobalamin to ileal mucosal homogenates was not impaired, indicating that membrane receptors were intact. Since patients with this disorder have normal serum levels of transcobalamin II (129, 148), it would appear that the defect in familial selective cobalamin malabsorption occurs at some point after the IF•Cbl complex attaches to the surface of ileal absorptive cells but before the absorbed vitamin binds to transcobalamin II. Until the transcellular transport of cobalamin can be directly examined, however, it will not be possible to identify the precise nature of this defect.

Drug-Induced Cobalamin Malabsorption

Malabsorption of cobalamin has been observed in some, but by no means all, patients receiving paraaminosalicylic acid (149), biguanides (150), colchicine (151), and Neomycin (152). Although it is not at all clear how these agents impair absorption of the vitamin in susceptible individuals, cobalamin absorption clearly increases when these drugs are discontinued.

In the case of paraaminosalicylic acid, an absorptive defect apparently specific for cobalamin has been described (153). Both colchicine (151) and Neomycin (152) appear to cause an intestinal mucosal lesion, however, and other nutrients in addition to cobalamin are poorly absorbed. Biguanides are known to inhibit intestinal transport of glucose (153) in addition to causing cobalamin malabsorption. Thus, the mechanisms involved in drug-induced cobalamin malabsorption are likely to be complex and have not been directly examined in any detail. Similarly, the administration of alcohol appears to decrease cobalamin absorption (155) but there is no information as to how alcohol interferes with intestinal absorption of the vitamin.

Zollinger-Ellison Syndrome

Impaired absorption of cobalamin which does not improve with IF administration has been observed in patients with massive gastric hypersecretion associated with gastrin-producing islet cell tumors of the pancreas (155, 156). In these cases the absorptive defect is corrected by total gastrectomy or by vigorous suppression of gastric acid secretion, although after total gastrectomy IF must be given. Since a pH greater than 5.5 is required for attachment of IF•Cbl to its intestinal receptor (145), the defect in Zollinger-Ellison syndrome may result from a low intraluminal pH caused by the massive secretion of gastric acid. On the other hand, in one carefully studied case the intraluminal pH at the level of the ileum was not sufficiently low to explain cobalamin malabsorption on this basis (155). Furthermore, although acid-induced mucosal lesions were present in the jejunum of this patient, the ileal mucosa was structurally intact. When the extremely acid contents present in the duodenum were neutralized by administration of bicarbonate, cobalamin absorption substantially increased. On the basis of one case report, therefore, it seems possible that acid-induced inactivation of pancreatic proteolytic enzymes may account for impaired absorption of cobalamin in the Zollinger-Ellison syndrome. A more extensive experience will be required, however, before this mechanism can be considered to be established.

Congenital Deficiency of Transcobalamin II

This disorder is described in detail in Chapter 8. It is necessary here only to point out that impaired absorption of cobalamin has been described in one child with isolated deficiency of transcobalamin II (157). This malabsorption was not corrected by IF. Although it is possible that transcobalamin II may therefore be required for normal absorption of the vitamin, such a conclusion must at present be considered premature.

Only one child has been studied thus far, and it was not possible to determine whether repletion with transcobalamin II corrected the absorptive defect. Since patients with this disorder are unable to deliver cobalamin to tissues, and since massive and frequent doses of the vitamin are required to ovecome the disorder, it remains possible that the observed malabsorption of cobalamin resulted from ileal mucosal dysfunction due to deprivation of the vitamin. Although the finding of cobalamin malabsorption in one case of transcobalamin II deficiency is of considerable interest, further studies are needed to determine whether transcobalamin II really plays an essential role in normal cobalamin absorption.

REFERENCES

1. Cooper, B., and Castle, W. B. (1960). Sequential mechanisms in the enhanced absorption of vitamin B_{12} by intrinsic factor in the rat. *J. Clin. Invest.* **30**, 199.
2. Schade, S. G., and Schilling, R. F. (1967). Effect of pepsin on the absorption of food vitamin B_{12} and iron. *Am. J. Clin. Nutr.* **20**, 636.
3. Reizenstein, R. (1959). Effect of digestive enzymes on bound vitamin B_{12}. *Acta Med. Scand.* **165**, 481.
4. Mahmud, K., Ripley, D., and Doscherholmen, A. (1971). Vitamin B_{12} absorption tests. Their unreliability in post gastrectomy states. *JAMA* **216**, 1167.
5. Chang, S., Ripley, D., Swain, W., and Doscherholmen, A. (1971). Gastric achlorhydria and hypochlorhydria with low serum B_{12} levels and normal B_{12} absorption. *Clin. Res.* **19**, 655.
6. Doscherholmen, A., and Swain, W. R. (1973). Impaired assimilation of egg Co^{57} vitamin B_{12} in patients with hypochlorhydria and achlorhydria and after gastric resection. *Gastroenterology* **64**, 913.
7. Deller, D. J., and Witts, L. J. (1962). Changes in the blood after partial gastrectomy with special reference to vitamin B_{12}. *Q. J. Med.* **31**, 71.
8. Hines, J. D., Hoffbrand, A. V., and Mollin, D. L. (1967). The hematologic complications following partial gastrectomy. A study of 292 patients. *Am. J. Med.* **43**, 555.
9. Sullivan, L. W., Herbert, V., and Reizenstein, P. (1962). Evidence against preferential intestinal absorption of physiological quantities of liver bound vitamin B_{12} by patients with pernicious anemia. *Am. J. Clin. Nutr.* **11**, 568.
10. Heyssell, R. M., Bozian, R. C., Darby, W. J., et al. (1966). Vitamin B_{12} turnover in man. The assimilation of vitamin B_{12} from natural food stuff by man and estimates of minimal daily dietary requirements. *Am. J. Clin. Nutr.* **18**, 176.
11. Schade, S. G., and Schilling, R. F. (1967). Effect of pepsin on the absorption of food vitamin B_{12} and iron. *Am. J. Clin. Nutr.* **20**, 636.
12. MacDonald, R. M., Ingelfinger, F. J., and Belding, H. W. (1947). Late effects of total gastrectomy in man. *N. Engl. J. Med.* **237**, 887.
13. Castle, W. B. (1970). Current concepts of pernicious anemia. *Am. J. Med.* **48**, 541.

14. Castle, W. B. (1953). Development of knowledge concerning the gastric intrinsic factor and its relation to pernicious anemia. *N. Engl. J. Med.* **249**, 603.

15. Witts, L. J. (1966). *The Stomach and Anemia,* Athlone Press, London.

16. Addison, T. (1855). On the constitutional and local effects of disease of the suprarenal capsules. London.

17. Fenwick, S. (1870). On atrophy of the stomach. *Lancet* **2**, 78.

18. Castle, W. B. (1929). Observations on etiologic relationship of achylia gastrica to pernicious anemia. I. Effect of administration to patients with pernicious anemia of contents of normal human stomach recovered after ingestion of beef muscle. *Am. J. Med. Sci.* **178**, 748.

19. Minot, G. R., and Murphy, W. P. (1926). Treatment of pernicious anemia by a special diet. *JAMA* **87**, 470.

20. Schilling, R. F. (1953). Intrinsic factor studies. II. The effect of gastric juice on the urinary excretion of radioactivity after the oral administration of radioactive vitamin B_{12}. *J. Lab. Clin. Med.* **42**, 860.

21. Rubin, W., Ross, L. L., Jeffries, G. H., and Sleisenger, M. H. (1966). Intestinal heterotopia: A fine structural study. *Lab. Invest.* **15**, 1024.

22. te Velde, K., Hoedemaeker, P. J., Anders, G. J. P. A., Arends, A., and Nieweg, H. O. (1966). A comparative morphological and functional study of gastritis with and without antibodies. *Gastroenterology*. **51**, 138.

23. Glass, G. B. J., Speer, F. D., Nieburgs, H. E., Ishimori, A., Jones, L. E., Baker, H., Schwartz, S., and Smith, R. (1960). Gastric atrophy, atrophic gastritis, and gastric secretory failure. *Gastroenterology* **39**, 429.

24. Strickland, R. G., and Mackay, I. R. (1973). A reappraisal of the nature and significance of chronic atrophic gastritis. *Am. J. Dig. Dis.* **18**, 426.

25. Poliner, I. J., and Spiro, H. M. (1958). The independent secretion of acid, pepsin and "intrinsic factor" by the human stomach. *Gastroenterology* **34**, 196.

26. Jeffries, G. H., and Sleisenger, M. H. (1965). The pharmacology of intrinsic factor secretion in man. *Gastroenterology* **48**, 444.

27. Ardeman, S., and Chanarin, I. (1965). Assay of gastric intrinsic factor in the diagnosis of Addisonian pernicious anemia. *Brit. J. Haematol.* **11**, 305.

28. Siurala, M., Vuorinen, Y., and Seppala, K. (1961). Follow-up studies of patients with atrophic gastritis. *Acta Med. Scand.* **170**, 151.

29. McIntyre, O. R., Sullivan, L. W., Jeffries, G. H., and Silver, R. H. (1965). Pernicious anemia in childhood. *N. Engl. J. Med.* **272**, 981.

30. Jeffries, G. H. (1971). Pernicious anemia and atrophic gastritis. *Immunological Diseases,* Vol. II, Samter, M., Ed., Little Brown, Boston, Massachusetts, p. 1228.

31. Callender, S. T., and Denborough, M. A. (1957). A family study of pernicious anaemia. *Brit. J. Haematol.* **3**, 88.

32. McIntyre, P. A., Hahn, H., Conley, C. L., and Glass, B. (1959). Genetic factors in predisposition to pernicious anemia. *Johns Hopkins Hosp.* **104**, 309.

33. Whittingham, S., MacKay, I. R., Ungar, B., and Matthews, J. D. (1969). The genetic factor in pernicious anaemia. A family study in patients with gastritis. *Lancet* **1**, 951.

34. Balcerzak, S. P., Westerman, M. P., and Heinie, E. W. (1968). Discordant occurrence of pernicious anemia in identical twins. *Blood* **32**, 701.

35. Schwartz, M. (1958). Intrinsic-factor-inhibiting substance in serum of orally treated patients with pernicious anaemia. *Lancet* **2**, 61.

36. Taylor, K. B. (1959). Inhibition of intrinsic factor by pernicious anaemia sera. *Lancet* **2**, 106.

37. Schade, S. G., Abels, J., and Schilling, R. F. (1967). Studies on antibody to intrinsic factor. *J. Clin. Invest.* **46**, 615.

38. Ardeman, S., and Chanarin, I. (1963). A method for the assay of human gastric intrinsic factor and for the detection and titration of antibodies against intrinsic factor. *Lancet* **2**, 1350.

39. Abels, J., Bouma, W., Jansz, A., Woldring, M. G., Bakker, A., and Nieweg, H. O. (1963). Experiments on the intrinsic factor antibody in serum from patients with pernicious anemia. *J. Lab. Clin. Med.* **61**, 893.

40. Jeffries, G. H., and Sleisenger, M. H. (1965). Studies of parietal cell antibody in pernicious anemia. *J. Clin. Invest.* **44**, 2021.

41. Carmel, R., Rosenberg, A., Lau, K. S., Streiff, R. R., and Herbert, V. (1969). Vitamin B_{12} uptake in human small bowel homogenate and its enhancement by intrinsic factor. *Gastroeterology* **56**, 548.

42. Hines, J. D., Rosenberg, A., and Harris, J. W. (1968). Intrinsic factor-mediated radio-B_{12} uptake in sequential incubation studies using everted sacs of guinea pig small intestine: Evidence that IF is not absorbed into the intestinal cell. *Proc. Soc. Exp. Biol. Med.* **129**, 653.

43. Gullberg, R. (1971). Sensitive test for antibody Type I to intrinsic factor. *Clin. Exp. Immunol.* **9**, 833.

44. Samloff, I. M., Kleinman, M. S., and Turner, M. D. (1968). Blocking and binding antibodies to intrinsic factor and parietal cell antibody in pernicious anemia. *Gastroenterology* **55**, 575.

45. Barditan, K. D., Hall, J. R., Spray, C. H., and Callender, S. T. E. (1968). Blocking and binding autoantibody to intrinsic factor. *Lancet* **1**, 62.

46. Ungar, B., Stocks, A., Martin, F., Whittingham, S., and Mackay, I. (1967). Intrinsic factor antibody in diabetes mellitus. *Lancet* **2**, 77.

47. Meecham, J., and Jones, E. W. (1967). Addison's disease and Addisonian anemia. *Lancet* **1**, 535.

48. Ardeman, S., Chanarin, I., Krafchik, B., and Singer, W. (1966). Addisonian pernicious anemia and intrinsic factor antibodies in thyroid disorders. *Q. J. Med.* **35**, 421.

49. Schade, S. C., Feick, P., Muckerheide, M., and Schilling, R. F. (1966). Occurrence in gastric juice of antibody to a complex of intrinsic factor and vitamin B_{12}. *N. Engl. J. Med.* **275**, 528.

50. Goldberg, L. G., Shuster, J., Stuckey, M., and Fudenberg, H. H. (1968). Secretory immunoglobulin A: Autoantibody in gastric juice. *Science* **160**, 1240.

51. Carmel, R., and Herbert, V. (1967). Intrinsic-factor antibody in the saliva of a patient with pernicious anemia. *Lancet* **1**, 80.

52. Hoedemaeker, P. J., Abels, J., Wachters, J. J., Arends, A., and Nieweg, H. O. (1964). Investigations about the site of production of Castle's gastric intrinsic factor. *Lab. Invest* **13**, 1394.

53. Markson, J. L., and Moore, J. M. (1962). Autoimmunity in pernicious anemia and iron-deficiency anemia: A complement-fixation test using human gastric mucosa. *Lancet* **2**, 1240.

54. Doniach, S., Roitt, I. M., and Taylor, K. B. (1965). Autoimmunity in pernicious anemia and thyroiditis: A family study. *Ann. N. Y. Acad. Sci.* **124**, 606.

55. Twomey, J. J., Jordan, P. H., Jarrold, T., Trubowitz, J., Ritz, N. D., and Conn, H. O. (1969). The syndrome of immunoglobin deficiency and pernicious anemia. *Am. J. Med.* **47**, 340.

56. McGuigan, J. E. (1967). *In vitro* behavior of lymphocytes from patients with pernicious anemia. *J. Clin. Invest.* **46**, 1094.

57. Fixa, B., and Thiele, H. G. (1969). Delayed hypersensitivity to intrinsic factor in patients with pernicious anemia. *Med. Exp. (Basel)* **19**, 231.

58. Rose, M. S., Chanarin, I., Doniach, D., Brostoff, J., and Ardeman, S. (1970). Intrinsic factor antibodies in absence of pernicious anemia. *Lancet* **2**, 9.

59. Jeffries, G. H., Todd, J. E., and Sleisenger, M. H. (1966). The effect of prednisolone on gastric mucosal histology, gastric secretion and vitamin B_{12} absorption in patients with pernicious anemia. *J. Clin. Invest.* **45**, 803.

60. Samloff, I. M., Kleinman, M. S., Turner, M. D., Sobel, M. V., and Jeffries, G. H. (1968). Blocking and binding antibody to intrinsic factor and parietal cell antibody in pernicious anemia. *Gastroenterology* **55**, 575.

61. Wangel, A. G., and Schiller, K. F. R. (1966). Diagnostic significance of antibody to intrinsic factor. *Brit. Med. J.* **1**, 1274.

62. Doniach, D., Roitt, I. M., and Taylor, K. B. (1963). Autoimmune phenomena in pernicious anaemia: Serologic overlap with thyroiditis, thyrotoxicosis and systemic lupus crythematosus. *Brit. Med. J.* **1**, 1374.

63. Bar-Shany, S. and Herbert, V. (1967). Transplacentally acquired antibody to intrinsic factor with vitamin B_{12} deficiency. *Blood* **30**, 777.

64. Fisher, J. M., and Taylor, K. B., (1965). A comparison of autoimmune phenomena in pernicious anemia and chronic atrophic gastritis. *N. Engl. J. Med.* **272**, 499.

65. Ashurst, P. M. (1968). Parietal cell antibodies in patients undergoing gastric surgery. *Brit. Med. J.* **1**, 647.

66. te Velde, K., Abels, J., Anders, G. J. P. A., Arends, A., Hoedemaeker, P. J., and Nieweg, H. A. (1964). A family study of pernicious anemia by an immunologic method. *J. Lab. Clin. Med.* **64**, 177.

67. Yates, T., and Cooper, B. A. (1967). Failure to demonstrate that antibody to intrinsic factor is a significant cause of vitamin B_{12} malabsorption in pernicious anemia. *Can. Med. Assoc. J.* **97**, 950.

68. Kaplan, M. E., Zalusky, R., Remington, J. and Herbert, V. (1963). Immunological studies with intrinsic factor in man. *J. Clin. Invest* **42**, 368.

69. Rose, M. S., and Chanarin, I. (1971). Intrinsic factor antibody and the absorption of vitamin B_{12} in pernicious anemia. *Brit. Med. J.* **1**, 25.

70. Rose, M. S., and Chanarin, I. (1969). Dissociation of intrinsic factor from its antibody: Application to study of pernicious anemia gastric juice specimens. *Brit. Med. J.* **1**, 468.

71. Goldberg, L. S., and Bluestone, R. (1970). Hidden gastric autoantibodies to intrinsic factor in pernicious anemia. *J. Lab. Clin. Med.* **75**, 449.
72. Ardeman, S., and Chanarin, I. (1965). Intrinsic factor antibodies and intrinsic factor mediated vitamin B_{12} absorption in pernicious anemia. *Gut* **6**, 436.
73. Donaldson, R. M. (1967). Role of enteric microorganisms in malabsorption. *Fed. Proc.* **26**, 1426.
74. Drasar, B. S., Shiner, M., and McLeod, G. M. (1969). Studies on the intestinal flora. I. The bacterial flora of the gastrointestinal tract in healthy and achlorhydria persons. *Gastroenterology* **56**, 71.
75. Sherwood, W. C., Goldstein, F., Haurani, F. I., and Wirts, C. W. (1964). Studies of the small intestinal flora and of intestinal absorption in pernicious anemia. *Am. J. Dig. Dis.* **9**, 416.
76. Haurani, F., Sherwood, W., and Goldstein, F. (1964). Intestinal malabsorption of vitamin B_{12} in pernicious anemia. *Metabolism* **13**, 1342.
77. Carmel, R., and Herbert, V. (1967). Correctable intestinal defect of vitamin B_{12} absorption in pernicious anemia. *Ann. Intern. Med.* **67**, 1201.
78. Foroozan, R., and Trier, J. S. (1967). Mucosa of the small intestine in pernicious anemia. *N. Engl. J. Med.* **227**, 553.
79. Spurling, C. L., Sacks, M. S., and Jiji, R. M. (1964). Juvenile pernicious anemia. *N. Engl. J. Med.* **271**, 995.
80. Lillibridge, S. B., Brandborg, L. L., and Rubin, C. E. (1967). Childhood pernicious anemia: Gastrointestinal secretory, histological, and electron microscopic aspects. *Gastroenterology* **52**, 792.
81. Miller, D. R., Bloom, G. E., Streiff, R. R., LoBuglio, A. F., and Diamond. L. K. (1966). Juvenile "congenital" pernicious anemia. Clinical and immunologic studies. *N. Engl. J. Med.* **275**, 978.
82. Katz, M., Lee, S. K., and Cooper, B. A. (1972). Vitamin B_{12} malabsorption due to a biologically inert intrinsic factor. *N. Engl. J. Med.* **287**, 425.
83. Katz, M., Mehlman, C. S., and Allen, R. H. Isolation and characterization of abnormal human intrinsic factor. *J. Clin. Invest.* **53**, 1274.
84. Halsted, J. A., Gasster, M., and Drenick, E. J. (1954). Absorption of radioactive vitamin B_{12} after total gastrectomy: Relation to macrocytic anemia and to the site of origin of Castle's intrinsic factor. *N. Engl. J. Med.* **251**, 161.
85. Watson, G. M., and Florey, H. W. (1955). The absorption of vitamin B_{12} in gastrectomized rats. *Brit. J. Exp. Pathol.* **36**, 479.
86. Adams, J. F., and Seaton, D. A. (1961). Reproducibility and reliability of the Schilling test. *J. Lab. Clin. Med.* **58**, 67.
87. Lowenstein, F. (1958). Absorption of cobalt^{60}-labeled vitamin B_{12} after subtotal gastrectomy. *Blood* **13**, 339.
88. Stelsberg, H., Sigbjorn, T. (1971). Stomach cancer following gastric surgery for benign conditions. *Lancet* **2**, 1175.
89. DuPlessis, D. J. (1965). Pathogenesis of gastric ulceration. *Lancet* **1**, 974.
90. Capper, W. M. (1967). Factors in the pathogenesis of gastric ulcer. *Ann. Roy. Coll. Surg. Eng.* **40**, 21.
91. Resnick, R. H., Colman, R., London, A., and Richter, H. (1963). Abnormal Schilling test corrected by intestinal juice. *N. Engl. J. Med.* **268**, 926.

92. McLean, L. D. (1957). Incidence of megaloblastic anemia after subtotal gastrectomy. *N. Engl. J. Med.* **257**, 262.

93. Donaldson, R. M. (1970). Small bowel bacterial overgrowth. *Adv. Intern. Med.* **16**, 191.

94. Tabaqchali, S. (1970). The pathophysiological role of small intestinal bacterial flora. *Scand. J. Gastroenterol. Suppl.* **6**, 139.

95. Bishop, R. F. (1963). Bacterial flora of the small intestine of dogs and rats with intestinal blind loops. *Brit. J. Exp. Pathol.* **44**, 189.

96. Donaldson, R. M., McConnel, C., and Defner, N. (1967). Bacteriological studies in clinical and experimental blind loop syndromes. *Gastroenterology* **52**, 1082.

97. Donaldson, R. M. (1962). Malabsorption of ^{60}Co-labeled cyanocobalamin in rats with intestinal diverticula. I. Evaluation of possible mechanisms. *Gastroenterology* **43**, 271.

98. Halstead, J. A., Lewis, P. M., and Gasster, M. (1956). Absorption of radioactive vitamin B_{12} in the syndrome of megaloblastic anemia associated with intestinal stricture or anastomosis. *Am. J. Med.* **20**, 52.

99. Ament, M. E., Shimoda, S. S., Saunders, D. R., and Rubin, C. F. (1972). Pathogenesis of steatorrhea in three cases of small intestinal stasis syndrome. *Gastroenterology* **63**, 728.

100. Schjonsby, H., and Tagaqchali, S. (1971). Uptake of rat gastric juice-bound vitamin B_{12} by intestinal brush borders isolated from blind-loop rats. *Scand. J. Gastroenterol.* **6**, 515.

101. Donaldson, R. M., Corrigan, H., and Natsios, G. (1962). Malabsorption of ^{60}Co-labeled cyanocobalamin in rats with intestinal diverticula. II. Studies on contents of the diverticula. *Gastroenterology* **43**, 282.

102. Giannella, R. A., Broitman, S. A., and Zamcheck, N. (1971). Vitamin B_{12} uptake by intestinal microorganisms: Mechanism and relevance to syndromes of intestinal bacterial overgrowth. *J. Clin. Invest.* **50**, 1100.

103. Schjonsby, H. (1972). The effect of bacteria on intestinal uptake of vitamin B_{12}. I. Effect of cultures of blind-loop contents. *Scand. J. Gastroenterol.* **7**, 119.

104. Booth, C. C., and Heath, J. (1962). The effect of *E. coli* on the absorption of vitamin B_{12}. *Gut* **3**, 70.

105. Dack, G. M., and Retran, E. (1934). Bacterial activity in different levels of the intestine and in isolated segments of small and large bowel in monkeys and in dogs. *J. Infect. Dis.* **54**, 204.

106. Dixon, J. M. S. (1960). The fate of bacteria in the small intestine. *J. Pathol. Bacteriol.* **79**, 131.

107. von Bonsdorff, B. (1957). Pathogenesis of vitamin B_{12} deficiency, with special reference to tapeworm pernicious anemia. *Vitamin B_{12} and Intrinsic Factor,* Heinvich, H. C., Ed., Enke, Stuttgart.

108. Brante, G., and Ernberg, T. (1957). The *in vitro* uptake of vitamin B_{12} by *Diphyllobothrium latum* and its blockage by intrinsic factor. *Scand. J. Clin. Lab. Invest.* **9**, 313.

109. Nyberg, W. (1960). The influence of *Diphyllobothrium latum* on the vitamin B_{12}-intrinsic factor complex. I. *In vivo* studies with Schilling test technique. *Acta Med. Scand.* **167**, 185.

110. Nyberg, W. (1960). The influence of *Diphyllobothrium latum* on the vitamin B_{12}-intrinsic factor complex. II. *In vitro* studies. *Acta Med. Scand.* **167**, 189.

111. Salokannel, J. (1970). Intrinsic factor in tapeworm anemia. *Acta Med. Scand.* **188** (Suppl. 517).

112. McIntyre, P. A., Sachs, M. V., Krevans, J. R., and Conley, C. L. (1956). Pathogenesis and treatment of macrocytic anemia: Information obtained with radioactive vitamin B_{12}. *Arch. Intern. Med.* **98**, 541.

113. Veeger, W., Abels, J., Hellemans, N., and Nieweg, H. O. (1962). Effect of sodium bicarbonate and pancreatin on the absorption of vitamin B_{12} and fat in pancreatic insufficiency. *N. Engl. J. Med.* **267**, 1341.

114. Toskes, P. P., and Deren, J. J. (1973). Vitamin B_{12} absorption and malabsorption. *Gastroenterology* **65**, 662.

115. Toskes, P. P., Hansell, J., Cerda, J., and Deren, J. J. (1971). Vitamin B_{12} malabsorption in chronic pancreatic insufficiency. Studies suggesting the presence of a pancreatic intrinsic factor. *N. Engl. J. Med.* **284**, 627.

116. Toskes, P. P., and Deren, J. J. (1972). The role of the pancreas in vitamin B_{12} absorption: Studies of vitamin B_{12} absorption in partially pancreatectomized rats. *J. Clin. Invest.* **51**, 216.

117. Toskes, P. P., Ginsberg, A., Conrad, M. E., and Deren, J. J. (1971). The physical chemical characteristics and mode of action of pancreatic intrinsic factor. *Blood* **38**, 609.

118. Henderson, J. T., Warwick, R. R. G., Simpson, J. D., and Shearman, J. C. (1972). Does malabsorption of vitamin B_{12} ocur in chronic pancreatitis? *Lancet* **II**, 241.

119. Johnson, P. C., Driscoll, T. B., and Honska, W. L. (1961). Inhibition of vitamin B_{12} absorption by bile. *Proc. Soc. Exp. Biol. Med.* **106**, 181.

120. Booth, C. C., and Mollin, D. L. (1959). The site of absorption of vitamin B_{12} in man. *Lancet* **1**, 18.

121. Smith, A. N., Falconer, C. W. A., and Small, W. P. (1969). Malabsorption in Crohn's disease. *Malabsorption,* Girdwood, R. H., and Smith, A. N., Eds., Edinburgh Press, Edinburgh.

122. Fromm, H., Thomas, P. J., and Hofman, A. F. (1973). Sensitivity and specificity in tests of distal ileal function. *Gastroenterology* **64**, 1077.

123. Katz, J. H., DiMase, J., and Donaldson, R. M. (1963). Simultaneous administration of gastric juice bound and free radioactive cyanocobalamin. *J. Lab. Clin. Med.* **61**, 266.

124. Booth, C. C. (1961). Metabolic effects of intestinal resection in man. *Postgrad. Med. J.* **37**, 725.

125. Donaldson, R. M. (1959). Absorption of vitamin B_{12}. *Lancet* **1**, 790.

126. Booth, C. C. (1968). Effect of location along the small intestine on absorption of nutrients. *Handbook of Physiology,* Sect. 6, Vol. III, Code, C. F., and Weidel, W., Eds., Williams & Wilkins, Baltimore, p. 1513.

127. Donaldson, R. M., Mackenzie, I. L., and Trier, J. S. (1967). Intrinsic factor-mediated attachment of vitamin B_{12} to brush borders and microvillous membranes of hamster intestine. *J. Clin. Invest.* **46**, 1215.

128. Rosenberg, A. H., Lau, K., and Herbert, V. (1965). Enhancement by intrinsic factor of vitamin B_{12} uptake by human ileal homogenate. *Clin. Res.* **13**, 281.

129. Mackenzie, I. L., Donaldson, R. M., Trier, J. S., and Mathan, V. I. (1972). Ileal mucosa in familial selective vitamin B_{12} malabsorption. *N. Engl. J. Med.* **286**, 1021.

130. Booth, C. C., MacIntyre, I., and Mollin, D. L. (1964). Nutritional problems associated with extensive lesions in the distal small intestine of man. *Q. J. Med.* **33**, 401.

131. Steinberg, F. (1961). The megaloblastic anemia of regional ileitis. *N. Engl. J. Med.* **264**, 186.

132. Rubin, C. E., Brandborg, L. L., Phelps, P. C., and Taylor, H. C. (1960). Studies of celiac disease. *Gastroenterology* **38**, 28.

133. MacKenzie, I. L., Donaldson, R. M., Kopp, W. L., and Trier, J. S. (1968). Antibodies to intestinal microvillous membranes. I. Characterization and morphologic localization. *J. Exp. Med.* **128**, 357.

134. Stewart, J. J., Pollock, D. J., Hoffbrand, A. V., Mollin, D. L., and Booth, C. C. (1967). A study of proximal and distal intestinal structure and absorptive function in intestinal steatorrhea. *Q. J. Med.* **36**, 425.

135. Mathan, V. I. (1973). Tropical sprue. *Gastrointestinal Disease,* Sleisenger, M., and Fordtran, J., Eds., Saunders, Philadelphia, Pennsylvania, p. 978.

136. Brunser, O., Eidelman, S., and Klipstein, F. (1970). Intestinal morphology of rural Haitians. A comparison between overt tropical sprue and asmptomatic subjects. *Gastroenterology* **58**, 655.

137. Bayless, T. M., Wheby, M. S., and Swanson, V. L. (1968). Tropical sprue in Puerto Rico. *Am. J. Clin. Nutr.* **21**, 1030.

138. Gorbach, S. L., Banwell, J. G., and Jacobs, B. (1970). Tropical sprue and malnutrition in West Bengal. I. Intestinal microflora and absorption. *Am. J. Clin. Nutr.* **23**, 1545.

139. Klipstein, F. A. (1969). Tropical sprue. *Gastroenterology* **54**, 275.

140. Scott, R. B., Kammer, R. B., Burger, W. F., and Middleton, F. G. (1968). Reduced absorption of vitamin B_{12} in two patients with folic acid deficiency. *Ann. Intern. Med.* **69**, 111.

141. Hermos, J. A., Adams, W. H., Liu, Y. K., Sullivan, L. W., and Trier, J. S. (1972). Mucosa of the small intestine in folate-deficient alcoholics. *Ann. Intern. Med.* **76**, 957.

142. Schloesser, L., and Schilling, R. (1963). Vitamin B_{12} absorption studies in a vegetarian with megaloblastic anemia. *Am. J. Clin. Nutr.* **12**, 70.

143. Herbert, V. (1959). Studies on the role of intrinsic factor in vitamin B_{12} absorption, transport, and storage. *Am. J. Clin. Nutr.* **7**, 433.

144. Gräsbeck, R., and Nyberg, W. (1958). Inhibition of radiovitamin B_{12} absorption by ethylenediamine tetraacetate (EDTA) and its reversal by calcium ions. *Scand. J. Clin. Lab. Invest.* **10**, 448.

145. Mackenzie, I. L., and Donaldson, R. M. (1972). Effect of divalent cations and pH on intrinsic factor-mediated attachment of vitamin B_{12} to intestinal microvillous membranes. *J. Clin. Invest.* **51**, 2465.

146. Imerslund, O. (1960). Idiopathic chronic megaloblastic anemia in children. *Acta Pediatr. Suppl.,* 119.

147. Gräsbeck, R., Gordin, R., Kantero, I., and Kuhlback, B. (1960). Selective vitamin B_{12} malabsorption and proteinuria in young people. *Acta Med. Scand.* **167**, 289.

148. Gräsbeck, R. (1972). Familial selective vitamin B_{12} malabsorption. *N. Engl. J. Med.* **287**, 358.

149. Heinivaara, O., and Pavla, I. P. (1965). Malabsorption and deficiency of vitamin B_{12} caused by treatment with para-aminosalicylic acid. *Acta Med. Scand.* **177**, 337.

150. Tomkin, G. H., Hadden, D. R., Weaver, J. A., and Montgomery, D. A. D. (1971). Vitamin B_{12} status of patients on long term Metformin therapy. *Brit. Med. J.* **1**, 685.

151. Webb, D. I., Chodos, R. B., Mahar, C. Q., and Faloon, W. W. (1968). Mechanism of vitamin B_{12} malabsorption in patients receiving colchicine. *N. Engl. J. Med.* **279**, 845.

152. Jacobson, E. D., Chodos, R. B., and Faloon, W. W. (1960). An experimental malabsorption syndrome induced by Neomycin. *Am. J. Med.* **28**, 524.

153. Heinivaara, O., Palva, I., Siorala, M., and Pelkonen, R. (1964). Selectivity of the PAS-induced malabsorption of vitamin B_{12}. *Ann. Med. Intern. Fenn.* **53**, 75.

154. Arvanitakis, C., Lorenzsonn, V., and Olson, W. (1972). Effect of phenformin on glucose and water absorption in man. *Gastroenterology* **62**, 837.

155. Schimoda, S., Saunders, D. R., and Rubin, C. E. (1968). The Zollinger-Ellison syndrome with steatorrhea. II. The mechanism of fat and vitamin B_{12} malabsorption. *Gastroenterology* **55**, 705.

156. Shum, H. Y., O'Neill, B. J., and Streeter, A. M. (1971). Vitamin B_{12} absorption and the Zollinger-Ellison syndrome. *Lancet* **I**, 1303.

157. Hakami, N., Neiman, P. E., Canellos, G. P., and Lazerson, J. (1971). Neonatal megaloblastic anemia due to inherited transcobalamin II deficiency in two siblings, *N. Engl. J. Med.* **285**, 1163.

CHAPTER EIGHT

INBORN ERRORS OF COBALAMIN METABOLISM

MAURICE J. MAHONEY, M.D.
AND LEON E. ROSENBERG, M.D.

Department of Human Genetics
Yale University School of Medicine
New Haven, Connecticut

CONTENTS

INTRODUCTION

Man has left the complexities and problems of synthesizing the cobalamin vitamin (Cbl) to the ingenuity of microorganisms. Even so, he has saved worthy tasks for himself, and in the failure of these tasks lie the origins of several human diseases. After ingestion the vitamin is bound to intrinsic factor (IF), a gastric secretory protein, which carries it intact to specific receptor sites in the ileum. There it is transported into ileal mucosal cells. From these specialized absorbing cells the vitamin enters the bloodstream, first to be carried to the liver in the portal venous blood and then to circulate to the entire body. While in the bloodstream the ingested vitamin, as well as other circulating cobalamins, is bound by special transport proteins called transcobalamins (TC). Today it is thought that some form of the vitamin is utilized by all cells; thus cells throughout the body must recognize and absorb circulating cobalamins for their own needs. Within the cell the vitamin is moved from compartment to compartment, perhaps again with the assistance of special binding proteins, and the crucial, multistep conversions of vitamin to active coenzymes are carried out. Only as the coenzymes, adenosylcobalamin (AdoCbl) and methylcobalamin (MeCbl), does the cobalamin molecule function, and these coenzymes must interact with specific apoenzymes to provide appropriate holoenzyme activities. Cobalamin has a half-life in the human bloodstream of 9 to 10 days; body organs such as liver, kidney, and the circulating blood apparently serve as storage sites. Although catabolism of the cobalamin molecule has not been described in human tissues, it is known that it leaves the body in bile and urine. From bile, reabsorption can occur in the ileum. An efficient salvage process by way of reabsorption in the kidney tubule has been suggested for the rat (1) but data to support renal reabsorption in man have not been obtained.

Each of the steps mentioned above which participate in cobalamin utilization is under genetic control. Mutations affecting transport proteins, membrane recognition sites, or enzymes which metabolize the cobalamin molecule would be expected to lead to disease just as absence of the vitamin from the diet does. Several inherited disorders of cobalamin metabolism have been recognized in the past 20 years and undoubtedly more are awaiting discovery. In some ways these disorders mimic a cobalamin deficiency state but in other ways they show characteristic changes of their own. As expected, study of the human diseases is adding new knowledge to our understanding of normal physiology and metabolism.

Today we recognize the following seven disorders as being inherited defects primarily related to cobalamin metabolism (Table 8-1): juvenile pernicious anemia where intrinsic factor either is absent from otherwise normal gastric secretions or is altered so as to be nonfunctional; selective cobalamin malabsorption presumably due to failure of ileal cells to absorb or transport cobalamin to the bloodstream; deficiency of the circulating transport protein, transcobalamin I (TC I); deficiency of the other major transport protein, transcobalamin II (TC II); an abnormality in the conversion of the cobalamin vitamin to one of its two coenzymes, AdoCbl; and the failure to convert the vitamin to either coenzyme, AdoCbl or MeCbl. Pernicious anemia with classic adult onset very likely has an important genetic component to its etiology as well, for familial clustering of the disease and an increased susceptibility in close relatives have long been recognized (2–5).

Table 8-1 Inborn Errors of Cobalamin Metabolism

Disorder	Phase of Metabolism Affected	Nature of Defect
Juvenile pernicious anemia		
I.	Intestinal absorption	Intrinsic factor deficiency
II.	Intestinal absorption	Inactive intrinsic factor
III.	Intestinal absorption	Defective ileal transport
Transcobalamin I deficiency	Plasma transport	TC I deficiency
Transcobalamin II deficiency	Plasma transport	TC II deficiency
Methylmalonic aciduria	Tissue utilization	Defective AdoCbl synthesis
Methylmalonic aciduria and homocystinuria	Tissue utilization	Defective AdoCbl and MeCbl synthesis

The first three disorders listed above prevent absorption of dietary cobalamin and lead to a syndrome like that seen in adult pernicious anemia. They have been discussed in Chapter 7 and are not described further here.

In this chapter we address ourselves to cobalamin uptake by mammalian cells, apart from the intestinal absorption process, and to the intracellular conversion of vitamin to coenzymes. In this framework we then discuss the last four human diseases listed above, the two abnormalities of circulating proteins and the two blocks in converting the vitamin to its coenzymes.

COBALAMIN UPTAKE AND METABOLISM BY MAMMALIAN CELLS

After it became evident that a factor in normal gastric juice was necessary for cobalamin absorption by cells of the small intestine, a similar phenomenon was sought in other cells. Using bone marrow cultures from patients with pernicious anemia, Callender and Lajtha (6) in 1951 showed that *in vitro* maturation of megaloblasts to normoblasts could be accomplished in the presence of normal gastric juice and also in the presence of normal serum; this effect was partially abolished if the serum was preheated to 56° for 1½ to 2 hr. They hypothesized that an extragastric intrinsic factor, which combined with cobalamin to form hemopoietic factor, was instrumental in the maturation. Twelve years later Hall and Finkler (7) suggested that a protein from serum that bound cobalamin and had β mobility was the factor which mediated cobalamin uptake by peripheral tissues. This protein, later named transcobalamin II, is thermolabile and is extremely important to the uptake process, as detailed below.

Prior to investigations of the specific role of TC II, several observations had made clear the importance of a binding protein in cobalamin uptake. Hog intrinsic factor was shown to stimulate cobalamin uptake in rat liver slices (8, 9) and a dependence on calcium ion was noted (9). Cooper and Paranchych (10, 11) showed that HeLa cells took up cobalamin when bound to human serum but not when bound to hog intrinsic factor or human gastric juice. Mouse ascites cells also accumlated cobalamin bound to human or mouse serum or to mouse ascites fluid, but not cobalamin bound to intrinsic factor or serum from other species. This investigation of cobalamin uptake by human and mouse tumor cells led to the concept of a biphasic process. The first phase was a physicochemical adsorption of bound cobalamin to cell surface receptors which were specific for a protein in serum or ascites fluid; this phase occurred in less than a minute and was relatively insensitive to changes in pH and temperature or to metabolic inhibitors. A slower second phase was sensitive to pH and temperature, and uptake was partially inhibited by cyanide and dinitrophenol and totally inhibited by iodoacetate. Cooper and Paranchych suggested that this second phase represented transport of the adsorbed cobalamin into the cell and that the transport was energy and sulfhydryl group dependent. A parallel between this biphasic uptake process and pinocytosis in amoebae was noted leading to the hypothesis that a pinocytotic mechanism might be operative for cobalamin uptake in mammalian cells.

Transfer of cobalamin into cells in the absence of a binding protein has also been demonstrated. Using reticulocyte-rich blood, Retief et al. (12) presented evidence for two types of uptake. One was serum mediated and calcium dependent and led to localization of cobalamin in the erythrocyte stroma. This uptake corresponds to the primary adsorption phase described by Cooper, but the second, delayed phase of uptake was not seen in the red cell. Another mechanism for uptake by erythrocytes was shown in saline medium without any serum. This mechanism had characteristics of a simple diffusion process without dependence on calcium, and the cobalamin did not associate with membrane stroma where hypothetical protein receptor sites might logically be found.

Transcobalamin II and Cobalamin Transport into Cells

In an important series of experiments, Finkler and Hall have extensively examined the uptake of cyanocobalamin (CN-Cbl) by HeLa cells in monolayer culture. Their initial studies demonstrated the specificity of the uptake system for CN-Cbl bound to TC II (13). The TC II-bound cobalamin was taken up rapidly in sharp contrast to the minimal uptake of cobalamin bound to TC I or to binders from red blood cells, leukocytes, gastric juice, and saliva, or of unbound cobalamin in buffered saline. Figure 8-1 shows the difference between TC II-bound and TC I-bound cobalamin uptake. Similar differences were reported when primary cultures of human amnion and human kidney cells were used. Further work showed that an antibody to human TC II blocked the uptake but antibodies to TC I and salivary binder did not (14). Also, TC IIs from other species, if they cross-reacted with antihuman TC II, would stimulate HeLa cell uptake of cobalamin; TC IIs that did not react with antihuman antibody had no effect (15). Thus binding sites on the TC II molecule for antibody and for cell surface receptors apparently are related.

When inhibitors were evaluated in the HeLa system (16), cyanide, dinitrophenol, iodoacetamide, and dithiothreitol caused more than 70% inhibition of TC II-mediated uptake. If the cells were preincubated with puromycin or actinomycin D, inhibition was also seen, presumably due to an effect on protein synthesis. These inhibitor studies provided further evidence that cobalamin uptake was in part dependent on cell energy metabolism and on sulfhydryl groups and disulfide bonds.

We have recently examined CN-[^{57}Co]Cbl uptake by diploid human skin fibroblasts in monolayer culture (17) and have also evaluated the

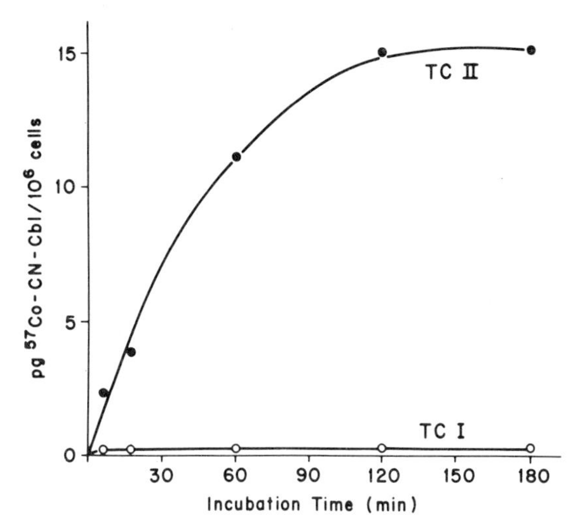

Fig. 8-1 Uptake by HeLa cells in monolayer culture of CN-[^{57}Co]Cbl bound to transcobalamin I (TC I) or to transcobalamin II (TC II). [Adapted from Finkler and Hall (13).]

effect of inhibitors. TC II-Cbl uptake over an 8-hr interval was 4 to 8 times greater than uptake of free cobalamin or cobalamin bound to leukocyte binder. Uptake was saturable for both TC II-Cbl and free cobalamin and the apparent affinity for each was the same, with a K_m of 1000 pg/ml. Maximum velocity for uptake of the bound vitamin, 100 pg/mg of cell protein each hour, was 10 times greater than with the free vitamin. We also witnessed striking inhibition by dithiothreitol but not by dinitrophenol or fluoride.

Other work has also supported an important role for TC II in cobalamin uptake. Retief and co-workers reported that β binder (TC II) was more effective than α binder (TC I) in mediating CN-Cbl uptake by reticulocyte-rich red blood cell suspensions (18) and by rat liver homogenates (19). Hall and Finkler (20) showed that when free CN-[^{57}Co]Cbl was injected intravenously or given orally, most of the cobalamin was immediately associated with TC II and disappeared from the plasma in a matter of hours. Only a small amount was found with TC I and this disappeared very slowly from plasma. This was further documented by Hom and Oleson (21). It has also been shown that, associated with the uptake of TC II-bound cobalamin from plasma, there is disappearance of TC II itself from the plasma (22). These several lines of experimentation have generated the concept that TC II mediates rapid cobalamin uptake in cells throughout the body much as IF does at the ileal mucosa and that this TC II-mediated uptake is the normal mechanism when physiologic amounts of cobalamin are present in the circulation. The uptake appears to have two phases: first, the adsorption of TC II-Cbl to a surface recognition site on the cell membrane, and second, entry of either the TC II-Cbl complex or the cobalamin molecule alone into the cell by a facilitated transport process. TC II is depicted as carrying recently absorbed cobalamin from the small intestine to the liver and peripheral tissues. TC I, on the other hand, is thought to play a storage role for cobalamins that will remain in the plasma for many days.

The roles of TC I, TC II, and possible additional plasma binders of cobalamin will undoubtedly be further defined. Earlier work presumed that, for the most part, TC II bound those cobalamins that had only recently gained access to the body. Evidence now has been published that TC II is also involved in plasma transport of cobalamins days to weeks after their entry into the body (23). Questions about the roles of the transport proteins also arise from the demonstration that several cobalamins circulate in the plasma. Lindstrand and Ståhlberg (24–26) and later Linnell and co-workers (27, 28) used chromatography in association with bioautography to identify the various forms. This work has suggested the presence of both of the vitamin forms OH-Cbl and CN-Cbl and both coenzymes AdoCbl and MeCbl in the circulation. Which ones are bound to TC I and which to TC II and how much exchange of cobalamins occurs between these two binders are not yet known. Answers to these questions should give a clearer picture of the relationships among several cobalamin pools: intracellular and extracellular cobalamins; newly absorbed and stored cobalamins; and vitamin and coenzyme forms of cobalamins.

Intracellular Metabolism of Cobalamins

Within the cell the cobalamin molecule must be transported to specific sites where vitamin forms are converted to coenzymes and where the attachment of coenzyme to apoenzyme is accomplished. Two cobalamin coenzymes have been found in human and other mammalian tissues; these are AdoCbl and MeCbl. Our knowledge of their synthesis is based largely on bacterial studies, the details of which have been presented in Chapters 2 and 3. It is important to our later discussion of inborn errors of cobalamin metabolism to review the coenzyme synthetic pathways; an outline is presented in Figure 8-2.

Cobalamin entering the cell as OH-Cbl contains trivalent cobalt.

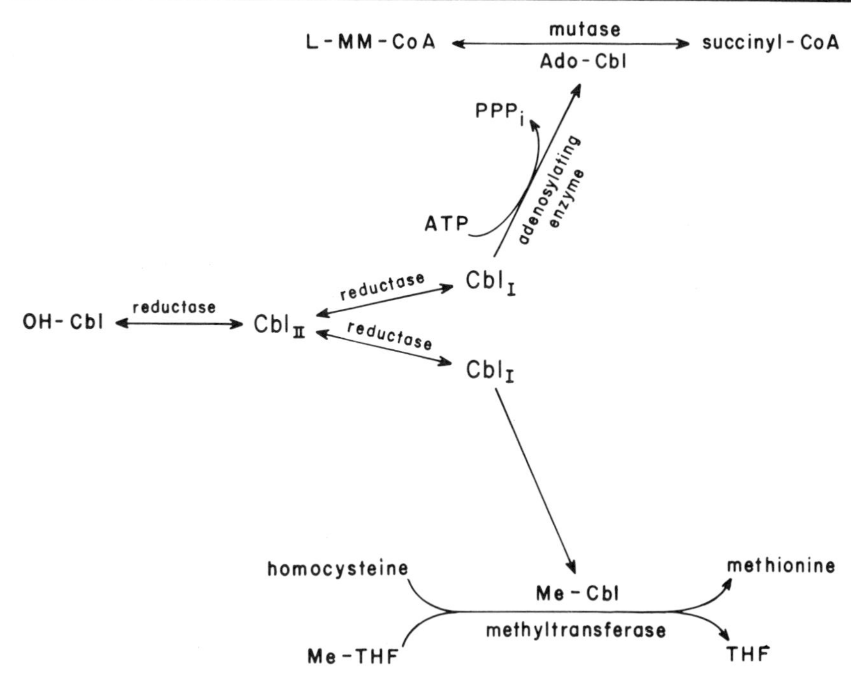

Fig. 8-2 Intracellular synthesis of cobalamin coenzymes. OH-Cbl is reduced in two steps to cob(I)alamin which is then adenosylated or methylated to form the coenzymes AdoCbl or MeCbl, respectively. AdoCbl is coenzyme for methylmalonyl CoA mutase and MeCbl for methyltetrahydrofolate homocysteine methyltransferase.

Since the synthesis of coenzymes requires reduced, monovalent cobalt, a two-step reduction process is required to reduce OH-Cbl first to cob(II)alamin and then to cob(I)alamin. Each reduction step appears to be catalyzed by a separate reductase. Cob(I)alamin is then adenosylated or methylated to form, respectively, AdoCbl or MeCbl. Whether one or both reduction steps are shared in the synthesis of AdoCbl and MeCbl is not known, nor has it been established whether the two coenzymes are synthesized in the same organelle. Since the equilibrium between cob(II)alamin and cob(I)alamin greatly favors the oxidized form, energy considerations would suggest that at least the final reduction step be closely linked spatially to adenosylation or methylation. Thus it is conceivable that different reductase enzymes exist in separate pathways to the two coenzymes. In Figure 8-2, we have arbitrarily drawn the first reductase as a single species common to the synthesis of both coenzymes, and have depicted the second reduction step as specific to the synthesis of each coenzyme. This scheme is entirely speculative with regard to the reduction steps and may be modified significantly as knowledge is acquired in this area.

In mammalian cells, each coenzyme functions with a single apoenzyme. AdoCbl is the coenzyme for methylmalonyl CoA mutase (MM-CoA mutase) and MeCbl is the coenzyme for methyltetrahydrofolate homocysteine methyltransferase. No other cobalamin-dependent enzymes have been found thus far in human or other mammalian tissues. Mammalian MM-CoA mutase is thought to be a mitochondrial enzyme (29, 30) and evidence is accumulating that AdoCbl is synthesized in the mitochondrion. The methyltransferase has most of its activity in the cytoplasm of rat liver cells (31) but the site of synthesis of MeCbl has not yet been defined. The two enzymes, MM-CoA mutase and methyltetrahydrofolate homocysteine methyltransferase, are discussed in detail elsewhere in this volume.

The first demonstration that human tissues could convert CN-Cbl to AdoCbl used liver and kidney homogenates; the tissues were incubated with CN-[^{60}Co]Cbl and labeled coenzyme was isolated by radiochromatography (32). Subsequently, we showed the synthesis of both AdoCbl and MeCbl by intact human skin fibroblasts growing in tissue culture medium containing OH-[^{57}Co]Cbl (33). Coenzymes were indentified in fibroblast extracts by radiochromatography.

The only part of the coenzyme synthetic pathway that has been studied in any detail in mammalian cell extracts is the final two steps in the synthesis of AdoCbl, the reduction of cob(II)alamin to cob(I)alamin followed by adenosylation. Because of the extreme lability of cob(I)alamin, the two steps have been assayed as one, without attempting to separate them. In extracts of HeLa cells, AdoCbl synthesis occurred in the presence of OH-Cbl, a reducing system, FAD, and ATP (34). The reducing system (dithiothreitol, mercaptoethanol, and hydrogen) bypassed the enzymatic reduction of OH-Cbl to cob(II)alamin. AdoCbl was identified by its ability to stimulate the dioldehydrase reaction (35) and by its photolability. In as yet unpublished studies from our own laboratory, we have used similar incubation conditions and a radioassay to demonstrate the conversion of OH-[^{57}Co]Cbl to Ado[^{57}Co]Cbl by extracts of human fibroblasts. Preliminary subcellular fractionation studies have placed this adenosylating activity in a mitochondrial fraction.

In vivo experiments have also studied the intracellular localization and metabolism of cobalamins. Newmark (36, 37) injected rats with CN-[^{57}Co]Cbl intramuscularly and examined subcellular fractions of their kidneys from 1 to 28 days later. Up to 70% of the injected cobalamin was found in the kidney in the first few days after injection; this fraction slowly declined to 20% after 4 weeks. During the entire 28 days, 50 to 65% of radioactive cobalamin in the kidney was found in a mitochondrial-lysosomal fraction, about 20% in the nucleus–cell debris fraction, 20% in the supernatant, and less than 3% in the microsomal fraction. Endogenous cobalamin in the kidney, assayed by chromatography-bioautography, had the same distribution. When the mitochondrial-lysosomal fraction was further fractionated, the lysosome-rich fraction had slightly larger amounts of radioactivity than did the mitochondria-rich fraction. Microbiological assay of endogenous cobalamins in those two fractions showed AdoCbl predominantly in the mitochondria-rich fraction and OH-Cbl in the lysosome-rich one.

Pletsch and Coffey used rat liver to study the intracellular distribution of ^{57}Co-labeled cobalamin (38, 39). They administered CN-[^{57}Co]Cbl by intracardiac injection and determined its hepatic subcellular distribution at intervals from 5 min to 72 hr later. Immediately after injection 98% of the CN-[^{57}Co]Cbl was bound to TC II. Five minutes after injection the fraction which contained plasma membrane vesicles and microsomes contained the highest amount (46%) of radioactivity. In this fraction the radioactivity was primarily with plasma membranes and was associated with a protein which had characteristics of TC II. At 30 min and 2 hr. the microsome–plasma membrane fraction continued to have the largest percentage of cobalamin, but during this period a lysosomal fraction also contained significant (15–20%) radioactivity. In this lysosomal fraction the cobalamin was 70% TC II bound and 30% free at the 30-min time point. When this fraction was further incubated *in vitro,* progressively less cobalamin was protein bound. At 24 and 72 hr after injection, one-third of the radioactivity was in the mitochondrial fraction and another third in the supernatant. Almost half of the mitochondrial cobalamin and 90% of the supernatant cobalamin was associated with a high-molecular-weight binder and none was associated with TC II.

These cellular fractionation studies, along with previously cited evidence, seem most compatible with the TC II-Cbl complex first being adsorbed to the surface of the cell and then entering intact into the cell. Pinocytosis and the formation of a secondary lysosomal vacuole, in which the cobalamin is split from the transport protein, have been suggested as the mechanism of entry (36, 38). The studies further suggest that, after being split from the carrier protein, the cobalamin molecule accumulates in mitochondria, the site of at least some of its further metabolism and function

Subcellular fractionation studies of ileal mucosal cells during IF-mediated cobalamin uptake have also been reported (40–42). These studies have not supported a pinocytotic ingestion of the IF•Cbl complex, to parallel the suggested pinocytosis of TC II-Cbl by kidney and liver cells, but they have garnered evidence for accumulation of the absorbed cobalamin in mitochondria. There is conversion of some of the absorbed vitamin to AdoCbl and perhaps some other processing of the cobalamin in that organelle before passage into the portal blood. Although the form or forms of the cobalamin that exit from the ileal cell remain uncertain, these studies provide further evidence that AdoCbl synthesis occurs in mitochondria.

In extracellular fluids, cobalamins in physiologic amounts are totally bound by specific binding proteins. Within the cells, at least some of the coenzymes must be bound to their respective apoenzymes; whether OH-Cbl and the remainder of the coenzymes have specific intracellular binders is not known. Cobalamins, including coenzymes, must also exit from cells. MeCbl is the major plasma cobalamin in the healthy individual and AdoCbl is probably also present in significant amounts in plasma, although the data on this point are conflicting (25, 28) (see also Chapter 6). The mode of exit of the cobalamins, for example by secretion across the cell membrane, by exocytosis, or by cell lysis, has not been studied. Finkler (43) showed the transfer of 95% of cellular [^{57}Co]Cbl from TC II to TC I after 16 hr in HeLa cell monolayers. This may mean that most cobalamins that exit from cells will be bound to TC I. This is by no means clear, however, for Rappazzo and Hall (44) have perfused isolated dog livers and kidneys and report that during one passage through the liver, free cobalamin comes out bound to an R binder, which could be TC I, but that during passage through the kidney, the free cobalamin is bound to TC II.

A general picture of cobalamin uptake and intracellular processing is emerging (Figure 8-3). It appears certain that TC II-Cbl is adsorbed to the cell surface and then gains entry, possibly by pinocytosis. At some early stage, perhaps at the cell surface but possibly within the cell, cobalamins are freed from TC II and transported to special intracellular sites. Some cobalamin molecules cross the mitochondrial membrane to enter the mitochondrion. The synthesis of AdoCbl almost certainly occurs within the mitochondion and the coenzyme is bound there to its apoenzyme, MM-CoA mutase. A recent study reports that 50 to 70% of labeled cobalamin within rat liver mitochondria is bound to MM-CoA mutase 3 to 7 days after CN-[^{57}Co]Cbl injection (45). Figure 8-3 does not suggest a site of synthesis of MeCbl in the mammalian cell. The methyltetrahydrofolate homocysteine methyltransferase may be an enzyme of the cytosol (31) and the association of MeCbl with its apoenzyme is shown in that location in the final diagram. In studies of *Escherichia coli,* the synthesis of MeCbl seems to be closely allied to the methyltrans-

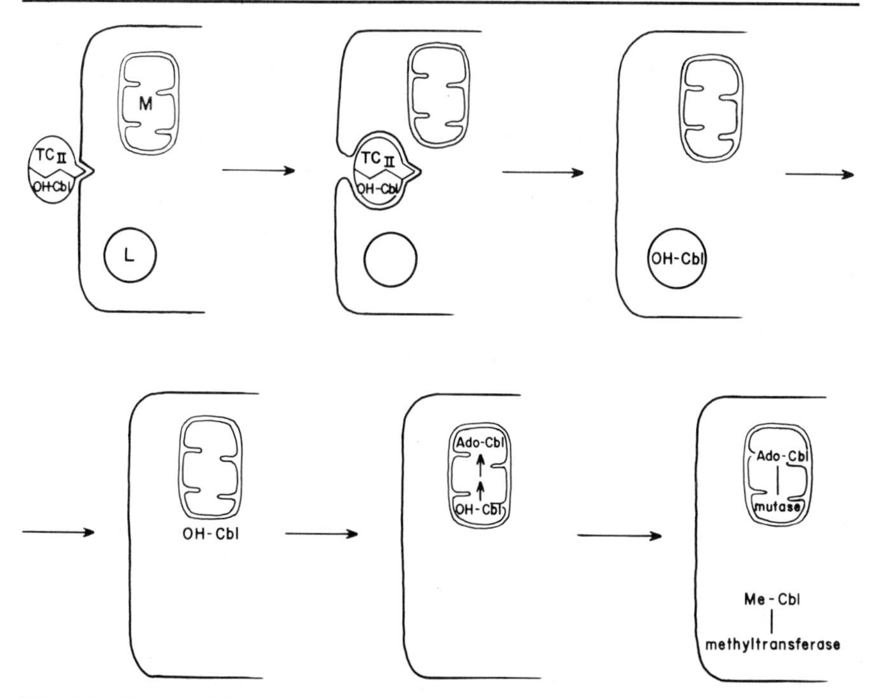

Fig. 8-3 Suggested intracellular processing of cobalamins. The circulating complex of TC II and cobalamin vitamin (OH-Cbl) binds to recognition sites on the cell membrane and is taken into the cell by a mediated process that suggests pinocytosis. The pinocytotic vacuole fuses with a primary lysosome (L) and OH-Cbl is split from TC II in the resultant secondary lysosome. From that organelle, some of the OH-Cbl moves to the mitochondrion (M), is transported across the mitochondrial membrane, and then converted to AdoCbl to function as coenzyme for MM-CoA mutase. Also, OH-Cbl is converted to MeCbl, at an unspecified site, and serves as coenzyme for the methyltetrahydrofolate homocysteine methyltransferase reaction in the cytosol.

ferase enzyme complex (46, 47). In the absence of similar information in mammalian cells, we have declined to speculate about the site or sites of MeCbl synthesis. Further intracellular phenomena, including exit from the cell, undoubtedly occur and details of these processes will very likely be clarified in the next several years.

INBORN ERRORS: INTRODUCTION

Two enzymatic reactions in man require a cobalamin coenzyme. These have been noted above and will be considered in detail in Chapter 9. Attempts to understand the pathophysiology and biochemical pathology of pernicious anemia based on blocks in these two reactions have been only partially successful. Similar problems and contradictions arise when one examines the inborn errors of cobalamin metabolism; these problems are highlighted in the following discussions. It may be that we do not yet have sufficient understanding of the pathophysiologic effects of a block in MM-CoA mutase or methyltetrahydrofolate homocysteine methyltransferase activity. Alternatively, the inability to explain all of the observed clinical and chemical findings may be due to the presence of still unrecognized cobalamin-dependent reactions or of other roles for the cobalamin coenzymes in human biochemistry.

Figure 8-4 shows the propionate-methylmalonate pathway indicating the precursors and major catabolites of these short-chain organic acids. AdoCbl is required in the last step for the isomerization of L-methylmalonyl CoA to succinyl CoA, which then enters the tricarboxylic acid cycle. A partial block in MM-CoA mutase activity leads to methylmalonic acidemia and methylmalonic aciduria as described in cobalamin deficiency states. A severe deficiency in mutase activity should have wider secondary effects on the related amino acid and lipid pathways. Figure 8-5 depicts part of the pathways of sulfur amino acid metabolism and the interaction with folate metabolism. The MeCbl-dependent methyltetrahydrofolate homocysteine methyltransferase is a key enzyme in the recycling of folates, specifically in the generation of tetrahydrofolate, and also provides one way of synthesizing methionine. A block in this reaction seriously interferes with folate metabolism, and is associated with deranged nucleic acid metabolism and megaloblastic changes in pernicious anemia. The block also causes homocystinuria, which has been seen in cobalamin deficiency. These effects of deficient mutase or methyltransferase activities should be kept in mind as we discuss the inborn errors.

INBORN ERRORS OF TRANSCOBALAMINS

Transcobalamin I Deficiency

Carmel and Herbert (48) described two brothers who had a partial deficiency of TC I (α binder), diagnosed at 46 and 47 years of age. These are the only two patients recognized with this condition. The index case was found because of a deficiency of serum cobalamin 3 years after subtotal gastrectomy for gastric ulcer disease. Repeated measurements of serum cobalamin in both brothers gave results less than 100 pg/ml (normal $>$150 pg/ml). Neither brother had any other evidence of cobalamin deficiency, however, and there was no methylmalonic aciduria or any hematologic change.

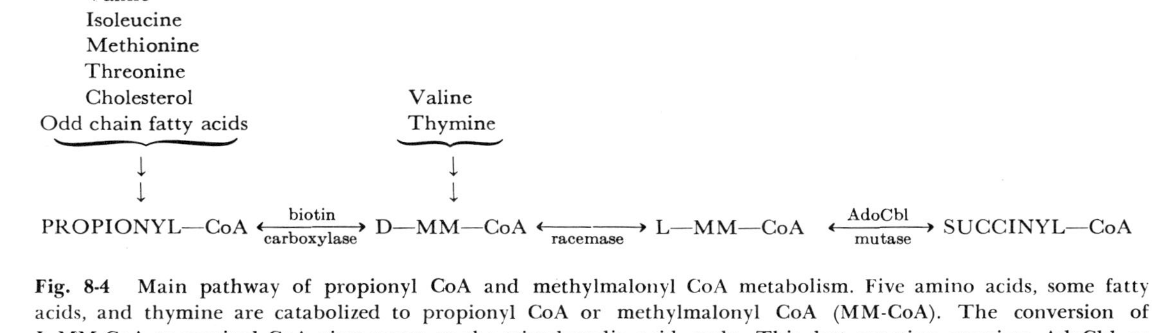

Fig. 8-4 Main pathway of propionyl CoA and methylmalonyl CoA metabolism. Five amino acids, some fatty acids, and thymine are catabolized to propionyl CoA or methylmalonyl CoA (MM-CoA). The conversion of L-MM-CoA to succinyl CoA gives access to the tricarboxylic acid cycle. This last reaction requires AdoCbl as coenzyme for MM-CoA mutase.

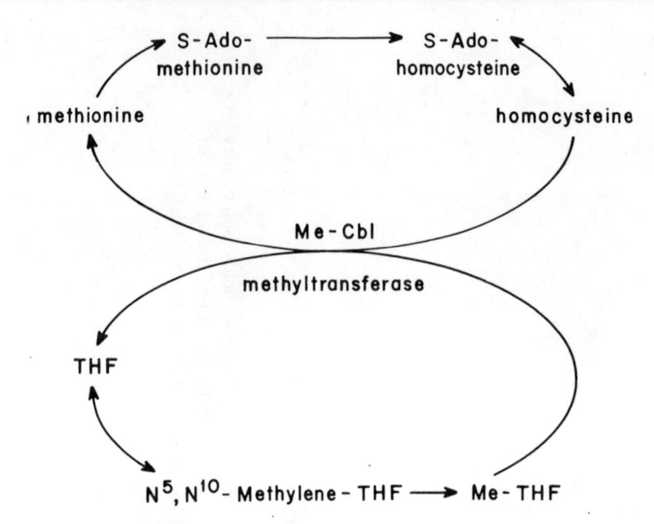

Fig. 8-5 Role of methylcobalamin (MeCbl) in sulfur amino acid and folate metabolism. Methyltetrahydrofolate homocysteine methyltransferase, with MeCbl as coenzyme and agent of the methyl group transfer, catalyzes the synthesis of methionine from homocysteine (homocysteine that is not reconverted to methionine is further metabolized to cystathionine, cysteine, and sulfate). Simultaneously, methyltetrahydrofolate (Me-THF) is converted to tetrahydrofolate (THF) to generate that compound for use in other folate-mediated reactions.

An examination of cobalamin binding proteins showed that TC I values in each brother were about 35% of the control mean. There was almost no detectable leukocyte or salivary binder. TC II levels were 70 to 80% of the control mean and intravenously administered [^{57}Co]Cbl disappeared rapidly in the manner typical of TC II-bound cobalamin. Monthly cobalamin injections (100 μg) for more than a year in the index case had no effect on the serum cobalamin level.

The parents of these two brothers were not living. Measurements of serum cobalamin and TC I in a brother and in children and grandchildren of the propositi all gave normal results. An inherited etiology was postulated but could not be further supported. These cases suggest that TC I deficiency of moderate degree, along with virtual absence of leukocyte and salivary cobalamin binders, is innocuous in its effect on cobalamin metabolism. Whether total TC I deficiency would be equally innocuous is entirely speculative.

Transcobalamin II Deficiency

As with TC I deficiency, only a single family, with two affected sisters, has been reported with the diagnosis of TC II deficiency (49). In striking contrast to the benign course of the first disorder, symptomatology in TC II deficiency was severe and appeared soon after birth. The affected infants presented at ages 3 and 5 weeks with failure to gain weight, vomiting, diarrhea, and infection. They had anemia, leukopenia, thrombocytopenia, and megaloblastic changes in their bone marrows. Serum cobalamin was 855 pg/ml in one of the girls. Treatment with high, pharmacologic amounts of intramuscular cobalamin, 1 mg twice weekly, induced and maintained full remission of symptoms and hematologic findings, and restored normal growth and development.

A study of the cobalamin binding proteins in the sisters revealed no TC II. TC I was present but was significantly decreased. In addition, oral absorption of [^{57}Co]Cbl with or without added IF was much reduced. Measurements of unsaturated TC II binding capacity were made in family members; several persons, including both parents, had low values that may reflect the heterozygous state. None of these individuals had any symptoms.

Biochemical parameters of cobalamin deficiency had not been sought in the affected girls during early infancy and the important experiment of withdrawing therapy to measure these was done when one of the patients was 12 months old (50). Therapy was stopped, for 6 weeks, at a time when the patient was in complete remission. Hematologic relapse occurred quickly and severe megaloblastic changes were present at the end of the 6 weeks. During that time no methylmalonic aciduria occurred and there was normal oxidation of propionate by peripheral leukocytes *in vitro,* a test of whether the propionate-methylmalonate pathway was intact. Levels of methionine, homocystine, and cystathionine in plasma and urine also remained normal. Thus no evidence of reduced activity of the known AdoCbl- and MeCbl-dependent reactions occurred despite severe megaloblastic changes.

These two infants have provided striking confirmation of the importance of TC II for delivery of cobalamin to body cells, especially to hematopoietic cells. The patients' cobalamin requirements can be met only by administering pharmacologic amounts of vitamin that maintain serum cobalamin levels at over 5000 pg/ml. Presumably this high concentration enables passive diffusion to deliver sufficient cobalamin to the cells. The finding of poor intestinal absorption in the presence of IF gives support to the proposed role of TC II as a serum recipient for newly absorbed cobalamin.

The important question that arises from studies of this family is the nature of the relationship between megaloblastic anemia and the known AdoCbl- and MeCbl-dependent enzymatic reactions. Without cobalamin therapy megaloblastic changes and severe anemia were present in a few weeks' time but methylmalonic aciduria and changes in sulfur amino acid metabolism had not occurred. Although direct measurements of enzyme activities were not made, the findings suggest that stores of holomutase and holomethyltransferase were not severely depleted at a time when megaloblastic changes were well advanced. Thus, the currently held hypothesis that a block in the MeCbl-dependent methyltransferase reaction leads to megaloblastic anemia by way of an effect on folate metabolism may not be a sufficient explanation.

INBORN ERRORS OF COENZYME METABOLISM

Methylmalonic Aciduria and Propionic Acidemia

In 1967 Oberholzer et al. (51) and Stokke et al. (52) described the first infants with congenital methylmalonic aciduria and proposed an inherited block in MM-CoA mutase activity and a possible association with cobalamin metabolism (Figure 8-4). The amount of methylmalonic acid excreted in the patients was several-fold higher than the amount excreted by adults with pernicious anemia. Stokke also noted the similarity of symptoms in these children with a previously reported condition known as "ketotic hyperglycinemia." Later work established that ketotic hyperglycinemia was due to propionyl CoA carboxylase deficiency (53) and the disorder was more appropriately named propionic acidemia. Soon after the first descriptions of congenital methylmalonic aciduria, Rosenberg (54), Lindblad (55), and co-workers reported a cobalamin responsive form of the disease. Morrow and colleagues (56) then provided the enzymatic evidence that there were two forms of congenital methylmalonic aciduria by studying liver extracts from patients who had died (Table 8-2). Mutase activity was barely detectable in the livers of three patients and showed no response when AdoCbl was added; these patients probably had a deficient or abnormal mutase apoenzyme. In a fourth patient mutase activity was restored to control levels by the addition of the coenzyme AdoCbl; it was suggested that this patient had an abnormality of cobalamin metabolism or a defect in coenzyme-apoenzyme interaction. Most recently a defect in methylmalonyl CoA racemase (57) has been added to the disorders of the propionate-methylmalonate pathway.

All of the patients with the above-mentioned blocks in propionate and methylmalonate metabolism have presented with a common symptomatology (58). Failure of normal growth and development, periodic

Table 8-2 Methylmalonyl CoA Mutase Activity in Liver Homogenates from Patients with Congenital Methylmalonic Aciduria

	cpm ^{3}H-Succinate Formed/mg Liver Protein	
Patient	No Added AdoCbl	Added AdoCbl
1	6	17
2	44	189
3	16	41
4	452	7770
Controls	3040–4920	4540–6010

Enzyme activity was assayed using racemic ^{3}H-methylmalonyl CoA for substrate and isolating the product, ^{3}H-succinate, by paper chromatography. Data from three controls are included. Adopted from Morrow et al. (56).

attacks of ketoacidosis including the presence of long-chain ketones, and intermittent hypoglycemia, hyperammonemia, and hyperglycinemia have been present in most of the patients. In propionyl CoA carboxylase deficiency, propionic acid is the major organic acid that accumulates in blood and urine. Methylmalonic acid characterizes the racemase and mutase blocks but propionic acid accumulates as well (59, 60). In severe cases of these diseases symptoms have appeared in the first days of life, after the infant has left the uterine environment where maternal metabolism could compensate for abnormal fetal metabolism. The introduction of protein feeding in the newborn period has at times been fatal and protein intolerance has been a common feature of the diseases. Many affected children have died or become severely retarded. Early diagnosis with prompt institution of overall protein restriction or specific amino acid restriction is now beginning to allow recovery and normal development in some children. Cobalamin therapy has been important for patients with a metabolic error of coenzyme synthesis; their disorders are discussed further in the next sections.

Defective Synthesis of Adenosylcobalamin

Methylmalonic aciduria, and later propionic aciduria, had been recognized as concomitants of cobalamin deficiency (61, 62). When congenital methylmalonic aciduria was described, cobalamin deficiency was sought but not found. Nonetheless, some of the patients responded to cobalamin therapy with a decrease in methylmalonate excretion and showed protection against the toxic effects of protein or of the amino acids which were catabolized to propionate or methylmalonate (63). These patients were not cobalamin deficient and had normal serum levels prior to therapy. They required massive, pharmacologic amounts of cobalamin (250–2000 μg/day) to derive maximal effect but even this maximal effect did not return methylmalonate excretion to normal. In one patient (R.P.) whom we have been treating with 1000 μg CN-Cbl intramuscularly each day, serum cobalamin rose to 19,000 pg/ml.

The finding of cobalamin responsiveness in patients with congenital methylmalonic aciduria was followed by demonstration of responsiveness in cells in culture. Skin fibroblasts from an affected child, grown in standard tissue culture medium containing 25 to 50 pg cobalamin/ml, could not oxidize ^{14}C-labeled propionate and were markedly deficient in intracellular coenzyme, AdoCbl (64). When cobalamin concentration, as either CN-Cbl or OH-Cbl, was raised to 10^6 pg/ml in the culture medium, the cells could oxidize propionate normally and AdoCbl content was measured at control levels.

These observations suggested an abnormality in the synthesis or storage of AdoCbl in patients with cobalamin-responsive methylmalonic aciduria. Direct confirmation of this defect used mutant fibroblasts from patient R.P. in tissue culture (65). During 5 days in culture, mutant cells took up and retained normal amounts of [^{57}Co]Cbl and synthesized normal amounts of Me[^{57}Co]Cbl; in contrast, accumulation of Ado[^{57}Co]Cbl was virtually absent (Figure 8-6). We have interpreted these results as a block in the synthesis of the single coenzyme AdoCbl.

After finding the evidence for a block in AdoCbl synthesis, we have attempted to localize the defect. As can be seen in Figure 8-2, the last steps in AdoCbl synthesis, that is, the final reduction step or the step involving the transfer of an adenosyl moiety from ATP to cob(I)alamin, seemed likely. Adapting the method of Kerwar et al. (34), we have assayed for the combined cob(II)alamin reductase and adenosylating activities in crude extracts of several fibroblast lines, all of which shared the whole cell phenotype of deficient AdoCbl but normal MeCbl synthesis in cells growing in culture. In three cell line extracts, including R.P. extract, we found no difference between mutant and control cell synthetic activities (66). In contrast, three other extracts had almost no detectable activity of the combined final two steps. Further evidence for the existence of two different intracellular defects leading to the same whole cell phenotype of deficient AdoCbl synthesis has been provided by Gravel's work in our laboratory (67). Using Sendai virus mediated heterokaryons of two mutant cell lines at a time, Gravel et al. identified

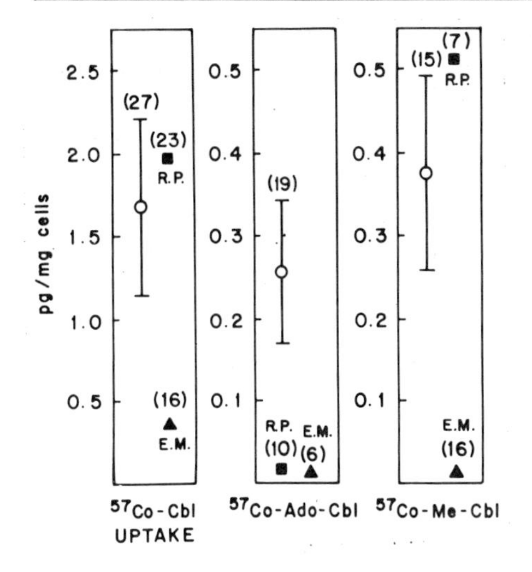

Fig. 8-6 Uptake of OH[^{57}Co]Cbl and conversion to coenzymes by skin fibroblasts in monolayer culture. R.P. cells came from a patient with methylmalonic aciduria, responsive *in vivo* to cobalamin therapy. E.M. cells were from an infant with methylmalonic aciduria, homocystinuria, and cystathioninuria. The cell cultures were grown with the labeled vitamin for 5 days, after which coenzymes were extracted from the cells and isolated by chromatography. Open circles represent mean ± S.D. for control cultures; numbers of observations are in parentheses (65).

two complementation groups. One included the cell lines whose extracts could synthesize normal amounts of AdoCbl in the assay measuring cob(II)alamin reductase and adenosylating activity, and the other included those cell lines whose extracts were severely deficient in that activity. Thus heterogeneity exists among the fibroblast lines where whole cells cannot synthesize adequate AdoCbl but do synthesize normal amounts of MeCbl. One class has a demonstrated defect in the final steps of AdoCbl synthesis; we have named these Cbl B mutants. The other class, Cbl A mutants, requires further work that will examine more proximal sites, for example initial reduction of OH-Cbl or cobalamin transport into mitochondria, in seeking an explanation for the defect.

The first patients who were identified as having a block in AdoCbl synthesis have also been clinically responsive to pharmacologic amounts of either CN-Cbl or OH-Cbl. Studies using cell culture or extracts of cells and tissues have demonstrated an *in vitro* parallel to the clinical responsiveness. This parallel of *in vivo* responsiveness with *in vitro* indications of defective AdoCbl synthesis has not been universal, however. Kaye et al. (68) reported studies of cultured fibroblasts from two patients who, *in vivo,* did not respond to cobalamin therapy with a decrease in methylmalonate excretion. Each patient received daily 1000 μg CN-Cbl parenterally, one for 5 days and the other for 8 days. The fibroblast lines in culture, however, did respond to added cobalamin by increasing propionate oxidation, and methylmalonyl CoA mutase activity in cell extracts was restored by adding AdoCbl. In our laboratory intact fibroblasts from both of these patients showed a severe deficiency in AdoCbl synthesis similar to that observed in R.P. cells (unpublished data). Thus it seems likely that some mutations affecting coenzyme metabolism cannot be overcome *in vivo* by providing very high concentrations of vitamin substrate. The use of the coenzyme AdoCbl for therapy might help in such situations although little is known about the ability of AdoCbl to cross plasma and mitochondrial membranes intact. We have treated one responsive patient (R.P.) with intramuscular AdoCbl and found that the same amounts were needed as with CN-Cbl.

Defective Synthesis of Adenosylcobalamin and Methylcobalamin

The combination of methylmalonic aciduria and abnormal sulfur amino acid metabolism in a male infant led Mudd and his associates (69, 70) to propose an abnormality of cobalamin metabolism which involved both cobalamin-dependent enzymes, MM-CoA mutase and methyltetrahydrofolate homocysteine methyltransferase. This infant boy presented with homocystinemia, cystathioninemia (cystathionine is a catabolite of homocysteine), and hypomethioninemia. Methylmalonate excretion was about 20 mg/day, considerably less than the 500 to 1000 mg/day excreted by infants described above with deficient synthesis of AdoCbl, but elevated five- to tenfold above normal excretion. He was critically ill almost from birth with anemia and seizures and died at $7\frac{1}{2}$ weeks of age. His bone marrow was hypoplastic but showed no evidence of megaloblastic change.

Study of postmortem tissues and of cultured fibroblasts gave further evidence of a defect in cobalamin metabolism (71, 72). Total cobalamin content of liver was normal but AdoCbl was less than 10% of control values. Cultured fibroblasts showed a block in propionate oxidation with accumulation of intracellular methylmalonate, while growth of the cells in OH-Cbl-supplemented medium corrected the defect in propionate oxidation. Holomethyltransferase activity was low in autopsy tissues and in extracts of cultured cells. However, methyltransferase apoenzyme was demonstrated in the cultured cells by adding MeCbl to the enzyme assay and partially restoring the methyltransferase activity. Growing the cells in medium supplemented with large amounts of OH-Cbl also restored activity.

Cell culture studies were used to substantiate the hypothesis of a failure of coenzyme accumulation (65). When the child's fibroblasts were examined for their ability to synthesize coenzyme (Figure 8-6, E.M. cells), almost no AdoCbl or MeCbl appeared, and the cells contained less than 25% of the total cobalamins accumulated by control cells. We have also assayed total endogenous cobalamin content of E.M. cells by radioassay and find only about one-fourth the amount present in control cells. Figure 8-6 contrasts the cellular coenzyme findings of the two coenzyme disorders. R.P. cells showed deficient synthesis only of AdoCbl just as the patient showed evidence of MM-CoA mutase deficiency alone without any sign of a sulfur amino acid disorder. R.P. cells synthesized normal amounts of MeCbl and had normal total cobalamin content. E.M. had deficient mutase and methyltransferase activities and his cells in culture failed to synthesize AdoCbl or MeCbl and had a low cobalamin content. We have termed this abnormality of cobalamin metabolism in cultured fibroblasts the Cbl C mutant. It forms another complementation group in the analysis using heterokaryons (67).

Analysis of the defect in E.M. cells has focused on transport of the vitamin into the cell or on an early step in intracellular metabolism. Possibilities would include binding of TC II-Cbl to the cell surface, cell uptake by pinocytosis or other mechanism, splitting of the cobalamin from TC II by a lysosomal hydrolase or other reaction, intracellular transport of the cobalamin molecule, and perhaps a common reductase in the pathways from vitamin to coenzymes (see Figures 8-2 and 8-3). Studies in our laboratory have identified no abnormality in the initial binding and uptake of labeled cobalamin by E.M. cells (17), but have found an important abnormality within the cells. A cobalamin binding protein, present in normal fibroblasts and in other cobalamin mutants, is absent in E.M. cells (73). This absence may prove to be the primary defect in the Cbl C mutant.

In addition to Mudd's case of methylmalonic aciduria and homocystinuria described above, two other reports with the same combination of biochemical abnormalities have appeared. One report (74) described two brothers, ages 14 years and 2 years, with consanguineous parents. The older boy was psychotic and retarded and had abnormal cerebellar-spinal cord function with nystagmus, intention tremor, and slightly hyperactive tendon reflexes; the younger boy was entirely well. Whether this marked difference in neurologic status is a consequence of 14 years with disease or is unrelated to the disease is a very important, but as yet unanswered, question. Each patient had normal serum cobalamin determinations and normal bone marrow examinations with no suggestion of megaloblastic change. When one of the brothers was treated with pharmacologic amounts of OH-Cbl (1000 μg/day parenterally), methylmalonic acid excretion decreased markedly (75). Studies with cultured fibroblasts suggested similar but less severe enzyme deficiencies than were present in E.M. cells. Cells from one of the brothers were used to study coenzyme synthesis in our laboratory and results very similar to those shown for E.M. in Figure 8-6 were obtained (unpublished data).

The most recent report of methylmalonic aciduria and homocystinuria is of a girl who died at age 7 (76). She had a recurrent megaloblastic anemia that none of the previous three cases had shown, and there was a suggestion on one occasion of improvement in the anemia after pharmacologic doses of CN-Cbl. Severe, progressive mental retardation was present. In addition, both clinical neurologic abnormalities and autopsy findings in the brain were typical of subacute combined degeneration of the cord as seen in cobalamin deficiency. Cobalamin assays of several tissues by chromatobioautography showed severe deficiencies of total cobalamin, AdoCbl, and MeCbl; in plasma total cobalamin content was normal but MeCbl content was very low. Fibroblast cell cultures showed the same deficiency and partial responsiveness of methyltransferase activity as noted with E. M. cells. In our laboratory cultured cells have behaved like E.M. cells also. They accumulate low amounts of total cobalamin and almost no AdoCbl or MeCbl; they belong to the complementation group of Cbl C mutants by heterokaryon analysis; and they lack the cellular cobalamin binding protein (unpublished data).

In these three families we have a congenital disorder that might well be expected to mimic acquired cobalamin deficiency. Neither coenzyme is present in normal amounts and deficiencies of the two cobalamin-dependent enzymes can be demonstrated. In cobalamin deficiency states, methylmalonic aciduria of moderate degree is common and at least one cobalamin-deficient patient has been reported with homocystinuria (77). Yet the major hallmark of cobalamin deficiency, a megaloblastic anemia, was present in only one of four children. E.M. was anemic and very sick throughout his life and his bone marrow was hypoplastic. Because of this, it has been suggested that megaloblastic changes had no opportunity to develop. The two brothers, although not as severely affected biochemically, had sufficient derangement of cobalamin metabolism that each one had methylmalonic aciduria and homocystinuria quantitatively as significant as in patients with pernicious anemia. But neither boy had any megaloblastic change. Only in the most recently reported patient, the girl who died of her disease, was megaloblastic anemia present. She also had findings of subacute degeneration of the cord, which is usually associated with prolonged cobalamin deficiency. Thus we have a disease where the failure to synthesize either cobalamin coenzyme is associated predictably with methylmalonic aciduria and homocystinuria but not always with megaloblastic anemia. Also, the effects on the nervous system and on growth and development have been quite varied. In contrast to the coenzyme synthesis defect, but with effects just as puzzling, is transcobalamin II deficiency, as described above in two sisters. Their disease might also mimic cobalamin deficiency because of failure to deliver cobalamin to the tissues. Severe megaloblastic anemia was present but no methylmalonic aciduria or homocystinuria occurred. These apparent conflicts are not understood at this time and point out the incomplete understanding of cobalamin metabolism at the cellular level.

Pathophysiology

Several direct sequelae of these inborn errors of coenzyme metabolism have already been described in the foregoing discussion but some further comments should be made. For the most part, the clinical and biochemical symptoms of all of the blocks in the propionate-methylmalonate pathway (Figure 8-4) have been very similar. This probably means that common pathophysiologic mechanisms can be found. Neither propionate nor methylmalonate accumulates in sufficient amounts to account for the acidosis that is often present. Several other organic acids build up behind the blocks to contribute to the acid load, and alternate pathways from propionate to lactate, pyruvate, and acetyl CoA are also operative (58). Oberholzer et al. (51) suggested that depletion of available CoA by the accumulation of MM-CoA or of propionyl CoA would divert acetyl CoA from fatty acid synthesis toward ketone body formation to provide free CoA. They also postulated that succinyl CoA might be in short supply due to the primary block, resulting in a slowing of the tricarboxylic acid cycle. Both of these phenomena, if they occur to a significant degree, would contribute to ketosis, acidosis, and hypoglycemia. The accumulation of long-chain ketones is thought to be partly a consequence of increased levels of ketoacid intermediates generated on the catabolic route from branched-chain amino acids to propionyl CoA and MM-CoA (78). Halperin et al. (79) considered possible ways that MM-CoA or methylmalonate might interfere with gluconeogenesis and cause ketoacidosis and hypoglycemia. Such interference might include inhibition of pyruvate carboxylase by MM-CoA, inhibition of succinic dehydrogenase by methylmalonate, and inhibition of mitochondrial transport of malate via the dicarboxylate carrier by methylmalonate. The authors favored the last mechanism as being operative from their studies in rat liver mitochondria.

Hyperammonemia is intermittently present in the blocks in the propionate-methylmalonate pathway, but has never been satisfactorily explained. Assay of urea cycle enzymes in the liver of a girl with methylmalonyl CoA racemase deficiency showed low, but not rate-limiting, activities for all the enzymes (57). The intermittent hyperglycinemia is likewise unexplained. Hillman et al. (80) suggested that glycine oxidation might be inhibited by the long-chain ketones and ketoacids that build up. Gompertz (81) offers the possibility that glycine condensation with acetyl CoA inside the mitochondrion may be inhibited by increased propionyl CoA. Neither mechanism has been demonstrated as yet in patients with propionic acidemia or methylmalonic aciduria.

A recurring hypothesis seeks to connect the consequences of increased intracellular MM-CoA and propionyl CoA on fatty acid synthesis with the chronic neurologic disease as seen in cobalamin deficiency. Propionyl CoA apparently rises when MM-CoA does (59, 60) and therefore the effects of both must be considered. Forward and Gompertz (82) showed that MM-CoA inhibits fatty acid synthetase from rat liver and brain; whether this occurs in the human disease state is not known. When propionyl CoA rather than acetyl CoA initiates fatty acid synthesis, odd-chain fatty acids rather than even-chain ones are made; and if MM-CoA substitutes for the structurally similar malonyl CoA, the normal substrate for fatty acid elongation, branched-chain fatty acids rather than straight-chain fatty acids would result. Barley et al. (83) demonstrated an increase in C_{15} and C_{17} odd-chain fatty acids in cultured rat glial cells as a consequence of cobalamin deficiency induced in culture. Kishimoto et al. (84) examined lipids from the brain and other tissues of the infant (E.M.) who died with homocystinuria and methylmalonic aciduria. They found a 6- to 13-fold increase in C_{15} and C_{17} odd-chain fatty acids in glycerolipids of the nervous system and a smaller increase in other tissues. They also identified significant amounts of branched-chain fatty acids in the nervous system from E.M., where only trace amounts were present in control tissues. Any of the above-mentioned effects on membrane lipids may have important and long-term consequences for the nervous system as well as for membrane functions elsewhere. Studies of membrane lipids in cobalamin deficiency states have not yet been done and should be very informative.

Genetics of the Coenzyme Defects

The authors know of more than 20 infants and children in whom there is some evidence to support a diagnosis of cobalamin-responsive methylmalonic aciduria, without homocystinuria. These cases have come from many areas of North America and Europe. A few families have had a second affected sibling and the sex ratio has been about equal. This evidence supports the presumption that the disease is an autosomal recessive trait. Parental consanguinity and the determination of a heterozygous state in parents, both of which would support the genetic hypothesis, have not yet been reported.

Only a few of the cases have had studies to establish a defect in AdoCbl synthesis. Basic enzymatic defects are just beginning to be identified and, since the coenzyme synthesis involves several steps, there may be considerable genetic heterogeneity among the patients currently referred to as having cobalamin-responsive methylmalonic aciduria. Also, as noted previously, a different type of heterogeneity has already appeared at the clinical level. Kaye et al. and ourselves have described two patients where cell studies show defective AdoCbl synthesis in cell culture and restoration of MM-CoA mutase activity when cell extracts are assayed in the presence of added AdoCbl, but the patients did not show response to cobalamin therapy *in vivo*. A complex set of different molecular abnormalities undoubtedly exists among these affected patients.

Thus far, only a small number of patients with the combination of congenital methylmalonic aciduria and homocystinuria have been recognized. As cited in the section on deficient synthesis of AdoCbl and MeCbl, the severity and symptomatology of the disorder have varied widely among the three boys and a girl, and parental consanguinity was reported in the family with two affected brothers. This information is very limited but would be consistent with autosomal recessive inheritance.

Prenatal Diagnosis

Cultured amniotic fluid cells, of fibroblast-like, epithelial-like, and intermediate morphologies, have MM-CoA mutase activity and can synthesize the cobalamin coenzymes. Activities are comparable to those seen in skin fibroblasts. Thus definitive diagnoses of the metabolic errors in coenzyme synthesis are possible. The cells also carry out propionate oxidation and methylmalonate oxidation. The use of an oxidation assay permits the diagnosis of any of the blocks in the propionate-methylmalonate pathway plus the blocks in cobalamin coenzyme metabolism. Differentiation between the disorders would then use further assays including those which measure coenzyme synthesis. Amniotic fluid methylmalonate determination can be used in conjunction with the cell assays to help substantiate the diagnosis. If a defect in cobalamin coenzyme synthesis is found, the infant may respond to cobalamin therapy. A thorough trial of therapy is then indicated for the newborn period and perhaps can be started prenatally. Since patients are now being recognized who do not respond to pharmacologic cobalamin therapy even though their cells show a block in AdoCbl synthesis, there must be caution in predicting responsiveness before the patient is actually treated.

Diagnosis of methylmalonic aciduria due to a MM-CoA mutase apoenzyme defect has been reported twice. Morrow et al. (85) made the first diagnosis and based it on the presence of methylmalonate in amniotic fluid and increased methylmalonate in maternal urine after 25 weeks' gestation; confirmatory studies were done postnatally. We recently reported a second diagnosis made in midtrimester from biochemical studies of cultured amniotic fluid cells, plus methylmalonate determinations in the amniotic fluid (86). Confirmation of the diagnosis was obtained by showing deficient MM-CoA mutase activity in cells cultured from the aborted fetus.

A prenatal diagnosis of an inborn error of coenzyme synthesis has also been made (87). This case is of special significance because of a successful attempt to treat the fetus. Diagnosis was based on biochemical studies of cultured amniotic fluid cells and on methylmalonate content of amniotic fluid and maternal urine. Huge doses of CN-Cbl were given to the mother during the last 9 weeks of pregnancy, first 10,000 μg/day orally and later 5000 μg/day parenterally. Methylmalonate excretion in maternal urine fell progressively after cobalamin therapy was begun, to an amount less than one-fifth of the daily excretion before therapy, despite a growing fetus. Total cobalamin concentration in the infant's blood was 19,000 pg/ml at birth, and methylmalonic content in blood and urine remained low during the first months of life. Studies of skin fibroblasts confirmed the prenatal diagnosis, and clinical investigation demonstrated that the infant's methylmalonic aciduria was responsive to cobalamin therapy. Although this case demonstrates the capability of prenatal diagnosis for one of the cobalamin errors and the feasibility of fetal therapy, questions remain about the necessity of prenatal therapy and the benefits one might expect from such therapy. Most infants with congenital methylmalonic aciduria appear well for the first day or more of life; nonetheless the possibility exists that development during the last fetal months has not been optimal because of tissue accumulations of methylmalonic acid.

SUMMARY

Seven inborn errors of cobalamin metabolism are now known (Table 8-1). Three of these affect the intestinal absorption phase of vitamin metabolism and cause a juvenile form of pernicious anemia with symptoms akin to an acquired cobalamin deficiency. The three disorders are discussed in detail in Chapter 7, but it should be noted here that dramatic clinical response is obtained when physiologic amounts of cobalamin (less than 5 μg/day) are given the patient, provided the vitamin is administered parenterally to circumvent intestinal absorption. Of the four other inborn errors, two are deficiencies of plasma transport proteins for cobalamin and two are errors of tissue utilization of the vitamin.

TC I apparently functions to provide cobalamin storage in plasma for relatively long periods of time, several days to a few weeks. One family, with two affected brothers, has been identified in which the probands have a partial deficiency of TC I and abnormally low plasma cobalamin content. No symptoms resulted from the partial deficiency and no therapy was required. TC II mediates the uptake of cobalamin from plasma by cells throughout the body. Deficiency of TC II has also been reported in only one family; but, in this case, deficiency was total and the affected infants had severe symptoms which required large, pharmacologic doses of cobalamin (over 100 μg/day) to control.

The two errors of tissue utilization are characterized by either a failure of cell synthesis of one cobalamin coenzyme, AdoCbl, or by failure of synthesis of both coenzymes, AdoCbl and MeCbl. The first one causes methylmalonic aciduria and the second, the combination of methylmalonic aciduria and homocystinuria. The disorders are due to abnormalities, just beginning to be defined, in the entry of cobalamin into cells or in the intracellular transport and metabolism of the vitamin. Where vitamin therapy has shown some benefit in affected patients, pharmacologic doses (250 μg/day or more) have been required, but only partial correction of biochemical abnormalities has occurred.

The inborn errors have raised questions about the relationship between megaloblastic anemia and known aspects of cobalamin metabolism. Cobalamin deficiency results in megaloblastic anemia, methylmalonic aciduria, and homocystinuria. The two patients with TC II deficiency had severe megaloblastic anemia but no methylmalonic aciduria or homocystinuria; and patients unable to synthesize AdoCbl or MeCbl have methylmalonic aciduria and homocystinuria but may or may not show megaloblastic changes. Our current knowledge of TC II function and of the two cobalamin-dependent reactions in man does not give sufficient insight to understand these inconsistencies. Future study of these diseases and of cobalamin metabolism will undoubtedly bring answers and will very probably expand the list of inborn errors.

REFERENCES

1. Newmark, P. A. (1972). The mechanism of uptake of vitamin B_{12} by the kidney of the rat *in vivo*. *Biochem. Biophys. Acta* **261**, 85.
2. Conner, H. M. (1930). Hereditary aspect of achlorhydria in pernicious anemia: Study of gastric acidity in 154 relatives of 109 patients having pernicious anemia. *JAMA* **94**, 606.
3. Callender, S. T., and Denborough, M. A. (1957). Family study of pernicious anaemia. *Brit. J. Haematol.* **3**, 88.

4. McIntyre, P. A., Hahn, R., Conley, C. L., and Glass, B. (1959). Genetic factors in predisposition to pernicious anemia. *Bull. Johns Hopkins Hosp.* **104**, 309.
5. Whittingham, S., MacKay, I. R., Ungar, B., and Matthews, J. D. (1969). The genetic factor in pernicious anaemia. A family study in patients with gastritis. *Lancet* **1**, 951.
6. Callender, S. T., and Lajtha, L. G. (1951). On the nature of Castle's hemopoietic factor. *Blood* **6**, 1234.
7. Hall, C. A., and Finkler, A. E. (1963). A second vitamin B_{12}-binding substance in human plasma. *Biochim. Biophys. Acta* **78**, 234.
8. Miller, O. N., Raney, J. L., and Hunter, F. M. (1957). Effect of intrinsic factor on uptake of radioactive vitamin B_{12} by slices of rat liver. *Fed. Proc.* **16**, 393.
9. Herbert, V. (1958). Studies of the mechanism of the effect of hog intrinsic factor concentrate on the uptake of vitamin B_{12} by rat liver slices. *J. Clin. Invest.* **37**, 646.
10. Cooper, B. A., and Paranchych, W. (1961). Selective uptake of specifically bound cobalt-58 vitamin B_{12} by human and mouse tumour cells. *Nature* **191**, 393.
11. Paranchych, W., and Cooper, B. A. (1962). Factors influencing the uptake of cyanocobalamin (vitamin B_{12}) by Ehrlich ascites carcinoma cells. *Biochim. Biophys. Acta* **60**, 393.
12. Retief, F. P., Gottlieb, C. W., and Herbert, V. (1966). Mechanism of vitamin B_{12} uptake by erythrocytes. *J. Clin. Invest.* **45**, 1907.
13. Finkler, A. E., and Hall, C. A. (1967). Nature of the relationship between vitamin B_{12} binding and cell uptake. *Arch. Biochem. Biophys.* **120**, 79.
14. Finkler, A. E., Green, P. D., and Hall, C. A. (1970). Immunological properties of human vitamin B_{12} binders. *Biochim. Biophys. Acta* **200**, 151.
15. Finkler, A. E., Hall, C. A., Green, P. D., and Young, C. C. (1969). Cross reactivity of mammalian B_{12} transport proteins. *Fed. Proc.* **28**, 700.
16. Hall, C. A., and Finkler, A. E. (1971). Protein-mediated uptake of vitamin B_{12} by cells in tissue culture. In *The Cobalamins,* Arnstein, H. R. V., and Wrighton, R. J., Eds. Churchill-Livingstone, Edinburgh, p. 49.
17. Rosenberg, L. E., Lilljeqvist, A., and Allen, R. H. (1973). Transcobalamin II-facilitated uptake of vitamin B_{12} by cultured fibroblasts: Studies in methylmalonicaciduria. *J. Clin. Invest.* **52**, 69a.
18. Retief, F. P., Gottlieb, C. W., and Herbert, V. (1967). Delivery of Co^{57} B_{12} to erythrocytes from α and β globulin of normal, B_{12}-deficient, and chronic myeloid leukemia serum. *Blood* **29**, 837.
19. Retief, F. P., Vandenplas, L., and Visser, H. (1969). Vitamin B_{12} binding proteins in liver disease. *Brit. J. Haematol.* **16**, 231.
20. Hall, C. A., and Finkler, A. E. (1965). The dynamics of transcobalamin II. A vitamin B_{12} binding substance in plasma. *J. Lab. Clin. Med.* **65** 459.
21. Hom, B. L., and Oleson, H. A. (1969). Plasma clearance of ^{57}cobalt-labelled vitamin B_{12} bound *in vitro* and *in vivo* to transcobalamin I and II. *Scand. J. Clin. Lab. Invest.* **23**, 201.
22. Tan, C. H., and Hansen, H. J. (1968). Studies on the site of synthesis of transcobalamin II. *Proc. Soc. Exp. Biol. Med.* **127**, 740.
23. Benson, R. E., Rappazzo, M. E., and Hall, C. A. (1972). Late transport of vitamin B_{12} by transcobalamin II. *J. Lab. Clin. Med.* **80**, 488.

24. Lindstrand, K., and Ståhlberg, K.-G. (1963). On vitamin B_{12} forms in human plasma. *Acta Med. Scand.* **174**, 665.
25. Ståhlberg, K.-G. (1967). Studies on methyl-B_{12} in man. *Scand. J. Haematol. Suppl.* 1.
26. Lindstrand, K. (1967). Vitamin B_{12} derivatives in the human organism. *Scand. J. Clin. Lab. Invest.* **19**, *Suppl.* 95.
27. Linnell, J. C., MacKenzie, H. M., and Matthews, D. M. (1969). Normal values for individual plasma cobalamins. *J. Clin. Pathol.* **22**, 506.
28. Linnell, J. C., Hoffbrand, A. V., Peters, T. J., and Matthews, D. M. (1971). Chromatographic and bioautographic estimation of plasma cobalamins in various disturbances of vitamin B_{12} metabolism. *Clin. Sci.* **40**, 1.
29. Gurnani, S., Mistry, S .P., and Johnson, B. C. (1960). Function of vitamin B_{12} in methylmalonate metabolism. I. Effect of a cofactor form of B_{12} on the activity of methylmalonyl-CoA isomerase. *Biochim. Biophys. Acta* **38**, 187.
30. Hegre, C. S., Miller, S. J., and Lane, M. D. (1962). Studies on methylmalonyl isomerase. *Biochim. Biophys. Acta* **56**, 538.
31. Wang, F. K., Koch, J., and Stokstad, E. L. (1967). Folate coenzyme pattern, folate linked enzymes and methionine biosynthesis in rat liver mitochondria. *Biochem. Z.* **346**, 458.
32. Pawełkiewicz, J., Gorna, M., Fenrych, W., and Magas, S. (1964). Conversion of cyanocobalamin *in vivo* and *in vitro* into coenzyme form in humans and animals. *Ann. N. Y. Acad. Sci.* **112**, 641.
33. Mahoney, M. J., and Rosenberg, L. E. (1971). Synthesis of cobalamin coenzymes by human cells in tissue culture. *J. Lab. Clin. Med.* **78**, 302.
34. Kerwar, S. S., Spears, C., McAuslan, B., and Weissbach, H. (1971). Studies on vitamin B_{12} metabolism in HeLa cells. *Arch. Biochem. Biophys.* **142**, 231.
35. Abeles, R. H., Myers, C., and Smith, T. A. (1966). An enzymatic assay for the determination of millimicrogram quantities of B_{12}-coenzyme. *Anal. Biochem.* **15**, 192.
36. Newmark, P., Newman, G. E., and O'Brien, J. R. P. (1970). Vitamin B_{12} in the rat kidney. Evidence for an association with lysosomes. *Arch. Biochem. Biophys.* **141**, 121.
37. Newmark, P. (1971). Metabolism of vitamin B_{12} in the kidney at a subcellular level. In *The Cobalamins,* Arnstein, H. R. V., and Wrighton, R. J., Eds., Churchill-Livingstone, Edinburgh, p. 79.
38. Pletsch, Q. A., and Coffey, J. W. (1971). Intracellular distribution of radioactive vitamin B_{12} in rat liver. *J. Biol. Chem.* **246**, 4619.
39. Pletsch, Q. A., and Coffey, J. W. (1972). Properties of the proteins that bind vitamin B_{12} in subcellular fractions of rat liver. *Arch. Biochem. Biophys.* **151**, 157.
40. Rosenthal, H. L., Cutler, L., and Sobieszczanska, W. (1970). Uptake and transport of vitamin B_{12} in subcellular fractions of intestinal mucosa. *Am. J. Physiol.* **218**, 358.
41. Peters, T. J., and Hoffbrand, A. V. (1970). Absorption of vitamin B_{12} by the guinea pig. I. Subcellular localization of vitamin B_{12} in the ileal enterocyte during absorption. *Brit. J. Haematol.* **19**, 369.
42. Peters, T. J., Linnell, J. C., Matthews, D. M., and Hoffbrand, A. V. (1971). Absorption of vitamin B_{12} in the guinea-pig. III. The forms of vitamin B_{12} in

ileal mucosa and portal plasma in the fasting state and during absorption of cyanocobalamin. *Brit. J. Haematol.* **20**, 299.

43. Finkler, A. E. (1972). Transfer of B_{12} from TC II to TC I in tissue culture. *Fed. Proc.* **31**, 723.

44. Rappazzo, M. E., and Hall, C. A. (1972). Transport function of transcobalamin II. *J. Clin. Invest.* **51**, 1915.

45. Whitaker, T. R., and Giorgio, A. J. (1973). Uptake of ^{57}Co-vitamin B_{12} by rat liver mitochondria and incorporation into methylmalonyl CoA mutase. *Fed. Proc.* **32**, 892.

46. Taylor, R. T., and Weissbach, H. (1969). *Escherichia coli* B N^5-methyltetrahydrofolate-homocysteine methyltransferase: Sequential formation of bound methylcobalamin with *S*-adenosyl-*L*-methionine and N^5-methyltetrahydrofolate. *Arch. Biochem. Biophys.* **129**, 728.

47. Foster, M. A., Dilworth, M. H., and Woods, D. D. (1964). Cobalamin and the synthesis of methionine by *Escherichia coli. Nature* **201**, 39.

48. Carmel, R., and Herbert, V. (1969). Deficiency of vitamin B_{12}-binding alpha globulin in two brothers. *Blood* **33**, 1.

49. Hakami, N., Neiman, P. E., Canellos, G. P., and Lazeron, J. (1971). Neonatal megaloblastic anemia due to inherited transcobalamin II deficiency in two siblings. *N. Engl. J. Med.* **285**, 1163.

50. Scott, C. R., Hakami, N., Teng, C. C., and Sagerson, R. N. (1972). Hereditary transcobalamin II deficiency: The role of transcobalamin II in vitamin B_{12}-mediated reactions. *J. Pediatr.* **81**, 1106.

51. Obenholzer, V. G., Levin, B., Burgess, E. A., and Young, W. F. (1967). Methylmalonic aciduria. An inborn error of metabolism leading to chronic metabolic acidosis. *Arch. Dis. Child.* **42**, 492.

52. Stokke, O., Eldjarn, J., Norum, K. R., Steen-Johnson, J., and Halvorsen, S. (1967). Methylmalonic acidemia. A new inborn error of metabolism which may cause fatal acidosis in the neonatal period. *Scand. J. Clin. Lab. Invest.* **20**, 313.

53. Hsia, Y. E., Scully, K. J., and Rosenberg, L. E. (1969). Defective propionate carboxylation in ketotic hyperglycinaemia. *Lancet* **1**, 757.

54. Rosenberg, L. E., Lilljeqvist, A-Ch., and Hsia, Y. E. (1968). Methylmalonic aciduria: Metabolic block localization and vitamin B_{12} dependency. *Science* **162**, 805.

55. Lindblad, B., Lindstrand, K., Svanberg, B., and Zetterström, R. (1969). The effect of cobamide coenzyme in methylmalonic acidemia. *Acta Paediatr. Scand.* **58**, 178.

56. Morrow, G. III, Barness, L. A., Cardinale, G. J., Abeles, R. H., and Flaks, J. G. (1969). Congenital methylmalonic acidemia: Enzymatic evidence for two forms of the disease. *Proc. Natl. Acad. Sci. U. S. A.* **63**, 191.

57. Kang, E. S., Snodgrass, P, J., and Gerald, P. S. (1972). Methylmalonyl coenzyme A racemase defect: Another cause of methylmalonic aciduria. *Pediatr. Res.* **6**, 875.

58. Rosenberg, L. E. (1972). Diseases of propionate, methylmalonate, and vitamin B_{12} metabolism. In *The Metabolic Basis of Inherited Disease,* 3rd ed., Stanbury, J. B., Wyngaarden, J. B., and Fredrickson, D. S., Eds., McGraw-Hill, New York, p. 440.

59. Ando, T., Rasmussen, K., Nyhan, W. L., Donnell, G. N., and Barnes, N. D. (1971). Propionic acidemia in patients with ketotic hyperglycinemia. *J. Pediatr.* **78**, 827.

60. Stokke, O., Jellum, E., Eldjarn, L., and Schnitler, R. (1973). The occurrence of β-hydroxy-*n*-valeric acid in a patient with propionic and methylmalonic acidemia. *Clin. Chim. Acta* **45**, 391.

61. Cox, E. V., and White, A. M. (1962). Methylmalonic acid excretion: An index of vitamin B_{12} deficiency. *Lancet* **2**, 853.

62. Cox, E. V., Robertson-Smith, D., Small, M., and White, A. M. (1968). The excretion of propionate and acetate in vitamin B_{12} deficiency. *Clin. Sci.* **35**, 123.

63. Hsia, Y. E., Scully, K., Lilljeqvist, A.-Ch., and Rosenberg, L. E. (1970). Vitamin B_{12}-dependent methylmalonicaciduria. Amino acid toxicity, long chain ketonuria, and protective effect of vitamin B_{12}. *Pediatrics* **46**, 497.

64. Rosenberg, L. E., Lilljeqvist, A.-Ch., Hsia, Y. E., and Rosenbloom, F. M. (1969). Vitamin B_{12}-dependent methylmalonicaciduria: Defective B_{12} metabolism in cultured fibroblasts. *Biochem. Biophys. Res. Commun.* **37**, 607.

65. Mahoney, M. J., Rosenberg, L. E., Mudd, S. H., and Uhlendorf, B. W. (1971). Defective metabolism of vitamin B_{12} in fibroblasts from children with methylmalonicaciduria. *Biochem. Biophys. Res. Commun.* **44**, 375.

66. Mahoney, M. J., Hart, A. C., and Rosenberg L. E. (1974). Defect in 5′-deoxyadenosylcobalamin synthesizing enzyme in methylmalonicacidemia. *Pediatr. Res.* **8**, 436.

67. Gravel, R. A., Mahoney, M. J., Ruddle, F. H., and Rosenberg, L. E. (1974). Genetic complementation in heterokaryons of human fibroblasts defective in cobalamin (vitamin B_{12}) metabolism. *Am. J. Hum. Genet.* **26**(6), 36a (abstr.).

68. Kaye, C. I., Morrow, III, G., and Nadler, H. L. (1974). *In vitro* "responsive" methylmalonic acidemia: a new variant. *J. Pediatr.* **85**, 55.

69. Mudd, S. H., Levy, H. L., and Abeles, R. H. (1969). A derangement in B_{12} metabolism leading to homocystinemia, cystathioninemia and methylmalonic aciduria. *Biochem. Biophys. Res. Commun.* **35**, 121.

70. Levy, H. L., Mudd, S. H., Schulman, J. D., Dreyfus, P. M., and Abeles, R. H. (1970). A derangement in B_{12} metabolism associated with homocystinemia, cystathioninemia, hypomethioninemia, and methylmalonic aciduria. *Am. J. Med.* **48**, 390.

71. Mudd, S. H., Uhlendorf, B. W., Hinds, K. R., and Levy, H. L. (1970). Deranged B_{12} metabolism: Studies of fibroblasts grown in tissue culture. *Biochem. Med.* **4**, 215.

72. Mudd, S. H., Levy, H. L., Morrow, G. III. (1970). Deranged B_{12} metabolism Effects on sulfur amino acid metabolism. *Biochem. Med.* **4**, 193.

73. Rosenberg, L. E., and Patel, L. (1975). Absence of a cobalamin binding protein in cultured fibroblasts from patients with defective cobalamin coenzyme synthesis. *J. Clin. Invest.* **55**(6) (abstr.), in press.

74. Goodman, S. I., Moe, P. G., Hammond, K. P., Mudd, S. H., and Uhlendorf, B. W. (1970). Homocystinuria with methylmalonic aciduria: Two cases in a sibship. *Biochem. Med.* **4**, 500.

75. Goodman, S. I., Keyser, A. J., Mudd, S. H., Schulman, J. D., Turse, H., and Lewy, J. (1972). Responsiveness of congenital methylmalonic aciduria to derivatives of vitamin B_{12}. *Pediatr. Res.* **6**, 138.

76. Dillon, M. J., England, J. M., Gompertz, D., Goodey, P. A., Grant, D. B., Hussein, H. A-A., Linnell, J. C., Matthews, D. M., Mudd, S. H., Newns, G. H., Seakins, J. W. T., Uhlendorf, B. W., and Wise, I. J. (1974). Mental retardation, megaloblastic anemia, methylmalonic aciduria and abnormal homocysteine metabolism due to an error in vitamin B_{12} metabolism. *Clin. Sci. Mol. Med.* **47**, 43.

77. Hollowell, J. G., Jr., Hall, W. K., Coryell, M. E., McPherson, J., Jr., Hahn, D. A. (1969). Homocystinuria and organic aciduria in a patient with vitamin B_{12} deficiency. *Lancet* **2**, 1428.

78. Rosenberg, L. E., Lilljeqvist, A.-Cr., and Hsia, Y. E. (1968). Methylmalonic aciduria. An inborn error leading to metabolic acidosis, long-chain ketonuria and intermittent hyperglycinemia. *N. Engl. J. Med.* **278**, 1319.

79. Halperin, M. L., Schiller, C. M., and Fritz, I. B. (1971). The inhibition by methylmalonic acid of malate transport by the dicarboxylate carrier in rat liver mitochondria. *J. Clin. Invest.* **50**, 2276.

80. Hillman, R. E., Feigen, R. D., Tennenbaum, S. M., and Keating, J. P. (1972). Defective isoleucine metabolism as a cause of the "ketotic hyperglycinemia" syndrome. *Pediatr. Res.* **6**, 394.

81. Gompertz, D. (1971). The metabolic effects of an impaired methylmalonyl CoA mutase. In *The Cobalamins,* Arnstein, H. R. V., and Wrighton, R. J., Eds., Churchill-Livingstone, Edinburgh, p. 101.

82. Forward, S. A., and Gompertz, D. (1970). The effects of methylmalonyl CoA on the enzymes of fatty acid biosynthesis. *Enzymologia* **39**, 379.

83. Barley, F. W., Sato, G. H., and Abeles, R. H. (1972). An effect of vitamin B_{12} deficiency in tissue culture. *J. Biol. Chem.* **247**, 4270.

84. Kishimoto, Y., Williams, M., Moser, H. W., Hignite, C., and Biemann, K. (1973). Branched-chain and odd-numbered fatty acids and aldehydes in the nervous system of a patient with deranged vitamin B_{12} metabolism. *J. Lipid Res.* **14**, 69.

85. Morrow, G. III, Schwartz, R. H., Hallock, J. A., and Barness, L. A. (1970). Prenatal detection of methylmalonic acidemia. *J. Pediatr.* **77**, 120.

86. Mahoney, M. J., Rosenberg, L. E., Waldenström, J., Linblad, B., and Zetterström R. (1975). Prenatal diagnosis of methylmalonicaciduria. *Acta Paediatr. Scand.* **64**, in press.

87. Ampola, M. G., Mahoney, M. J., Nakamura, E., and Tanaka, K. (1974). *In utero* treatment of methylmalonic acidemia with vitamin B_{12}. *Pediatr. Res.* **8**, 387.

CHAPTER NINE

METABOLIC FEATURES OF COBALAMIN DEFICIENCY IN MAN

WILLIAM S. BECK, M.D.

Department of Medicine, Harvard Medical School, and the
Hematology Research Laboratory of the Massachussetts General Hospital
Boston, Massachussetts

CONTENTS

INTRODUCTION

It is proposed to present in this chapter (a) a brief description of the major clinical features of the cobalamin deficiency syndrome in man; (b) a survey of cobalamin-dependent metabolic systems known to be present in human tissues; and (c) a summary of current views on the metabolic basis of the cobalamin deficiency syndrome in its several aspects.

We should note at the outset that cobalamin deficiency is an abnormal nutritional state having many causes. These include intrinsic factor deficiency due to Addisonian pernicious anemia or gastrectomy; a variety of intestinal disorders leading to defective absorption of the vitamin (e.g., selective malabsorption, ileal resection, ileitis, etc.); competitive parasites (fish tapeworm, bacterial overgrowth, etc.); increased nutrition requirements (pregnancy, neoplastic disease, hyperthyroidism, etc.); and on rare occasions severe malnutrition. Cobalamin deficiency can result from all of these pathologic states. When it does, its clinical manifestations are superimposed on those of the underlying primary disorder.

Much of what is known of the cobalamin deficiency syndrome in man has been learned from studies of patients with pernicious anemia. However, it should be borne in mind that this is but one cause of cobalamin deficiency—albeit a relatively common one—and that our concern here is primarily with the pathophysiology of the vitamin deficiency syndrome, not with the pathogenesis of pernicious anemia. Although the focus of our attention will be on changes occurring in human material, we will perforce refer to relevant data from other animal systems and from bacteria.

CLINICAL FEATURES OF THE COBALAMIN DEFICIENCY SYNDROME

Cobalamin deficiency in man leads to two broad classes of clinical phenomena: (a) megaloblastic anemia and related changes in other body cell lines; and (b) a variety of neurological disorders. The following is a brief description of the clinical and pathological features of each of these phenomena.

Megaloblastic Anemia and Related Phenomena

The term "megaloblastic anemia" designates a group of disorders of varying etiology having in common a characteristic pattern of morphologic and functional abnormalities in the blood and bone marrow. As we shall point out later, the pattern is thought to result from impairment of DNA synthesis, a phenomenon that may have many causes, of which cobalamin deficiency is but one.

In the present discussion, the adjective "megaloblastic" refers to certain morphologic and functional patterns in erythrocyte, granulocyte, and platelet precursors, while the term "megaloblast" specifically denotes any maturation stage of the megaloblastic erythroid series. The process wherein normoblastic cells become megaloblastic is termed "megaloblastic transformation."

Clinical Features

The patient with cobalamin deficiency typically displays megaloblastic anemia. The anemia may be severe, but because it develops slowly it may produce few symptoms until the hematocrit is profoundly depressed. Patients characteristically demonstrate a combination of pallor and slight jaundice, the often-mentioned lemon-yellow skin color of the older clinical literature. Perhaps because cobalamin deficiency today is less commonly allowed to become severe, this sign is rarely observed in modern practice. In addition to anemia, leukopenia and thrombocytopenia are also commonly seen in cobalamin deficiency.

Megaloblastic anemia is associated with two pathophysiologic abnormalities: ineffective erythropoiesis with intramedullary destruction of the abnormal erythrocyte precursors, and moderate hemolysis of the circulating erythrocytes. The occurrence of each of these abnormalities is indicated by several lines of evidence which are discussed in detail elsewhere (1–8). Although the mechanism of the leukopenia of megaloblastic anemia has received relatively little study, it seems reasonable to suppose that ineffective leukopoiesis accounts at least in part for this abnormality (9). This conclusion is compatible with the morphologic evidence of active but abnormal leukopoiesis in marrow aspirates. Ineffective thrombopoiesis also appears to be present (10).

The formed elements of the blood display characteristically abnormal morphology. Erythrocytes are generally normochromic and macrocytic, and show striking variations in size and shape. The reticulocyte count is lower than normal. Nuclei of neutrophils often have more than the usual three to four segments, a consequence not of cell age, as previously thought, but more likely of abnormalities of nuclear division or of chromatin structure.

The bone marrow is cellular or hypercellular, with megaloblastic changes in the erythrocyte, granulocyte, and platelet prescursors, though frequently major changes are confined to the erythroid series. In this series, the megaloblastic precursor is larger than the corresponding normoblastic cell and often has a higher than normal ratio of cytoplasmic area to nuclear area. As the cell matures, the chromatin retains its granular texture and forms coarse, deeply basophilic clumps only slowly. With the appearance of hemoglobin, the apparent maturity of the cytoplasm contrasts with the apparent immaturity of the nucleus, a feature referred to as nuclear-cytoplasmic asynchronism. Megaloblastic granulocyte precursors, when present, also display nuclear-cytoplasmic asynchronism and apparent enlargement. In the later stages of development (metamyelocyte and beyond), the cytoplasm appears more abnormally immature. The nuclei are large and sometimes bizarre in shape, with a characteristic uneven chromatin pattern.

Mechanism of Megaloblastic Transformations

Study of the biochemical features of megaloblasts has been impeded by many technical problems, among them difficulties in resolving different bone marrow cell types. Nevertheless, it has been established that megaloblasts contain a substantially increased amount of RNA and a normal or a slightly increased amount of DNA per cell (11, 12), the former presumably accounting for the cytoplasmic basophilia. Also, some DNA synthesis can occur if labeled DNA precursors are provided *in vitro* (13–15).

These data suggested that the megaloblast is a cell in a state of unbalanced growth owing to impaired synthesis of one or more deoxyribonucleotides, the precursors of DNA, (16, 17). As in the unbalanced growth patterns observed in other species (18), DNA replication and cell division are blocked, while synthesis of cytoplasm (RNA and protein) proceeds normally; hence the RNA/DNA ratio rises.

Studies of the unbalanced growth states in bacteria (19–22) and animal cells (23), whether due to lack of an essential nutrient or exposure to an inhibitor of DNA synthesis, indicate that the prolongation of the state results in permanent loss of the capacity for mitosis and eventual cell death. Such loss of cell viability occurs in animal cells in culture following exposure to an inhibitor of DNA synthesis for a period corresponding to about one generation time (24). It may be assumed that in megaloblastic bone marrow the degree of impairment of DNA synthesis varies from cell to cell, ranging from complete inhibition to none, and from cell series to cell series, being usually (but not always) more evident among erythrocyte precursors than granulocyte precursors.

Many data support the conclusion that megaloblastic cells are in a state of unbalanced growth. These include evidence that the maturation time of the promegaloblast is prolonged (25–28) and that DNA synthesis is impaired, as demonstrated by a variety of isotopic methods (29–33). It is perhaps simplistic to conclude that a deficit in DNA synthesis is the ultimate cause of all the cytokinetic abnormalities of megaloblastic anemia. There is evidence, for example, that DNA synthesis can be experimentally dissociated from the cell division cycle (34). Also, it is not known what consequences if any may flow from possible impairment of the synthesis of mitochondrial DNA as contrasted with nuclear DNA. Nonetheless, until these and other features in the chain of events in megaloblastic transformation are understood in greater detail, the postulate that in this disorder defective DNA synthesis blocks or delays cell division, and the resulting state of unbalanced growth predisposes to premature cell death or abnormal cell division with resulting dysplasia, continues to be the best biochemical rationalization of the megaloblastic changes produced by cobalamin deficiency.

Neurological Abnormalities

In its classic description, the neurologic syndrome of cobalamin deficiency consists of symmetrical paresthesias in feet and fingers, with associated disturbances of vibratory sense and proprioception, progressing to spastic ataxia with "subacute combined system" disease of spinal cord, that is, degenerative changes of the dorsal and lateral columns. Typically, sensory systems are affected early and motor systems late. In fact, the picture is more often chronic than subacute and more varied and complex. Clinical signs may include cerebral manifestations, irritability, somnolence, "megaloblastic madness" (35), perversion of taste, smell, and vision, and occasional optic atrophy. Tobacco amblyopia is a curious visual disorder in cobalamin-deficient smokers that has been attributed to the tendency of cyanide in tobacco smoke to convert a meager supply of coenzymatically active adenosylcobalamin to metabolically inert cyanocobalamin (36).

Early pathologic changes in the affected spinal cord consist of swelling of individual myelinated nerve fibers in small foci. Later the lesions coalesce into larger foci involving many fiber systems, but not in a systematic manner (37–39). Lesions bear no constant relationship to blood vessels. The pathologic process takes the form of diffuse uneven degeneration of white matter with multiple foci of spongy degeneration. Although neuropathologists consider the lesions distinctive and readily distinguishable from other myelopathies associated with metabolic abnormalities, they have obtained little evidence regarding their nature or their genesis. The major conclusions to come from this avenue of study are (a) the lesions are demyelinating, that is, myelin appears to diminish in quantity as judged by histologic criteria; (b) neurological changes are for a time reversible by cobalamin therapy, but if untreated they become irreversible; and (c) the neuropathy of cobalamin deficiency is also associated with defective nerve conduction (40). A puzzling feature of the neurologic picture is its apparent dissociation from megaloblastic anemia; the severity of the neurological disorder is often unrelated to the rest of the clinical picture, and the occurrence and progression of the neurologic syndrome in the absence of anemia is common (41).

Associated Metabolic Abnormalities

Changes Secondary to Megaloblastic Anemia per se

Changes in blood chemistry are relatively few. Plasma bilirubin and iron levels may be somewhat increased; serum lactic dehydrogenase level (isozymes 1 and 2) is markedly elevated (42–44). It is of interest that in erythrocytes of normal subjects and in those of patients with various nonmegaloblastic anemias, the level of the lactic dehydrogenase isozyme 2 (LDH-2) exceeds that of LDH-1, whereas in the erythrocytes of megaloblastic anemia, LDH-1 activity exceeds that of LDH-2 (45). This reversal of the LDH isozyme pattern is also observed in the serum in megaloblastic anemia.

Elevations are also found in the levels of serum muramidase (9), malic dehydrogenase, and 6-phosphogluconic dehydrogenase (46). Serum glutamic oxaloacetic transaminase is normal (46). Aminoaciduria reportedly occurs (47–49) but observers differ on its frequency and significance. Studies in the author's laboratory failed to detect measurable aminoaciduria in 10 patients with moderately severe megaloblastic anemia (50). Serum and uric acid are variably affected; often they are depressed.

Changes Secondary to Cobalamin Deficiency per se

DECREASED LEVEL OF SERUM COBALAMIN

A decreased serum cobalamin level remains the single most decisive datum in the diagnosis of cobalamin deficiency. Until recently standard assay methods were microbiologic. The cobalamin auxotrophs employed are *L. leichmannii* ATCC 7830 (51, 52), *Euglena gracilis,* strain Z (53), and *E. coli* 113-3 (54). Of these the *L. leichmannii* method is simplest and fastest. Recently an acceptable nonmicrobiologic procedure has been devised (55, 56). Several adaptations of this method, which is based on an isotope dilution principle, are now available commercially in kit form. Performance of the assay with most kits generally requires no special facilities other than a radioactivity scaler. Isotopic methods yield somewhat higher levels then microbiologic methods.

Ranges of normal serum cobalamin levels vary in different laboratories. The normal range in the writer's laboratory is believed to be 150 to 450 pg/ml. As is the case with most quoted ranges of normal, this one was not determined with statistical rigor. Nor does it take account of the several subcategories of normal—elderly subjects, children, pregnant women, and so on. In any event, we find (as do others) that signs and symptoms ordinarily begin to appear when the serum level is below 100.

Serum folate is often elevated when serum cobalamin is depressed, unless there is coexisting folate deficiency. A number of other abnormalities of folic acid metabolism are observed in cobalamin deficiency. These are discussed later in connection with the pathogenesis of clinical features of the deficiency syndrome.

METHYLMALONIC ACIDURIA

Methylmalonic aciduria has been considered a reliable and sensitive index of cobalamin deficiency (57–59), except in the rare cases in which it is due to an inborn metabolic error (60, 61). However, a recent study (62) reports no methylmalonic aciduria in a quarter of the cobalamin-deficient patients surveyed. It is shown that oral loading doses of valine and/or isoleucine (which, as described below, are metabolic precursors of methylmalonyl CoA) increase urinary excretion of methylmalonic acid, but a deficient group remains (albeit with higher serum cobalamin levels) in which methylmalonic aciduria fails to materialize. It would appear, therefore, that normal methylmalonic acid excretion does not rule out the presence of cobalamin deficiency.

In most series, normal subjects excrete only trace amounts of methylmalonic acid, that is, a mean of about 1.5 to 2.0 mg/24 hr with a range of 0 to about 10. Levels vary in cobalamin deficiency but the majority are elevated, sometimes to more than 300 mg/24 hr. Cobalamin therapy restores excretion patterns to normal in several days, though an occasional patient continues to excrete abnormal amounts of methylmalonic acid following vitamin repletion. Methylmalonic acid excretion is normal in folate deficiency. In the usual clinical setting, determination of urinary methylmalonic acid is rarely necessary as a diagnostic tool. Precise analysis continues to present difficulties for the routine laboratory.

RESPONSE TO THE SPECIFIC THERAPY

The clinical response to cobalamin therapy is another useful diagnostic datum. Following parenteral administration of cobalamin to deficient subjects, elevated plasma bilirubin, iron, and lactic dehydrogenase levels fall promptly (Figure 9-1). Decreasing plasma iron turnover and fecal urobilinogen excretion reflect cessation of ineffective erythropoiesis. Within 8 to 12 hr the morphologic appearance of aspirated bone marrow

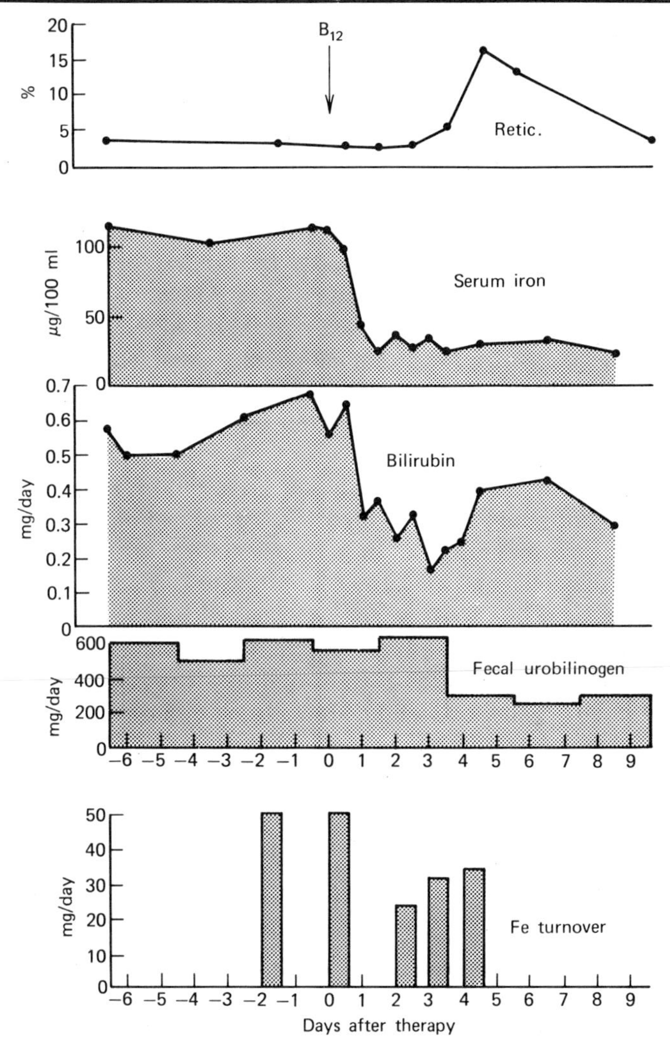

Fig. 9-1 Effect of therapy with cyanocobalamin on reticulocyte count, plasma iron, serum bilirubin, fecal urobilinogen, and plasma iron turnover, in a patient with the cobalamin deficiency of pernicious anemia. [Adapted from Finch et al. (62a). Reprinted by permission.]

begins to convert from megaloblastic to normoblastic. Transformation is complete in 48 to 72 hr.

It is currently thought that after therapy begins the population of accumulated megaloblasts is converted to normoblasts ineffectively—that is, a majority of the megaloblasts die within the marrow or soon after delivery into the circulation. Abrupt reticulocytosis begins on the third to fifth day, reaching a climax on the fourth to tenth day. Cells released in the reticulocyte crisis apparently come from new normoblasts, not from converted megaloblasts. The intensity of the reticulocyte crisis is roughly proportional to the severity of the anemia.

Hemoglobin levels begin to rise, though the percentage increase of hemoglobin during the reticulocyte crisis is smaller than percentage increase in circulating red cells, as judged by the reticulocyte count. The discrepancy has been attributed to the fact that the unusually young reticulocytes delivered after sudden remission of a megaloblastic process undergo prolonged maturation in the peripheral blood (63). Hence, the observed percentage at a given moment reflects the accumulated output of several days.

When the maturation delay in the marrow is corrected by cobalamin therapy a new condition is established in which a still severe anemia elicits intensive erythropoietin-mediated stimulation of erythropoiesis. In later stages of hemoglobin restoration, hypochromia and other signs of iron deficiency may appear. In such instances the plasma iron level decreases as the cobalamin level becomes normal. A second reticulocyte response may then be produced by iron administration.

Other changes that may follow repletion of cobalamin deficiency include the following: (a) striking and rapid improvement in the sense of well-being and capacity for mental effort; (b) rise in serum alkaline phosphatase (which is often depressed in cobalamin deficiency) (64); (c) nitrogen balance, previously negative, becomes positive; (d) sharp rise in serum and urine uric acid within 24 hr of the start of therapy, a peak occurring 24 hr before the peak of the reticulocyte crisis; (e) decrease in serum folate; (f) decrease in urine phosphorus after cobalamin administration, increase during reticulocytosis, and then return to normal (65); (g) sharp drop in serum potassium sometimes warranting replacement therapy (66) and on rare occasions leading to death (67); and (h) rise in serum cobalamin, either slow or rapid (68).

A patient deficient only in cobalamin fails to respond to a "physiologic" dose of folic acid (i.e., 100 to 400 μg/day) that is capable of producing a maximal response to folic acid deficiency (69). This is the basis for a little-used procedure for distinguishing cobalamin and folate deficiency. Large "pharmocologic" doses of folic acid for example, 5 to 15 mg/day, produce sizable reticulocyte responses and suboptimal hemoglobin regeneration in cobalamin deficiency. Treatment of cobalamin deficiency with folate also induces or accelerates the development of neurological abnormalities (70). These putative relationships between cobalamin and folate are discussed below.

ROLE OF COBALAMIN IN ANIMAL CELLS

Discussion of the role of cobalamin in animal cell physiology is conveniently conducted on two levels. Consideration must first be given to evidence for the dependence of individual enzyme reactions on one or more of the cobamide coenzymes. Claims of such dependence, to be acceptable, must rest on demonstrations of requirements in purified systems.

Attention must then be directed to longer metabolic sequences or pathways in which cobamide-dependent enzymes catalyze either individual steps within the sequence or related reactions that supply essential cofactors to enzyme reactions of the main pathway. For purposes of discussion we shall refer to the former category of dependence as "enzymologic" and to the latter as "metabolic." We are concerned at the metabolic level with larger systems that are cobamide dependent because an essential step is cobamide dependent and that become impaired en bloc in cobalamin deficiency. Because the cobamide dependence of these systems may be somewhat indirect it may be difficult to delineate them experimentally. As we shall later see some claims in this area rest entirely on fragmentary data obtained in clinical material displaying greater or lesser degrees of cobalamin deficiency.

The enzymologic role of cobalamin in animal cells is now reasonably well understood, though the prudent investigator is prepared at all times for imminent surprises. Earlier workers proposed many functions for the vitamin on the basis of suggestive data—including an unspecified role in protein synthesis (71, 72), a role in maintaining –SH compounds in the reduced state (73, 74), and so on. However, time has culled these claims to a relatively short list of adequately resolved enzyme reactions. Although the cobamide-dependent reactions of human metabolism are not among those that have been fully resolved enzymologically, two—deoxyadenosylcobalamin-dependent methylmalonyl CoA mutase and the methylcobalamin-dependent methionine synthetase system—have been well characterized in nonhuman animal cells. All other purified cobamide-dependent systems have been derived from bacteria and other protists.

The following section briefly summarizes present understanding of the nature and significance of cobamide-dependent systems in human material at both the enzymologic and metabolic levels—and in both cobalamin-repleted and cobalamin-deficient subjects. We shall then consider how the behavior of these systems in cobalamin deficiency might account for the clinical features of the deficiency syndrome.

Adenosylcobalamin-Dependent Reactions

As noted elsewhere in this volume, it appeared at first that the several reactions requiring adenosylcobalamin included enzymes with two types of mechanisms—one a mutation (or isomerization) with no net change in oxidation level and the other a net reduction. This apparent diversity was clarified with the recognition that (a) in all of its reactions, the adenosylcobalamin coenzyme is an acceptor-donor of hydrogen, the locus of the transfer being the 5′-deoxyadenosyl carbon atom that is attached to the cobalt atom, and (b) in some cobamide reactions hydrogen transfer can occur intramolecularly. A cobamide-mediated intramolecular transfer of hydrogen with formation of a new carbon-hydrogen bond occurs in mutase (or isomerase)-type reactions. A cobamide-mediated intermolecular transfer of hydrogen from outside reductant to substrate occurs in reductase-type reactions.

Methylmalonyl CoA Mutase: Enzymologic Aspects

The cobamide-dependent mutation (or isomerization) of methylmalonyl-CoA, the only adenosylcobalamin-dependent reaction demonstrated thus far in any animal tissue, is a step in the catabolic pathway of propionic acid catabolism (Figure 9-2) (75–79). As discussed in detail elsewhere in this volume, the methylmalonyl CoA mutase reaction occurs as follows:

```
  H  COOH                                      H  COOH
  |   |            adenosylcobalamin           |   |
H—C———C—H        ====================>      H—C———C—H
  |   |          <====================         |   |
 (H)  |                                       (H)  |
   (CO—S—CoA)                             (CO—S—CoA)

Methylmalonyl CoA                         Succinyl CoA
```

Following the discovery by Flavin and co-workers (75, 76) that propionic acid is metabolized in animal tissue by a conversion of propionyl CoA to succinyl CoA by way of the interesting intermediary compound, methylmalonyl CoA, the search began for a cofactor of the enzyme catalyzing the reversible conversion of methylmalonyl CoA to succinyl CoA. Two lines of investigation soon implicated cobalamin or a cobalamin derivative. Marston's observations (80) of propionate intolerance in cobalamin-deficient sheep led to the discovery (81) of depressed methylmalonyl CoA mutase activity in cobalamin-deficient rat tissues. Almost simultaneously, Barker's studies of the cobamide-dependent glutamate mutase of *Clostridium tetanomorphum* revealed the obvious similarity of the two mutases. Several laboratories then demonstrated cobamide coenzyme participation in the mutation of methylmalonyl CoA (82–85).

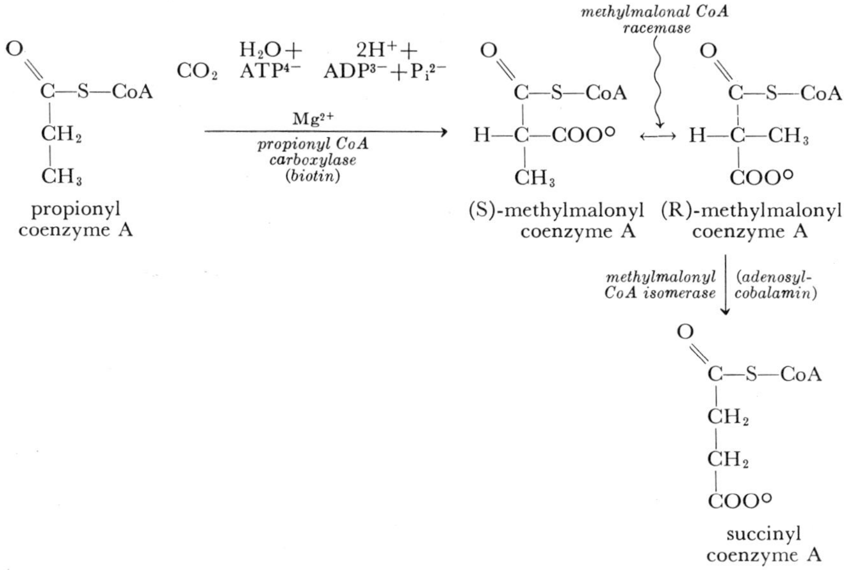

$$CH_3CH_2C(=O)\text{—S—CoA} + CO_2 + H_2O + ATP^{4-} \xrightarrow[\text{propionyl CoA carboxylase (biotin)}]{Mg^{2+}} \text{(S)-methylmalonyl CoA} + 2H^+ + ADP^{3-} + P_i^{2-}$$

propionyl coenzyme A

(S)-methylmalonyl coenzyme A: $O{=}C(\text{—S—CoA})\text{—}CH(COO^{\ominus})\text{—}CH_3$

methylmalonal CoA racemase ⟷

(R)-methylmalonyl coenzyme A: $O{=}C(\text{—S—CoA})\text{—}CH(CH_3)\text{—}COO^{\ominus}$

↓ *methylmalonyl CoA isomerase* (*adenosylcobalamin*)

succinyl coenzyme A: $O{=}C(\text{—S—CoA})\text{—}CH_2\text{—}CH_2\text{—}COO^{\ominus}$

Fig. 9-2 Pathway of propionic acid metabolism.

Interestingly, it was found (84–86) that charcoal readily removes adenosylcobamides from bacterial mutases—in contrast to the earlier experience with mammalian isomerase. The coenzyme is so strongly bound to sheep isomerase that it can be removed only by treating the enzyme with acid ammonium sulfate (85). Enzyme so treated can be reactivated by addition of adenosylcobalamin. Added coenzyme is no longer tightly bound. Enzyme-bound coenzyme is not inactivated by light under conditions that rapidly cleave cobamide coenzymes in solution.

Methylmalonyl CoA Mutase: Metabolic Aspects

As shown in Figure 9-2, propionic acid is metabolized in animal tissues by a biotin-dependent carboxylation of propionyl CoA to methylmalonyl CoA, an α-carboxy derivative of propionyl CoA. After a racemization step, methylmalonyl CoA mutase catalyzes the reversible conversion of methylmalonyl CoA to its β-carboxy isomer, succinyl CoA, which after deacylation enters the tricarboxylic acid cycle. In the main, propionyl CoA is a byproduct of the catabolism of odd-chain fatty acids. These, of course, comprise only a small fraction of the total fatty acid pool and even in them only the terminal three carbons appear as propionyl CoA. The metabolism of propionyl CoA is therefore not of great quantitative significance in fatty acid oxidation in man; it is of substantially greater importance in sheep and other ruminants. In these species, cellulose digestion in the rumen is facilitated by bacteria such as *Propionibacterium shermanii*. Propionate and other short-chain fatty acids are major products of cellulose digestion. Hence, in these species propionic acid is a nutrient, accounting for 25% of their metabolizable energy (79, 87, 88).

In nonruminant mammals, several other pathways can give rise to propionyl CoA and/or methylmalonyl CoA. One or the other (or both) may arise from the metabolism of several amino acids, including valine, isoleucine, threonine, and methionine. Also the catabolic pathway of propionyl CoA operates to some extent in the catabolism of uracil and thymine and perhaps in the catabolism of unsaturated fats by way of a proposed gamma-oxidation pathway (89). The propionate pathway is therefore of more general significance than a mere disposal system for uncommon fatty acids with odd numbers.

As noted above, ingestion of valine and isoleucine can increase the methylmalonic aciduria of cobalamin deficiency. The CoA derivatives of these branched-chain amino acids undergo oxidation, hydration, and a second oxidation as do ordinary fatty acid acyl CoA derivatives, except that the valine derivative loses its coenzyme A along the route. The methylacetoacetyl CoA derived from isoleucine is cleaved as is acetoacetyl CoA except that one of the products of the cleavage is propionyl CoA. The route of metabolism of the semialdehyde of methylmalonic acid, which is derived from valine, is less certain. It is evidently oxidized to methylmalonic acid and then activated to form the corresponding CoA derivative. Hence isoleucine and valine give rise to 1 mole of succinyl CoA by way of methylmalonyl CoA. Threonine is metabolized by removal of ammonia under the action of threonine dehydratase. The product, 2-ketobutyrate, undergoes an oxidative decarboxylation that yields CO_2 and propionyl CoA.

Many studies have been reported of factors affecting methylmalonic aciduria in cobalamin-deficient animals. We alluded earlier to comparable studies in humans (62). Cobalamin deficiency sharply increases urinary excretion of methylmalonic acid in rats, although excretion levels vary with the amount and type of food (90, 91). During starvation, excretion continues at a high level for about 16 hr and then declines rapidly. Measurements of methylmalonic acid excretion in cobalamin-deficient rats starved for 24 hr have been used to assay the metabolism of propionate or precursors of propionyl CoA or methylmalonyl CoA (90). Feeding of sodium propionate, for example, greatly increases methylmalonic acid excretion; leucine and valine produce only moderate increases. Methionine and threonine have little or no effect.

The extent of methylmalonic aciduria in cobalamin-deficient rats can be correlated with decreases in methylmalonyl CoA mutase activity in liver and kidney homogenates (91–93). An excretion of 1.5 μmoles or more of methylmalonic acid per gram of body weight each day reflects a decrease of about 50% in the hepatic mutase level. The tissue level of propionyl CoA carboxylase is not altered by cobalamin deficiency (94). Decrease in methylmalonyl CoA mutase activity in cobalamin-deficient rats and pigs can be correlated with decreased levels of adenosylcobalamin in various tissues (93). In cobalamin-repleted rats, coenzyme level is higher in kidney (0.320 μg/g tissue) than in liver (0.129 μg/g), whereas in the pig the reverse is true and the absolute levels are lower (0.059 μg/g liver). In cobalamin deficiency, coenzyme levels decrease to 12 to 15% of those in normal tissues. Comparable studies in human subjects show that cobalamin deficiency sharply depresses leukocyte methylmalonyl CoA mutase (95). Addition of adenosylcobalamin *in vitro* shows that deficiency leads to increased apoenzyme levels. Interestingly, no correlation was observed between leukocyte mutase activities, urinary methylmalonic acid levels, and severity of megaloblastic anemia in deficient subjects.

Even in cobalamin-repleted rats the level of adenosylcobalamin may be inadequate to saturate the methylmalonyl CoA mutase fully. This is indicated by the fact that addition of coenzyme to homogenates of normal liver and other tissues results in a two- to three-fold increase in the activity of methylmalonyl CoA mutase. However, in homogenates of cobalamin-deficient tissues, the increase in mutase activity following coenzyme addition is larger (3.6- to 7.6-fold). These small differences in response of normal and deficient homogenates to coenzyme additions may or may not be an accurate index of the actual degree of mutase saturation with coenzyme in intact tissue.

There is some evidence that rats and pigs respond to cobalamin deficiency by increasing the level of apoenzyme in liver. In homogenates of some cobalamin-deficient animals, total mutase activity, measured in the presence of excess coenzyme, was from 1.5- to 4-fold higher than that of normal controls. Elevated apoenzyme levels appear to compensate in part for coenzyme deficiency.

A number of interesting studies have dealt with the metabolism of propionic acid in cobalamin deficiency. Aside from the fact that administered propionate enhances methylmalonic aciduria in cobalamin-deficient rats (90, 96), it appears that perfused livers and kidney cortex slices of such rats display depressed gluconeogenesis from propionate (97). Deficient rats excrete more endogenous propionate (98) and more administered ^{3}H-propionate than normal controls (99).

It is noteworthy that the disappearance of methylmalonyl CoA added to animal tissue homogenate is surprisingly unaffected by cobalamin deficiency (93). This finding suggests that the fate of methylmalonyl CoA may well be determined by a reaction other than mutation to succinyl CoA, one that is cobamide-indepenednt.

Methylcobalamin-Dependent Reactions

Methylation of Homocysteine: Enzymologic Aspects

As described elsewhere in this volume, methylcobalamin participates in the cobamide-dependent synthesis of methionine according to the scheme shown in Figure 9-3 (100). Careful studies of the pathway in animal tissues (101–103) indicate that the net reaction occurs as follows:

$$N^5\text{-methyltetrahydrofolate} + \text{homocysteine} \rightarrow \text{tetrahydrofolate} + \text{methionine}$$

The enzyme (N^5-methyltetrahydrofolate: homocysteine methyltransferase) requires *S*-adenosylmethionine (102), a reducing system (103), and a cobamide derivative. The animal holoenzyme has not yet been highly purified and thus it has not been studied as extensively as the corresponding *E. coli* enzyme. For example, details of its cobamide content remain to be elucidated. Enzyme-bound methylcobalamin has been found in a partially purified transferase preparation from porcine kidney (104). This enzyme preparation accumulates labeled methyl groups from [^{14}C-methyl]-N^5-methyltetrahydrofolate only in the presence of *S*-adenosylmethionine.

Much of the literature dealing with the metabolic implications of this pathway rests on the assumption that human or animal enzyme closely resembles the well-studied enzyme from *E. coli*.

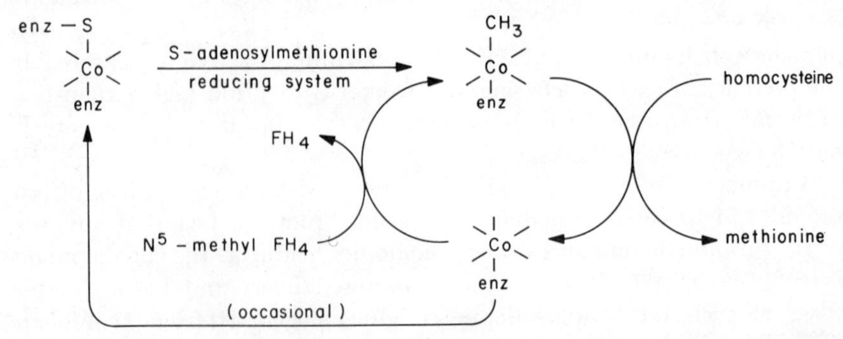

Fig. 9-3 Methylcobalamin-dependent pathway of methionine synthesis. [Adapted from Weissbach and Taylor (100). Reprinted by permission.]

Methylation of Homocysteine: Metabolic Aspects

Assuming that a cobamide-dependent pathway of homocysteine methylation occurs in man, it would then appear that it is one of several means by which the body acquires methionine.[1] At least two pathways would exist for the methylation of homocysteine. The one under consideration requires folate and a cobamide derivative. In the other betaine supplies the transferable methyl group in a reaction that is independent of cobamide coenzymes. This pathway accounts for well-known nutritional data showing that methionine-deficient rats respond to administered homocysteine and betaine in the complete absence of folic acid and cobalamin. The findings of Du Vigneaud (106) on the methylation of homocysteine by choline in rats doubtless involved betaine as an intermediate. Choline is dehydrogenated in the liver—first to betaine aldehyde and then to betaine, which then surrenders a methyl group to homocysteine to form methionine and dimethylglycine. Rats on a methionine-deficient diet lacking folic acid and cobalamin respond well to choline. Hence, folic acid and cobalamin do not participate in these reactions.

In view of the apparent fact that the methylation of homocysteine can occurs by way of a cobamide-dependent pathway and a betaine-dependent pathway, it would seem that methionine can be amply synthesized (via the betaine pathway) when there is a deficiency of cobalamin. If true, this would suggest that the cobamide-dependent pathway may have as its major function not the synthesis of methionine but the conversion of N^5-methyltetrahydrofolate to tetrahydrofolate. In any event, the cobamide-dependent pathway of homocysteine methylation does offer a locus for the interaction of cobamide and folate coenzymes.

Other points of interest relating to the mammalian transferase system concern its regulatory behavior. Supplementation of a deficient diet with cobalamin elevates the transferase activity in chicken (107) and rat (108) liver. Even greater rises occur when methionine is removed from a cobalamin-containing diet (107–109). Detection of transferase activity in cultured cells (110–112) led to a number of interesting observations. Studies of mammalian cells grown in tissue culture have identified relatively few enzymes whose synthesis is derepressed or induced in response to changes in concentration of metabolites. However, transferase activity in baby hamster kidney cells cultured in cobalamin-deficient media is sharply increased following addition of cobalamin or replacement of methionine in the medium with homocysteine (113). These results imply that trans-

[1] It is of biologic interest, according to recent work (105), that the synthesis of methionine in the prebiotic era could have resulted from the action of a spark on a primitive atmosphere containing CH_4, N_2, NH_3, H_2O, and H_2S.

ferase is subject to repression control arising from both cofactor and product. Such mechanisms undoubtedly account for the striking changes in transferase activity observed in the livers and kidneys of rats given diets varying in content of methionine and homocysteine (114).

Pending Claims

Three old and new proposals concerning other possible roles for cobalamin in animal cells merit brief mention. Though tentative or unconfirmed, all are provocative.

Possible Role of Cobalamins in Membrane Transport of Folate

In light of the fact that many patients with pernicious anemia have elevated levels of serum folate, depressed levels of red cell folate, and the other manifestations of altered folate metabolism to be discussed below, Das and Hoffbrand investigated the uptake of N^5-methyltetrahydrofolate (the major component of serum folate) by dividing lymphocytes (115). They found that uptake rates were subnormal in cells from patients with untreated pernicious anemia but quite normal in cells from folate-deficient patients. Similar experiments in bone marrow cells (116) yielded similar results and led Tisman and Herbert to propose that the cellular uptake of N^5-methyltetrahydrofolate is dependent on cobalamin. In this view, diminution of such uptake accounts for the elevated serum folate and depressed red cell folate commonly seen in cobalamin deficiency.

These results, though of great interest, need to be viewed with reserve. It is not clear whether the techniques employed are measuring transport across, absorption onto, or entry into cells membranes. It is not certain whether the molecular species transported is N^5-methyltetrahydrofolate, tetrahydrofolate, or some other bearer of labeled atoms, whether or not the cells responsible for observed data are hematopoietic marrow cells, whether the active principle is the cyanocobalamin added or another cobamide derivative, whether the effect of the vitamin is on the membrane or on some intracellular system, or whether the effects are specific and not simply the consequences of a general improvement in cellular integrity. A final argument—perhaps a weak one—is that this would be a surprising role for a cobamide compound in view of present understanding of their enzymologic role. Further work in this area is awaited with interest.

Possible Role of Cobalamins in Cyanide Metabolism

A fascinating if fragmented literature has been accumulating for some years on the possible participation of cobalamin in the metabolism of cyanide in animal cells. The subject has been reviewed recently (117) and is summarized here only briefly.

The suspicion that cobalamins participate in cyanide metabolism has rested on three considerations: (a) the presence of a cyanide group in cyanocobalamin and the avidity of the cobalamins for cyanide; (b) the fact that chronic cyanide exposure has certain neuropathological effects (as does cobalamin deficiency); and (c) data suggesting that cobalamin is necessary for the conversion of cyanide to thiocyanate (118). It was found, for example, that urinary thiocyanate excretion in cobalamin-deficient patients is lower than in healthy controls and that cobalamin treatment rapidly increases this excretion. These early notions on the connection of cobalamin and cyanide led, among other things, to a proposal for the use of large doses of hydroxocobalamin in the therapy of cyanide poisoning (119). An interaction between hydroxocobalamin, cyanide, and the enzyme rhodanese was also proposed (118).

In fact, the body has to deal with a continuing burden of cyanide from three main sources: (a) smoke from both cigarette and pipe tobacco, the cyanide arising from combustion of tobacco (as of any vegetable matter) at relatively low temperatures; (b) free cyanide arising, for example, from certain fungi and microorganisms and from industrial exposure, and (c) certain cyanogenetic glycosides (amygdalin, linamarin, etc.) in fruits, beans, and nuts that yield HCN on hydrolysis. Specific hydrolases occur with these glycosides. These are released by bruising or crushing and they in turn release HCN.

The metabolism of cyanide follows at least three pathways: (a) conversion to thiocyanate; (b) conversion to 2-aminothiazoline-4-carboxylic acid; and (c) incorporation into the metabolic pool of "one-carbon" units. Most cyanide injected in toxic concentrations is converted to thiocyanate. However, the importance of cobalamins in this conversion is debatable. Matthews and Wilson (117) have argued that since most of the thiocyanate present in body fluids of nonsmokers probably arose not from cyanide, but from dietary thiocyanate, the reduced excretion of thiocyanate in cobalamin deficiency may be due to anorexia and poor diet. The increased excretion after treatment begins may be due initially to transient diuresis and later to increased dietary intake. The question awaits biochemical evidence.

Other proposals have attempted to link cobalamins and cyanide. Studies of normal subjects, heavy smokers, and patients with multiple sclerosis and other ills have shown inverse relationships between the concentrations of serum cobalamin and cyanide (120–123). Of particular interest is an evident tendency for cyanocobalamin to appear more frequently in the plasma, and in larger amounts, in smokers than in nonsmokers.

This phenomenon has been thought to account for higher "extractability" of serum cobalamin. About 75% of the cobalamin in serum is normally in the form of methylcobalamin. The rest is a mixture of adenosylcobalamin and hydroxocobalamin (124–126). Cyanocobalamin is ordinarly undetectable and when present is normally less than 5% of the total. This figure rises in smokers. It has been suggested that exogenous cyanide in smokers may convert cobamide coenzymes to cyanocobalamin, which is metabolically inert. Also, since cyanocobalamin is less firmly bound to plasma proteins than methylcobalamin or adenosylcobalamin, it is more readily excreted by the kidney and thus there may occur a small continuing drain on total body cobalamin. Another possibility is that individuals with low serum cobalamin levels may have a reduced ability to detoxify cyanide and this may predispose to the association between low serum cobalamin level and high plasma cyanide levels. These considerations may be of little moment in normal subjects. But in persons with cobalamin deficiency, the conversion of even a small proportion of the body's cobalamin supply to cyanocobalamin could produce adverse effects. Cyanocobalamin is not only metabolically useless, it acts in higher concentrations as a competitive in inhibitor of certain cobamide-dependent enzyme reactions. We shall consider these matters further in connection with the neurological disorder of cobalamin deficiency.

Possible Role of Cobalamins in the Biosynthesis of Thymidylate Synthetase

Results have appeared recently that appear to give cobalamin a role in determining the level of thymidylate synthetase activity in cells (127). It was found that phytohemagglutinin-stimulated cultured lymphocytes of patients with cobalamin deficiency had little or no demonstrable thymidylate synthetase activity, even though these cells did appear to undergo blastogenesis. Ample activity was found in lymphocytes from normal individuals and patients with folic acid deficiency or pernicious anemia in remission. The data were statistically significant.

The author's inference that cobalamin participates specifically in the synthesis of thymidylate synthetase is hardly justified in the absence of data on whether or not other enzymes are affected and whether the observed depression of thymidylate synthetase is due to lack of enzyme or lack of enzyme activity. Also, if the speculative proposal were correct, it would confer upon the vitamin an entirely novel and unique role.

Important Negative Conclusions

Direct or indirect participation of a cobamide coenzyme in the pathway of DNA synthesis has been postulated on the basis of convincing evidence that DNA synthesis is impaired in cobalamin deficiency. Some of that evidence was summarized above. At this point it seems appropriate to note those enzymatic steps in the pathway that have been shown by conclusive biochemical study to be cobamide-independent.

Cobamide-Independence of Ribonucleoside Diphosphate Reductase

A schematic diagram of the known pathways of RNA and DNA synthesis is shown in Figure 9-4 (128). The pathways in the upper portion of the figure are those known to occur in *E. coli* (129) and Novikoff hepatoma (130) and suspected to occur in other animal cells. The pathways in the lower portion of the figure occur in *Lactobacillus leichmannii* (and other cobamide-requiring lactobacilli) (131, 132) and *Euglena gracilis* (133, 134).

In both systems the *de novo* synthesis of purines and pyrimidines follows established pathways, yielding ribonucleotides that are then phosphorylated to degrees appropriate to their further fate—to the triphosphate level for those serving as substrates of RNA polymerase and to the di- or triphosphate level for those serving as precursors of deoxyribonucleotides. Those in the latter category undergo a reduction that is catalyzed by a ribonucleotide reductase system that differs notably in the two groups of organism. In *E. coli* and Novikoff hepatoma, ribonucleotide reduction occurs at the *di*phosphate level; in *L. leichmannii* reduction occurs at the *tri*phosphate level. In both systems, the hydrogen transferred to the ribosyl moiety arises from reduced thioredoxin, a low-molecular-weight protein bearing two active thiol groups (135). In lactobacilli thioredoxin functions as a reductant for adenosylcobamide, which then transfers hydrogen to that substrate (136, 137). No cobamide derivative participates in the *E. coli* reductase, and the uncertainty over which agency serves the function exercised by cobamide coenzyme in the lactobacillus system led to the discovery that *E. coli* reductase contains nonheme iron (138). Addition of complexing agents such as hydroxyurea binds the iron and inactivates the enzyme.

Other major differences between the two systems include an absolute requirement for Mg^{2+} by *E. coli* reductase and a relative requirement by *L. leichmannii* reductase. Both enzymes are subject to elaborate mechanisms of allosteric control but the systems differ considerably in detail. For example, with enzymes from *E. coli* (or Novikoff hepatoma), reduction of CDP is stimulated by ATP and dTTP (139). The "prime effec-

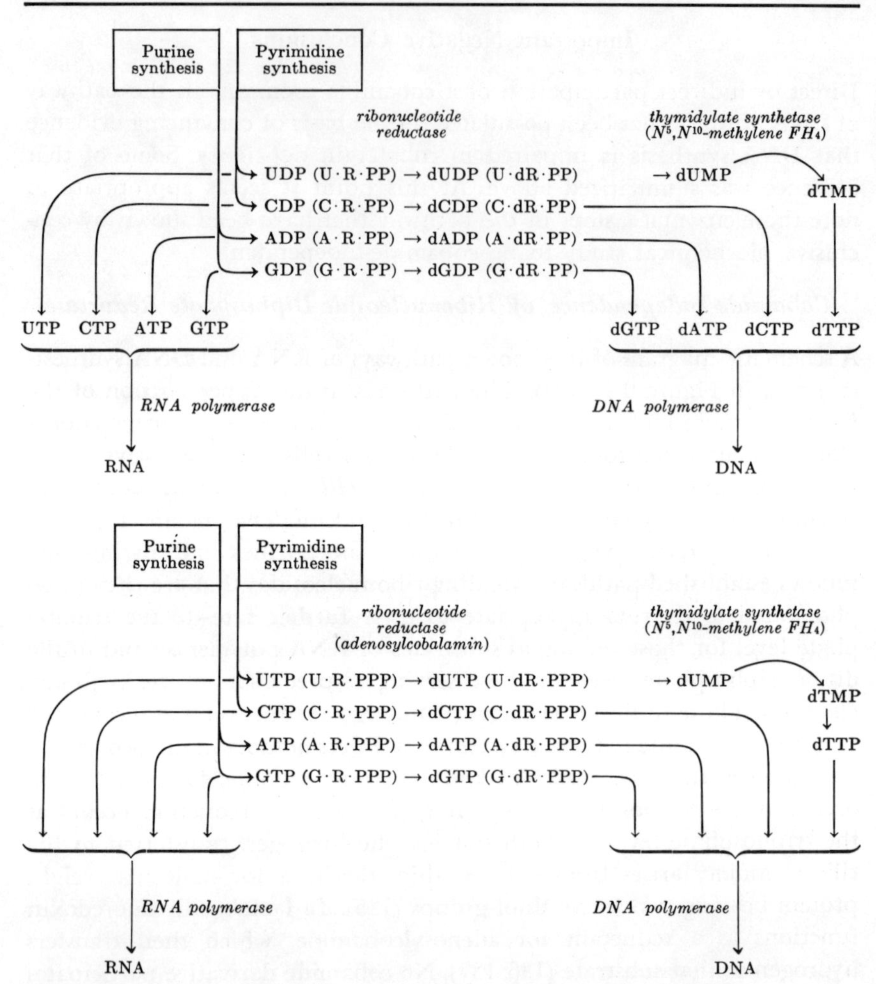

Fig. 9-4 Schematic diagram of pathways of nucleotide and nucleic acid synthesis. Upper diagrams shows pattern in *Escherichia coli* and Novikoff hepatoma. Lower diagram shows pattern in *Lactobacillus leichmannii*. [From Beck (128). Reprinted by permission.]

tor" for reduction of CTP by enzyme from *L. leichmannii* is dATP (140). But dATP is a negative effector for CDP reduction with the ribonucleoside diphosphate reductase of *E. coli* and Novikoff hepatoma. In view of the existence of two strikingly different microbial systems—one cobamide-dependent and the other cobamide-independent—it had been anticipated that the former would prove to be the relevant model for bone marrow cells and other mammalian cells, especially in light of the fact that these animal cells have similar requirements for cobalamin and folic acid. Also, it was found that cobamide-starved *L. leichmannii* and other cobamide-requiring organisms undergo unbalanced growth with resulting elongation (17, 19, 141, 142). These filamentous bacteria are analogous to the megaloblastic cells of human cobalamin deficiency.

Despite the compelling character of the bacterial model, efforts to demonstrate a cobamide-dependent ribonucleoside diphosphate reductase in bone marrow and other animal cells proved unsuccessful. Moreover, evidence has appeared suggesting that bone marrow reductase is, in fact, cobamide-independent (143, 144). Until the reductase of bone marrow and other animal cells is fully purified and characterized, the question requires further study. Nevertheless, the weight of present evidence makes it necessary to seek other explanations for the impairment of DNA synthesis in cobalamin deficiency. These are discussed later.

Cobamide-Independence of Thymidylate Synthetase

Despite the many claims that cobalamin deficiency impairs the conversion of deoxyuridylate to thymidylate, no evidence has implicated a cobamide derivative as a cofactor in the thymidylate synthetase reaction (145, 146). This critical enzyme requires only the presence of its substrate. dUMP, and its cofactor, $N^{5,10}$-methylenetetrahydrofolate, which serves as donor of both the methyl carbon and two hydrogens. Purified enzymes from bacteria (e.g., *E. coli, L. casei*) and other animal cells do not contain cobamide derivatives and are not stimulated by them. Moreover, there is no need to postulate participation of a cobamide derivative on the basis of present knowledge of the mechanisms of the thymidylate synthetase reaction.

This argument does not, of course, rule out the possibility that the net level of thymidylate synthetase activity *in vivo* may be affected by diminished cobalamin availability. Indeed, evidence was cited above suggesting that cobalamin deficiency may lead to decreased levels of thymidylate synthetase in stimulated lymphocytes (127). Also, at least one hypothesis concerning the role of cobalamin in DNA synthesis posulates mechanisms that act indirectly to reduce the level of thymidylate synthetase activity in tissues. This view, the so-called methylfolate trap hypothesis, is considered below.

METABOLIC BASIS OF THE COBALAMIN DEFICIENCY SYNDROME

We turn now to a discussion of known or postulated metabolic changes underlying the two main categories of clinical phenomena in human cobalamin deficiency: (a) megaloblastic anemia and related phenomena; and (b) certain neurological abnormalities. It is worth repeating that these are the two major modes of clinical abnormality in human cobalamin deficiency and that for reasons unknown one may occur in the absence of the other. It may be noted in the context of this discussion that the clinical picture of folic acid deficiency in man includes megaloblastic anemia and not neurological disease. Moreover, the megaloblastic anemia of folate deficiency is in every way indistinguishable from that of cobalamin deficiency. Surely this signifies in some preliminary sense that cobalamin deficiency leads to at least one category of biochemical sequelae that are unrelated to those responsible for the megaloblastic anemia.

It should be stated at the outset that despite vast amounts of work over a period of many years we still lack unimpeachable explanations for megaloblastic hematopoiesis and neurological disease in cobalamin deficiency. Several factors account for this state of affairs. Perhaps the most significant is the great difficulty in producing analogs of the human deficiency syndrome in experimental animals. Although it is possible with some difficulty to produce reliably deficient animals of several species—as judged by serum cobalamin levels, coenzyme levels in tissue extracts, assays of methylmalonyl CoA mutase levels, and so on—it is well known to most investigators in the field that such animals do not develop a clinical picture resembling that of human cobalamin deficiency. For example, rats, pigs, and other laboratory animals do not display clear-cut megaloblastic anemia when severely depleted of cobalamin nor do they show characteristic neurological disorders, either from the clinical or neuropathological viewpoint. A second source of difficulty is the relative rarity of human cobalamin deficiency and especially its fully developed neurological component.

Another factor is the inability of investigators to sample relevant tissues such as bone marrow or neural tissue in quantities sufficient to permit decisive biochemical study. Only recently has it become possible to develop cobalamin-dependent tissues culture systems. As will be noted later, cobalamin-deficient cultures have begun to reveal certain interesting biochemical phenomena.

A final factor that has undoubtedly slowed progress in the explication of the deficiency syndrome is the high probability that the chain of events between the nutritional deficiency and its clinical manifestations is a very long one indeed, which may be analyzable only in kinetic terms. It may be, for example, that cobalamin deficiency slows one pathway, which in turn slows another pathway, which in turn slows DNA synthesis. If this is the case, study of the system could proceed adequately only in the whole animal, and even under the best of circumstances the rigor of the evidence might leave something to be desired.

Megaloblastic Anemia and Related Phenomena

Nature of the Basic Defect

As noted in the foregoing discussion, it is generally agreed that the metabolic derangement that leads to megaloblastic anemia is impairment of DNA synthesis unassociated with impairment of RNA synthesis. The evidence for this statement, as was mentioned briefly above, consists in part of observations indicating an elevated RNA/DNA ratio in megaloblastic bone marrow (11, 12). The fact that megaloblastic bone marrow cells are capable of incorporating exogenously supplied precursors into DNA (13–15) suggests that the primary difficulty is in the endogenous synthesis of one or more of the nucleotide precursors of DNA.

Other data have supported this view of the basic defect in megaloblastic anemias of whatever cause. For example, labeling studies employing ^{3}H-thymidine (^{3}H-TdR) revealed a sharp increase in the number of unlabeled S-phase cells in bone marrow from subjects with megaloblastic anemia due to cobalamin deficiency (147–150) or methotrexate therapy (151). These are taken to reflect depression or arrest of the DNA synthesis observed in the normal S phase. Also, there reportedly occurs an accumulation of cells in the G_2 phase (152) and an increase in hypertetraploid cells (150, 152).

It bears repeating that although megaloblastosis occurs most commonly in deficiencies of cobalamin or folic acid—with resulting impairment of pathways dependent on those vitamins—impairment of any pathway essential in the biosynthesis of any portion of the deoxyribonucleoprotein complex may produce megaloblastosis. For example, megaloblastosis refractory to therapy with cobalamin or folic acid but responsive to therapy with adenine occurs in the Lesch-Nyhan syndrome in which there is a defect of the salvage pathway for converting hypoxanthine and guanine to purine nucleotides (153). Similarly, megaloblastosis refractory to cobalamin or folic acid but reversible by therapy with the pyrimidine nucleotides, cytidylic and uridylic acids, occurs in hereditary orotic aciduria, in which there is a defect in the biosynthesis of pyrimidine nucleotides (154, 155).

The problem requiring solution, then, within the purview of this essay, is to identify the locus of cobalamin action in the pathways of DNA synthesis. We have already summarized the attempts made to determine whether the role of cobalamin in animal cells is analogous to that in cobamide-dependent lactobacilli. Despite the strong presumptions set by the bacterial model—and the detailed evidence available on its use of adenosylcobalamin—efforts to demonstrate a cobamide-dependent ribonucleotide reductase in bone marrow and other animal cells have, as we have seen, been unsuccessful. Hence, it became necessary to devise other theories to account for the observable fact that cobalamin deficiency leads to impairment of DNA synthesis.

The "Methylfolate Trap" Hypothesis

The hypothesis that has acquired the jargon name "methylfolate trap" was first proposed in 1961 by Noronha and Silverman to account for certain abnormalities of folate metabolism appearing to arise as sequelae of cobalamin deficiency in rats (156). Following experiments showing that methionine supplementation causes an apparent shift of the forms of folate in rat liver, the proportion of N^5-methyltetrahydrofolate declining from 80 to 90% of the total to about 50%, they suggested that a similar shift might follow suppression of the cobamide-dependent pathway of methionine synthesis in cobalamin deficiency. This postulation was extrapolated to the situation in human cobalamin (157) and the further suggestion was made that accumulation of folate as N^5-methyltetrahydrofolate—the methylfolate trap—would have the effect of sequestering folate in this form, thereby preventing its conversion to tetrahydrofolate, a precursor of $N^{5,10}$-methylenetetrahydrofolate, the essential cofactor of thymidylate synthetase. Hence, the conversion of dUMP to dTMP— and consequently the synthesis of DNA—would be impaired.

This hypothesis rests on certain assumptions, none of which has been rigorously proved beyond all doubt. They include the following: (a) no cobamide-independent avenue exists in man for the conversion of N^5-methyltetrahydrofolate to tetrahydrofolate—indeed, there can exist no significant cobamide-independent pathway for the release of the methyl group from N^5-methyltetrahydrofolate; (b) N^5-methyltetrahydrofolate cannot be directly converted to $N^{5,10}$-methylenetetrahydrofolate; and (c) such interconversions of folate derivatives can occur only by way of processes in which folate temporarily releases its one-carbon substituent.

This hypothesis has been the subject of much debate, a good deal of which has been centered on the merits of data that in any case would not provide proof. It would simplify discussion if we were merely to recount the several pros and cons now on the record. The evidence tending to support the view that sequestration of folate as N^5-methyltetrahydrofolate eventually interferes with thymidylate synthesis includes the following: (a) changes in folate metabolism that often (but, significantly, not always) occur in cobalamin deficiency, among them the altered partitioning of tissue folate compounds referred to above (156), elevation of the serum N^5-methyltetrahydrofolate (157), increased urinary excretion of formiminoglutamic acid after an oral histidine loading dose (158), and elevated urinary excretion of aminoimidazole carboxamide (159); (b) the presence of megaloblastosis in a patient with congenital methyltransferase deficiency (160); (c) evidence of diminished methyltransferase activity in cobalamin-deficient rats (161); and (d) certain isotopic studies of Killmann (162) and others that are discussed below. The increases in urinary excretion of both formiminoglutamic acid and aminoimidazole carboxamide bespeak a tissue deficiency of folate coenzymes in cobalamin deficiency—though the possible presence of associated folate deficiency in reported cases of cobalamin deficiency is almost never rigorously ruled out.

The following evidence tends to deny the validity of the methylfolate trap hypothesis: (a) the lack of urinary excretion of formiminoglutamic acid and aminoimidazole carboxamide in many cobalamin-deficient subjects (163); (b) clearance studies by some workers (164, 165)—but not by all (166)—showing normal rates of N^5-methyltetrahydrofolate utilization in cobalamin deficiency and suggesting that the rise in serum N^5-methyltetrahydrofolate is due to translocation rather than entrapment; (c) normal rates of $^{14}CO_2$ production from [^{14}C-methyl]-N^5-methyltetrahydrofolate in cobalamin-deficient rats (167); (d) the recent discovery that pathways exist for the release of methyl groups from N^5-methyltetrahydrofolate—in the *N*- and *O*-methylation of biogenic amines (168, 169)—that appear to be cobamide-independent, though that matter has not been exhaustively studied; (e) the depression of total red cell folate in cobalamin deficiency (170), which is not predicted by the methylfolate trap hypothesis; (f) the disproportionate decrease in plasma methylcobalamin (171, 172), which would seem surprising if N^5-methyltetrahydrofolate were present in abundance; (g) the fact that therapy with methylcobalamin produces good responses in cobalamin deficiency (173), but is less effective than adenosylcobalamin (174); and (h) finally, the seemingly important recorded observations of three children with severe homocystinuria, cystathioninuria, hypomethioninemia, and methylmalonic aciduria (175–178), who displayed evidence of defective formation of both adenosylcobalamin and methylcobalamin, resulting perhaps from a defect in membrane transport of cobalamin or binding by an essential cobalamin binding protein. Significantly, these children had no megaloblastic anemia. Experiments with cultured fibroblasts from skin biopsies showed a depressed net level of methyltransferase activity and a correspondingly impaired conversion of homocysteine to methionine—precisely the defect envisioned in the methylfolate trap hypothesis.

A defect in the conversion of homocysteine to methionine thus seems insufficient to account for the megaloblastic hematopoiesis of cobalamin deficiency. Nor does it account for puzzling phenomena such as the failure of folate therapy to produce full remissions in cobalamin deficiency, the fact that rises in serum folate are greatest in the least anemic patients (179), and the curious ability of administered methionine to diminish the formiminoglutamic aciduria of cobalamin deficiency without correcting the megaloblastosis (180, 181). In sum, it would appear that no clear-cut evidence is available to prove that failure of the cobamide-dependent conversion of homocysteine to methionine "traps" folate as N^5-methyltetrahydrofolate.

Validity of Conclusions Based on the "Killmann Experiment"

The pathophysiologic problem for which the methylfolate trap hypothesis was a proposed solution is the relative failure of DNA synthesis in cobalamin deficiency. That hypothesis sought to link a postulated cobamide-dependent dislocation in the metabolism of folate coenzymes and a postulated failure in the conversion of dUMP to dTMP. Until the recent and still unconfirmed studies of Haurani on thymidylate synthetase activity in cobalamin deficiency (127), the only evidence for the presence of a defect in the conversion of dUMP to dTMP came from certain isotopic data derived from what the reviewer prefers to term the "Killmann experiment"—though others have termed it the "deoxyuridine (or dU) suppression test." These workers, following the pioneer work of Killmann (162) have utilized the following experimental protocol employing incubations with allegedly intact cells. In "tube A" cells were incubated with ^{3}H-thymidine (^{3}H-TdR). The rate of DNA synthesis was judged by the rate of incorporation of ^{3}H into DNA, which was determined either by grain counts in a radioautographic preparation or by radioassay of extracted DNA. "Tube B" contained a similar aliquot of cells and ^{3}H-TdR and, in addition, nonradioactive deoxyuridine (UdR). It has been regularly observed that incorporation of radioactivity into DNA was markedly lower (e.g., 90% lower) in tube B than in Tube A.

This result was interpreted by Killmann and successors to mean that exogenous nonradioactive UdR leads to the intracellular formation of nonradioactive dTMP, which substantially dilutes the pool of radioactive dTMP arising from radioactive TdR and thereby diminishes the labeling of DNA. When this experiment was repeated by Killman with bone marrow cells from cobalamin-deficient subjects, the percentage of "dilution" of DNA labeling of nonradioactive UdR was notably decreased. It was concluded that the conversion of dUMP to dTMP is directly or indirectly cobalamin dependent, since in cobalamin deficiency "dilution" of the radioactive dTMP pool by nonradioactive dTMP arising from nonradioactive dUMP is decreased.

This procedure was promptly and widely utilized in one or another modification by many workers in subsequent years (182–185). Under the name "dU suppression test" it has been recommended as a routine clinical test for cobalamin deficiency (186) and as a tool for investigating folate metabolism, drug effects, amino acid metabolism, and other phenomena that may directly or indirectly affect the synthesis of dTMP. In general, these experiments involve tests of the effects of various additions to bone marrow cell incubations on the "percent suppression" of the uptake of ^{3}H-TdR by nonradioactive UdR. In none of these reports are data given to show that added drugs, metabolites, or whatever, actually enter cells or that they enter as such.

In view of the heavy reliance being placed on this methodology, it seems appropriate that it be closely scrutinized. When this is done it becomes evident that the procedure raises searching questions that are in need of sound answers. For example, it seems surprising that variations in the concentration of an exogenous deoxyribonucleoside would have such drastic effects on the intracellular concentration of a given nucleotide, especially one as critically involved in regulatory networks as dTMP. Also, if the interpretation of the Killman experiment is correct, the reciprocal experiment (with radioactive UdR and nonradioactive TdR) should give comparable results, yet that study seems not to have been adequately reported. Controls also are lacking on the effects of exogenous nucleosides other than UdR. Finally, it should be noted that users of this procedure usually report results in terms of "percent suppression" of radioactive TdR incorporation by nonradioactive UdR rather than in terms of the actual incorporation rate of TdR into DNA (which is presumably a measure of the rate of DNA synthesis). This masks the possibility that the major effect of cobalamin deficiency may be depression of DNA synthesis per se and not a decrease in the suppression of TdR incorporation by UdR resulting from a putative decrease in the conversion of dUMP to dTMP.

With a view toward clarifying this matter certain experiments are currently being performed in the writer's laboratory (187). Though this work is still in progress, a number of relevant preliminary conclusions may be offered. First, we repeated the essential elements of the original Killmann protocol in experiments with normal and megaloblastic bone marrow cells employing a rigorous procedure of DNA extraction (188) and purification to the point of essential freedom from RNA (189). Careful determinations were made of the specific radioactivity of ^{3}H-TdR and other labeled nucleosides and data were expressed in terms of millimicomoles of the TdR incorporated into DNA per hour per milligram of total DNA, the latter being assayed by a highly reproducible modification (190) of the diaminobenzoic acid method (191). Incorporation rates in duplicate incubations agreed within 5%. When the rate of incorporation of ^{3}H-TdR was determined as a function of the number of cells incubated, rates were identical over a tenfold range of cell counts (0.3–3.0 × 10^{8} cells per incubation).

Experiments based on these procedures yielded several features of interest. The original observation of Killmann was confirmed. Addition of nonradioactive deoxycytidine (CdR) had a similar effect, conceivably by leading to the production of dUMP through a deamination reaction (192, 193). However, in the reciprocal experiments, nonradioactive TdR failed completely to decrease the incorporation of ^{3}H-UdR or ^{3}H-CdR into DNA. Also, nonradioactive deoxyadenosine and deoxyguanosine depressed incorporation of TdR to some extent. These results seem as likely to reflect inhibition of DNA synthesis by added exogenous nucleosides as hypothetical dilution effects on the intracellular pool of dTMP.

The experiments also confirmed that addition of UdR has a smaller relative effect on the incorporation of ^{3}H-TdR in megaloblastic bone marrow. But when these results were reported in absolute terms rather than as "percent suppression" it was found that the absolute rate of DNA synthesis in megaloblastic marrows was already quite low. Addition of nonradioactive UdR further decreases the incorporation of ^{3}H-TdR, but with the control incorporation rate low to begin with the UdR effect could as reasonably be attributed to further inhibition of DNA synthesis as to ineffective dilution of the ^{3}H-dTMP pool. The megaloblastic state similarly decreased ^{3}H-UdR incorporation.

Efforts have also been made to assay directly the intracellular pools of dTMP, dTTP, and dATP in these systems. Procedures were developed wherein extracts containing intracellular dTMP were treated with a protein fraction from L cells (194) or from *L. leichmannii* that contained sufficient deoxymononucleoside kinase activity to convert dTMP to dTTP quantitatively. This (along with endogenous dATP) was assayed by the precise enzymatic method of Lindberg and Skoog (195), which is based on the effects of limiting concentrations of dATP or dTTP in limiting the incorporation, respectively, of excess ^{3}H-dTTP or ^{3}H-dATP into poly-d(A-T) under the influence of DNA polymerase. Preliminary results indicate that exogenous UdR in several concentrations had only negligible effects on the intracellular pool of dTMP and dTTP. We are not yet prepared to comment on the possible differences in deoxyribonucleotide concentration of normoblastic and megaloblastic bone marrows. That question is still being studied.

If correct, these results suggest that the original observation of Killmann (which we repeated without difficulty) is to be explained either by an isotope dilution such as Killman postulated, but one occurring in a sequestered pool of thymidylates, or by some other phenomenon. The latter possibility is perhaps enhanced by our experiments with deoxynucleoside other than UdR.

In sum, these results, though preliminary, cast some doubt on the premise of the Killman experiment and on conclusions that have been based on it. While it is entirely possible that cobalamin deficiency directly and indirectly impairs the conversion of dUMP to dTMP, we consider that this proposition is not finally established by the Killmann experiment.

Neurological Abnormalities

The search for an explanation of the neurological consequences of cobalamin deficiency has been especially frustrating. Impediments to direct study of this facet of the deficiency syndrome have been even more tiresome than those hindering investigation of the mechanism of megaloblastic anemia. The main ones have been the unavailability of affected tissues and useful animal models.

Clinical and pathologic manifestations of the neurological syndrome of cobalamin deficiency were described above. In the following discussion, we attempt first to reconstruct a model of the basic process as it stands revealed by the methods of neuropathology. We then consider two speculative approaches to a theory of pathogenesis.

Nature of the Basic Defect

The earliest change observed in the spinal cords of patients with cobalamin deficiency has been swelling of the neuronal fibers, with distention of the myelin sheath. The change is not specific, similar patterns having been observed in several rapidly evolving degenerations of spinal cord white matter and in experimental cyanide encephalopathy. However, in the myelopathy of cobalamin deficiency, fibers swell in a multifocal, asymmetric, uneven fashion. The result is a topographic distribution that gives a characteristic appearance to the lesions. Specific fiber systems distal to the lesions undergo secondary degeneration in the course of time, so that regional and nonsystematized series of "empty spaces" develop that give the cord an open or spongy texture. If the deficiency state is not corrected, the changes become irreversible. Eventually they become associated with glial responses that occur in proportion to the extent of tissue breakdown.

It is usually said that a major feature of this picture is "demyelination." This term presumably signifies a decrease in the amount of myelin present. It is not clear, however, whether this state of affairs came about through impairment of myelin synthesis, abnormal destruction of existing myelin, or some other process. For example, a possible underlying process might be loss of myelin resulting from the synthesis of defective myelin. This has been termed "dysmyelination," though in the view of some (196) that term should be restricted to diseases of development since little myelin synthesis is believed to occur in adult life.

Neuropathologic evidence does not resolve these mechanistic uncertainties. These data do not even ascertain whether the primary target of the disease process is the myelin sheath or the axon. The myelin sheath seems more severely involved in early lesions, but the limitations of light microscopy hinder evaluation of the axons in these lesions. Since secondary degeneration of fiber tracts occurs distal to the characteristic regional spongiform lesion, significant axonal involvement must occur. If degeneration distal to spongiform lesions were greater than could be accounted for by secondary change, it would then be reasonable to infer that disease occurs earliest in the distal parts of the axons, but this has not been the observed pattern. On the other hand, distal tract degeneration is never completely absent and there is little if any focal demyelination with preservation of axons. Thus, it is an open question whether axon or myelin sheath is attacked preferentially in cobalamin deficiency.

In general, heavily myelinated fibers are affected first and most severely (37). Since the spinal cord fibers with the thickest myelin sheaths are in the posterior columns, this may explain why the lesions appear first in them. This explanation fails to account for the preferential localization of lesions in the cervical and the upper thoracic portions of the spinal cord. It may be that supporting tissues in these regions have unusually heavy metabolic requirements, but evidence for that supposition is also lacking.

The myelin sheath of peripheral nerves is believed to consist of multiple membrane layers made up of many windings of the extended plasma membranes of Schwann cells (197, 198). Some have suggested that in the central nervous system myelin is a cellular constituent of the oligodendrocytes (199–201). In several studies, the developing myelin sheath has appeared to indent the oligodendrocyte, which gives off a short process that is contiguous with the lamellae of the myelin sheath. There are many fewer neuroglical cells per myelinated nerve fiber in the central than in the peripheral nervous system. Probably in the former a myelin-forming cell can form more than one myelin segment. The role of the oligodendrocyte, then, remains uncertain. If the myelin sheath is attacked in cobalamin deficiency, they could be the locus of the disorder. It is agreed, in any case, that destruction of the myelin sheath—wherever it occurs and by whatever means—would lead to neurological dysfunction.

Virtually the entire dry weight of myelin is accounted for by proteins and lipids that form a continuous spiral of lipoproteic units arranged so that two layers of lipid point their fatty acid chains inward (i.e., toward one another) and their hydrophilic ends outward, where they associate closely with layers of protein. There are interesting differences between central and peripheral myelin that have not yet been given biological meaning. For example, proteolipids are plentiful in the former and absent in the latter (202) and the amino acid compositions of the several proteins differ (203).

Myelin is synthesized at a rapid rate in the early life of the animal (from the twelfth to twenty-fifth day of life in the rat). The rate of deposition then slowly declines. The material persists for a long period of time and has a relatively low rate of turnover. Reviewers (204–207) have emphasized the metabolic stability and inertness of myelin. However, recent workers have uncovered several areas of definite metabolic activity (208, 209) which conceivably could be weak links in a chain of pathophysiologic events.

Three myelin lipid constituents—inositol phosphatide, lecithin, and serine phosphatide—have certain metabolic characteristics which differentiate them from another group of myelin lipids including cholesterol, ethanolamine phosphatide, cerebroside, cerebroside sulfate, and sphingomyelin. The latter group accumulates rapidly during myelination; the former remains constant or decreases proportionately. Cholesterol, plasmalogen, galactolipids, and sphingomyelin have also been found to disappear most rapidly in Wallerian degeneration and in central demyelinating lesions. The lipids in the first group were not originally classified as myelin lipids. They were so identified only when methods were developed for purification of myelin. When the fatty acids were labeled with acetate-1-^{14}C, inositol phosphatide, lecithin, and serine phosphatide were found to have a higher rate of turnover than the other myelin lipids. The turnover pattern of lipds of myelin and other brain membrane fractions is similar. Nonetheless, the metabolic rate of myelin constituents is relatively sluggish compared with that of other brain fractions.

Studies of lipids in cobalamin deficiency were performed decades ago. However, these dealt with lipids of plasma and red cells. Williams and co-workers (210) found that remission was accompanied by a rise in the levels of plasma lecithin and sphingomyelin, and a decrease in cephalin. Kirk (211) found that the feeding of liver to pernicious anemia patients raised the sphingomyelin level of plasma and the cerebrosides of red cells. However, few reports have correlated the neurological syndrome of cobalamin deficiency with changes in the lipids of central and peripheral neural tissue. One study in albino rats (212) showed that although the cobalamin deficiency induced led to no overt clinical or pathologic manifestations of neurological disease, significant changes in the concentration of lipids in peripheral nerves did occur. For example, triglyceride concentrations decreased markedly and progressively over several months. Phospholipid levels remained unchanged.

Any theory of the pathogenesis of the neurological lesions of cobalamin deficiency ought to take account of the following considerations: (a) the "demyelinating" character of the lesions, as described above; (b) the lack of correlation in some patients between the presence or severity of the neurological and megaloblastic syndrome; (c) the apparent predilection of cobalamin-deficient ruminants for neurological rather than megaloblastic lesions; (d) the fact that folate therapy in cobalamin deficiency initiates or aggravates neurological abnormalities; (e) the several threads of evidence, to be described below, suggesting a connection between chronic cyanide intoxication, demyelination, and cobalamin deficiency; and (f) the argument that different cobamide coenzyme functions are impaired in the two clinical syndromes, which is based on the homocystinuric children discussed above (175, 176) who have methylmalonic aciduria and a measure of neurological involvement (mental retardation, increased tendon reflexes, and impaired proprioceptive sensation), but no megaloblastic anemia.

The reviewer is aware of only two coherent proposals to connect known enzymologic functions of cobamide coenzymes with postulated mechanisms of neurological involvement. One envisions certain abnormalities in the metabolism of myelin lipids as a consequence of impairment of the methylmalonyl CoA mutase reaction. The other postulates certain effects arising from an occult chronic cyanide intoxication. Underlying both schemes is the tidy but unproven assumption that the cobamide coenzyme active in hematopoiesis (methylcobalamin perhaps) is not the one active in pathways essential to neural integrity (adenosylcobolamin perhaps).

Possible Defects in the Metabolism of Myelin Lipids

Although methylmalonyl CoA is an intermediate in the catabolism of a short-chain fatty acid, propionyl CoA, two considerations seem to have deprived this pathway of direct involvement in the process under discussion. One is the relatively low level of proprionate metabolism in man. The other has been the widely confirmed observation that the degree of methylmalonic aciduria cannot be correlated with the presence or severity of neurological disease. In fact, the only metabolic observations in human material that do display such a correlation are the recently described studies of urinary excretion of acetic acid in cobalamin deficiency (213). Surprisingly few extensions of these data have been reported.

Nevertheless, there is a provocative association between the neurological lesions of cobalamin deficiency in sheep and the fact that propionate metabolism is quantitatively and qualitatively of greater importance in ruminants than in man. This, among other considerations, has led to many ingenious efforts to connect these facts. These have followed two lines. One postulates that accumulated methylmalonyl CoA (or propionyl CoA) exercises an inhibitory effect on the synthesis of myelin lipids. The other postulates that methylmalonyl CoA is incorporated into the lipids synthesized which thereby become abnormal. The resulting myelin is defective and "demyelination" follows.

The evidence bearing on these hypotheses may be briefly summarized. Cardinale et al. (214) examined the possibility that methylmalonyl CoA, because of its structural analogy with malonyl CoA, either inhibits fatty acid synthesis or partially replaces malonyl CoA as a substrate for fatty acid synthesis with the resulting formation of abnormal branched-chain fatty acids. Both effects were observed in *in vitro* experiments with rat liver extract. Addition of 0.4 mM methylmalonyl CoA to a reaction mixture containing acetyl CoA in similar concentrations inhibited fatty acid synthesis by 47 to 68%. Tritium from ^{3}H-methylmalonyl CoA was incorporated with low efficiency into long-chain fatty acids in the presence but not in the absence of both acetyl CoA and malonyl CoA. Interestingly, examination by gas-liquid chromatography of fatty acids formed in the presence of methylmalonyl CoA revealed components not formed in the absence of methylmalonyl CoA, which were saturated and had branched-chain structures. It was concluded that methylmalonyl residues are incorporated into fatty acids in nonterminal positions. Propionyl CoA is not an intermediate in the process. Branched-chain fatty acids similar to those formed *in vitro* from methylmalonyl CoA could not be detected in liver lipids of cobalamin-deficient pigs, but suggestive evidence was obtained of an increased accumulation of C_{18} monosaturated fatty acids.

A recent report presented evidence from tissue culture studies that tends to support the hypothesis that methylmalonyl CoA may enter into reactions leading to fatty acid elongation with resulting formation of fatty acids with one-carbon branches (215). Rat glial cells were grown in culture in cobalamin-deficient media. As intracellular levels of adenosylcobalamin decreased (as shown by direct assay), cells synthesized increasing amounts of two unbranched fatty acids which were unusual in that they contained 15 and 17 carbon atoms (Figure 9-5). They appeared in increasing amounts as cells progressively lost the ability to metabolize 1-^{14}C-propionic acid through the methylmalonyl CoA mutase reaction. Levels of the 15- and 17-carbon acids returned to normal when cultures were supplemented with cobalamin. Similar preliminary results were obtained by us some years ago in extracts of red cell and plasma lipids from patients with pernicious anemia (216). One wonders whether the acetic aciduria of cobalamin (213) might reflect the displacement of acetyl CoA from the fatty acid synthesis pathway by methylmalonyl CoA or propionyl CoA.

Further evidence has been reported by Frenkel who showed that nerve biopsy slices from patients with pernicious anemia incorporate ^{14}C-propionate into branched-chain fatty acids (217, 218). Similar experiments

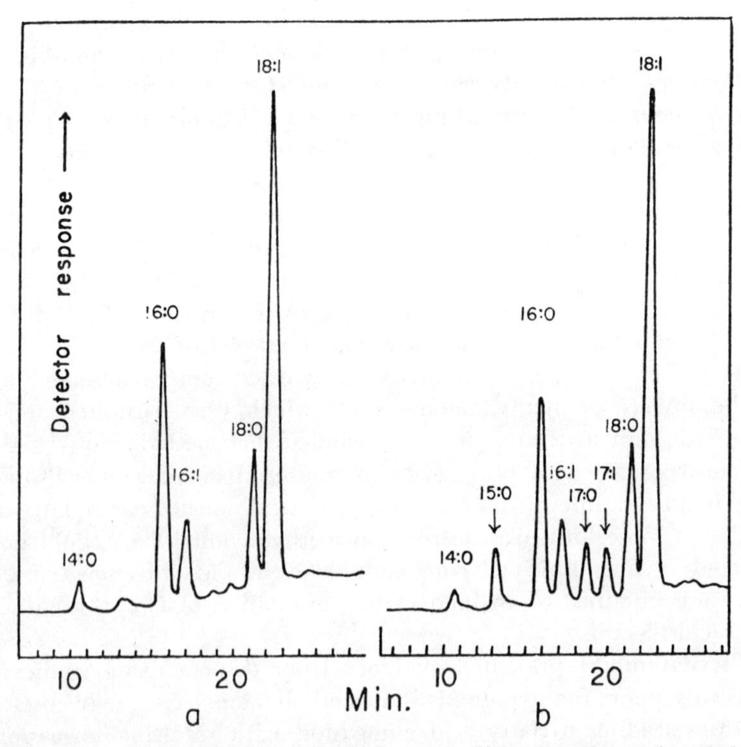

Fig. 9-5 Fatty acid composition of lipids from rat glial cells cultured in cobalamin-deficient media for 35 days. (*a*) Culture was supplemented with hydroxocobalamin and grown for an additional 7 days before isolation of fatty acids. (*b*) No hydroxocobalamin supplement. Curves show detector response observed in gas chromatography of methyl esters of fatty acids [From Barley, et al. (215). Reprinted by permission.]

with normal material revealed no such acids. Moreover, propionate was found to inhibit incorporation of ^{14}C-acetate into normal fatty acids. These results, when considered alongside the fact that homocystinuric children with marked methylmalonic aciduria have only modest neurological changes, suggest that the major culprit may be accumulated propionyl CoA rather than methylmalonyl CoA.

Aspects of the tissue culture studies of Barley et al. (215) merit comment. There was no evidence in cobalamin-deficient cell cultures of methyl-branched fatty acids. However, differences were observed in the fatty acids of normal and cobalamin-deficient cells. Deficient cells made increased amounts of straight-chain fatty acids with an odd number of carbon atoms and the level of these odd-chain acids returned to normal when the cells were repleted with cobalamin. Odd-chain fatty acids were readily formed when propionyl CoA initiated fatty acid synthesis. Propionyl CoA was an effective initiator of fatty acid synthesis (214). The normal low level of odd-chain fatty acids in animals may very well reflect the low level of propionyl CoA *in vivo.*

Since propionyl CoA is metabolized by carboxylation of methylmalonyl CoA, conditions that increase concentrations of methylmalonyl CoA would be expected to increase the level of propionyl CoA also. Propionic acid did increase in the media of cobalamin-deficient cell cultures as it does in the urine of cobalamin-deficient patients (219), and experiments in which the cells were grown in the presence of exogenous propionate showed that odd-chain fatty acids arise when the level of propionic acid in the medium increases.

The tissue culture system may, of course, be deceptive. Cells *in vitro* are not required to preserve the same functions they display *in vivo.* For example, they do not transmit nerve impulses. The fact that an increased level of odd-chain fatty acids was not observed in cobalamin-deficient pigs (214) suggests that either a rise in odd-chain fatty acids might be lethal before it is large enough to be detected, or alternatively, that these changes may not occur significantly in the pig. In light of the variable ability of tissue culture cells to metabolize propionic acid, it is possible that another cell type is even more sensitive to propionate levels than glial cells are, and that the fatty acids from these cells were not examined in the studies performed in whole animals.

Such tissue culture experiments do not eliminate the possibility that methylmalonyl CoA is incorporated into fatty acids of cell types that were not tested. Similar studies in L cells yielded similar results although the magnitude of the fatty acid changes was smaller (215). This work has several interesting implications. First, it shows the potential usefulness of tissues culture techniques in the study of metabolic consequences of cobalamin deficiency. Second, it provides support of a plausible hypothsis concerning the pathogenesis of neural symptoms in human cobalamin deficiency. Further explorations along these lines are awaited with interest.

Possible Role of Chronic Cyanide Intoxication

We have earlier cited evidence pointing to a possible role for cobalamin in the metabolism of cyanide. Aside from the suggestiveness of biochemical arguments relating to the conversions of cyanide to thiocyanate and of adenosylcobalamin to cyanocobalamin, a number of diverse clinical and pathologic observations have enlivened discussion.

Several studies have revealed a relatively high incidence of retrobulbar neuritis and optic atrophy in cobalamin-deficient tobacco smokers and in cobalamin-deficient males who were presumed more likely to be smokers than observed females (118, 220, 221). We have already seen that plasma cobalamin of smokers contains a relatively elevated cyanocobalamin fraction (123); see also Chapter 6. Comparable incidences of retrobulbar neuritis and optic atrophy, often in association with other neuropathies, have been found in West Africa and the West Indies where native populations eat cassava and other foods rich in cyanide or cyanogenetic glycosides (222). Malnourished prisoners of war have been similarly afflicted (223).

Perhaps the most suggestive of the clinical arguments is based on the disorder known as tobacco amblyopia. This is a disease of insidious onset that predominantly affects older men. Visual disability occurs, often with central scotomata and poor acuity for red objects. Although it has occurred mainly in smokers of pipe and cigarette tobaccos, some have given an etiological role to alcohol. The condition usually clears on cessation of smoking, but irreversible optic atrophy may occur. Although some observers have equated tobacco amblyopia with the retrobulbar neuritis and optic atrophy of cobalamin deficiency, there has been no consistent pattern of cobalamin depletion (as judged by serum levels) in patients thought to be suffering from tobacco amblyopia. Heaton and co-workers (224) pointed out the association of tobacco amblyopia and cobalamin deficiency, and Wokes (221) suggested that the cyanide in tobacco smoke might lead to the optic neuropathy of patients whose detoxification mechanisms had been impaired by cobalamin deficiency. Smith (225) speculated on a possible relation between tobacco amblyopia and visual failure of pernicious anemia on the one hand and cyanide exposure and cobalamin deficiency on the other. Finally, recent work (226) has suggested that the therapeutic efficacy of hydroxocobalamin is superior to that of cyanocobalamin in the treatment of tobacco amblyopia. However, the precise role of cobalamin deficiency remains to be established and it remains uncertain as to whether or not this disorder may be considered a special case of the neuropathy of cobalamin deficiency.

A variety of other neurological disorders provide some circumstantial evidence that cyanide exposure or abnormalities of cyanide metabolism may be of pathogenetic significance. As well reviewed by Matthews and Wilson (117), these include tropical ataxic neuropathy, Leber's hereditary optic atrophy, "dominantly inherited" optic atrophy, a variety of optic atrophies of obscure origin, and the subacute combined spinal cord degeneration of pernicious anemia itself. In Leber's disease, there is consistent depression of serum methylcobalamin, though total serum cobalamin is normal (227). This pattern resembles the serum cobalamin pattern observed in a child with an inborn error of cobalamin metabolism, who has methylmalonic aciduria unresponsive to cobalamin therapy, mental deficiency and convulsions, and megaloblastic anemia (228).

We have earlier taken note of the several hypotheses that have sought to account for the neural effects of cyanide and the role of cobalamin deficiency. According to one, an additional exogenous cyanide load tends to convert coenzyme derivatives of cobalamin to cyanocobalamin. This phenomenon occurs in normal subjects, but only on a small scale. Cyanocobalamin, which is less firmly bound to plasma proteins, is more readily excreted than the other forms of cobalamin. This conversion thus might account for an increased urinary excretion of cobalamin in smokers. The increment is small, however, and seems not to be enough to deplete a normal individual of cobalamin. It is possible that in persons with relatively low, though not necessarily subnormal, serum cobalamin concentrations the capacity to detoxify cyanide is decreased. These changes may contribute to the association between low serum cobalamin and high plasma cyanide. In this view, the moderate increases in cyanide intake experienced by smokers lead to disturbances in cobalamin-deficient individuals in whom the conversion of even a small proportion of body cobalamin stores is detrimental.

A facet of the neurological syndrome that none of the available hypotheses just listed has explained is the effect of therapy with folate on the cobalamin deficiency syndrome. As noted above, this commonly leads in time either to an appearance of neurological symptoms where none existed before or to frank neurological relapse. The mechanism of this effect is also without explanation. It has always seemed to the writer that a plausible explanation could be a translocation of cobamide coenzymes away from neural tissues to other sites—perhaps to bone marrow where folate may have stimulated hematopoiesis previously slowed by megaloblastosis. The validity of that concept also remains to be tested.

SUMMARY AND PROSPECTUS

The status of the problem discussed in this essay is clearly unsatisfactory. Despite recent successes in the assignment of precise enzymologic and metabolic roles to cobalamin and its coenzymes, it is still not possible to write a coherent account of the pathophysiology of the megaloblastic anemia and neuropathy associated with cobalamin deficiency in man. Following the discovery that animal cells do not mimic cobamide-dependent lactobacilli in their method of harnessing cobalamin into the pathway of DNA synthesis, one theory after another has sought to fill the void. The methylfolate trap hypothesis seems now to have collapsed and we await the testing of other ideas—ideas such as a role for cobalamin in promoting cellular uptake of folate, a role for cobalamin in promoting the synthesis (or activity) of thymidylate synthetase, and so on.

The apparent fact that human metabolism has been shown to make use of cobalamin in only two enzyme reactions—N^5-methyltetrahydrofolate homocysteine methyltransferase and methylmalonyl CoA mutase—has generated an understandable impulse to associate defects of one (the former) with megaloblastic anemia and defects of the other (the latter) with neurological disease. To date, this view remains only a hope, though promising recent results suggest that impairment of the mutase system may lead to abnormalities of lipid metabolism that could conceivably affect the integrity of myelin. Also, there is accumulating evidence that a number of neurological and ophthalmological syndromes may represent an accumulation of the neurotoxic effects of chronic cyanide exposure. In certain hereditary conditions, a failure of cyanide detoxification may occur as an inborn error of metabolism. In cobalamin deficiency, so the speculation goes, defects of cyanide detoxification may occur as an acquired abnormality.

One anticipates that new approaches to these problems will materialize soon. It is possible, for example, that more than two enzymes in the human body are cobamide-dependent and these must be searched out and found. It is possible, too, that megaloblastosis is due to a combination of factors, of which impairment of DNA synthesis is only one. One wonders, for example, if mitochondrial integrity is altered in cobalamin deficiency, especially since mitochondrial duplication necessarily includes replication of mitochondrial DNA. Studies now in progress in our laboratory (229) confirm that mitochondria are rich in cobalamin in several tissues but show that mitochondrial DNA synthesis is dependent on extramitochondrial sources of deoxyribonucleotides. More also needs to be done in animal cells on the mechanisms of conversion of cobalamin to cobamide coenzymes.

It is hard to avoid the suspicion that important light will be shed on the metabolic role of cobalamin in man by more careful examination of the biology of this extraordinary vitamin. Its chemical structure and biological sources bespeak its evolutionary antiquity. This inference is supported also by the fact that many and diverse species apparently required its presence before they could begin to synthesize DNA. One cannot keep from doubting that this entire remarkable situation arose as a byproduct of a minor pathway of methionine synthesis.

REFERENCES

1. Finch, C. A., Coleman, D. H., Motulsky, A. G., Donohue, D. M., and Reiff, R. H. (1956). *Blood* **11**, 807.
2. Myhre, E. (1964). *Scand. J. Clin. Lab. Invest.* **16**, 391.
3. Heller, P., Weinstein, H. G., West, M., and Zimmerman, H. J. (1960). *Ann. Intern. Med.* **53**, 898.
4. Elliot, B. A., and Fleming, A. F. (1965). *Brit. Med. J.* **1**, 626.
5. Libnoch, H. A., Yakulis, V. J., and Heller, P. (1966). *Am. J. Clin. Pathol.* **45**, 302.
6. London, I. M., and West, R. (1950). *J. Biol. Chem.* **184**, 359.
7. White, P., Coburn, R. F., Williams, W. J., Goldwein, M. I., Rother, M. L., and Shafer, B. C., (1967). *J. Clin. Invest.* **46**, 1986.
8. Lynch, E. C., and Alfrey, C. P., Jr., (1966). *Texas Rep. Biol. Med.* **24**, 180.
9. Perillie, P. E., Kaplan, S. S., and Finch, S. C. (1967). *N. Engl. J. Med.* **277**, 10.
10. Harker, L. A., and Finch, C. A. (1969). *J. Clin. Invest.* **48**, 963.
11. Glazer, H. S., Mueller, J. F., Jarrold, T., Sakurai, K., Will, J. J., and Vilter, R. R. (1954). *J. Lab. Clin. Med.* **43**, 905
12. White, J. C., Leslie, I., and Davidson, J. N. (1953). *J. Pathol. Bacteriol.* **66**, 291.
13. Lessner, H. E., and Friedkin, M. (1959). *Clin. Res.* **7**, 207.
14. Wichramasinghe, S. N., Cooper, E. H., and Chalmers, D. E. (1968). *Blood* **31**, 304.
15. Yoshida, Y., Todo, A., Shirakawa, S., Wakisaka, G., and Uchino, H. (1968). *Blood* **31**, 292.
16. Beck, W. S. (1962). *N. Engl. J. Med.* **266**, 708, 814.
17. Beck, W. S. (1964). *Medicine (Balt.)* **43**, 715.
18. Cohen, S. S., and Barner, H. E. (1954). *Proc. Natl. Acad. Sci. U. S. A.* **40**, 885.
19. Beck, W. S., Hook, S., and Barnett, B. H. (1962). *Biochem. Biophys. Acta* **55**, 455.
20. Freifelder, D., and Maaløe, O. (1964). *J. Bacteriol.* **88**, 987.
21. Smith, B. J., and Burton, K. (1965). *Biochem. J.* **97**, 240.
22. Freifelder, D. (1969). *J. Mol. Biol.* **45**, 1.
23. Ruekert, R. R., and Mueller, G. C. (1960). *Cancer Res.* **20**, 1584.

24. Kim, J. H., Perez, A. G., and Djordjevic, B. (1968). *Cancer Res.* **28**, 2443.
25. Myhre, E. (1964). *Scand. J. Clin. Lab. Invest.* **16**, 320.
26. Nathan, D. G., and Gardner, F. H. (1962). *J. Clin. Invest.* **41**, 1086.
27. Rondanelli, E. G., Gorini, P., Magliulo, E., and Fiori, G. P. (1964). *Blood* **24**, 542.
28. Wichramasinghe, S. N., Chalmers, D. G., and Cooper, E. H. (1967). *Nature (London)* **215**, 189.
29. Williams, A. M., Chosy, J. J., and Schilling, R. F. (1963). *J. Clin. Invest.* **42**, 670.
30. Lajtha, L. G., and Kumatori, T. (1957). *Nature (London)* **180**, 991.
31. Bock, H. E., Hartje, J., Müller, D., and Wilmanns, W. (1967). *Klin. Wochenschr.* **45**, 176.
32. Rotherham, J., Price, F. M., Otani, T. T., and Evans, V. J. (1971). *J. Natl. Cancer Inst.* **47**, 277.
33. Waxman, S., Metz, J., and Herbert, V. (1969). *J. Clin. Invest.* **48**, 284.
34. Lark, K. G. (1963). In *Molecular Genetics,* Taylor, J. H., Ed., Pt. 1, Academic Press, New York, p. 153.
35. Smith, A. D. M. (1960). *Brit. Med. J.* **2**, 1840.
36. Smith, A. D. M., and Duckett, S. (1965). *Brit. J. Exp. Pathol.* **46**, 615.
37. Pant, S. S., Asbury, A. K., and Richardson, E P. (1968). *Acta Neurol. Scand. (Suppl. 35)* **44**, 7,
38. Adams, R. D., and Kubik, C. S. (1944). *N. Engl. J. Med.* **231**, 2.
39. Greenfield, J. G., and O'Flynn, E. (1933). *Lancet* **2**, 62.
40. Mayer, R. F. (1965). *Arch. Neurol* **13**, 355.
41. Victor, M., and Lear, A. (1956). *Am. J. Med.* **20**, 896.
42. Hess, B., and Gehm, E. (1955). *Klin. Wochenschr.* **33**, 91.
43. Anderssen, N. (1964). *Scand. J. Haematol.* **1**, 212.
44. Emerson, P. M., and Wilkinson, J. H. (1966). *Brit. J. Haematol.* **12**, 678.
45. Winston, R. M., Warburton, F. G., and Stott, A. (1970). *Brit. J. Haematol.* **19**, 587.
46. Heller, P., Weinstein, H. G., West, M., and Zimmerman, H. J. (1960). *J. Lab. Clin. Med.* **55**, 425.
47. Keeley, K. J., and Politzer, W. M. (1956). *J. Clin. Pathol.* **9**, 142.
48. Todd, D. (1959). *J. Clin. Pathol.* **12**, 238.
49. Fowler, D., Cox, E. V., Cooke, W. T., and Meynell, M. J. (1960). *J. Clin. Pathol.* **13**, 230.
50. Beck, W. S., and Girey, G. J. D. Unpublished results.
51. Thompson, H. T., Dietrich, L. S., and Elvehjem, C. A. (1950). *J. Biol. Chem.* **184**, 175.
52. *United States Pharmacopeia* **15**, 885 (1955).
53. Shinton, N. K. (1959). *Clin. Sci.* **18**, 389.
54. Burkholder, P. R. (1951). *Science* **114**, 459.
55. Kelly, A., and Herbert, V. (1967). *Blood* **29**, 139.
56. Raven, J. L., Walker, P. L., and Barkhan, P. (1966). *J. Clin. Pathol.* **19**, 610.

57. Cox, E. V., and White, A. M. (1962). *Lancet* **2**, 853.

58. Kahn, S. B., Williams, W. J., Barness, L. A., Young, D., Shafer, B., Vicacqua, R. J., and Beaupre, E. M. (1965). *J. Lab. Clin. Med.* **66**, 75.

59. Bashir, H. V., Hinterberger, H., and Jones, B. P. (1966). *Brit. J. Haematol.* **12**, 704.

60. Rosenberg, L. E., Lilljeqvist, A.-C., and Hsia, Y. E. (1969). *N. Engl. J. Med.* **278**, 1319.

61. Mudd, S. H., Levy, H. L., and Abeles, R. H. (1969). *Biochem. Biophys. Res. Commun.* **35**, 121.

62. Chanarin, I., England, J. M., Mollin, C., and Perry, J. (1973). *Brit. J. Haematol.* **25**, 45.

62a. Finch, C. A., Coleman, D. H., Motulsky, A. G., Donohue, D. M., and Reiff, R. H. (1956). *Blood,* **11**, 807.

63. Hillman, R. S., Adamson, J., and Burka, E. (1968). *Blood* **31**, 419.

64. van Dommelen, C. K. V., and Klaassen, C. H. L. (1964). *N. Engl. J. Med.* **271**, 541.

65. James, G. W. III, and Abbott, L. D., Jr. (1952). *Metabolism* **1**, 259.

66. Lawson, D. H., Murray, R. M., Parker, J. L. W., and Hay, G. (1970). *Lancet* **2**, 588.

67. Lawson, D. H., Murray, R. M., and Parker, J. L. W. (1972). *Q. J. Med.* **41**, 1.

68. Adams, J. F., Hume, R., Kennedy, E. H., Pirrie, T. G., and Whitelaw, J. W. (1968). *Brit. J. Nutr.* **22**, 575.

69. Marshall, R. A., and Jandl, J. H. (1960). *Arch. Intern. Med. (Chicago)* **105**, 352.

70. Vilter, R. W., Vilter, C. F., and Hawkins, R. (1947). *J. Lab. Clin. Med.* **32**, 1426.

71. Wagle, S. R., Mehta, R., and Johnson, B. C. (1957). *J. Am. Chem. Soc.* **79**, 4249.

72. Arnstein, H. R. V., and Simkin, J. L. (1959). *Nature (London)* **183**, 523–525.

73. Dubnoff, J. W. (1950). *Arch. Biochem.* (1950). **27**, 466.

74. Ling, C. T., and Chow, B. F. (1962). In *Vitamin B_{12} and Intrinsic Factor, Second European Symposium,* Heinrich, H. C., Ed., Enke, Stuttgart, p. 127.

75. Flavin, M., Ortiz, P. J., and Ochoa, S. (1955). *Nature (London)* **176**, 823.

76. Beck, W. S., Flavin, M., and Ochoa, S. (1957). *J. Biol. Chem.* **229**, 997.

77. Beck, W. S., and Ochoa, S. (1958). *J. Biol. Chem.* **232**, 931.

78. Retey, J., and Arigoni, D. (1966). *Experientia* **22**, 783.

79. Kaziro, Y., and Ochoa, S. (1964). *Adv. Enzymol.* **26**, 283.

80. Marston, H. R. (1959). *Med. J. Aust.* **46**, 105.

81. Smith, R. M., and Monty, K. J. (1959). *Biochem. Biophys. Res. Commun.* **1**, 105.

82. Rabinowitz, J. C. (1960). In *The Enzymes,* 2nd ed., Vol. 2, Boyer, P. D., Lardy, H., and Mÿrback, K. Eds., Academic Press, New York, p. 185.

83. Eggerer, H., Stadtman, E. R., Overath, P., and Lynen, F. (1960). *Biochem. Z.* **333**, 1.

84. Wood, H. G., and Stjernholm, R. (1961). *Proc. Natl. Acad. Sci. U. S. A.* **47**, 303.

85. Lengyel, P., Mazumder, R., and Ochoa, S. (1960). *Proc. Natl. Acad. Sci. U. S. A.* **46**, 1312.

86. Stadtman, E. R., Overath, P., Eggerer, H., and Lynen, F. (1960). *Biochem. Biophys. Res. Commun.* **2**, 1.

87. Smith, R. M., and Osborne-White, W. S. (1965). *Biochem. J.* **95**, 411.

88. Marston, H. R. (1961). *Nature (London)* **190**, 1085.

89. Dupont, J., and Matthias, M. M. (1969). *Lipids* **4**, 478.

90. Williams, D. L., Spray, G. H., Newman, G. E., and O'Brien, J. R. P. (1969). *Brit. J. Nutr.* **23**, 343.

91. Reed, E. B., and Tarver, H. (1970). *J. Nutr.* **100**, 935.

92. Venkataraman, S., Biswas, D. K., and Johnson, B. C. (1967). *J. Nutr.* **93**, 131.

93. Cardinale, G. J., Dreyfus, P. M., Auld, P., and Abeles, R. H. (1969). *Arch. Biochem. Biophys.* **131**, 92.

94. Weidemann, M. J., Hems, R., Williams, D. L., Spray, G. H., and Krebs, H. A. (1970). *Biochem. J.* **117**, 177.

95. Contreras, E., and Giorgio, A. J. (1972). *Am. J. Clin. Nutr.* **25**, 695.

96. Barness, L., Young, D., Nocho, R., and Kahn, B. (1963). *J. Clin. Invest.* **42**, 915.

97. Weidemann, M. J., Hems, R., Williams, D. L., Spray, G. H., and Krebs, H. A. (1970). *Biochem. J.* **117**, 177.

98. Williams, D. L., and Spray, G. H. (1972). *Brit. J. Nutr.* **28**, 263.

99. Stokstad, E. L. R., Webb, R. E., and Shah, E. (1966). *J. Nutr.* **88**, 225.

100. Weissbach, H., and Taylor, R. T. (1968). *Vitam. Horm. (N. Y.)* **26**, 395.

101. Sakami, W., and Ukstins, I. (1961). *J. Biol. Chem.* **235**, PC50.

102. Mangum, J. H., and Scrimgeour, K. G. (1962). *Fed. Proc.* **21**, 242.

103. Buchanan, J. M., Elford, H. C., Loughlin, R. E., McDougall, B. M., and Rosenthal, S. (1964). *Ann. N. Y. Acad. Sci.* **112**, 756.

104. Burke, G. T., Mangum, J. H., and Brodie, J. D. (1970). *Biochem.* **9**, 4297.

105. Van Trump, J. E., and Miller, S. L. (1972). *Science* **178**, 859.

106. Du Vigneaud, V., Chandler, J. P., Moyer, A. W., and Keppel, D. M. (1939). *J. Biol. Chem.* **131**, 57.

107. Dickerman, H., Redfield, B. G., Bieri, J. G., and Weissbach, H. (1964). *J. Biol. Chem.* **239**, 2545.

108. Kutzbach, C., Galloway, E., and Stokstad, E. L. R. (1967). *Proc. Soc. Exp. Biol. Med.* **124**, 801.

109. Finkelstein, J. D., Kyle, W. E., and Harris, B. J. (1971). *Arch. Biochem. Biophys.* **146**, 84.

110. Mangum, J. H., and North, J. A. (1968). *Biochem. Biophys. Res. Commun.* **32**, 105.

111. Mangum, J. H., Murray, B. K., and North, J. A. (1969). *Biochem.* **8**, 3496.

112. Kerwar, S. S., Spears, C., McAuslan, B., and Weinbach, H. (1971). *Arch. Biochem. Biophys.* **142**, 231.

113. Kamely, D., Littlefield, H. W., and Erbe, R. W. (1973). *Proc. Natl. Acad. Sci. U. S. A.* **70**, 2585.

114. Sauer, H. J., Howell, J. N., and Jaenicke, L. (1973). *Res. Exp. Med.* **160**, 171.

115. Das, K. C., and Hoffbrand, A. Y. (1970). *Brit. J. Haematol.* **19**, 203.
116. Tisman, G., and Herbert, V. (1973). *Blood* **41**, 465.
117. Matthews, D. M., and Wilson, J. (1971). In *The Cobalamins,* Arnstein, H. R. V., and Weighton, R. J., Eds. Churchill-Livingstone, London, p. 115.
118. Wokes, F., and Picard, G. W. (955). *Am. J. Clin. Nutr.* **3**, 383.
119. Mushett, C. W., Kelley, K. L., Boxer, G. E., and Rickards, J. C. (1952). *Proc. Soc. Exp. Biol. Med.* **81**, 234.
120. Basil, W., Brown, J. K., and Matthews, D. M. (1965). *J. Clin. Pathol.* **18**, 317.
121. Matthews, D. M., Wilson, J., and Zilkha, K. J. (1965). *J. Neurol. Neurosurg. Psychiat.* **28**, 426.
122. Wilson, J., and Matthews, D. M. (1966). *Clin. Sci.* **31**, 1.
123. Linnell, J. C., Wilson, J., and Matthews, D. M. (1969). *Clin. Sci.* **37**, 878.
124. Lindstrand, K., and Stahlberg, K. C. (1963). *Acta Med. Scand.* **174**, 665.
125. Linnell, J. C., Mackenzie, H. M., and Matthews, D. M. (1969). *J. Clin. Pathol.* **22**, 506.
126. Linnell, J. C., Mackenzie, H. M., Wilson, J., and Matthews, D. M. (1969). *J. Clin. Pathol.* **22**, 545.
127. Haurani, F. I. (1973). *Science* **182**, 78.
128. Beck, W. S. (1968). *Vitam. Horm.* **26**, 413.
129. Reichard, P. (1962). *J. Biol. Chem.* **237**, 3513.
130. Moore, E. C., and Hurlbert, R. (1966). *J. Biol. Chem.* **241**, 4802.
131. Blakley, R. L., and Barker, H. A. (1964). *Biochem. Biophys. Res. Commun.* **16**, 391.
132. Beck, W. S., and Hardy, J. (1965). *Proc. Natl. Acad. Sci. U. S. A.* **54**, 286.
133. Gleason, F. K., and Hogenkamp, H. P. C. (1970). *J. Biol. Chem.* **245**, 4894.
134. Beck, W. S., McCabe, J. T., and Sutherland, J. A. Unpublished results.
135. Laurent, T. C., Moore, E. C., and Reichard, P. (1964). *J. Biol. Chem.* **239**, 3436.
136. Beck, W. S., Goulian, M., Larsson, A., and Reichard, P. (1966). *J. Biol. Chem.* **241**, 2177.
137. Abeles, R. H., and Beck, W. S. (1967). *J. Biol. Chem.* **242**, 3589.
138. Brown, N. C., Eliasson, R., and Reichard, P. (1969). *Eur. J. Biochem.* **9**, 512.
139. Larsson, A., and Reichard, P. (1966). *J. Biol. Chem.* **241**, 2533, 2540.
140. Beck, W. S. (1967). *J. Biol. Chem.* **242**, 3148.
141. Beck, W. S., Goulian, M., and Hook, S. (1962). *Biochem. Biophys. Acta* **55**, 470.
142. Beck, W. S., and Kashket, S. (1964). *Medicine* **43**, 195.
143. Fujioka, S., and Silber, R. (1969). *Biochem. Biophys. Res. Commun.* **35**, 759.
144. Hopper, S. (1972). *J. Biol. Chem.* **247**, 3336.
145. Blakely, R. L. J. (1963). *J. Biol. Chem.* **238**, 2113.
146. Friedkin, M. (1972). *Adv. Enzymol.* **38**, 235.
147. Menzies, R. C., Crossen, P. E., Fitzgerald, P. H., and Gunz, F. W. (1966). *Blood* **28**, 281.
148. Messner, H., Fliedner, T. M., and Cronkite, E. P. (1969). *Haematology* **2**, 44.

149. Wickramasinghe, S. N., Cooper, E. H., and Chalmers, D. E. (1968). *Blood* **31**, 304.

150. Yoshida, Y., Todo, A., Shirakawa, S., Wakisaka, G., and Uchino, H. (1968). *Blood* **31**, 292.

151. Young C. W., and Hodas, S. (1964). *Science* **146**, 1172.

152. Cooper, E. N., and Wickramasinghe, S. N. (1969). *Ser. Haematol.* **2**, 65.

153. Van Der Zee, S. P. M., Lommen, E. J. P., Trijbels, J. M. F., and Schretlen, E. D. A. M. (1970). *Acta Paediatr. Scand.* **59**, 259.

154. Fallon, H. J., Smith, L. H., Graham, J. B., and Burnett, C. H. (1964). *N. Engl. J. Med.* **270**, 878.

155. Huguley, C. M., Jr., Bain, J. A., and Rivers, S. L. (1959). *Blood* **14**, 615.

156. Noronha, J. M., and Silverman, M. (1962). In *Vitamin B_{12} and Intrinsic Factor, Second European Symposium,* Heinrich, H., Ed., Enke, Stuttgart, p. 728.

157. Herbert, V., and Zalusky, R. (1962). *J. Clin. Invest.* **41**. 1263.

158. Silverman, M., and Pitney, A. S. (1958). *J. Biol. Chem.* **233**, 1179.

159. Luhby, A. L., and Cooperman, J. M. (1962). *Lancet* **2**, 1381.

160. Arakawa, I. (1970). *Am. J. Med.* **48**, 594.

161. Kutzbach, C., Galloway, E., and Stokstad, E. L. R. (1967). *Proc. Soc. Exp. Biol. Med.* **124**, 801.

162. Killmann, S.-Aa. (1964). *Acta Med. Scand.* **175**, 483.

163. Beck, W. S. Unpublished results.

164. Chanarin, I., and Perry, J. (1968). *Brit. J. Haematol* **14**, 297.

165. Nixon, P. F., and Bertino, J. R. (1972). *J. Clin. Invest.* **51**, 1431.

166. Herbert, V. (1971). In *The Cobalamins,* Arnstein, H. R. V., and Wrighton, R. J., Eds., Churchill-Livingstone, London, p. 20.

167. Thenen, S. W., Gawthorne, J. M., and Stokstad, E. L. R. (1970). *Proc. Soc. Exp. Biol. Med.* **134**, 199.

168. Laduron, P. (1972). *Nature New Biol.* **238**, 212.

169. Banerjee, S. P., and Snyder, S. H. (1973). *Science* **182**, 74.

170. Cooper, B. A., and Lowenstein, L. (1964). *Blood* **24**, 502.

171. Linnell, J. C., Mackenzie, H. M., Wilson, J., and Matthews, D. M. (1969). *J. Clin. Pathol.* **22**, 545.

172. Linnell, J. C., Hoffbrand, A. V., Peters, T. J., and Matthews, D. M. (1971). *Clin. Sci.* **40**, 1.

173. Abe, T., Nahajima, T., and Okamura, H. (1970). *Japanese Clin. Hematology* **11**, 473.

174. Hamilton, H. E., Falabella, F., Hogenkamp, H. P. C., and Sheets, R. F. (1969). *J. Lab. Clin. Med.* **74**, 881.

175. Mudd, S. H., Levy, H. L., and Abeles, R. H. (1969). *Biochem. Biophys. Res. Commun.* **35**, 121.

176. Levy, H. L., Mudd, S. H., Schulman, J. D., Dreyfus, P. M., and Abeles, R. H. (1970). *Am. J. Med.* **48**, 390.

177. Mudd, S. H., Uhlendorf, B. W., Hinds, K. R., and Levy, H. L. (1970). *Biochem. Med.* **4**, 215.

178. Goodman, S. I., Moe, P. G., Hammond, K. B., Mudd, S. H., and Uhlendorf, B. W. (1970). *Biochem. Med.* **4**, 500.

179. Waters, A. H., and Mollin, D. L. (1963). *Brit. J. Haematol.* **9**, 319.

180. Chanarin, I. (1963). *Brit. J. Haematol.* **9**, 141.

181. Herbert, V., and Sullivan, L. W. (1963). *Proc. Soc. Exp. Biol. Med.* **112**, 304.

182. Metz, J., Kelly, A., Sweth, V. C., Waxman, S., and Herbert, V. (1968). *Brit. J. Haematol.* **14**, 575.

183. Waxman, S., Metz, J., and Herbert, V. (1969). *J. Clin. Invest.* **48**, 284.

184. Stebbins, R., Scott, J., and Herbert, V. (1972). *Blood* **40**, 927.

185. Van Der Weyden, M. B., Cooper, M., and Firkin, B. G. (1973). *Blood* **41**, 299.

186. Herbert, V. (1971). *Ann. Clin. Lab. Sci.* **1**, 193.

187. Beck, W. S., Ribas-Mundo, M., and Hinrichsen, M. Unpublished results.

188. Morris, N. R., and Cramer, J. W. (1966). *Mol. Pharmacol.* **2**, 1.

189. Klamerth, O. (1965). *Nature* **208**, 1318.

190. Beck, W. S. (1967). *J. Biol. Chem.* **242**, 3148.

191. Kissane, J. M., and Robins, E. (1958). *J. Biol. Chem.* **233**, 184.

192. Maley, F., and Maley, G. F. (1960). *J. Biol. Chem.* **235**, 2968.

193. Gelbard, A. S., Kim, J. H., and Perez, A. G. (1969). *Biochim. Biophys. Acta* **182**, 564.

194. Griffith, T. J., and Helleiner, C. W. (1965). *Biochim. Biophys. Acta* **108**, 114.

195. Lindberg, U., and Skoog, L. (1970). *Anal. Biochem.* **34**, 152.

196. Davison, A. N. (1970). In *Myelination,* Davison, A. N., and Peters, A., Eds. Thomas, Springfield, p. 162.

197. Fernández-Morán, H. (1950). *Exp. Cell Res.* **1**, 143.

198. Geren, B. B. (1954). *Exp. Cell Res.* **7**, 558.

199. Maturana, H. R. (1960). *J. Biophys. Biochem. Cytol.* **7**, 197.

200. Peters, A. (1962). *J. Biophys. Biochem. Cytol.* **7**, 121.

201. Bunge, M. B., Bunge, R. P., and Pappas, G. D. (1962). *J. Cell. Biol.* **12**, 448.

202. Johnson, A. C., McNabb, A. R., and Rossiter, R. J. (1949). *Biochem. J.* **45**, 500.

203. Wolfgram, F., and Kotorii, K. (1968). *J. Neurochem.* **15**, 1291.

204. Adams, C. W. M., and Davison, A. N. (1965). In *Neurochemistry,* Adams, C. W. M., Ed., Elsevier, Amsterdam, p. 332.

205. Smith, M. E. (1967). In *Advances in Lipid Research,* Vol. 5, Paoletti, R., and Kritchevsky, D., Eds., Academic Press, New York, p. 2410.

206. Davison, A. N. (1970). In *Myelination,* Davison, A. N., and Peters, A. Eds., Thomas, Springfield, p. 80.

207. Wolman, M. (1971). In *Progress in Neuropathology,* Vol. 1, Zimmerman, H. M., Ed., Grune & Stratton, New York, p. 62.

208. Smith, M. E., and Eng, L. F. (1965). *J. Am. Oil Chem. Soc.* **42**, 1013.

209. Eichberg, J., and Davidson, R. M. C. (1965). *Biochem. J.* **96**. 644.

210. Williams, H. M., Erickson, B. N., Bernstein, S. S., and Macy, I. G. (1940). *Proc. Soc Exp Biol. Med.* **45**, 151.

211. Kirk, E. (1938). *Am. J. Med. Sci.* **196**, 648.

212. Turner, D. A., and Cevallos, W. H. (1968). *Clin. Biochem.* **2**, 1.

213. Cox, E. V., Robertson-Smith, D., Small, M., and White, A. M. (1968). *Clin. Sci.* **35**, 123.

214. Cardinale, G. J., Carty, T. J., and Abeles, R. H. (1970). *J. Biol. Chem.* **245**, 3771.

215. Barley, F. W., Sato, G. H., and Abeles, R. H. (1972). *J. Biol. Chem.* **247**, 4270.

216. Kashket, S., and Beck, W. S. Unpublished results.

217. Frenkel, E. P. (1971). *J. Clin. Invest.* **50**, 332.

218. Frenkel, E. P. (1971). *Clin. Res.* **19**, 74.

219. White, A. M. (1965). *Biochem. J.* **95**, 17.

220. Hamilton, H. E., Ellis, P. P., and Sheets, R. F. (1959). *Blood* **14**, 378.

221. Wokes, F. (1958). *Lancet* **2**, 526.

222. Montgomery, R. D. (1965). *Am. J. Clin. Nutr.* **17**, 103.

223. Denny-Brown, D. (1947). *Medicine (Balt.)* **26**, 41.

224. Heaton, J. M., McCormick, A. J. A., and Freeman, A. G. (1958). *Lancet* **2**, 286.

225. Smith, A. D. M. (1961). *Lancet* **1**, 1001.

226. Chisolm, J. A., Bronte-Stewart, J., and Foulds, W. S. (1967). *Lancet* **2**, 450.

227. Wilson, J., Linnell, J. C., and Matthews, D. M. (1971). *Lancet* **1**, 259.

228. Matthews, D. M., and Linnell, J. C. (1971). In *The Cobalamins,* Arnstein, H. R. V., and Wrighton, R. J., Eds., Churchill-Livingstone, London, p. 23.

229. Beck, W. S., Biswas, C., and Cohen, R. Unpublished results.

Afterword

BERNARD M. BABIOR

As is apparent from the foregoing chapters, a great deal of time, effort, and money has been invested in the study of vitamin B_{12}, the factor postulated by Minot to be the antipernicious anemia principle of liver. Much has been learned about the chemistry of the vitamin and its relatives; significantly less is known about the biology of the vitamin; while the relationships between cobalamin-requiring metabolic processes and the clinical manifestations of cobalamin deficiency remain for the most part completely obscure.

Work currently in progress in many laboratories promises to yield substantial growth in knowledge about cobamides. Investigations of the chemistry of cobamides are being pursued with a view to understanding the mechanisms involved in axial ligand exchange and rearrangements. Biochemical studies are concerned with the relationship between the chemistry of the axial ligands and the mechanisms of cobalamin-requiring enzymatic reactions, as well as with more general aspects of the mechanisms of these reactions and the structures of the enzymes that catalyze them. Studies in whole animals are turning from questions of absorption per se to questions dealing with cobamide pool kinetics in various tissues, and with the interconversions of the various forms of cobamide in one tissue or another, while studies of absorption are focusing more and more on events at the molecular level. Transport of cobamides across bacterial and mammalian cell membranes and their uptake by mitochondia are becoming areas of intense research interest, partly because of the insights such investigations may provide regarding the biochemical lesions in familial errors of cobalamin metabolism and partly because of the possibility that cobamide analogs, known to inhibit the growth of cobamide-dependent microorganisms, may if taken up inhibit the growth of rapidly dividing tumor cells as folate antagonists do. Finally, the whole question of the nature of the metabolic aberrations responsible for the clinical manifestations of cobalamin deficiency is now beginning to be reevaluated.

It is clear that after almost 50 years of investigation, only the surface of the cobamide problem has been scratched. It is likely, however, that the great advances of recent years in the general understanding of biological systems as well as in the methodology available for the study of such systems will lead after a substantially shorter interval of time to a deep understanding of the role played by cobamides in living systems.

APPENDIX I

THE NOMENCLATURE OF CORRINOIDS[1] (1973 RECOMMENDATIONS)

IUPAC-IUB COMMISSION ON BIOCHEMICAL NOMENCLATURE[2]

1. The corrinoids are a group of compounds containing four reduced pyrrole rings joined into a macrocyclic ring by links between their α positions; three of these links are formed by a one-carbon unit (methylidyne radicals) and the other by a direct $C\alpha$–$C\alpha$ bond. They include various B_{12} vitamins, factors, and derivatives based on the skeleton of *corrin,* $C_{19}H_{22}N_4$ (structure I). The atoms are numbered and the rings are lettered as shown in structure I. The numbering is thus the same as

[1] Reprinted with permission from IUPAC-IUB Commission on Biochemical Nomenclature. The Nomenclature of Corrinoids (1973 Recommendations) (1974). *Biochemistry* **13,** 1555–1560. Copyright by the American Chemical Society.

[2] Revision of the 1965 document (1) of the Commission on Biochemical Nomenclature (CBN) of the International Union of Pure and Applied Chemistry (IUPAC) and the International Union of Biochemistry (IUB), approved by CBN, IUPAC, and IUB in 1973 and published by permission of IUPAC and IUB. Significant changes from the 1965 version (1) are indicated by ▲ in the margin. An appendix on abbreviations has been added.

Comments on and suggestions for future revisions of these Recommendations may be sent to any member of CBN. Reprints of this publication may be obtained from W. E. Cohn, Director, NAS–NRC Office of Biochemical Nomenclature, Oak Ridge National Laboratory, Box Y, Oak Ridge, Tenn., U.S.A., 37830. Members of the Commission are O. Hoffmann-Ostenhof (Chairman), W. E. Cohn (Secretary), A. E. Braunstein, B. L. Horecker, P. Karlson, B. Keil, W. Klyne, C. Liébecq, E. C. Webb, and W. J. Whelan.

that of the porphyrin nucleus, number 20 being omitted to preserve the identity.

▲ *Note.* The name "corrin" was proposed by those who established its structure because it is the *core* of the vitamin B_{12} molecule; the letters "co" of corrin are *not* derived from the fact that vitamin B_{12} contains cobalt. However, this does not apply to the "cob" terms below, all of which do contain "co" for cobalt.

▲ 2. Some important corrinoids that are more unsaturated than corrin itself are derivatives of octadehydrocorrin. This has sometimes been called tetradehydrocorrin[3] because it has four additional double bonds. Although this could be indicated by the prefix "tetrakis(didehydro)," "octadehydro" is preferred.

The octadehydrocorrin system IA has the trivial name *corrole.*

3. Many important corrinoids have a regular pattern of substituents on the methylene carbon atoms of the reduced pyrrole rings and a cobalt atom in the center of the macrocylic ring. The heptacarboxylic acid II is named *cobyrinic acid.* The carboxyl groups are designated by the locants *a* to *g,* as shown in II.[4] Cobyrinic *a,b,c,d,e,g*-hexaamide, formerly sometimes referred to as *Factor* V_{1a}, is named *cobyric acid.*[4] Substituents on the side chains may be designated by appropriate locants, for example, $7\beta^1$-methylcobyrinic acid, if $-CH(CH_3)CO_2H$ replaces $-CH_2CO_2H$ at C-7β of cobyrinic acid.

4. The compound III (R = OH, R′ = H), which is the amide formed by combination of cobyrinic acid with D-1-amino-2-propanol at position *f,* is named *cobinic acid*[4]; its hexaamide (III; R = NH_2, R′ = H) is named *cobinamide.*[4]

5. The compound III (R = OH, R′ = structure V) in which cobinic acid is further substituted at the 2 position of the aminopropanol by an *α-D-ribofuranose 3-phosphate* residue (V) is named *cobamic acid*[4]; its hexaamide (III; R = NH_2, R′ = V) is named *cobamide.*[4]

▲ 6. Glycosyls and nucleotides (which are *N*-glycosyl derivatives at C-1 of the ribofuranose unit) of cobamides are named by adding the name of the appropriate aglycon radical (ending in "yl") as a prefix to the name of the corrinoid allotted according to 1–5, for example, aglyconylcobamide (VI).

[3] "Pentadehydrocorrin" in the previous document (1) was an error.

▲ [4] The names cobyrinic acid, cobinic acid, cobamic acid, and cobalamin, and names derived from them, imply the relative and absolute configurations shown in the structural formulas. α and β are used as in Steroid Nomenclature (2) and the IUPAC E Rules (3) to indicate stereochemical configuration. Epimers at C-3, C-8, and C-13 may be designated as, for example, 13-epicobalamin.

▲ **7.** Most of the important natural products in this series have aglycon radicals containing an imidazole nucleus, one N of the latter being covalently bonded to the ribose while the other is coordinately bonded to what is, by this attachment, defined as the cobalt-α position. *The latter situation (VII) is assumed to exist unless otherwise indicated.* When another ligand occupies the cobalt-α position, it and its locant may be indicated by, for example, (Coα-ligand)-aglyconylcobamide (VIII). The absence of a "*Coα*-ligand" term, as in the cobalamins (see **9**), indicates that the aglycon radical attached to the ribose occupies the cobalt-α position as well.

8. Cobamides bearing a ligand in the cobalt-β position [which implies Co(III)] may be named as follows:

(*Coα*-ligandyl)- (*Coβ*-ligandyl)-(aglyconylcobamide) (IX)

or, if the aglycon is attached to the cobalt-α position, as indicated in **7**

aglyconyl-(*Coβ*-ligandyl)cobamide (Xa)

In a cobalamin (see **9**), the latter becomes simply

ligandylcobalamin (Xb)

See also **15**.

9. Cobalamins.[4] A cobalamin is a cobamide in which 5,6-dimethylbenzimidazole is the aglycon attached by a glycosyl link from its N-1 to the C-1 of the ribose and additionally linked, as stated in **7**, by a bond between the N-3 and the cobalt (in position α).[4] They may be named as cobamides, as above, or according to the pattern:

(ligand in Coβ position, if any)-cobalamin (Xb)

Examples:

Coα-[α-(5,6-Dimethylbenzimidazolyl)]-*Coβ*-cyanocobamide, also known as vitamin B_{12}, is termed *cyanocobalamin.*

Coα-[α- (5,6-Dimethylbenzimidazolyl)]-*Coβ*-aquacobamide, also known as vitamin B_{12a}, is termed *aquacobalamin.*

Coα-[α- (5,6-Dimethylbenzimidazolyl)]-*Coβ*-hydroxocobamide, also known as vitamin B_{12b}, is termed *hydroxocobalamin.*

(*Note.* Aquacobalamin is the conjugate acid of hydroxocobalamin.)

Coα-[α-(5,6-Dimethylbenzimidazolyl]-*Coβ*-nitritocobamide, also known as vitamin B_{12c}, is termed *nitritocobalamin.*

10. Anion(s) associated with the corrinoids is (are) stated in the usual way after the name of the (cationic) corrinoid, for example, cobamic dichloride (not dichlorocobamic acid).

11. The state of oxidation of the cobalt may be specified, when necessary, as follows:

vitamin B_{12} cyanocob(III)alamin
▲ cob(II)alamin[5]
▲ cob(I)alamin[5]

▲ **12.** Displacement of the ribosyl-bound aglycon base from its normal coordinate bonding to position α of the cobalt by another ligand (or by water) may be indicated by placing the name and locant of the replacing ligand before the corrinoid name and enclosing the modified corrinoid name (see **6**) in parentheses. (See also **7**.)

Example:

Coα-Aqua-*Co*β-methyl(2-methyladenylcobamide), in which the 2-methyladenyl residue is attached to the ribose residue but is not coordinately bound to the cobalt atom, having been displaced by water. Methyl occupies the *Co*β position.

13. Modified, derived, or related compounds are named systematically from the largest of the compounds I, II, or III that is contained in them.

Examples:

Cobyrinic acid *a,b,c,d,e,g*-hexaamide *f*-2-hydroxyethylamide 3,8,13,17-tetraethyl-1,2,2,5,7,7,12,12,15,17,18-undecamethylcobalticorrin dichloride (for the dichloride of fully decarboxylated cobyrinic acid).

$12\alpha^1$-Carboxycobyrinic acid (for cobyrinic acid in which the 12α-methyl group has been replaced by $-CH_2CO_2H$).

14. Replacement of the cobalt atom in compound II or III by another metal or by hydrogen is indicated by replacing "co" in the "cob" part of the name with the name or the root of the name of the replacing metal followed by "o" or "i" according to its valence (e.g., cupri, cupro, zinco). When cobalt is replaced by hydrogen, "hydrogeno" replaces "co."

Examples: ferrobamic acid; hydrogenobamic acid.

See Note to **1** concerning corrin. This replacement nomenclature does not apply to corrole (see **2**).

▲ **15. Cofactor Forms.** The coenzymatically active forms of the B_{12} vitamins (see **12**) and their analogs possess an organic ligand, either

[5] The previous document (1) erred in prefixing "cyano" to these names.

methyl or 5′-deoxy-5′-adenosyl,[6] attached to the β position of the cobalt by a carbon-to-cobalt bond, that is, in the position of the CN in formula IV. These adducts (*4*) should be named according to the pattern:

Coα-(radical in α position)-*Co*β-(ligand in β position)(corrinoid name) or (ligand in β position)cobalamin, if the radical in the α position is dimethylbenzimidazole

Examples:

Coα-[α-(5,6-Dimethylbenzimidazolyl)]-*Co*β-adenosylcobamide,[6] or adenosylcobalamin,[6] for the compound formerly known as "coenzyme B_{12}."

Coα-[α-(5,6-Dimethylbenzimidazolyl)]-*Co*β-methylcobamide or methylcobalamin, for the compound involved in several reactions, including methionine biosynthesis, where a methyl group is ligated to the cobalt in the β position

Coα[α-(7-adenyl)]-*Co*β-adenosylcobamide,[6] the coenzymatically active form of "pseudovitamin B_{12}," capable of replacing adenosylcobalamin in many systems.

16. Summary. The trivial names applied to corrinoids of varying complexity are perhaps confusing to the nonspecialist, and it seems desirable to tabulate (in outline) how they are interrelated (Table A-1).

▲ [6] For brevity, 5′-deoxy-5′-adenosyl may be replaced by adenosyl, with definition, as is commonly seen in *S*-adenosylmethionine. The intermediate form, 5′-deoxyadenosyl, should not be used.

NOMENCLATURE OF CORRINOIDS

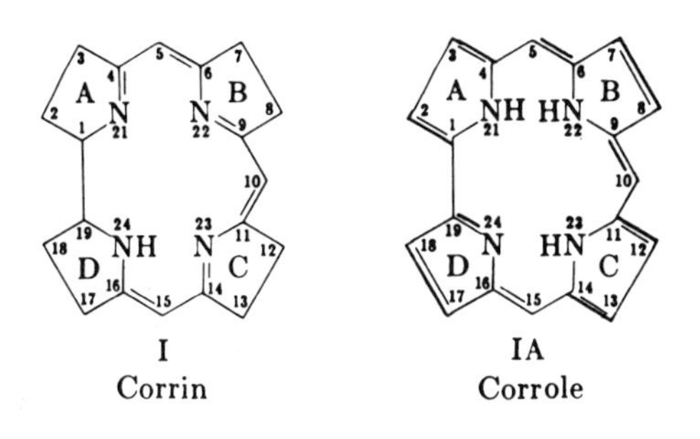

I
Corrin

IA
Corrole

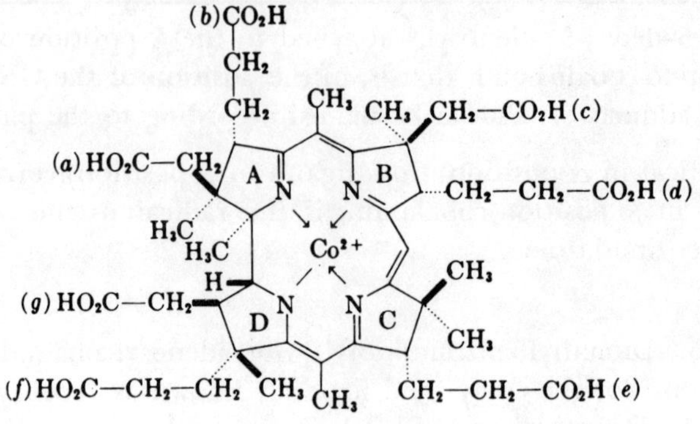

(b) CO_2H
CH_2
CH_2 CH_3 CH_3 CH_2—CO_2H (c)
(a) HO_2C—CH_2 A N N B CH_2—CH_2—CO_2H (d)
H_3C H_3C Co^{2+} CH_3
H
(g) HO_2C—CH_2 D N N C CH_3
(f) HO_2C—CH_2—CH_2 CH_3 CH_3 CH_2—CH_2—CO_2H (e)

II
Cobyrinic acid

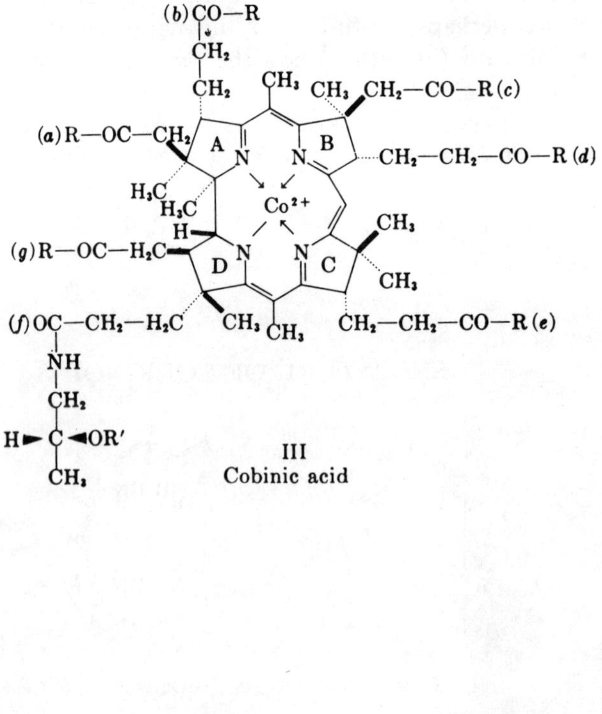

(b) CO—R
CH_2
CH_2 CH_3 CH_3 CH_2—CO—R (c)
(a) R—OC—CH_2 A N N B CH_2—CH_2—CO—R (d)
H_3C H_3C Co^{2+} CH_3
H
(g) R—OC—H_2C D N N C CH_3
(f) OC—CH_2—H_2C CH_3 CH_3 CH_2—CH_2—CO—R (e)
NH
CH_2
H—C—OR′
CH_3

III
Cobinic acid

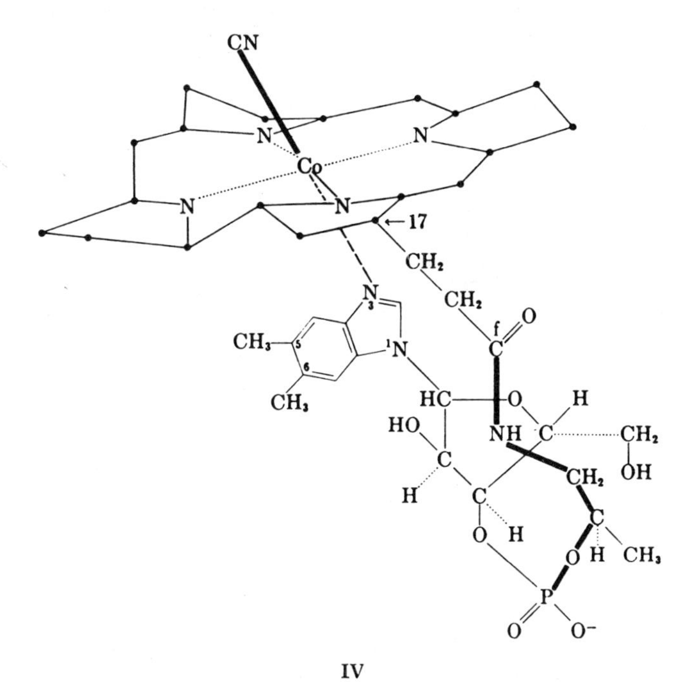

IV

Vitamin B_{12}

Sketch based on Hodgkin et al. (5). Detail of substituents on corrin nucleus (except side chain at C-17) is omitted for the sake of clarity.

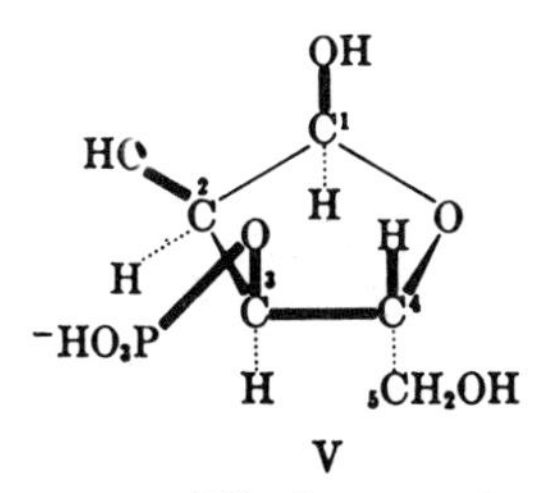

V

α-D-Ribofuranose 3-phosphate residue

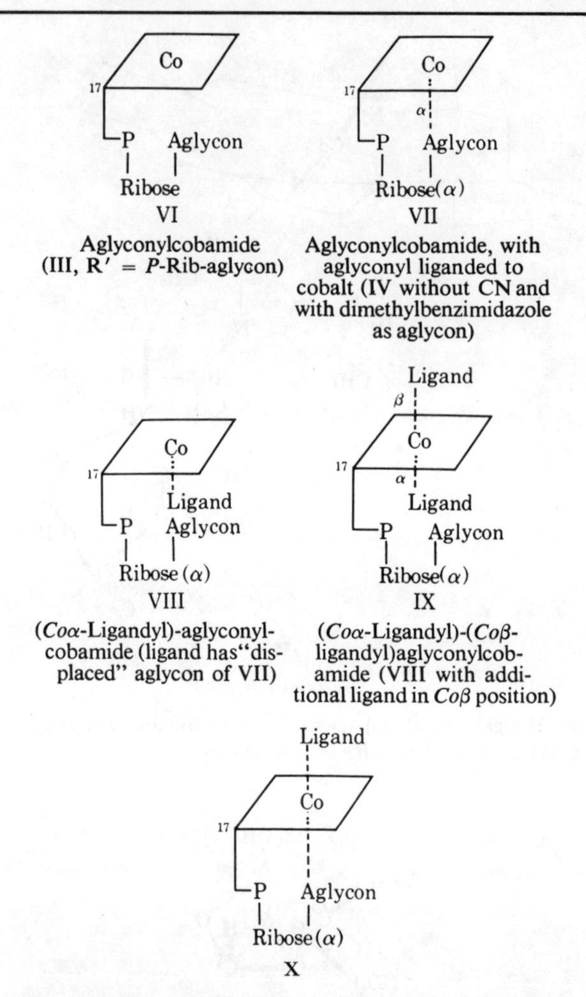

Xa, Aglyconyl-(*Coβ*-ligandyl)cobamide (VII with additional ligand in *Coβ* position; IV with dimethylbenzimidazole as aglycon, CN as *Coβ* ligand)

Xb, Ligandylcobalamin (if aglycon is dimethylbenzimidazole as in IV, and with CN as *Coβ* ligand)

Notes on Formulas

1. In formulas II and III, the corrin nucleus is represented as being roughly in the plane of the paper. Bonds joining peripheral substituents to the nucleus are shown by the same convention as in the steroid series (2) that is, full (heavy) lines are bonds lying *above* the plane of the ring system, while dashed (broken) lines are bonds lying *below* this plane.

Table A-1

Section	Description	Specific Names, in Increasing Order of Complexity	
1	Skeleton (porphyrin nucleus minus C-20)	Corrin (I)	
		Heptaacid	Heptaacid, hexaamide
3	**1**, with standard side chains and with cobalt	Cobyrinic acid (II)	Cobyric acid
4	**3**, with D-1-amino-2-propanol at position *f*	Cobinic acid (III)	Cobinamide
5	**4**, with D-ribofuranose 3-phosphate at position 2 of the aminopropanol	Cobamic acid (III–V)	Cobamide
7	**5**, with heterocyclic base attached by *N*-glycosyl link at position 1 of ribose and attached as an α ligand to cobalt (see **6**, also)		Aglyconylcobamide (VI)
9	Many "B_{12}" vitamins and derivatives, in which the heterocyclic base is 5,6-dimethylbenzimidazole, are given the trivial name "cobalamin"		Cobalamin
15	"B_{12} coenzymes," compounds in which a further organic group (X-yl) is β-ligated to cobalt (see **9** also)		X-ylcobalamin; (*Co*α-ligandyl)-(*Co*β-X-yl) cobamide (X)

2. Formulas II, III, and IV represent the true absolute stereochemical configuration of the structures as determined by X-ray work (*5, 6*). The CN is in the β position of the cobalt, the ribose-bound heterocyclic base in the α position, in formula IV. When adenine is the heterocyclic base, it is usually bound to the ribose by its N-7, the opposite from what is seen in the nucleic acids.

3. Formula V represents the absolute stereochemical configuration of the ribofuranose residue. For convenience in comparing it with formula IV, it is written with the α substituent at C-1 *above* the plane of the ring (i.e., the reverse of the usual carbohydrate method) (7).

REFERENCES

1. IUPAC–IUB (1966). *J. Biol. Chem.* **241**, 2992; *Biochem. J.* **102**, 19; and elsewhere.
2. IUPAC–IUB (1969). *Biochemistry* **8**, 2227; **10**, 4994; *Biochim. Biophys. Acta* **164**, 453 (**248**, 387); and elsewhere.
3. IUPAC (1971). *Eur. J. Biochem.* **18**, 151; *J. Org. Chem.* **35**, 2899; and elsewhere.
4. Ljungdahl, L., Irion, E., and Wood, H. G. (1966). *Fed. Proc. Fed. Am. Soc. Exp. Biol.* **25**, 1642.
5. Hodgkin, D. C., Kamper, J., Lindsey, J., MacKay, M., Pickworth, J., Robertson, J. H., Shoemaker, C. B., White, J. E., Prosen, R. J., and Trueblood, K. N. (1957) *Proc. Roy. Soc. Ser. A* **242**, 228.
6. Hodgkin, D. C. (1965). *Science* **150**, 979.
7. IUPAC–IUB (1971). *Biochemistry* **10**, 3983; *Eur. J. Biochem.* **21**, 455; and elsewhere.

APPENDIX: ABBREVIATIONS FOR CORRINOIDS

This appendix was inspired by the burgeoning literature concerning corrinoid compounds, many of which have long and unwieldy names—a fact that has led to a variety of *ad hoc* abbreviations that in turn has led to difficulties for the reader.

In accordance with several preceding CBN documents (1–3), as well as with standard chemical practice, the abbreviations are constructed by assembling symbols representing the various radicals involved, rather than from combinations of letters drawn haphazardly from the complete names of the compounds. The use of symbols reflects the actual structure of a compound and facilitates the writing of equations for its chemical transformations. In particular, the use of DBC, DMBC, and so on, is discouraged, as is the use of B_{12} (except as vitamin B_{12}), coenzyme B_{12}, and "factor" terms.

Names and Symbols

Names		Symbols
Corrin		Crn
of Free Acid	of Hexaamide	(for the Hexaamide)
Cobyrinic acid	Cobyric acid	Cby
Cobinic acid	Cobinamide	Cbi
Cobamic acid	Cobamide	Cba
	Cobalamin	Cbl[7]

Designation of Substituents Attached to Cobalt

Ligands coordinated to the α and β position of the cobalt (below and above the plane of the corrin residue, respectively, as shown in structure IV) are represented by terms that precede the symbol for the corrinoid residue. If the positions of the ligands are *unknown or not specified,* the two terms representing the ligands in the α and β position are enclosed in one set of parentheses and are separated by a *comma.* If the positions are *known and specified,* the α ligand is set apart by parentheses; the β is enclosed (separately) only if its complexity may make it ambiguous. If the ligands are identical, a single term followed by the subscript 2 is used.

Anion Substituents

The chemical symbol for the anion is used. Aqua is abbreviated aq. Examples:

(Me)aqCbi	(methyl)aquacobinamide (4) (methyl in α position)
(CN)MeCbi	(cyano)methylcobinamide (methyl in β position)
(CN,aq)Cbi or (aq,CN)Cbi	cyanoaquacobinamide (ligand location unspecified)

Alkyl Substituents

1. Primary substituents are designated by naming the alkyl group[8] without denoting the position attached to the cobalt, as it is always 1.

[7] A cobalamin is a cobamide in which 5,6-dimethylbenzimidazole is covalently bonded to the ribose in α-glycosidic linkage; it is thus a dimethylbenzimidazolylcobamide and can be symbolized as such. However, it is often convenient to have a short symbol for this complex, hence Cbl. Cbl is recommended in place of the former B_{12} or B-12 for chemical use.

[8] Symbols for alkyl groups are Me, Et, Pr, Pr^i, Bu^s, Bu^t, Pe, Hx, Hp, ...

Examples:

(aq)EtCbi	(aqua)ethylcobinamide
(CN)(2-OAcBu)Cbi	(cyano)(2-acetoxybutyl)cobinamide

(2) Secondary substituents are named similarly, except that the position attached to the cobalt is given by a locant suffixed to the name of the alkyl group (as in the -X-yl name). Examples:

(aq)(Bu-2)Cbi or (aq)Bu^{s}Cbi	(aqua)(*sec*-butyl)cobinamide (*5*)
(aq)(3-OAcBu-2)Cby	(aqua)(3-acetoxybut-2-yl)-cobyric acid

(3) Alicyclic groups are indicated by a small "c" before the symbol for the alkyl residue. In these compounds, cobalt is always assumed to be substituted in position 1 of the ring. Examples:

(aq)cHxCbi	(aqua)cyclohexylcobinamide (*5*)
(CN)(2-HOcPe)Cby	(cyano)(2-hydroxycyclopentyl)-cobyric acid

(4) 5′-Deoxy-5′-adenosyl in the β coordination position, as in "coenzyme B_{12}," is represented by the symbol Ado for "adenosyl"[6]; a 2′-deoxyadenosine residue by dAdo (3). Unusual deoxyadenosyl residues can be indicated by superscripts (e.g., d^3Ado, $d^{2,3}$Ado). See the following section.

Cobamides of the Cobalamin[6] Type

As the symbol Cbl designates α-(5,6-dimethylbenzimidazolyl)cobamide [cob(III)alamin], only those cobamides having this base utilize Cbl. Those containing another base are named as cobamides utilizing the symbol Cba. Hence, Cbl is preceded by only a single term, the one representing the β substituent. Examples are given in Table A-2.

Notes: (i) The hyphenation in the case of secondary alkyl substituents and similar situations of potential confusion may make it necessary to enclose the β substituent in parentheses or set it off by a hyphen.

(ii) If the replacing base (in α position) is unspecified, the term (?) is used, for example, (?)MeCba. The term (OH/base) indicates that the ribose residue is not attached to the Coα-linked base.

Table A-2

CN-Cbl	Cyanocob(III)alamin (vitamin B_{12})
AdoCbl	Adenosylcob(III)alamin (6,7)
PrCbl	*n*-Propylcob(III)alamin; methyl-, etc., similarly (8,9)
(Ade)(Pr-2)Cba or (Ade)Pr^{i}-Cba[a]	*Coα*-[α-Aden-9-yl)]-*Coβ*-isopropylcobamide
(Bza)MeCba[b]	*Coα*-(α-Benzimidazolyl)-*Coβ*-methylcobamide
2-(MeOOC)EtCbl	(2-Methoxycarbonylethyl)cob(III)alamin (8)
(Ade-7)AdoCba[a]	*Coα*-[α-(Aden-7-yl)]-*Coβ*-adenosylcobamide (10)
(2-SHAde-7)AdoCba[a]	*Coα*-[α-(2-Thiaaden-7-yl)]-*Coβ*-adenosylcobamide (11)
(5-MeOBza)MeCba	*Coα*-(5-Methoxybenzimidazolyl)-*Coβ*-methylcobamide[d] (12,13)
(2-MeAde-7)CN-Cba[a]	*Coα*-[α-(2-Methyladen-7-yl)]-*Coβ*-cyanocobamide (11,14)
(Ade)CN-Cba[a]	*Coα*-[α-(Aden-9-yl)]-*Coβ*-cyanocobamide (pseudovitamin B_{12})
(Ade)OH-Cba[a]	*Coα*-[α-(Aden-9-yl)]-*Coβ*-hydroxocobamide (hydroxopseudovitamin B_{12})
(Ade)MeCba[a]	*Coα*-[α-(Aden-9-yl)]-*Coβ*-methylcobamide
[4-(Ade-9)Bu]Cbl[c]	[4-(Aden-9-yl)butyl]cob(III)alamin (6)
(6MeSPur)AdoCba	*Coα*-(α-6-Methylthiopurinyl)-*Coβ*-adenosylcobamide (15)

[a] Ade alone represents adenine bonded to the ribosyl moiety through its 7 position (*i.e.*, a 7-α-D-ribofuranosyladenine). Bonding to the cobalt is thus through N-9. When these positions are reversed, Ade-7 and aden-7-yl are used (*i.e.*, the locant specifies that N linked to cobalt).

[b] Bza = benzimidazolyl.

[c] As this is a cobalamin, the adenine residue is not in the *Coα* position, but is attached (-9-yl) to a but-4-yl residue that is in turn linked to the β position of the cobalt. Named as a cobamide, it would be (Me_2Bza) [4-(Ade-9)Bu]Cba.

[d] Factor III_m (12,13).

(iii) If the α substituent (the "base") is displaced from the cobalt by another ligand, but remains attached to the ribosyl residue, the same system is used. Example:

(2-MeAde/aq)MeCba	*Coα*-aqua-*Coβ*-methyl(2-methyladenylcobamide)
(Ade/CN)CN-Cba	*Coα*-cyano-*Coβ*-cyano(adenylcobamide) or dicyanoadenylcobamide

In abbreviating cobalamin derivatives, the base need not be specified. Replacement of the base by another *Coα* ligand is indicated by merely adding to the abbreviation a term corresponding to the replacing ligand. Examples

(OH)MeCbl	*Coα*-hydroxo-*Coβ*-methylcobalamin or *Coα*-hydroxo-Coβ-methyl (dimethylbenzimidazolylcobamide)
$(CN)_2Cbl$	*Coα*-cyano-*Coβ*-cyanocobalamin or dicyanocobalamin

(iv) Cobalt valences of II or I may be indicated by superscripts (e.g., Cbl^{II}).

Designation of Alterations and Substitutions on the Corrin Ring

Substituents on the ring itself are represented by symbols following the symbol for the corrinoid, with locants indicating the positions of the substituents. Epimerization is indicated in a smiliar manner. Symbols representing replacements on the carboxyl groups at the periphery of the corrin residue follow those that designate substituents directly on the ring. The location of the substituent is indicated by the letter corresponding to the carboxyl group that carries it. Examples:

CN-Cbl(13-epi)	cyano(13-epi)cobalamin
CN-Cbl(13epi-*e*OH)	*Coα*-(α-5,6-dimethylbenzimidazolyl)-*Coβ*-cyano- (13-epi)cobamic *a,b,c,d,g*-pentaamide
AdoCbl (10-Cl)	adenosyl-10-chlorocobalamin
(aq)AdoCbi (*e*-PhNH)	*Coα*-aqua-*Coβ*-adenosylcobinic *a,b,c,d,g*-pentaamide e-anilide
$(CN)_2Cby(OMe)_7$	dicyanocobyrinic heptamethyl ester (*16*)

If the location of the carboxyl substituent(s) is unknown, a term of the following structure should be used:

$$(a:g\text{-X})_n$$

where *a*:*g* indicates a substitution at the periphery of the ring, X is the replacing group, and *n* refers to the number of carboxyl groups substituted. Examples:

(CN,aq)Cby[*a*:*g*-$(NH_2)_5$]	cyanoaquacobyrinic acid pentaamide (*17*)
(CN,aq)Cby[10-Cl-*a* : *g*-$(NH_2)_5$]	10-chloro derivative of the above

Replacement of Cobalt by Other Metals (18, 19)

Corrinoids containing metals other than cobalt are symbolized by placing the symbol of the replacing metal in square brackets preceding and attached to the symbol of the corrinoid. Thus, a hydrogenocobamide utilizes [H]Cba, a nickelocobalamin [Ni]Cbl^{II}, a zincocobinamide [Zn]Cbi^{II}, and so on, Phenylcupribamide *(19, 20)* could be indicated as (Ph)[Cu]Cba^{II}. I, II, and III may be added as superscripts when needed.

Isotopic Labeling

A labeled position is indicated in the usual fashion *(21)*, for example,

(Bza)Me[^{57}Co]Cba	*Co*α-(α-benzimidazolyl)-*Co*β-methyl-[^{57}Co]cobamide
(Bza)[^{14}C]MeCba	*Co*α-(α-benzimidazolyl)-*Co*β-[^{14}C]-methylcobamide
([4-^{3}H]Bza)MeCba	*Co*α-(α-[4-^{3}H]benzimidazolyl)-*Co*β-methylcobamide

Metallocorrins

As corrin contains no metal (the name "corrin" being derived from "core," not "cobalt"), complexes of metals with corrin require specification of both terms. Example: Cu^{II}Crn for copper(II) corrin.

REFERENCES

1. IUPAC–IUB Commission on Biochemical Nomenclature (1972). *Biochemistry* **11**, 1726, and in other journals.
2. IUPAC–IUB Commission on Biochemical Nomenclature (1972). *Biochemistry* **11**, 942, and in other journals.
3. IUPAC–IUB Commission on Biochemical Nomenclature (1970). *Biochemistry* **9**. 4022.
4. Pailes, W. H., and Hogenkamp, H. P. C. (1968). *Biochemistry* **7**, 4160.
5. Brodie, J. D. (1969). *Proc. Natl. Acad. Sci. U. S. A.* **62**, 461.
6. Babior, B. M. (1969). *J. Biol. Chem.* **244**, 2917.
7. Cardinale, G. J., and Abeles, R. H. (1967). *Biochim. Biophys. Acta* **132**, 517.
8. Hogenkamp, H. P. C., Rush, J. E., and Swenson, C. A. (1965). *J. Biol. Chem.* **240**, 3641.
9. Stavrianopoulos, J., and Jaenicke, L. (1967). *Eur. J. Biochem.* **3**, 95.
10. Overath, P., Stadtman, E. R., Kellerman, G. M., and Lynen, F. (1962). *Biochem. Z.* **336**, 77.
11. Bonnett, R. (1963). *Chem. Rev.* **63**, 573.
12. Irion, E., and Ljungdahl, L. (1968). *Biochemistry* **7**, 2350.
13. Ljungdahl, L., Irion, E., and Wood, H. G. (1966). *Fed. Proc., Fed. Am. Soc. Exp. Biol.* **25**, 1642.

14. Hayashi, M., and Kamikubo, T. (1971). *FEBS (Fed. Eur. Biochem. Soc.) Lett.* **15**, 213.
15. Uchida, Y., Hayashi, M., and Kamikubo, T. (1973). *Vitamins (Jap.)* **47**, 27; (1973). *Chem. Abstr.* **78**, 94046e.
16. Werthemann, L. (1968). Dissertation, Zürich.
17. Bernhauer, K., Vogelmann, H., and Wagner, F. (1968). *Z. Physiol. Chem.* **349**, 1281.
18. Toohey, J. I. (1966). *Fed. Proc., Fed. Am. Soc. Exp. Biol.* **25**, 1628.
19. Koppenhagen, V. B., and Pfiffner, J. J. (1971). *J. Biol. Chem.* **246**, 3075.
20. Koppenhagen, V. B., and Pfiffner, J. J. (1970). *J. Biol. Chem.* **245**, 5865.
21. Instructions to Authors (1973). *J. Biol. Chem.* **248**, 4; and elsewhere.

APPENDIX II

IUPAC-IUB RECOMMENDATIONS FOR ABBREVIATIONS FOR SOME CORRINOIDS FREQUENTLY REFERRED TO BY "FACTOR" OR OTHER TRIVIAL TERMINOLOGY

Terminology or Abbreviation[a]	Recommended Abbreviation
B_{12a}	OH-Cbl
B_{12b}	OH-Cbl
B_{12d}	OH-Cbl
B_{12f}	(Ade)CNCba
B_{12m}	(2-MeAde)CNCba
B_{12p}	$(CN)_2Cbi$
B_{12r}	Cbl^{II}
B_{12s}	Cbl^{I}
ψB_{12}	(Ade)CNCba
ψB_{12d}	(2-MeAde)CNCba
Factor I	$(CN)_2Cbi$ or (5-OHBza)CNCba
Factor Ib	$(CN)_2Cbi$
Factor II	CNCbl
Factor III	(5-OHBza)CNCba
Factor III_m	(5-OMeBza)CNCba
Factor IV	(Ade)CNCba
Factor V_{1a}	$(CN)_2Cby$

Terminology or Abbreviation[a]	Recommended Abbreviation
Factor A	(2-MeAde)CNCba
Factor B	$(CN)_2Cbi$
Factor G	(Ino)CNCba
Factor H	(2-MeIno)CNCba
Pseudo-B_{12}	(Ade)CNCba
Coβ-Adenosylcorrins	
DBCC	AdoCbl
DBC	AdoCbl
DMBC	AdoCbl
CoB_{12}	AdoCbl
Coenzyme B_{12}	AdoCbl
Pseudo-B_{12} coenzyme	(Ade)AdoCba
ψB_{12} coenzyme	(Ade)AdoCba
AC	(Ade)AdoCba
BC	(Bza)AdoCba
SB_{12p}	(OH)AdoCbi

[a] Taken largely from Bonnett, R. (1963). *Chem Rev.* **63,** *573.*

INDEX